INTRODUCTION TO

RADAR ANALYSIS

SECOND EDITION

Advances in Applied Mathematics

Series Editor: Daniel Zwillinger

Published Titles

Advanced Engineering Mathematics with MATLAB, Fourth Edition
Dean G. Duffy

CRC Standard Curves and Surfaces with Mathematica®, Third Edition
David H. von Seggern

CRC Standard Mathematical Tables and Formulas, 33rd Edition
Dan Zwillinger

Dynamical Systems for Biological Modeling: An Introduction
Fred Brauer and Christopher Kribs

Fast Solvers for Mesh-Based Computations *Maciej Paszyński*

Green's Functions with Applications, Second Edition *Dean G. Duffy*

Handbook of Peridynamic Modeling *Floriin Bobaru, John T. Foster, Philippe H. Geubelle, and Stewart A. Silling*

Introduction to Financial Mathematics *Kevin J. Hastings*

Introduction to Radar Analysis, Second Edition *Bassem R. Mahafza*

Linear and Complex Analysis for Applications *John P. D'Angelo*

Linear and Integer Optimization: Theory and Practice, Third Edition
Gerard Sierksma and Yori Zwols

Markov Processes *James R. Kirkwood*

Pocket Book of Integrals and Mathematical Formulas, 5th Edition
Ronald J. Tallarida

Stochastic Partial Differential Equations, Second Edition *Pao-Liu Chow*

Quadratic Programming with Computer Programs *Michael J. Best*

Advances in Applied Mathematics

INTRODUCTION TO
RADAR ANALYSIS
SECOND EDITION

BASSEM R. MAHAFZA

GEORGIA TECH RESEARCH INSTITUTE

CRC Press is an imprint of the
Taylor & Francis Group, an **informa** business
A CHAPMAN & HALL BOOK

Book Dedication

To

Hamsa, Zachary, Joseph, Jacob, Jordan

Bassem R. Mahafza
Huntsville, Alabama
United States of America

Table of Contents

Chapter 8: Pulse Compression, 207

PART III: Special Topics in Radar Systems

Chapter 9: Radar Clutter, 225

Chapter 10: Moving Target Indicator, 243

Chapter 11: Radar Detection, 267

Preface

I developed the first edition of *Introduction to Radar Analysis* as a textbook on radar systems analysis that would be suitable for college courses. In my opinion, there existed a need for a radar book that can serve as a textbook, and yet be a good source of information for engineers working in the field. Over the past three decades, numerous books have been published in the field of radar systems. Some of those books place emphasis on radar technology, while most others present the subject based on its practical aspects. Little or no emphasis has been placed on presenting this subject in the context of a textbook suitable for college courses. To this end, a good textbook must concentrate on the fundamentals and principles; it must also include examples and exercise problems. And finally, it must include detailed mathematical derivations.

Indeed, since its publication in 1998, the first edition has been adopted by many professors as either a primary, or as a supporting, textbook for university level courses taught on the subject; and the time has come to introduce an update to this bestseller textbook. For this purpose, the second edition includes, expands upon, and in some cases rewrites almost all topics presented in the first edition. Just like its predecessor, it is written primarily as a textbook for use as a senior level and/or a first graduate level college course on radar systems. It is written within the context of communication theory as well as the theory of signals and noise. In addition to the new edition being developed as a textbook it will also serve as a valuable tool for practicing radar engineers. Users of this textbook will be able to analyze and understand the many issues of radar systems design and analysis. Each chapter provides the necessary mathematical and analytical coverage required for good understanding of the theory. Unlike many other books available on the market, the second edition of this text provides comprehensive, easy to follow mathematical derivations of all key equations and formulas. The second edition contains new and updated end-of-chapter problems. And finally, a solutions manual has been developed to benefit instructors who adopt the book as text and will be made available to adopting professors through the publisher.

For many years, I have taught courses (at different collegiate levels) on the subject and I have acquired extensive industry experience in the field. Here are some of my observations. As a teacher, one of the most challenging tasks I faced was choosing a suitable textbook for my students. Yes, a seemingly endless list of publications exists on the subject; yet, most of these publications emphasize technology and/or practical aspects of radar systems rather than fundamentals and basic operations. As a professional working in the industry, I observed that the knowledge acquired by most engineers working in the field appears to be spiky in nature. Most engineers seem to have detailed knowledge of few aspects of radar systems and limited capabilities throughout the rest of the field.

The second edition of this book builds upon its first edition's intent, and that is, to complement existing books on the market by presenting the fundamentals of radar systems and providing rigorous mathematical derivations. The subject of radar systems should not be complicated to understand nor difficult to analyze. By emphasizing principles and fundamentals, this book serves as a stepping stone for students and radar engineers so that they can move towards tackling advanced radar topics.

This second edition is divided into three distinct but complementary parts. Part I covers the necessary background in signals and systems as well as communication theory. Topics like Amplitude Modulation (AM) and Frequency Modulation (FM) are discussed as well as continuous and discrete time domain signals and systems. The treatments of these topics are such that they naturally lead to Continuous Wave (CW) and pulsed radar systems; thus, bridging the gap between communication theory, signals and noise on one end, and radar system analysis on the other. The radar equation is developed into a general form and then translated into its unique CW and pulsed radars forms. A detailed presentation of radar waveforms in the context of signals and systems comprises a significant portion of Part I of this textbook.

Part II of this second edition continues the build up of material presented in Part I and is primarily concerned with the treatment of radar signal processing including pulse compression. A detailed analytical description of CW and pulsed radar systems is presented. Then the basics of the radar matched receiver are analyzed. In this part, the radar ambiguity function is first developed in its general form; then, specific ambiguity functions for the most common radar waveforms are introduced. Finally, this part closes with an overview of pulse compression.

Alternatively, Part III presents chapters on the different radar subsystems/systems. To this end, the subject of radar clutter is in Chapter 9. Area clutter as well as volume clutter are defined and the radar equation is re-derived to reflect the importance of clutter, where in this case, the signal to interference ratio becomes more critical than the signal to noise ratio. A step-by-step mathematical derivation of clutter RCS is presented, and the statistical models for the clutter backscatter coefficient are also presented. Next, the Moving Target Indicator (MTI) chapter discusses how delay line cancelers can be used to mitigate the impact of clutter within the radar signal processor. Pulse repetition frequency staggering is analyzed in the context of blind speeds and in the context of resolving range and Doppler ambiguities. Finally, pulsed Doppler radars are briefly analyzed.

Target detection is presented in Chapter 11, and it includes target fluctuation where the Swerling target models are discussed. Detailed discussion of coherent and noncoherent integration in the context of a square law detector are in this chapter. Cumulative probability of detection and M-out-of-N detection are also discussed. Chapter 12 discusses target tracking radar systems. The first part of this chapter covers the subject of single target tracking. Topics such as sequential lobing, conical scan, monopulse, and range tracking are discussed in detail. The second part of this chapter introduces multiple target tracking techniques. Fixed gain tracking filters such as the $\alpha\beta$ and the $\alpha\beta\gamma$ filters are presented in detail. The concept of the Kalman filter is introduced. Special cases of the Kalman filter are analyzed. The next chapter focuses on Phased Array Antennas; it starts by developing the general array formulation. Linear arrays and several planar array configurations such as rectangular, circular, rectangular with circular boundaries, and concentric circular arrays are discussed.

Synthetic Aperture Radars is the subject of Chapter 14; topics of this chapter include SAR signal processing, SAR design considerations, and the SAR radar equation. Finally, Chapter 15 presents a new and very unique feature of this book and that is, a Radar Design Case Study, referred to as *Textbook Radar Design Case Study*. The approach taken in this chapter is such that the design case study mirrors the chapter sequence of the book. In this case, my view of the design process is detailed and analyzed using an approach suitable for an open literature publication and is commensurable with academic classroom settings. Suitable and comprehensive trade off analyses and calculations are performed.

Based on my own teaching experience, the following breakdown can be utilized by professors using this book as a text:

1. Option I: Chapters 1-5, and parts of Chapter 15 can be used as a senior-level course. Chapters 4-8 as well as the remainder of Chapter 15 can be used as a first graduate level course. Finally, Chapters 9-14 can be used as a second advanced graduate-level course.
2. Option II: Chapters 1-8 can be used as an introductory graduate-level course. Chapters 9-14 can be used as a second graduate-level course. In both cases, suitable sections of Chapter 15 should be included.

Bassem R. Mahafza
bassemrm@gmail.com
Huntsville, Alabama
United States of America

Part I

Signals, Systems, Modulation Techniques, and Basic Radar Concepts

Chapter 1

Introduction

In this chapter a top level overview of communication systems and their similarities to radar systems is discussed. The presentation is very general in nature and is suitable for an introductory chapter. The intent is to introduce a bird's-eye view discussion of both systems that is commensurable with causal and non-expert readers interested in learning more about radar systems. In this context, we will demonstrate that much of the top level system components of either system are similar in form and function. To this end, both systems use almost identical modulation/demodulation techniques as well as similar signal processing algorithms to transmit and or extract information from a signal. As a matter of fact, early radar systems were named after the type of signal (waveform) they used, such as pulsed and continuous-wave radars. While many readers might find the material presented in this chapter to be useful, well-versed individuals in either system may skip over this chapter and move on to subsequent chapters based on their particular interest.

1.1. The General Communication System

Communication systems are used to transmit information from one point to another. Information can be either analog or digital. Examples include radio, telephone and television. In today's modern digital world, computers and smart phones are used to transfer a huge amount of data, in many different forms, almost instantaneously using the Internet. Figure 1.1 shows a block diagram of a general communication system. The system comprises an information source, a transmitter, a transmission medium, a receiver, and an output transducer. The time varying signal $x(t)$ represents the information that needs to be transmitted by the system. This signal may take on any number of things, such as speech, music, data, or photographs to name a few. Almost in all cases, the signal $x(t)$ is not suitable for transmission in its raw form. Thus, the signal must be prepared (i.e., pre-processed) before it is emitted into the channel or medium of transmission. Such signal preparation is done inside the transmitter.

The transmitter subsystem comprises a few electrical devices whose primary function is to pre-process (prepare) the raw data (signal) and turn it into a form suitable for transmission. In this case, the transmitter knows the nature of the signal as well as the characteristics of the medium of transmission. Generally, the information being transmitted is used to modulate a carrier signal. For example, the modulated carrier can be a sine wave whose amplitude, phase or frequency is completely known and vary linearly as a function of time.

The primary motivations behind modulation include:

1. Ease of transmission of signal via the antenna into the transmission medium
2. Reduction of the effects of noise and interference in the channel
3. Better utilization of bandwidth assignments
4. Transmission of more than one simultaneous message over one channel, and
5. Overcoming of hardware or electrical devices limitations.

The antenna is the device that acts as the coupler (transducer) between the transmitter and the transmission medium. Antennas convert the modulated electrical signal from the transmitter into an electromagnetic wave capable and suitable to travel through the channel. The design shape, size and other specific characteristics of the antenna are closely coupled to the carrier frequency being used and to the distance that the signal must travel between the transmitter and receiver.

The channel designates the transmission medium between the transmitter and receiver. Although in some cases the transmission medium may be made of an electrical cable or a fiber optics medium, it is the atmosphere in most cases. In almost all cases, the channel adds unwanted or undesirable effects upon the signal being transmitted. Simply put, the channel introduces uncertainty upon the signal. This uncertainty is due to the presence of unwanted signal perturbations, which is widely referred to as noise. It is usually sufficient to approximate the perturbations as additive noise. Additionally, the noise can be assumed to be Gaussian with a zero mean as a result of the central limit theorem.

Noise is random and unpredictable in nature; hence, the design and analysis of communication systems require strong knowledge of statistical techniques. Interfering signals (deliberate or otherwise) as well as signal perturbations are other key characteristics of the transmission medium. The signal typically goes through degradation while traveling through the channel. This degradation includes:

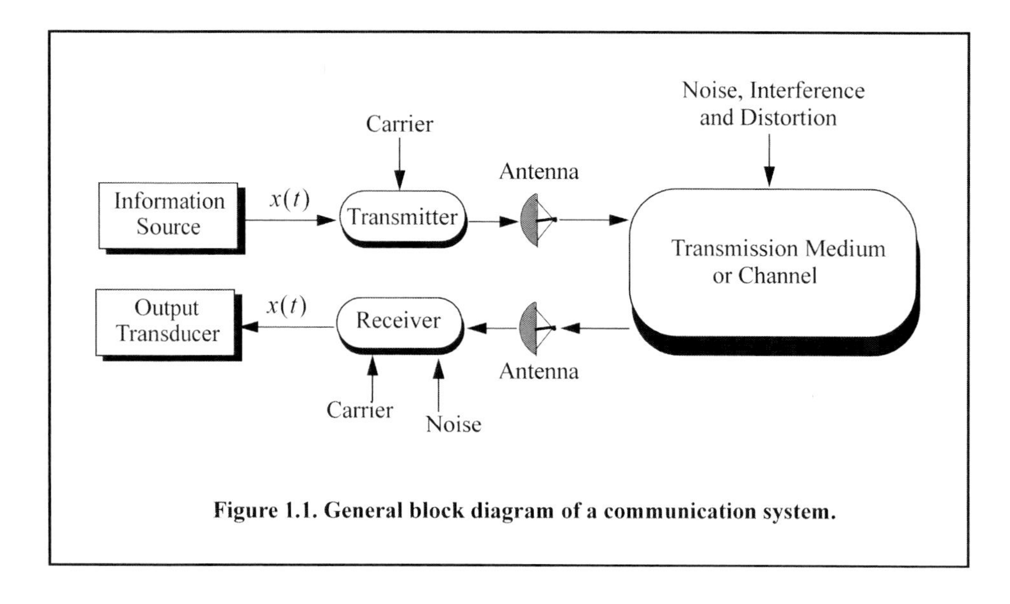

Figure 1.1. General block diagram of a communication system.

1. Attenuation in amplitude
2. Phase distortion
3. Contamination by noise and other interfering signals
4. Other forms of distortion due to effects such as multiple transmission paths and filtering.

The receiver is the subsystem whose primary function is to extract the original information from the received signal and deliver it in a suitable form to the output transducer. Typically, the receiver employs filtering and amplification to the received signal prior to performing the demodulation function (sometimes called detection). The output transducer converts the electrical signal outputted by the receiver into its desired form per the design of the system.

1.2. The General Radar System

The word *radar* is an abbreviation for *RAdio Detection And Ranging*. In general, radar systems use modulated waveforms and directive antennas to transmit electromagnetic energy into a specific volume in space to search for targets. Objects (targets) within a search volume will reflect portions of this energy (radar returns or echoes) back to the radar. These echoes are then processed by the radar receiver to extract target information such as range, velocity, angular position, and other target identifying characteristics. Figure 1.2 shows a block diagram for the general radar system. At first glance, one observes that a radar system is similar, at least functionally, to a communication system. That is not surprising, since radar systems make extensive use of communication theory. Having said that, one should not conclude that the two systems are the same.

Just like the case in a communication system, the radar transmitter prepares and sends a signal to the antenna that is suitable for transmission into the channel. In this case, the channel affects the signal twice, once as it travels from transmitting antenna to the target and once on its way back to the receiving antenna after reflection from a target. One big difference between a communication system's channel and that of a radar system is the presence of a clutter signal in the radar echoes.

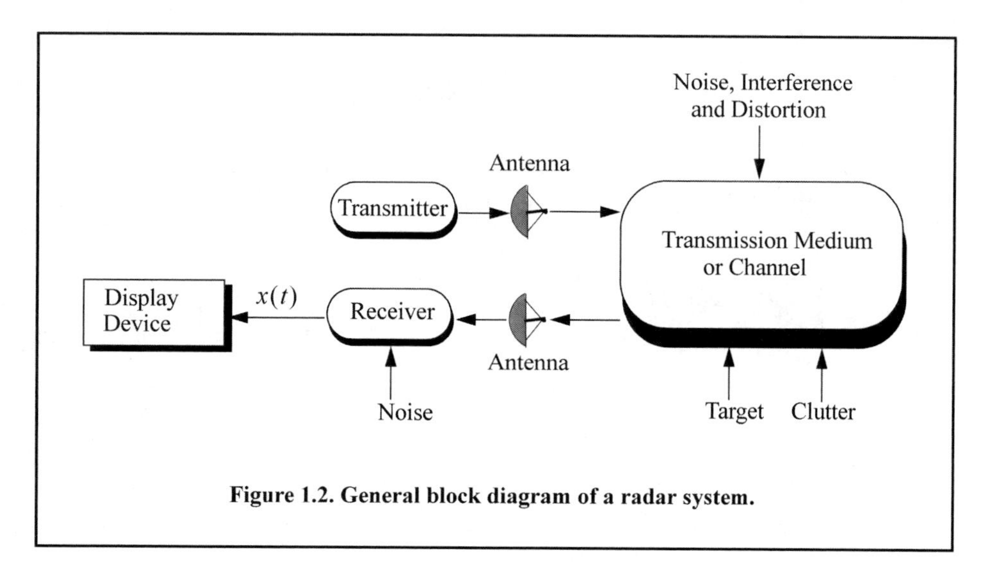

Figure 1.2. General block diagram of a radar system.

Clutter is a term used to describe radar returns from any object that may generate unwanted radar returns that may interfere with normal radar operations. Clutter echoes are random and have noise-like characteristics because the individual clutter components (scatterers) have random phases and amplitudes. In many cases, the clutter signal level is much higher than the receiver noise level.

The radar receiver demodulates the radar echo signal so that the signal processor can extract target relevant information from the returned signal. In this case, the radar receiver knows exactly the nature of the transmitted signal. The radar antenna, again, acts as the coupler between the transmitter and the transmission medium. It also captures the radar echo signal and converts it from electromagnetic wave into an electrical signal suitable for processing by the receiver.

1.3. Comparison of the Two Systems

The functional similarities between the two systems are rather obvious as observed from their respective block diagrams. The differences between the two systems can be vast, however. To demonstrate these differences, let us take a closer look at the antennas used in both systems. As mentioned before, the main purpose of the antenna in a communication system is to convert electrical signals into electromagnetic waves and vice versa.

Typically, antennas are not thoroughly analyzed in communication systems, and in most cases, they are either replaced by a filter or even completely ignored. This is true because one does not need extensive knowledge of antenna theory to fully understand communication theory. Alternatively, that is far from being true in the case of radar systems. The antenna is a major contributor to the overall performance of a radar system. Because of that, it has been a very common practice to name ceratin radar systems after the type of antenna they use.

Other differences between the two systems include:

1. In a radar system, the source of information (i.e., the target) is located within the channel rather than at the input of the transmitter as in the case of a communication system.
2. In most cases, radar systems use the same antenna to transmit and receive the signal, while communication systems have distinct transmitting and receiving antennas.
3. The information sought after in a radar system is not a message but simply the presence of the source signal (target).
4. In a radar system, the receiver is in close proximity to the transmitter, but the two subsystems are typically located far away from each other in a communication system.
5. While the Radio Frequency (RF) portion of signal processing is similar in both systems, the signal processing employed by both systems differs significantly once demodulation has been completed.
6. Finally, the definition of the interfering signals can be very different in both systems because it includes clutter returns in the case of a radar.

Chapter 2

Signals and Systems

In the first chapter, we addressed the similarities between communication and radar systems. A key element in this comparison was the signal type used by either system and the associated time domain as well as frequency domain characteristics. In a communication system, a modulated signal carries the information being transmitted from one point to another. Alternatively, in the case of a radar system, a completely known signal, generated within the transmitter, is emitted into space by the radar using a directive antenna. The antenna is also used to collect a modified version of the emitted signal reflected from a target within the radar field of view; the reflected signal is referred to as echo or return signal. Simply put, the echo signal is a modified, in amplitude and phase, version of the signal emitted by the radar. The radar receiver extracts the targets' information, such as range, velocity and target type, from the echo signal. The extraction mechanism is known as signal and data processing. To this end, different signal types can be superior to others. This is the reason why it is critical to first study signals as well as their relevant characteristics in the context of a radar system.

In this chapter an overview of elements of signals and systems that are relevant to radar systems is presented. It is assumed that the reader has sufficient and adequate background in signals and systems as well as in Fourier transform and its associated properties.

2.1. Signal Classifications

In general, electrical signals can represent either current or voltage and may be classified into two main categories: energy signals and power signals. Energy signals can be deterministic or random, while power signals can be periodic or random. A signal is said to be random if it is a function of a random parameter (such as random phase, random amplitude, or both). Additionally, signals may be divided into lowpass or bandpass signals. Signals that contain very low frequencies (close to DC) are called lowpass signals; otherwise they are referred to as bandpass signals. Through modulation, lowpass signals can be mapped into bandpass signals.

The average power P for the current or voltage signal $x(t)$ over the interval (t_1, t_2) across a 1Ω resistor is

$$P = \frac{1}{t_2 - t_1}\int_{t_1}^{t_2} |x(t)|^2 \ dt . \qquad \text{Eq. (2.1)}$$

The signal $x(t)$ is said to be a power signal over a very large interval $T = t_2 - t_1$, if and only if it has finite power and satisfies the relation:

$$0 < \lim_{T \to \infty} \frac{1}{T} \int_{-T/2}^{T/2} |x(t)|^2 \ dt < \infty . \qquad \text{Eq. (2.2)}$$

Using Parseval's theorem, the energy E dissipated by the current or voltage signal $x(t)$ across a 1Ω resistor, over the interval (t_1, t_2), is

$$E = \int_{t_1}^{t_2} |x(t)|^2 \ dt . \qquad \text{Eq. (2.3)}$$

The signal $x(t)$ is said to be an energy signal if and only if it has finite energy,

$$E = \int_{-\infty}^{\infty} |x(t)|^2 \ dt \quad < \infty . \qquad \text{Eq. (2.4)}$$

A signal $x(t)$ is said to be periodic with period T if and only if

$$x(t) = x(t + nT) \qquad \textit{for all } t \qquad \text{Eq. (2.5)}$$

where n is an integer.

Example:

Classify each of the following signals as an energy signal, a power signal, or neither. All signals are defined over the interval $(-\infty < t < \infty)$*:*

$$x_1(t) = \cos t + \cos 2t$$

$$x_2(t) = \exp(-\alpha^2 t^2) .$$

Solution:

$$P_{x_1} = \frac{1}{T} \int_{-T/2}^{T/2} (\cos t + \cos 2t)^2 dt = 1 \Rightarrow \textit{power signal} .$$

Note that since the cosine function is periodic, the limit is not necessary.

$$E_{x_2} = \int_{-\infty}^{\infty} (e^{-\alpha^2 t^2})^2 dt = 2\int_0^{\infty} e^{-2\alpha^2 t^2} dt = 2\frac{\sqrt{\pi}}{2\sqrt{2}\alpha} = \frac{1}{\alpha}\sqrt{\frac{\pi}{2}} \Rightarrow \textit{energy signal} .$$

****

2.2. The Fourier Transform

The Fourier Transform (FT) of the time-domain signal $x(t)$ is

$$F\{x(t)\} = X(\omega) = \int_{-\infty}^{\infty} x(t)e^{-j\omega t}\, dt \qquad \text{Eq. (2.6)}$$

where, in general, t represents time in seconds, while ω and f represent frequency in radians per second and Hertz, respectively.

Using the relationship $\omega = 2\pi f$, then the FT can also be expressed by

$$F\{x(t)\} = X(f) = \int_{-\infty}^{\infty} x(t)e^{-j2\pi ft}\, dt \qquad \text{Eq. (2.7)}$$

and the Inverse Fourier Transform (IFT) is

$$F^{-1}\{X(\omega)\} = x(t) = \frac{1}{2\pi}\int_{-\infty}^{\infty} X(\omega)e^{j\omega t}\, d\omega \qquad \text{Eq. (2.8)}$$

$$F^{-1}\{X(f)\} = x(t) = \int_{-\infty}^{\infty} X(f)e^{j2\pi ft}\, df. \qquad \text{Eq. (2.9)}$$

A detailed table of the FT pairs is listed in Appendix 2-A. Some of the most important FT properties are listed below,

$$\begin{aligned}
\frac{d}{dt}x(t) &\Leftrightarrow j\omega X(\omega) \\
a_1x_1(t) + a_2x_2(t) &\Leftrightarrow a_1X_1(\omega) + a_2X_2(\omega) \\
x(t - t_0) &\Leftrightarrow e^{-j\omega t_0}X(\omega) \\
e^{j\omega_0 t}x(t) &\Leftrightarrow X(\omega - \omega_0) \\
\cos(\omega_0 t)x(t) &\Leftrightarrow \frac{1}{2}[X(\omega - \omega_0) + X(\omega + \omega_0)]
\end{aligned} \qquad \text{Eq. (2.10)}$$

2.3. Systems Classification

Any system can be represented mathematically as a transformation of an input signal into an output signal. This transformation relationship between the input signal $x(t)$ and the corresponding output signal $y(t)$ can be written as

$$y(t) = f[x(t);\ (-\infty < t < \infty)]\,. \qquad \text{Eq. (2.11)}$$

The relationship described in Eq. (2.11) can be linear or nonlinear, time invariant or time varying, causal or noncausal, and stable or nonstable systems. When the input signal is a unit impulse (*Dirac delta function)* $\delta(t)$, the output signal is referred to as the system's impulse response $h(t)$.

2.3.1. Linear and Nonlinear Systems

A system is said to be linear if superposition holds true. More specifically, if

$$\begin{aligned} y_1(t) &= f[x_1(t)] \\ y_2(t) &= f[x_2(t)] \end{aligned} \qquad \text{Eq. (2.12)}$$

then for a linear system

$$f[ax_1(t) + bx_2(t)] = ay_1(t) + by_2(t) \qquad \text{Eq. (2.13)}$$

for any constants (a, b). If the relationship in Eq. (2.13) is not true, the system is said to be nonlinear.

2.3.2. Time Invariant and Time Varying Systems

A system is said to be time invariant (or shift invariant) if a time shift at its input produces the same time shift at its output. That is if

$$y(t) = f[x(t)] \qquad \text{Eq. (2.14)}$$

then

$$y(t - t_0) = f[x(t - t_0)]; -\infty < t_0 < \infty . \qquad \text{Eq. (2.15)}$$

If the above relationship is not true, the system is called a time varying system.

Any Linear Time Invariant (LTI) system can be described using the convolution integral between the input signal and the system's impulse response, as

$$y(t) = \int_{-\infty}^{\infty} x(t-u)h(u)\ du = x \otimes h \qquad \text{Eq. (2.16)}$$

the operator $\otimes$ is used to symbolically describe the convolution integral. In the frequency domain, convolution translates into multiplication. That is

$$Y(f) = X(f)H(f) , \qquad \text{Eq. (2.17)}$$

where $H(f)$ is the FT for $h(t)$ and is typically referred to as the system transfer function.

2.3.3. Stable and Nonstable Systems

A system is said to be stable if every bounded input signal produces a bounded output signal. From Eq. (2.16)

$$|y(t)| = \left| \int_{-\infty}^{\infty} x(t-u)h(u)\ du \right| \le \int_{-\infty}^{\infty} |x(t-u)||h(u)|\ du . \qquad \text{Eq. (2.18)}$$

If the input signal is bounded, then there is some finite constant K such that

$$|x(t)| \le K < \infty . \qquad \text{Eq. (2.19)}$$

Therefore,

$$y(t) \le K \int_{-\infty}^{\infty} |h(u)| \; du \qquad \text{Eq. (2.20)}$$

which can be finite if and only if

$$\int_{-\infty}^{\infty} |h(u)| \; du < \infty . \qquad \text{Eq. (2.21)}$$

Thus, the requirement for stability is that the impulse response must be absolutely integrable. Otherwise, the system is said to be unstable.

2.3.4. Causal and Noncausal Systems

A causal (or physically realizable) system is one whose output signal does not begin before the input signal is applied. Thus, the following relationship is true when the system is causal:

$$y(t_0) = f[x(t); t \le t_0]; -\infty < t, t_0 < \infty . \qquad \text{Eq. (2.22)}$$

A system that does not satisfy Eq. (2.22) is said to be noncausal which means it cannot exist in the real world.

2.4. The Fourier Series

A set of functions $S = \{\varphi_n(t) \;\; ; \; n = 1, \ldots, N\}$ is said to be orthogonal over the interval (t_1, t_2) if and only if

$$\int_{t_1}^{t_2} \varphi_i^*(t)\varphi_j(t)dt = \int_{t_1}^{t_2} \varphi_i(t)\varphi_j^*(t)dt = \begin{Bmatrix} 0 & i \ne j \\ \lambda_i & i = j \end{Bmatrix}; \qquad \text{Eq. (2.23)}$$

the asterisk indicates complex conjugation and λ_i are constants. If $\lambda_i = 1$ for all i, then the set S is said to be an orthonormal set. A signal $x(t)$ can be expressed over the interval (t_1, t_2) as a weighted sum of a set of orthogonal functions as

$$x(t) \approx \sum_{n=1}^{N} X_n \varphi_n(t) \qquad \text{Eq. (2.24)}$$

where X_n are, in general, complex constants and the orthogonal functions $\varphi_n(t)$ are called basis functions. If the integral-square error over the interval (t_1, t_2) is equal to zero as N approaches infinity, i.e.,

$$\lim_{N \to \infty} \int_{t_1}^{t_2} \left| x(t) - \sum_{n=1}^{N} X_n \varphi_n(t) \right|^2 dt = 0 , \qquad \text{Eq. (2.25)}$$

then the set $S = \{\varphi_n(t)\}$ is said to be complete, and Eq. (2.24) becomes an equality. The constants X_n are computed as

$$X_n = \frac{\int_{t_1}^{t_2} x(t)\varphi_n^*(t)dt}{\int_{t_1}^{t_2} |\varphi_n(t)|^2 dt} . \qquad \text{Eq. (2.26)}$$

Let the signal $x(t)$ be periodic with period T, and let the complete orthogonal set S be

$$S = \left\{ e^{\frac{j2\pi nt}{T}} \; ; \; n= -\infty, \infty \right\} . \qquad \text{Eq. (2.27)}$$

Then the complex exponential Fourier series of $x(t)$ is

$$x(t) = \sum_{n=-\infty}^{\infty} X_n e^{\frac{j2\pi nt}{T}} . \qquad \text{Eq. (2.28)}$$

Using Eq. (2.27) in Eq. (2.26) yields

$$X_n = \frac{1}{T} \int_{-T/2}^{T/2} x(t) e^{\frac{-j2\pi nt}{T}} dt . \qquad \text{Eq. (2.29)}$$

The FT of Eq. (2.28) is given by

$$X(\omega) = 2\pi \sum_{n=-\infty}^{\infty} X_n \delta\left(\omega - \frac{2\pi n}{T}\right) \qquad \text{Eq. (2.30)}$$

where $\delta(\ )$ is delta function.

When the signal $x(t)$ is real, we can compute its trigonometric Fourier series from Eq. (2.28) as

$$x(t) = a_0 + \sum_{n=1}^{\infty} a_n \cos\left(\frac{2\pi nt}{T}\right) + \sum_{n=1}^{\infty} b_n \sin\left(\frac{2\pi nt}{T}\right) \qquad \text{Eq. (2.31)}$$

$$a_0 = X_0 \qquad \text{Eq. (2.32)}$$

$$a_n = \frac{1}{T} \int_{-T/2}^{T/2} x(t) \cos\left(\frac{2\pi nt}{T}\right) dt \qquad \text{Eq. (2.33)}$$

$$b_n = \frac{1}{T} \int_{-T/2}^{T/2} x(t) \sin\left(\frac{2\pi nt}{T}\right) dt . \qquad \text{Eq. (2.34)}$$

The coefficients a_n are all zeros when the signal $x(t)$ is an odd function of time. Alternatively, when the signal is an even function of time, then all b_n are equal to zero.

Consider the periodic energy signal defined in Eq. (2.31). The total energy associated with this signal is then given by

$$E = \frac{1}{T}\int_{t_0}^{t_0+T} |x(t)|^2 dt = \frac{a_0^2}{4} + \sum_{n=1}^{\infty}\left(\frac{a_n^2}{2} + \frac{b_n^2}{2}\right). \qquad \text{Eq. (2.35)}$$

2.5. Convolution and Correlation Integrals

The convolution $\rho_{xh}(t)$ between the signals $x(t)$ and $h(t)$ is defined by

$$\rho_{xh}(t) = x(t) \otimes h(t) = \int_{-\infty}^{\infty} x(\tau)h(t-\tau)d\tau \qquad \text{Eq. (2.36)}$$

where τ is a dummy variable. Convolution is commutative, associative, and distributive. More precisely,

$$\begin{aligned} x(t) \otimes h(t) &= h(t) \otimes x(t) \\ x(t) \otimes \{h(t) \otimes g(t)\} &= \{x(t) \otimes h(t)\} \otimes g(t) = x(t) \otimes \{h(t) \otimes g(t)\} \end{aligned}. \qquad \text{Eq. (2.37)}$$

For the convolution integral to be finite at least one of the two signals must be an energy signal. The convolution between two signals can be computed using the FT as

$$\rho_{xh}(t) = F^{-1}\{X(\omega)H(\omega)\}. \qquad \text{Eq. (2.38)}$$

Consider an LTI system with impulse response $h(t)$ and input signal $x(t)$. It follows that the output signal $y(t)$ is equal to the convolution between the input signal and the system impulse response,

$$y(t) = \int_{-\infty}^{\infty} x(\tau)h(t-\tau)d\tau = \int_{-\infty}^{\infty} h(\tau)x(t-\tau)d\tau. \qquad \text{Eq. (2.39)}$$

The cross-correlation function between the signals $x(t)$ and $g(t)$ is

$$R_{xg}(t) = \int_{-\infty}^{\infty} x^*(\tau)g(t+\tau)d\tau = R^*_{gx}(-t) = \int_{-\infty}^{\infty} g^*(\tau)x(t+\tau)d\tau. \qquad \text{Eq. (2.40)}$$

Again, at least one of the two signals should be an energy signal for the correlation integral to be finite. The cross-correlation function measures the similarity between the two signals.

The peak value of $R_{xg}(t)$ and its spread around this peak are an indication of how good this similarity is. This similarity is measured by a factor called *the correlation coefficient*, denoted by C_{xg}. For example, consider the signals $x(t)$ and $g(t)$; the correlation coefficient is

$$C_{xg} = \frac{\left|\int_{-\infty}^{\infty} x(t)\ g^*(t) dt\right|^2}{\int_{-\infty}^{\infty} |x(t)|^2 dt \int_{-\infty}^{\infty} |g(t)|^2 dt} = C_{gx} . \quad \text{Eq. (2.41)}$$

Clearly the correlation coefficient is limited to $0 \le C_{xg} = C_{gx} \le 1$, with $C_{xg} = 0$ indicating no similarity while $C_{xg} = 1$ indicates 100% similarity between the signals $x(t)$ and $g(t)$. The cross-correlation integral can be computed as

$$R_{xg}(t) = F^{-1}\{X^*(\omega)G(\omega)\} . \quad \text{Eq. (2.42)}$$

When $x(t) = g(t)$, we get the autocorrelation integral,

$$R_{xx}(t) = \int_{-\infty}^{\infty} x^*(\tau)x(t+\tau)d\tau . \quad \text{Eq. (2.43)}$$

When the signals $x(t)$ and $g(t)$ are power signals, the correlation integral becomes infinite, and thus time averaging must be included. More precisely,

$$\bar{R}_{xg}(t) = \lim_{T \to \infty} \frac{1}{T} \int_{-T/2}^{T/2} x^*(\tau)g(t+\tau)d\tau . \quad \text{Eq. (2.44)}$$

2.5.1. Energy and Power Spectrum Densities

Consider an energy signal $x(t)$. From Parseval's theorem, the total energy associated with this signal is

$$E = \int_{-\infty}^{\infty} |x(t)|^2 dt = \frac{1}{2\pi} \int_{-\infty}^{\infty} |X(\omega)|^2 d\omega . \quad \text{Eq. (2.45)}$$

When $x(t)$ is a voltage signal, the amount of energy dissipated by this signal when applied across a network of resistance R is

$$E = \frac{1}{R} \int_{-\infty}^{\infty} |x(t)|^2 dt = \frac{1}{2\pi R} \int_{-\infty}^{\infty} |X(\omega)|^2 d\omega . \quad \text{Eq. (2.46)}$$

Alternatively, when $x(t)$ is a current signal, we get

$$E = R \int_{-\infty}^{\infty} |x(t)|^2 dt = \frac{R}{2\pi} \int_{-\infty}^{\infty} |X(\omega)|^2 d\omega . \quad \text{Eq. (2.47)}$$

The quantity $\int |X(\omega)|^2 d\omega$ represents the amount of energy spread per unit frequency across a 1Ω resistor; therefore, the Energy Spectrum Density (ESD) function for the energy signal $x(t)$ is defined as

$$ESD = |X(\omega)|^2 . \quad \text{Eq. (2.48)}$$

The ESD at the output of an LTI system when $x(t)$ is at its input is

$$|Y(\omega)|^2 = |X(\omega)|^2 |H(\omega)|^2 \quad \text{Eq. (2.49)}$$

where $H(\omega)$ is the FT of the system impulse response, $h(t)$. It follows that the energy present at the output of the system is

$$E_y = \frac{1}{2\pi} \int_{-\infty}^{\infty} |X(\omega)|^2 |H(\omega)|^2 d\omega . \quad \text{Eq. (2.50)}$$

Example:

The voltage signal $x(t) = e^{-5t};\ t \geq 0$ *is applied to the input of a lowpass LTI system. The system bandwidth is* $5Hz$, *and its input resistance is* 5Ω. *If* $H(\omega) = 1$ *over the interval* $(-10\pi < \omega < 10\pi)$ *and zero elsewhere, compute the energy at the output.*

Solution:

From Eq. (2.50) one computes

$$E_y = \frac{1}{2\pi R} \int_{\omega = -10\pi}^{10\pi} |X(\omega)|^2 |H(\omega)|^2 d\omega .$$

Using Fourier transform tables and substituting $R = 5$ *yields*

$$E_y = \frac{1}{5\pi} \int_{0}^{10\pi} \frac{1}{\omega^2 + 25} d\omega .$$

Completing the integration yields

$$E_y = \frac{1}{25\pi}[\operatorname{atanh}(2\pi) - \operatorname{atanh}(0)] = 0.01799\ Joules .$$

Note that an infinite bandwidth would give $E_y = 0.02$, *only 11% larger.*

****

The total power associated with a power signal $g(t)$ is

$$P = \lim_{T \to \infty} \frac{1}{T} \int_{-T/2}^{T/2} |g(t)|^2 dt . \quad \text{Eq. (2.51)}$$

The Power Spectrum Density (PSD) function for the signal $g(t)$ is $S_g(\omega)$, where

$$P = \lim_{T \to \infty} \frac{1}{T} \int_{-T/2}^{T/2} |g(t)|^2 dt = \frac{1}{2\pi} \int_{-\infty}^{\infty} S_g(\omega) d\omega . \quad \text{Eq. (2.52)}$$

It can be shown that

$$S_g(\omega) = \lim_{T \to \infty} \frac{|G(\omega)|^2}{T}. \quad \text{Eq. (2.53)}$$

Let the signals $x(t)$ and $g(t)$ be two periodic signals with period T. The complex exponential Fourier series expansions for those signals are, respectively, given by

$$x(t) = \sum_{n=-\infty}^{\infty} X_n e^{j\frac{2\pi nt}{T}} \quad \text{Eq. (2.54)}$$

$$g(t) = \sum_{m=-\infty}^{\infty} G_m e^{j\frac{2\pi mt}{T}}. \quad \text{Eq. (2.55)}$$

The power cross-correlation function $\bar{R}_{gx}(t)$ was given in Eq. (2.44) and is repeated here as Eq. (2.56),

$$\bar{R}_{gx}(t) = \frac{1}{T} \int_{-T/2}^{T/2} g^*(\tau)x(t+\tau)d\tau. \quad \text{Eq. (2.56)}$$

Note that since both signals are periodic the limit is no longer necessary. Substituting Eqs. (2.54) and (2.55) into Eq. (2.56), collecting terms, and using the definition of orthogonality, yields

$$\bar{R}_{gx}(t) = \sum_{n=-\infty}^{\infty} G_n^* X_n e^{j\frac{2n\pi t}{T}}. \quad \text{Eq. (2.57)}$$

When $x(t) = g(t)$, Eq. (2.57) becomes the power autocorrelation function,

$$\bar{R}_{xx}(t) = \sum_{n=-\infty}^{\infty} |X_n|^2 e^{j\frac{2n\pi t}{T}} = |X_0|^2 + 2\sum_{n=1}^{\infty} |X_n|^2 e^{j\frac{2n\pi t}{T}}. \quad \text{Eq. (2.58)}$$

The power spectrum and cross-power spectrum density functions are then computed as the FT of Eqs. (2.58) and (2.57), respectively. More precisely,

$$\bar{S}_{xx}(\omega) = 2\pi \sum_{n=-\infty}^{\infty} |X_n|^2 \delta\left(\omega - \frac{2n\pi}{T}\right) \quad \text{Eq. (2.59)}$$

$$\bar{S}_{gx}(\omega) = 2\pi \sum_{n=-\infty}^{\infty} G_n^* X_n \delta\left(\omega - \frac{2n\pi}{T}\right). \quad \text{Eq. (2.60)}$$

The line (or discrete) power spectrum is defined as the plot of $|X_n|^2$ versus n, where the lines are $\Delta f = 1/T$ apart. The DC power is $|X_0|^2$, and the total power is

$$\sum |X_n|^2. \quad \text{Eq. (2.61)}$$

Consider a signal $x(t)$ and its FT $X(f)$. The corresponding autocorrelation function and power spectrum density are, respectively, $\bar{R}_{xx}(t)$ and $\bar{S}_{xx}(f)$. A few very useful relations that will be utilized often in this book include

$$x(0) = \int_{-\infty}^{\infty} X(f)df \qquad \text{Eq. (2.62)}$$

$$\int_{-\infty}^{\infty} x(t)dt = X(0) \qquad \text{Eq. (2.63)}$$

$$\bar{R}_{xx}(0) = \int_{-\infty}^{\infty} |x(t)|^2 dt = \int_{-\infty}^{\infty} |X(f)|^2 df = \bar{S}_{xx}(0) \qquad \text{Eq. (2.64)}$$

$$\int_{-\infty}^{\infty} |\bar{R}_{xx}(t)|^2 dt = \int_{-\infty}^{\infty} |X(f)|^4 df. \qquad \text{Eq. (2.65)}$$

Note that Eq. (2.62) or Eq. (2.63) represents the total DC power (in the case of a power signal) or voltage (in the case of an energy signal). Equation (2.64) represents the signal's total power (for power signals) or total energy (for energy signals).

2.6. Lowpass, Bandpass Signals and Quadrature Components

Signals that contain significant frequency composition at a low frequency band including DC are called lowpass (LP) signals. Signals that have significant frequency composition around some frequency away from the origin are called bandpass (BP) signals. A real BP signal $x(t)$ can be represented mathematically by

$$x(t) = r(t)\cos(2\pi f_0 t + \phi_x(t)) \qquad \text{Eq. (2.66)}$$

where $r(t)$ is the amplitude modulation or envelope, $\phi_x(t)$ is the phase modulation, f_0 is the carrier frequency, and both $r(t)$ and $\phi_x(t)$ have frequency components significantly smaller than f_0. The frequency modulation is

$$f_m(t) = \frac{1}{2\pi}\frac{d}{dt}\phi_x(t) \qquad \text{Eq. (2.67)}$$

and the instantaneous frequency is

$$f_i(t) = \frac{1}{2\pi}\frac{d}{dt}(2\pi f_0 t + \phi_x(t)) = f_0 + f_m(t). \qquad \text{Eq. (2.68)}$$

If the signal bandwidth is B and f_0 is very large compared to B, then the signal $x(t)$ is referred to as a narrow bandpass signal.

Bandpass signals can also be represented by two lowpass signals known as the quadrature components; in this case Eq. (2.66) can be rewritten as

$$x(t) = x_I(t)\cos 2\pi f_0 t - x_Q(t)\sin 2\pi f_0 t \qquad \text{Eq. (2.69)}$$

where $x_I(t)$ and $x_Q(t)$ are real LP signals referred to as the quadrature components and are given, respectively, by

$$\begin{aligned} x_I(t) &= r(t)\cos\phi_x(t) \\ x_Q(t) &= r(t)\sin\phi_x(t) \end{aligned}. \qquad \text{Eq. (2.70)}$$

2.6.1. The Analytic Signal (Pre-Envelope)

Given a real-valued signal $x(t)$, its Hilbert transform is

$$H\{x(t)\} = \hat{x}(t) = \frac{1}{\pi}\int_{-\infty}^{\infty} \frac{x(u)}{t-u}\, du. \qquad \text{Eq. (2.71)}$$

Observation of Eq. (2.71) indicates that the Hilbert transform is computed as the convolution between the signals $x(t)$ and $h(t) = 1/(\pi t)$. More precisely,

$$\hat{x}(t) = x(t) \otimes \frac{1}{\pi t}. \qquad \text{Eq. (2.72)}$$

The Fourier transform of $h(t)$ is

$$FT\{h(t)\} = FT\left\{\frac{1}{\pi t}\right\} = H(\omega) = e^{-j\frac{\pi}{2}}\,\mathrm{sgn}(\omega) \qquad \text{Eq. (2.73)}$$

where the function $\mathrm{sgn}(\omega)$ is given by

$$\mathrm{sgn}(\omega) = \frac{\omega}{|\omega|} = \left\{\begin{array}{l} 1\ ;\ \omega > 0 \\ 0;\ \omega = 0 \\ -1\ ;\ \omega < 0 \end{array}\right\}. \qquad \text{Eq. (2.74)}$$

Thus, the effect of the Hilbert transform is to introduce a phase shift of $\pi/2$ on the spectra of $x(t)$. It follows that,

$$FT\{\hat{x}(t)\} = \hat{X}(\omega) = -j\,\mathrm{sgn}(\omega)X(\omega). \qquad \text{Eq. (2.75)}$$

The analytic signal $\psi(t)$ corresponding to the real signal $x(t)$ is obtained by cancelling the negative frequency contents of $X(\omega)$. Then, by definition

$$\Psi(\omega) = \left\{\begin{array}{ll} 2X(\omega) & ;\omega > 0 \\ X(\omega) & ;\omega = 0 \\ 0 & ;\omega < 0 \end{array}\right\} \qquad \text{Eq. (2.76)}$$

or equivalently,

$$\Psi(\omega) = X(\omega)(1 + \mathrm{sgn}(\omega)). \qquad \text{Eq. (2.77)}$$

It follows that

$$\psi(t) = FT^{-1}\{\Psi(\omega)\} = x(t) + j\hat{x}(t). \qquad \text{Eq. (2.78)}$$

The analytic signal is often referred to as the pre-envelope of $x(t)$ because the envelope of $x(t)$ can be obtained by simply taking the modulus of $\psi(t)$. Note that the Hilbert transform for the bandpass signal defined in Eq. (2.69) is

$$\hat{x}(t) = x_I(t)\sin 2\pi f_0 t + x_Q(t)\cos 2\pi f_0 t. \qquad \text{Eq. (2.79)}$$

Using Eq. (2.69) and Eq. (2.79) into Eq. (2.78) and collecting terms yields

$$\psi(t) = [x_I(t) + jx_Q(t)]e^{j2\pi f_0 t} = \tilde{x}(t)e^{j2\pi f_0 t} . \qquad \text{Eq. (2.80)}$$

The signal $\tilde{x}(t) = x_I(t) + jx_Q(t)$ is the complex envelope of $x(t)$. Thus, the envelope signal and associated phase deviation are given by

$$a(t) = |\tilde{x}(t)| = |x_I(t) + jx_Q(t)| = |\psi(t)| \qquad \text{Eq. (2.81)}$$

$$\phi(t) = \arg(\tilde{x}(t)) = \angle\tilde{x}(t). \qquad \text{Eq. (2.82)}$$

In summary, a bandpass signal $x(t)$ and its corresponding pre-envelope (analytic signal) and complex envelope are

$$x(t) = x_I(t)\cos 2\pi f_0 t - x_Q(t)\sin 2\pi f_0 t \qquad \text{Eq. (2.83)}$$

$$\psi(t) = x(t) + j\hat{x}(t) \equiv \tilde{x}(t)e^{j2\pi f_0 t} \qquad \text{Eq. (2.84)}$$

$$\tilde{x}(t) = x_I(t) + jx_Q(t). \qquad \text{Eq. (2.85)}$$

Obtaining the complex envelope for any bandpass signal requires extraction of the quadrature components. Figure 2.1 shows how the quadrature components can be extracted from a bandpass signal. First, the bandpass signal is split into two parts; one part is mixed with $2\cos 2\pi f_0 t$, and the other is mixed with $-2\sin 2\pi f_0 t$. From the figure, the two signals $x_1(t)$ and $x_2(t)$ are,

$$x_1(t) = 2x_I(t)(\cos 2\pi f_0 t)^2 - 2x_Q(t)\cos(2\pi f_0 t)\sin(2\pi f_0 t) \qquad \text{Eq. (2.86)}$$

$$x_2(t) = -2x_I(t)\cos(2\pi f_0 t)\sin(2\pi f_0 t) + 2x_Q(t)(\sin 2\pi f_0 t)^2. \qquad \text{Eq. (2.87)}$$

Finally, using the appropriate trigonometry identities and after lowpass filtering the quadrature components are extracted.

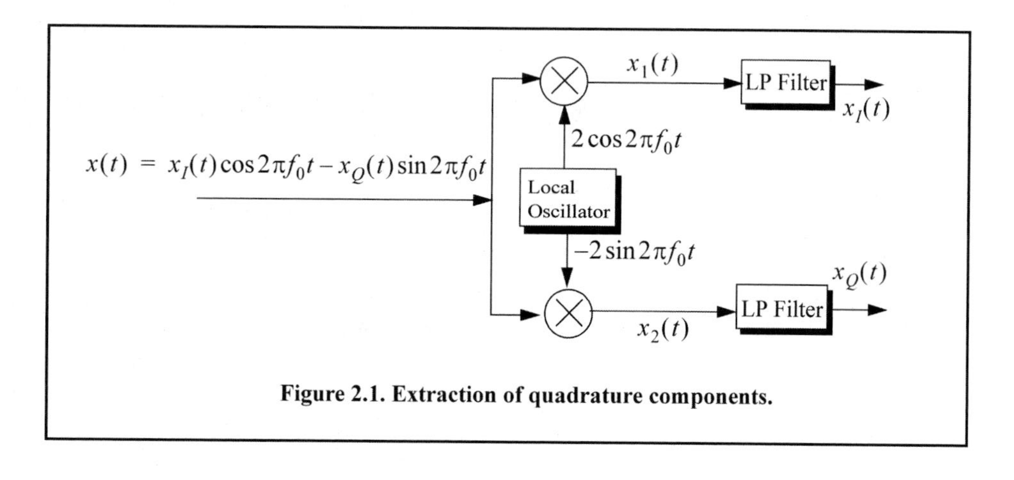

Figure 2.1. Extraction of quadrature components.

2.7. Random Variables

Consider an experiment with outcomes defined by a certain sample space. The rule or functional relationship that maps each point in this sample space into a real number is called the "random variable". Random variables are designated by capital letters (e.g., $X, Y, \ldots$), and a particular value of a random variable is denoted by a lowercase letter (e.g., $x, y, \ldots$). The Cumulative Distribution Function (*cdf*) associated with the random variable X is denoted as $F_X(x)$, and is interpreted as the total probability that the random variable X is less or equal to the value x. More precisely,

$$F_X(x) = Pr\{X \le x\} . \qquad \text{Eq. (2.88)}$$

The probability that the random variable X is in the interval (x_1, x_2) is then given by

$$F_X(x_2) - F_X(x_1) = Pr\{x_1 \le X \le x_2\} . \qquad \text{Eq. (2.89)}$$

The *cdf* has the following properties,

$$\begin{gathered} 0 \le F_X(x) \le 1 \\ F_X(-\infty) = 0 \\ F_X(\infty) = 1 \\ F_X(x_1) \le F_X(x_2) \Leftrightarrow x_1 \le x_2 \end{gathered} \qquad . \qquad \text{Eq. (2.90)}$$

It is often practical to describe a random variable by the derivative of its *cdf*, which is called the Probability Density Function *(pdf)*. The *pdf* of the random variable X *is*

$$f_X(x) = \frac{\mathrm{d}}{\mathrm{d}x} F_X(x) \qquad \text{Eq. (2.91)}$$

or, equivalently,

$$F_X(x) = Pr\{X \le x\} = \int_{-\infty}^{x} f_X(\lambda) d\lambda . \qquad \text{Eq. (2.92)}$$

The probability that a random variable X has values in the interval (x_1, x_2) is

$$F_X(x_2) - F_X(x_1) = Pr\{x_1 \le X \le x_2\} = \int_{x_1}^{x_2} f_X(x) dx . \qquad \text{Eq. (2.93)}$$

Define the *nth* moment for the random variable X as

$$E[X^n] = \overline{X^n} = \int_{-\infty}^{\infty} x^n f_X(x) dx . \qquad \text{Eq. (2.94)}$$

The first moment, $E[X]$, is called the mean value, while the second moment, $E[X^2]$, is called the mean squared value. When the random variable X represents an electrical signal across a 1Ω resistor, then $E[X]$ is the DC component, and $E[X^2]$ is the total average power.

The *nth* central moment is defined as

$$E[(X-\bar{X})^n] = \overline{(X-\bar{X})^n} = \int_{-\infty}^{\infty} (x-\bar{x})^n f_X(x)dx \qquad \text{Eq. (2.95)}$$

and thus, the first central moment is zero. The second central moment is called the variance and is denoted by the symbol σ_X^2,

$$\sigma_X^2 = \overline{(X-\bar{X})^2}. \qquad \text{Eq. (2.96)}$$

Appendix 2-B has some common *pdf*s and their means and variances.

In practice, the random nature of an electrical signal may need to be described by more than one random variable. In this case, the joint *cdf* and *pdf* functions need to be considered. The joint *cdf* and *pdf* for the two random variables X and Y are respectively defined by,

$$F_{XY}(x, y) = Pr\{X \le x; Y \le y\} \qquad \text{Eq. (2.97)}$$

$$f_{XY}(x, y) = \frac{\partial^2}{\partial x \partial y} F_{XY}(x, y). \qquad \text{Eq. (2.98)}$$

The marginal *cdf*s are obtained as follows,

$$\begin{aligned} F_X(x) &= \int_{-\infty}^{\infty}\int_{-\infty}^{x} f_{UV}(u, v)dudv = F_{XY}(x, \infty) \\ F_Y(y) &= \int_{-\infty}^{\infty}\int_{-\infty}^{y} f_{UV}(u, v)dvdu = F_{XY}(\infty, y) \end{aligned}. \qquad \text{Eq. (2.99)}$$

If the two random variables are statistically independent, then the joint *cdf*s and *pdf*s are respectively given by

$$F_{XY}(x, y) = F_X(x)F_Y(y) \qquad \text{Eq. (2.100)}$$

$$f_{XY}(x, y) = f_X(x)f_Y(y). \qquad \text{Eq. (2.101)}$$

Let us now consider a case when the two random variables X and Y are mapped into two new variables U and V through some transformations T_1 and T_2 defined by

$$\begin{aligned} U &= T_1(X, Y) \\ V &= T_2(X, Y) \end{aligned}. \qquad \text{Eq. (2.102)}$$

The joint *pdf*, $f_{UV}(u, v)$, may be computed based on the invariance of probability under the transformation. One must first compute the matrix of derivatives; then the new joint *pdf* is computed as

$$f_{UV}(u, v) = f_{XY}(x, y)|\mathbf{J}|, \qquad \text{Eq. (2.103)}$$

where

$$|\mathbf{J}| = \begin{vmatrix} \frac{\partial x}{\partial u} & \frac{\partial x}{\partial v} \\ \frac{\partial y}{\partial u} & \frac{\partial y}{\partial v} \end{vmatrix} . \qquad \text{Eq. (2.104)}$$

Note that the determinant of the matrix of derivatives $|\mathbf{J}|$ is called the Jacobian.

The characteristic function for the random variable X is defined as

$$C_X(\omega) = E[e^{j\omega X}] = \int_{-\infty}^{\infty} f_X(x)e^{j\omega x}dx . \qquad \text{Eq. (2.105)}$$

The characteristic function can be used to compute the *pdf* for a sum of independent random variables. More precisely, let the random variable Y be equal to

$$Y = X_1 + X_2 + \ldots + X_N \qquad \text{Eq. (2.106)}$$

where $\{X_i \ ; \ i = 1, \ldots N\}$ is a set of independent random variables. It can be shown that

$$C_Y(\omega) = C_{X_1}(\omega)C_{X_2}(\omega)\ldots C_{X_N}(\omega) \qquad \text{Eq. (2.107)}$$

and the *pdf* $f_Y(y)$, is computed as the inverse Fourier transform of $C_Y(\omega)$ (with the sign of y reversed),

$$f_Y(y) = \frac{1}{2\pi}\int_{-\infty}^{\infty} C_Y(\omega)e^{-j\omega y}d\omega . \qquad \text{Eq. (2.108)}$$

The characteristic function may also be used to compute the nth moment for the random variable X as

$$E[X^n] = (-j)^n \frac{\mathrm{d}^n}{\mathrm{d}\omega^n} C_X(\omega)\bigg|_{\omega = 0} . \qquad \text{Eq. (2.109)}$$

2.8. Random Processes

A random variable X is by definition a mapping of all possible outcomes of a random experiment to numbers. When the random variable becomes a function of both the outcomes of the experiment as well as time, it is called a random process and is denoted by $X(t)$. Thus, one can view a random process as an ensemble of time domain functions that are the outcome of a certain random experiment, as compared to single real numbers in the case of a random variable.

Since the *cdf* and *pdf* of a random process are time dependent, we will denote them as $F_X(x;t)$ and $f_X(x;t)$, respectively. The nth moment for the random process $X(t)$ is

$$E[X^n(t)] = \int_{-\infty}^{\infty} x^n f_X(x;t)dx . \qquad \text{Eq. (2.110)}$$

A random process $X(t)$ is referred to as stationary to order one if all its statistical properties do not change with time. Consequently, $E[X(t)] = \bar{X}$, where $\bar{X}$ is a constant. A random process $X(t)$ is called stationary to order two (or wide sense stationary) if

$$f_X(x_1, x_2; t_1, t_2) = f_X(x_1, x_2; t_1 + \Delta t, t_2 + \Delta t), \qquad \text{Eq. (2.111)}$$

for all t_1, t_2 and Δt. Define the statistical autocorrelation function for the random process $X(t)$ as

$$\Re_X(t_1, t_2) = E[X(t_1)X(t_2)]. \qquad \text{Eq. (2.112)}$$

The correlation $E[X(t_1)X(t_2)]$ is, in general, a function of (t_1, t_2). As a consequence of the wide sense stationary definition, the autocorrelation function depends on the time difference $\tau = t_2 - t_1$, rather than on absolute time; and thus, for a wide sense stationary process we have

$$\begin{aligned} E[X(t)] &= \bar{X} \\ \Re_X(\tau) &= E[X(t)X(t+\tau)] \end{aligned}. \qquad \text{Eq. (2.113)}$$

If the time average and time correlation functions are equal to the statistical average and statistical correlation functions, the random process is referred to as an ergodic random process. The following is true for an ergodic process

$$\lim_{T \to \infty} \frac{1}{T} \int_{-T/2}^{T/2} x(t)dt = E[X(t)] = \bar{X} \qquad \text{Eq. (2.114)}$$

$$\lim_{T \to \infty} \frac{1}{T} \int_{-T/2}^{T/2} x^*(t)x(t+\tau)dt = \Re_X(\tau). \qquad \text{Eq. (2.115)}$$

The covariance of two random process $X(t)$ and $Y(t)$ is defined by

$$Cov_{XY}(t, t+\tau) = E[\{X(t) - E[X(t)]\}\{Y(t+\tau) - E[Y(t+\tau)]\}], \qquad \text{Eq. (2.116)}$$

which can be written as

$$Cov_{XY}(t, t+\tau) = \Re_{XY}(\tau) - \bar{X}\bar{Y}. \qquad \text{Eq. (2.117)}$$

2.9. Discrete Time Systems and Signals

Advances in computer hardware and in digital technologies completely revolutionized radar systems signal and data processing techniques. Virtually all modern radar systems use some form of a digital representation (signal samples) of their received signals for the purposes of signal and data processing. These samples of a time-limited signal are nothing more than a finite set of numbers (thought of as a vector) that represents discrete values of the continuous time domain signal. These samples are typically obtained by using Analog-to-Digital (A/D) conversion devices.

Since in the digital world the radar receiver is now concerned with processing a set of finite numbers, its impulse response will also compose a set of finite numbers. Consequently, the radar receiver is now referred to as a discrete system. All input/output signal relationships are now carried out using discrete time samples. It must also be noted that just as in the case of continuous time-domain systems, the discrete systems of interest to radar applications must also be causal, stable, and linear time invariant.

Consider a continuous lowpass signal that is essentially time-limited with duration τ and band-limited with bandwidth B. This signal (as will be shown in the next section) can be completely represented by a set of $\{2\tau B\}$ samples. Since a finite set of discrete values (samples) is used to represent the signal, it is common to represent this signal by a finite dimensional vector of the same size. This vector is denoted by $\mathbf{x}$, or simply by the sequence $x[n]$,

$$\mathbf{x} \equiv x[n] = [x(0)\ \ x(1)\ \ \dots x(N-2)\ \ x(N-1)]^t \quad \text{Eq. (2.118)}$$

where the superscript t denotes transpose operation. The value N is at least $2\tau B$ for a real lowpass essentially limited signal $x(t)$ of duration τ and bandwidth B. If, however, the signal is complex, then N is at least τB and the components of the vector $\mathbf{x}$ are complex. The samples defined in Eq. (2.118) can be obtained from pulse-to-pulse samples at a fixed range (i.e., delay) of the radar echo signal. The PRF is denoted by f_r and the total observation interval is T_0; then N would be equal to $T_0 f_r$. Define the radar receiver transfer function as the discrete sequence $h[n]$ and the input signal sequence as $x[n]$; then the output sequence $y[n]$ is given by the convolution sum

$$y[n] = \sum_{m=0}^{M-1} h(m)x(n-m) \quad \text{Eq. (2.119)}$$

where $\{h[n] = [h(0)\ \ h(1)\ \ \dots h(M-2)\ \ h(M-1)];\ M \le N\}$.

2.9.1. Sampling Theorem

In general, it is required to determine the necessary condition such that a signal can be fully reconstructed from its samples by filtering, or data processing. The answer to this question lies in the sampling theorem, which may be stated as follows: let the signal $x(t)$ be real-valued, essentially band-limited by the bandwidth B; this signal can be fully reconstructed from its samples if the time interval between samples is no greater than $1/(2B)$. Figure 2.2 illustrates the sampling process concept. The sampling signal $p(t)$ is periodic with period T_s, which is called the sampling interval. The Fourier series expansion of $p(t)$ and the sampled signal $x_s(t)$ expressed using this Fourier series definition are, respectively, given by

$$p(t) = \sum_{n=-\infty}^{\infty} P_n e^{j\frac{2\pi nt}{T_s}} \quad \text{Eq. (2.120)}$$

$$x_s(t) = p(t) \cdot x(t) \quad \text{Eq. (2.121)}$$

$$x_s(t) = \sum_{n=-\infty}^{\infty} x(t) P_n e^{j\frac{2\pi nt}{T_s}} . \quad \text{Eq. (2.122)}$$

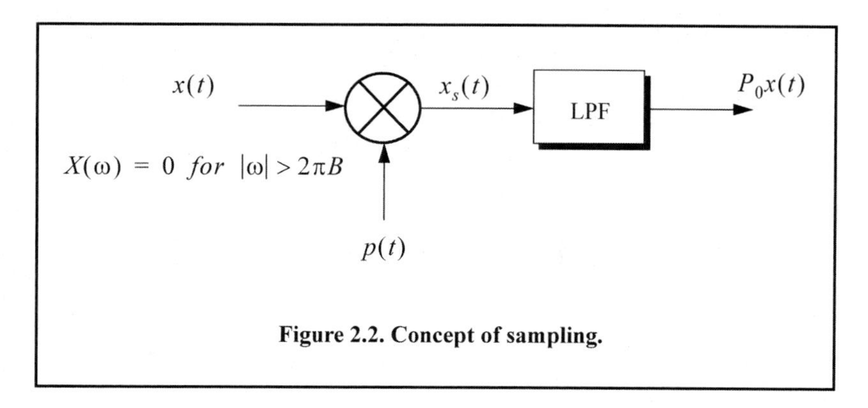

Figure 2.2. Concept of sampling.

Taking the FT of Eq. (2.122) yields

$$X_s(\omega) = \sum_{n=-\infty}^{\infty} P_n\, X\left(\omega - \frac{2\pi n}{T_s}\right) = P_0 X(\omega) + \sum_{\substack{n=-\infty \\ n \neq 0}}^{\infty} P_n\, X\left(\omega - \frac{2\pi n}{T_s}\right) \qquad \text{Eq. (2.123)}$$

where $X(\omega)$ is the FT of $x(t)$. Therefore, we conclude that the spectral density, $X_s(\omega)$, consists of replicas of $X(\omega)$ spaced $(2\pi/T_s)$ apart and scaled by the Fourier series coefficients P_n. A lowpass filter (LPF) of bandwidth B can then be used to recover the original signal $x(t)$.

When the sampling rate is increased (i.e., T_s decreases), the replicas of $X(\omega)$ move farther apart. Alternatively, when the sampling rate is decreased (i.e., T_s increases), the replicas get closer to one another. The value of T_s such that the replicas are tangent to one another defines the minimum required sampling rate so that $x(t)$ can be recovered from its samples by using an LPF. It follows that,

$$\frac{2\pi}{T_s} = 2\pi(2B) \Leftrightarrow T_s = \frac{1}{2B}. \qquad \text{Eq. (2.124)}$$

The sampling rate defined by Eq. (2.124) is known as the Nyquist sampling rate. When $T_s > (1/2B)$, the replicas of $X(\omega)$ overlap, and thus $x(t)$ cannot be recovered cleanly from its samples. This is known as aliasing. In practice, ideal LPF cannot be implemented; hence, practical systems tend to oversample in order to avoid aliasing.

Example:

Assume that the sampling signal $p(t)$ is given by

$$p(t) = \sum_{n=-\infty}^{\infty} \delta(t - nT_s).$$

Compute an expression for $X_s(\omega)$.

Solution:

The signal $p(t)$ is called the Comb function, with exponential Fourier series

$$p(t) = \sum_{n=-\infty}^{\infty} \frac{1}{T_s} e^{2\frac{\pi n t}{T_s}}.$$

It follows that

$$x_s(t) = \sum_{n=-\infty}^{\infty} x(t) \frac{1}{T_s} e^{2\frac{\pi n t}{T_s}}.$$

Taking the Fourier transform of this equation yields

$$X_s(\omega) = \frac{2\pi}{T_s} \sum_{n=-\infty}^{\infty} X\left(\omega - \frac{2\pi n}{T_s}\right).$$

****

It is desired to develop a general expression from which any lowpass signal can be recovered from its samples. In order to do that, let $x(t)$ and $x_s(t)$ be the desired lowpass signal and its corresponding samples, respectively. Then an expression for $x(t)$ in terms of its samples can be derived as follows. First, obtain $X(\omega)$ by filtering the signal $X_s(\omega)$ using an ideal LPF whose transfer function is

$$H(\omega) = T_s Rect\left(\frac{\omega}{4\pi B}\right). \quad \text{Eq. (2.125)}$$

Thus,

$$X(\omega) = H(\omega) X_s(\omega) = T_s Rect\left(\frac{\omega}{4\pi B}\right) X_s(\omega). \quad \text{Eq. (2.126)}$$

The signal $x(t)$ is now obtained from the inverse FT of Eq. (2.126) as

$$x(t) = FT^{-1}\left\{ T_s Rect\left(\frac{\omega}{4\pi B}\right) X_s(\omega) \right\} = 2BT_s Sinc(2\pi B t) \otimes x_s(t). \quad \text{Eq. (2.127)}$$

The sampled signal $x_s(t)$ can be represented using an ideal sampling signal

$$p(t) = \sum_n \delta(t - nT_s) \quad \text{Eq. (2.128)}$$

thus,

$$x_s(t) = \sum_n x(nT_s) \delta(t - nT_s). \quad \text{Eq. (2.129)}$$

Substituting Eq. (2.129) into Eq. (2.127) yields an expression for the signal $x(t)$ in terms of its samples

$$x(t) = 2BT_s \sum_n x(nT_s) \; Sinc(2\pi B(t - nT_s)) \;\; ;T_s \le \frac{1}{2B}. \quad \text{Eq. (2.130)}$$

It was established earlier that any bandpass signal can be expressed using the quadrature components. It follows that it is sufficient to construct the bandpass signal $x(t)$ from samples of the quadrature components $\{x_I(t), x_Q(t)\}$. Let the signal $x(t)$ be essentially bandlimited with bandwidth B; then each of the lowpass signals $x_I(t)$ and $x_Q(t)$ is also bandlimited, each with bandwidth $B/2$. Hence, if either of these lowpass signals is sampled at a rate $f_s \le 1/B$, then the Nyquist criterion is not violated. Assume that both quadrature components are sampled synchronously, that is

$$x_I(t) = BT_s \sum_{n=-\infty}^{\infty} x_I(nT_s)\ Sinc(\pi B(t - nT_s)) \qquad \text{Eq. (2.131)}$$

$$x_Q(t) = BT_s \sum_{n=-\infty}^{\infty} x_Q(nT_s)\ Sinc(\pi B(t - nT_s)) \qquad \text{Eq. (2.132)}$$

where if the Nyquist rate is satisfied, then $BT_s = 1$ (unity time bandwidth product).

2.10. The Z-Transform

The Z-transform is a transformation that maps samples of a discrete time-domain sequence into a new domain known as the z-domain. It is defined as

$$Z\{x(n)\} = X(z) = \sum_{n=-\infty}^{\infty} x(n)z^{-n} \qquad \text{Eq. (2.133)}$$

$z = re^{j\omega}$, and for most cases, $r = 1$. It follows that Eq. (2.133) can be rewritten as

$$X(e^{j\omega}) = \sum_{n=-\infty}^{\infty} x(n)e^{-jn\omega}. \qquad \text{Eq. (2.134)}$$

The region over which $X(z)$ is finite is called the Region of Convergence (ROC).

Example:

Show that $Z\{nx(n)\} = -z\frac{d}{dz}X(z)$.

Solution:

Starting with the definition of the Z-transform,

$$X(z) = \sum_{n=-\infty}^{\infty} x(n)z^{-n}.$$

Taking the derivative, with respect to z, of the above equation yields

$$\frac{d}{dz}X(z) = \sum_{n=-\infty}^{\infty} x(n)(-n)z^{-n-1} = (-z^{-1})\sum_{n=-\infty}^{\infty} nx(n)z^{-n}.$$

It follows that

$$Z\{nx(n)\} = (-z)\frac{d}{dz}X(z).$$

****

A discrete LTI system has a transfer function $H(z)$ that describes how the system operates on its input sequence $x(n)$ in order to produce the output sequence $y(n)$. The output sequence $y(n)$ is computed from the discrete convolution between the sequences $x(n)$ and $h(n)$:

$$y(n) = \sum_{m=-\infty}^{\infty} x(m)h(n-m). \qquad \text{Eq. (2.135)}$$

However, since practical systems require the sequence $x(n)$ and $h(n)$ to be of finite length, we can rewrite Eq. (2.135) as

$$y(n) = \sum_{m=0}^{N} x(m)h(n-m); \qquad \text{Eq. (2.136)}$$

where N denotes the input sequence length. The Z-transform of Eq. (2.136) is

$$Y(z) = X(z)H(z) \qquad \text{Eq. (2.137)}$$

where $H(z)$ is the transfer function. The transfer function can be written as

$$H(z)\big|_{z=e^{j\omega}} = \left|H(e^{j\omega})\right| e^{\angle H(e^{j\omega})} \qquad \text{Eq. (2.138)}$$

where $\left|H(e^{j\omega})\right|$ is the amplitude response, and $\angle H(e^{j\omega})$ is the phase response. A detailed table of the FT pairs is listed in Appendix 2-C.

2.11. The Discrete Fourier Transform

The Discrete Fourier Transform (DFT) is a mathematical operation that transforms a discrete sequence, usually from the time domain into the frequency domain, in order to explicitly determine the spectral information for the sequence. The time-domain sequence can be real or complex.

The DFT has finite length N and is periodic with period N. The discrete Fourier transform pairs for the finite sequence $x(n)$ are defined by

$$X(k) = \sum_{n=0}^{N-1} x(n)e^{-j\frac{2\pi nk}{N}} \quad ; \; k = 0, \ldots, N-1 \qquad \text{Eq. (2.139)}$$

$$x(n) = \frac{1}{N}\sum_{k=0}^{N-1} X(k)e^{j\frac{2\pi nk}{N}} \quad ; \; n = 0, \ldots, N-1. \qquad \text{Eq. (2.140)}$$

The Fast Fourier Transform (FFT) is not a new kind of transform that is different from the DFT. Instead, it is an algorithm used to compute the DFT more efficiently. There are numerous FFT algorithms that can be found in the literature. In this book we will interchangeably use the terms DFT and the FFT to mean the same thing. Furthermore, we will assume a radix-2 FFT algorithm, where the FFT size is equal to $N = 2^m$ for some integer m.

2.11.1. Discrete Power Spectrum

Practical discrete systems utilize DFTs of finite length as a means of numerical approximation for the Fourier transform. The input signals must be truncated to a finite duration (denoted by T) before they are sampled. This is necessary so that a finite length sequence is generated prior to signal processing. Unfortunately, this truncation process may cause serious spectral leakage problems.

To demonstrate this difficulty, consider the time-domain signal $x(t) = \sin 2\pi f_0 t$. The spectrum of $x(t)$ consists of two spectral lines at $\pm f_0$. Now, when $x(t)$ is truncated to length T seconds and sampled at a rate $T_s = T/N$, where N is the number of desired samples, we produce the sequence $\{x(n);\ n = 0, 1, \ldots, N-1\}$.

The spectrum of $x(n)$ would still be composed of the same spectral lines if T is an integer multiple of T_s and if the DFT frequency resolution Δf is an integer multiple of f_0. Unfortunately, those two conditions are rarely met, and as a consequence, the spectrum of $x(n)$ spreads over several lines (normally the spread may extend up to three lines). This is known as spectral leakage. Since f_0 is normally unknown, this discontinuity caused by an arbitrary choice of T cannot be avoided. Windowing techniques can be used to mitigate the effect of this discontinuity by applying smaller weights to samples close to the edges.

A truncated sequence $x(n)$ can be viewed as one period of some periodic sequence with period N. The discrete Fourier series expansion of $x(n)$ is

$$x(n) = \sum_{k=0}^{N-1} X_k e^{j\frac{2\pi nk}{N}} . \qquad \text{Eq. (2.141)}$$

It can be shown that the coefficients X_k are given by

$$X_k = \frac{1}{N}\sum_{n=0}^{N-1} x(n) e^{-j\frac{2\pi nk}{N}} = \frac{1}{N}X(k) \qquad \text{Eq. (2.142)}$$

where $X(k)$ is the DFT of $x(n)$. Therefore, the Discrete Power Spectrum (DPS) for the band-limited sequence $x(n)$ is the plot of $|X_k|^2$ versus k, where the lines are Δf apart,

$$P_0 = \frac{1}{N^2}|X(0)|^2 \qquad \text{Eq. (2.143)}$$

$$P_k = \frac{1}{N^2}\{|X(k)|^2 + |X(N-k)|^2\} \quad ; \ k = 1, 2, \ldots, \frac{N}{2} - 1 \qquad \text{Eq. (2.144)}$$

$$P_{N/2} = \frac{1}{N^2}|X(N/2)|^2. \qquad \text{Eq. (2.145)}$$

Before proceeding to the next section, we will show how to select the FFT parameters. For this purpose, consider a band-limited signal $x(t)$ with bandwidth B. If the signal is not band-limited, an LPF can be used to eliminate frequencies greater than B. In order to satisfy the sampling theorem, one must choose a sampling frequency $f_s = 1/T_s$, such that

$$f_s \geq 2B. \qquad \text{Eq. (2.146)}$$

The truncated sequence duration T and the total number of samples N are related by

$$T = NT_s \qquad \text{Eq. (2.147)}$$

or equivalently,

$$f_s = N/T. \qquad \text{Eq. (2.148)}$$

It follows that

$$f_s = \frac{N}{T} \geq 2B \qquad \text{Eq. (2.149)}$$

and the frequency resolution is

$$\Delta f = \frac{1}{NT_s} = \frac{f_s}{N} = \frac{1}{T} \geq \frac{2B}{N}. \qquad \text{Eq. (2.150)}$$

2.11.2. Windowing Techniques

Truncation of the sequence $x(n)$ can be accomplished by computing the product

$$x_w(n) = x(n)w(n) \qquad \text{Eq. (2.151)}$$

where

$$w(n) = \begin{Bmatrix} f(n) & ; n = 0, 1, \ldots, N-1 \\ 0 & otherwise \end{Bmatrix} \qquad \text{Eq. (2.152)}$$

where $f(n) \leq 1$. The finite sequence $w(n)$ is called a windowing sequence, or simply a window. The windowing process should not impact the phase response of the truncated sequence. Consequently, the sequence $w(n)$ must retain linear phase. This can be accomplished by making the window symmetrical with respect to its central point.

If $f(n) = 1$ for all n, we have what is known as the rectangular window. It leads to the Gibbs phenomenon, which manifests itself as an overshoot and a ripple before and after a discontinuity. Figure 2.3 shows the amplitude spectrum of a rectangular window. Note that the first sidelobe is at $-13.46dB$ below the main lobe. Windows that place smaller weights on the samples near the edges will have less overshoot at the discontinuity points (lower sidelobes); hence, they are more desirable than a rectangular window. However, reduction of the sidelobes is offset by a widening of the main lobe. Therefore, the proper choice of a windowing sequence is a continuous trade-off between sidelobe reduction and mainlobe widening. Table 2.1 gives a summary of some commonly used windows with the corresponding impact on main beam widening and peak reduction. Table 2.2 lists the mathematical expressions for some common windows.

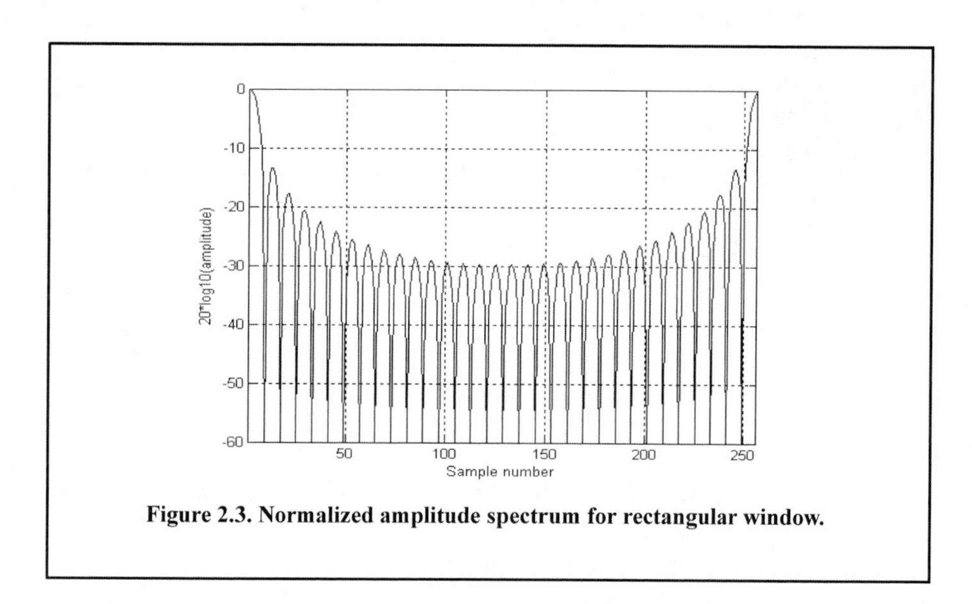

Figure 2.3. Normalized amplitude spectrum for rectangular window.

Table 2.1. Common Windows.

Window	Beamwidth reference to rectangular window	Peak reduction
Rectangular	*1*	*1*
Hamming	*2*	*0.73*
Hanning	*2*	*0.664*
Blackman	*6*	*0.577*
Kaiser $(\beta = 6)$	*2.76*	*0.683*
Kaiser $(\beta = 3)$	*1.75*	*0.882*

Table 2.2 Some common windows, $n = 0, \ldots, N-1$.

Window	Expression	First side lobe	Main lobe width
Rectangular	$w(n) = 1$	$-13.46dB$	1
Hamming	$w(n) = 0.54 - 0.46\cos\left(\frac{2\pi n}{N-1}\right)$	$-41dB$	2
Hanning	$w(n) = 0.5\left[1 - \cos\left(\frac{2\pi n}{N-1}\right)\right]$	$-32dB$	2
Kaiser	$w(n) = \frac{I_0[\beta\sqrt{1-(2n/N)^2}]}{I_0(\beta)}$; I_0 is the zero-order modified Bessel function of the first kind	$-46dB$ *for* $\beta = 2\pi$	$\sqrt{5}$ *for* $\beta = 2\pi$

The multiplication process defined in Eq. (2.151) is equivalent to cyclic convolution in the frequency domain. It follows that $X_w(k)$ is a smeared (distorted) version of $X(k)$. To minimize this distortion, we would seek windows that have a narrow main lobe and small side-lobes. Additionally, using a window other than a rectangular window reduces the power by a factor P_w, where

$$P_w = \frac{1}{N}\sum_{n=0}^{N-1} w^2(n) = \sum_{k=0}^{N-1} |W(k)|^2 . \qquad \text{Eq. (2.153)}$$

It follows that the DPS for the sequence $x_w(n)$ is now given by

$$P_0^w = \frac{1}{P_w N^2}|X(0)|^2 \qquad \text{Eq. (2.154)}$$

$$P_k^w = \frac{1}{P_w N^2}\{|X(k)|^2 + |X(N-k)|^2\} \qquad ; k = 1, 2, \ldots, \frac{N}{2} - 1 \qquad \text{Eq. (2.155)}$$

$$P_{N/2}^w = \frac{1}{P_w N^2}|X(N/2)|^2 \qquad \text{Eq. (2.156)}$$

where P_w is defined in Eq. (2.153). Figures 2.4 through 2.6 show the frequency domain characteristics for these windows.

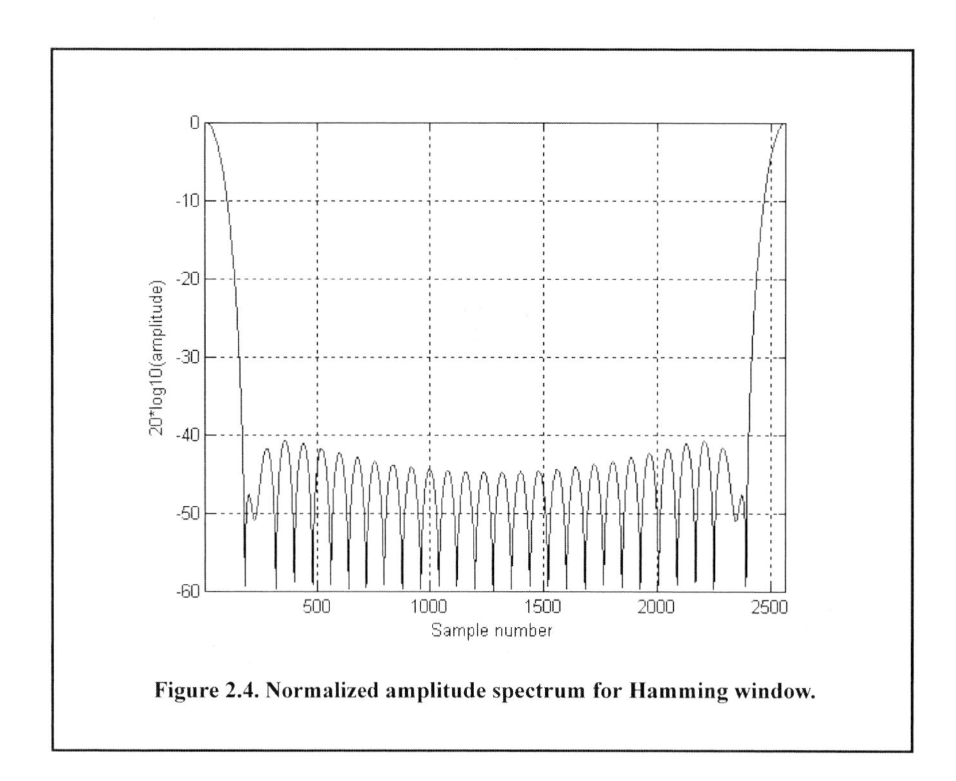

Figure 2.4. Normalized amplitude spectrum for Hamming window.

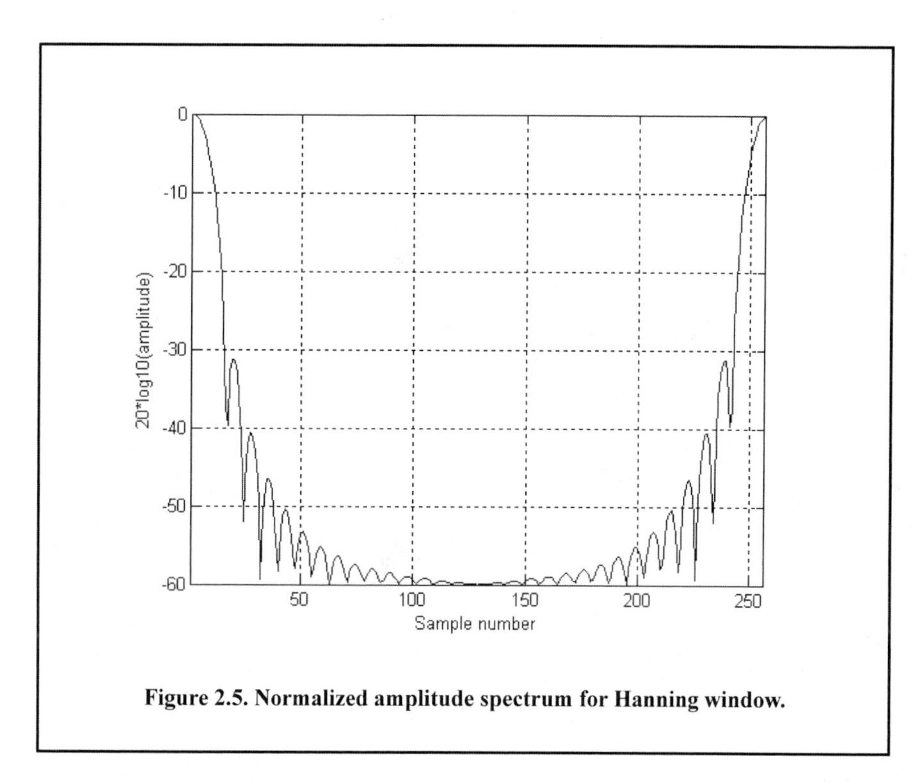

Figure 2.5. Normalized amplitude spectrum for Hanning window.

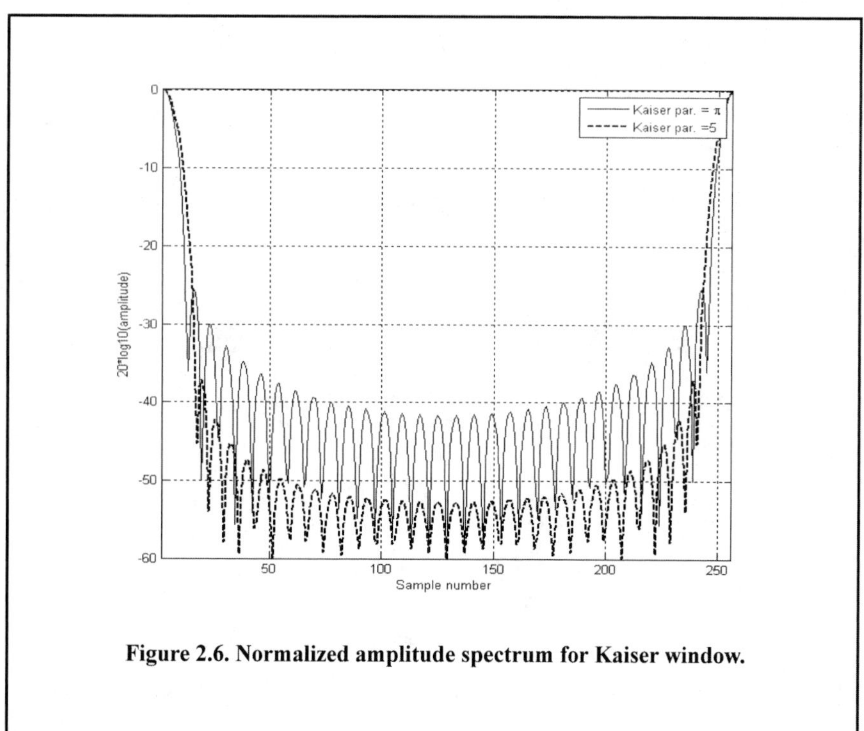

Figure 2.6. Normalized amplitude spectrum for Kaiser window.

Problems

2.1. Classify each of the following signals as an energy signal, a power signal, or neither.

(a) $\exp(0.5t)$ $(t \geq 0)$; (b) $\exp(-0.5t)$ $(t \geq 0)$; (c) $\cos t + \cos 2t$ $(-\infty < t < \infty)$; (d) $e^{-a|t|}$ $(a > 0)$.

2.2. Compute the energy associated with the signal $x(t) = ARect(t/\tau)$.

2.3. (a) Prove that $\varphi_1(t)$ and $\varphi_2(t)$, shown in the figure below, are orthogonal over the interval $(-2 \leq t \leq 2)$. (b) Express the signal $x(t) = t$ as a weighted sum of $\varphi_1(t)$ and $\varphi_2(t)$ over the same time interval.

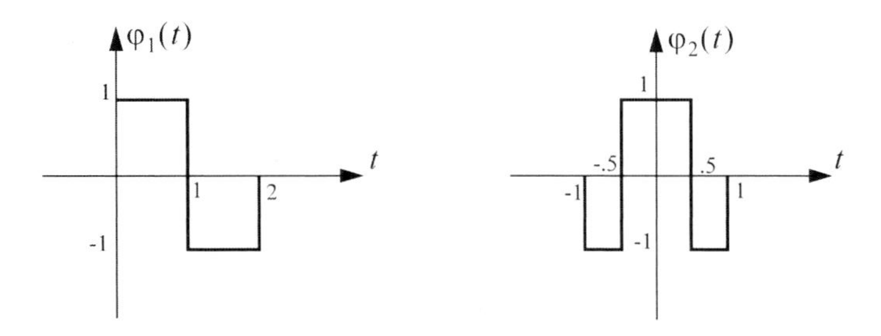

2.4. Derive Eq. (2.53).

2.5. If the Fourier series is

$$x(t) = \sum_{n=-\infty}^{\infty} X_n e^{j2\pi nt/T},$$

define $y(t) = x(t - t_0)$. Compute an expression for the complex Fourier series expansion of $y(t)$.

2.6. A definition for the instantaneous frequency was given in Eq. (2.68). A more general definition is

$$f_i(t) = \frac{1}{2\pi} Im\left\{\frac{d}{dt}\ln\psi(t)\right\}$$

where $Im\{.\}$, indicates imaginary part and $\psi(t)$ is the analytic signal. Using this definition, calculate the instantaneous frequency for

$$x(t) = Rect\left(\frac{t}{\tau}\right)\cos\left(2\pi f_0 t + \frac{B}{2\tau}t^2\right).$$

2.7. Consider the two bandpass signals $x(t) = r_x(t)\cos(2\pi f_0 t + \phi_x(t))$ and $h(t) = r_h(t)\cos(2\pi f_0 t + \phi_h(t))$. Derive an expression for the complex envelope for the signal $s(t) = x(t) + h(t)$.

2.8. Consider the bandpass signal $x(t)$ whose complex envelope is equal to $\tilde{x}(t) = x_I(t) + jx_Q(t)$. Derive an expression for the autocorrelation function and the power spectrum density for $x(t)$ and $\tilde{x}(t)$. Assume that the signal $x(t)$ is the input to an LTI filter whose impulse response is $h(t)$; give an expression for the output's autocorrelation and power spectrum density.

2.9. Find the autocorrelation integral of the pulse train

$$y(t) = Rect(t/T) - Rect\left(\frac{t-T}{T}\right) + Rect\left(\frac{t-2T}{T}\right);$$

where,

$$Rect\left(\frac{t}{\tau_0}\right) = \left\{\begin{array}{ll} 1 & -\frac{\tau_0}{2} \le t \le \frac{\tau_0}{2} \\ 0 & otherwise \end{array}\right\}.$$

2.10. Compute the discrete convolution $y(n) = x(m) \bullet h(m)$ where

$\{x(k), k = -1, 0, 1, 2\} = [-1.9, 0.5, 1.2, 1.5]$ $\{h(k), k = 0, 1, 2\} = [-2.1, 1.2, 0.8]$.

2.11. Define $\{x_I(n) = 1, -1, 1\}$ and $\{x_Q(n) = 1, 1, -1\}$. (a) Compute the discrete correlations: R_{x_I}, R_{x_Q}, $R_{x_I x_Q}$, and $R_{x_Q x_I}$. (b) A certain radar transmits the signal $s(t) = x_I(t)\cos 2\pi f_0 t - x_Q(t)\sin 2\pi f_0 t$. Assume that the autocorrelation $s(t)$ is equal to

$$y(t) = y_I(t)\cos 2\pi f_0 t - y_Q(t)\sin 2\pi f_0 t.$$

Compute and sketch $y_I(t)$ and $y_Q(t)$.

2.12. A periodic signal $x_p(t)$ is formed by repeating the pulse $x(t) = 2\Delta((t-3)/5)$ every 10 seconds. (a) What is the Fourier transform of $x(t)$? (b) Compute the complex Fourier series of $x_p(t)$. (c) Give an expression for the autocorrelation function $\bar{R}_{x_p}(t)$ and the power spectrum density $\bar{S}_{x_p}(\omega)$.

2.13. An LTI system has impulse response

$$h(t) = \left\{\begin{array}{ll} \exp(-2t) & t \ge 0 \\ 0 & t < 0 \end{array}\right\}.$$

(a) Find the autocorrelation function $R_h(\tau)$. (b) Assume the input of this system is $x(t) = 3\cos(100t)$. What is the output?

2.14. Show that (a) $\bar{R}_x(-t) = \bar{R}_x^*(t)$. (b) If $x(t) = f(t) + m_1$ and $y(t) = g(t) + m_2$, show that $\bar{R}_{xy}(t) = m_1 m_2$, where the average values for $f(t)$ and $g(t)$ are zeroes.

2.15. What is the power spectral density for the signal $x(t) = A\cos(2\pi f_0 t + \theta_0)$?

2.16. A certain radar system uses linear frequency modulated waveforms of the form

$$x(t) = Rect\left(\frac{t}{\tau}\right)\cos\left(\omega_0 t + \mu\frac{t^2}{2}\right).$$

What are the quadrature components? Give an expression for both the modulation and instantaneous frequencies.

2.17. Consider the signal $x(t) = Rect(t/\tau)\cos(\omega_0 t - Bt^2/2\tau)$ and let $\tau = 15\mu s$ and $B = 10MHz$. What are the quadrature components?

2.18. Determine the quadrature components for the signal

$$h(t) = \delta(t) - \left(\frac{\omega_0}{\omega_d}\right)e^{-2t}\sin(\omega_0 t)$$

for $t \geq 0$.

2.19. If $x(t) = x_1(t) - 2x_1(t-5) + x_1(t-10)$, determine the autocorrelation functions $R_{x_1}(t)$ and $R_x(t)$ when $x_1(t) = \exp(-t^2/2)$.

2.20. Write an expression for the autocorrelation function $R_y(t)$, where

$$y(t) = \sum_{n=1}^{5} Y_n Rect\left(\frac{t-n5}{2}\right)$$

and $\{Y_n\} = \{0.8, 1, 1, 1, 0.8\}$. Give an expression for the density function $S_y(\omega)$.

2.21. Suppose you want to determine an unknown DC voltage v_{dc} in the presence of additive white Gaussian noise $n(t)$ of zero mean and variance σ_n^2. The measured signal is $x(t) = v_{dc} + n(t)$. An estimate of v_{dc} is computed by making three independent measurements of $x(t)$ and computing the arithmetic mean, $\underset{\sim}{v}_{dc} \approx (x_1 + x_2 + x_3)/3$. (a) Find the mean and variance of the random variable $\underset{\sim}{v}_{dc}$. (b) Does the estimate of v_{dc} get better by using ten measurements instead of three? Why?

2.22. Assume the X and Y miss distances of darts thrown at a bulls-eye dart board are Gaussian with zero mean and variance σ^2. (a) Determine the probability that a dart will fall between 0.8σ and 1.2σ. (b) Determine the radius of a circle about the bull's-eye that contains 80% of the darts thrown. (c) Consider a square with side s in the first quadrant of the board. Determine s so that the probability that a dart will fall within the square is 0.07.

2.23. (a) A random voltage $v(t)$ has an exponential distribution function $f_V(v) = a\exp(-av)$, where $(a > 0);(0 \leq v < \infty)$. The expected value $E[V] = 0.5$. Determine $Pr\{V > 0.5\}$.

2.24. Consider the network shown in the figure below, where $x(t)$ is a random voltage with zero mean and autocorrelation function $\Re_x(\tau) = 1 + \exp(-a|t|)$. Find the power spectrum $S_x(\omega)$. What is the transfer function? Find the power spectrum $S_v(\omega)$.

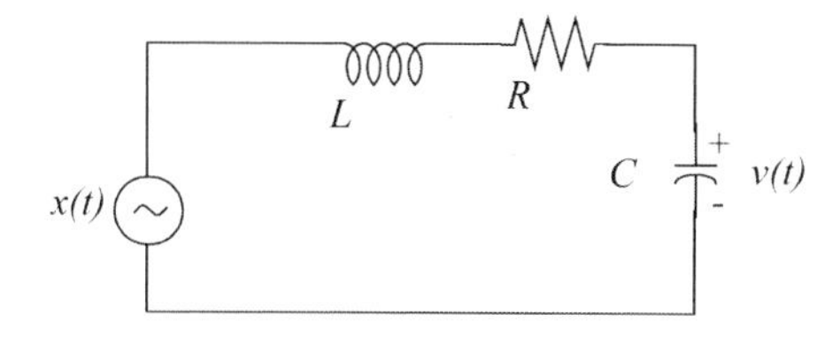

2.25. Let $\bar{S}_X(\omega)$ be the PSD function for the stationary random process $X(t)$. Compute an expression for the PSD function of

$$Y(t) = X(t) - 2X(t-T).$$

2.26. Let X be a random variable with

$$f_X(x) = \begin{Bmatrix} \frac{1}{\sigma}t^3 e^{-t} & t \geq 0 \\ 0 & elsewhere \end{Bmatrix}.$$

(a) Determine the characteristic function $C_X(\omega)$. (b) Using $C_X(\omega)$, validate that $f_X(x)$ is a proper *pdf.* (c) Use $C_X(\omega)$ to determine the first two moments of X. (d) Calculate the variance of X.

2.27. Let the random variable Z be written in terms of two other random variables X and Y as follows: $Z = X + 3Y$. Find the mean and variance for the new random variable in terms of the other two.

2.28. Suppose you have the following sequences of statistically independent Gaussian random variables with zero means and variances σ^2. if

$$X_1, X_2, \ldots, X_N \; ; \; X_i = A_i \cos\Theta_i \text{ and } Y_1, Y_2, \ldots, Y_N \; ; \; Y_i = A_i \sin\Theta_i .$$

Define $Z = \sum_{i=1}^{N} A_i^2$. Find an expression where Z exceeds a threshold value v_T.

2.29. Repeat the previous problem when two single delay line cancelers are cascaded to produce a double delay line canceler. Let $X(t)$ be a stationary random process, $E[X(t)] = 1$ and the autocorrelation $\Re_x(\tau) = 3 + \exp(-|\tau|)$. Define a new random variable Y as

$$Y = \int_0^2 x(t)dt .$$

Compute $E[Y(t)]$ and σ_Y^2.

2.30. Consider the single delay line canceler in the figure below. The input $x(t)$ is a wide-sense stationary random process with variance σ_x^2 and mean μ_x and a covariance matrix Λ. Find the mean and variance and the autocorrelation function of the output $y(t)$.

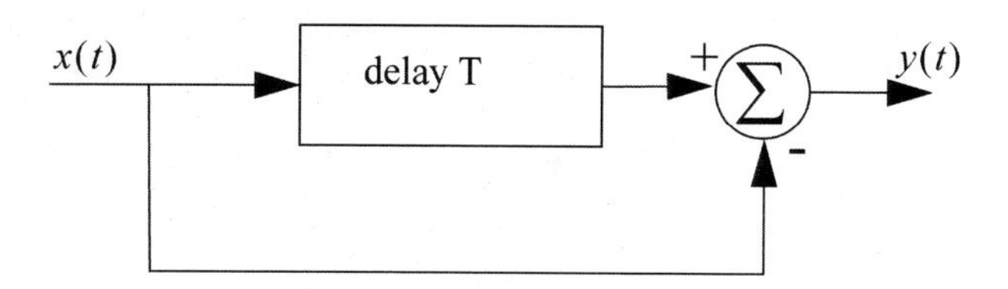

2.31. Compute the Z-transform for

(a) $x_1(n) = \frac{1}{n!}u(n)$, (b) $x_2(n) = \frac{1}{(-n)!}u(-n)$.

2.32. (a) Write an expression for the FT of $x(t) = Rect(t/3)$. (b) Assume that you want to compute the modulus of the FT using a DFT of size *512* with a sampling interval of *1* second. Evaluate the modulus at frequency $(80/512)Hz$. Compare your answer to the theoretical value and compute the error.

2.33. In Figure 2.2, let

$$p(t) = \sum_{n=-\infty}^{\infty} ARect\left(\frac{t-nT}{\tau}\right).$$

Give an expression for $X_s(\omega)$.

2.34. Generate *512* samples of the signal $x(t) = 2.0e^{-5t}\sin(4\pi t)$, using a sampling interval equal to 0.002. Compute the resultant spectrum and then truncate the spectrum at 15*Hz*. Generate the time-domain sequence for the truncated spectrum. Determine the sampling rate of the new sequence.

2.35. A certain band-limited signal has bandwidth $B = 20KHz$. Find the FFT size required so that the frequency resolution is $\Delta f = 50Hz$. Assume radix-2 FFT and a record length of 1 second.

2.36. Assume that a certain sequence is determined by its FFT. If the record length is $2ms$ and the sampling frequency is $f_s = 10KHz$, find N.

Appendix 2-A: Fourier Transform Pairs

$x(t)$	$X(\omega)$
$ARect(t/\tau)$; rectangular pulse	$A\tau Sinc(\omega\tau/2)$
$A\Delta(t/\tau)$; triangular pulse	$A\frac{\tau}{2}Sinc^2(\tau\omega/4)$
$\frac{1}{\sqrt{2\pi}\sigma}\exp\left(-\frac{t^2}{2\sigma^2}\right)$; Gaussian pulse	$\exp\left(-\frac{\sigma^2\omega^2}{2}\right)$
$e^{-at}u(t)$	$1/(a+j\omega)$
$e^{-a\|t\|}$	$\frac{2a}{a^2+\omega^2}$
$e^{-at}\sin\omega_0 t\ \ u(t)$	$\frac{\omega_0}{\omega_0^2+(a+j\omega)^2}$
$e^{-at}\cos\omega_0 t\ \ u(t)$	$\frac{a+j\omega}{\omega_0^2+(a+j\omega)^2}$
$\delta(t)$	1
1	$2\pi\delta(\omega)$
$u(t)$	$\pi\delta(\omega)+\frac{1}{j\omega}$
$\mathrm{sgn}(t)$	$\frac{2}{j\omega}$
$\cos\omega_0 t$	$\pi[\delta(\omega-\omega_0)+\delta(\omega+\omega_0)]$
$\sin\omega_0 t$	$j\pi[\delta(\omega+\omega_0)-\delta(\omega-\omega_0)]$
$u(t)\cos\omega_0 t$	$\frac{\pi}{2}[\delta(\omega-\omega_0)+\delta(\omega+\omega_0)]+\frac{j\omega}{\omega_0^2-\omega^2}$
$u(t)\sin\omega_0 t$	$\frac{\pi}{2j}[\delta(\omega+\omega_0)-\delta(\omega-\omega_0)]+\frac{\omega_0}{\omega_0^2-\omega^2}$
$\|t\|$	$\frac{-2}{\omega^2}$

Appendix 2-B: Common Probability Densities

Chi-Square with N degrees of freedom

$$f_X(x) = \frac{x^{(N/2)-1}}{2^{N/2}\Gamma(N/2)} \exp\left\{\frac{-x}{2}\right\} \quad ; \; x > 0$$

$$\bar{X} = N$$

$$\sigma_X^2 = 2N$$

$$gamma \; function = \Gamma(z) = \int_0^{\infty} \lambda^{z-1} e^{-\lambda} d\lambda$$

$$Re\{z\} > 0$$

Exponential

$$(f_X(x) = a \exp\{-ax\}) \quad ; \; x > 0$$

$$\bar{X} = \frac{1}{a}$$

$$\sigma_X^2 = \frac{1}{a^2}$$

Gaussian

$$f_X(x) = \frac{1}{\sqrt{2\pi}\sigma} \exp\left\{-\frac{1}{2}\left(\frac{x - x_m}{\sigma}\right)^2\right\}$$

$$\bar{X} = x_m$$

$$\sigma_X^2 = \sigma^2$$

Laplace

$$f_X(x) = \frac{\sigma}{2} \exp\{-\sigma|x - x_m|\}$$

$$\bar{X} = x_m$$

$$\sigma_X^2 = \frac{2}{\sigma^2}$$

Log-Normal

$$f_X(x) = \frac{1}{x\sigma\sqrt{2\pi}} \exp\left(-\frac{(\ln x - \ln x_m)^2}{2\sigma^2}\right) \quad ; \; x > 0$$

$$\bar{X} = \exp\left\{\ln x_m + \frac{\sigma^2}{2}\right\}$$

$$\sigma_X^2 = [\exp\{2\ln x_m + \sigma^2\}][\exp\{\sigma^2\} - 1]$$

Rayleigh

$$f_X(x) = \frac{x}{\sigma^2}\exp\left\{\frac{-x^2}{2\sigma^2}\right\} \quad ; \; x \geq 0$$

$$\bar{X} = \sqrt{\frac{\pi}{2}}\sigma$$

$$\sigma_X^2 = \frac{\sigma^2}{2}(4 - \pi)$$

Uniform

$$f_X(x) = \frac{1}{b - a} \quad ; \; a < b$$

$$\bar{X} = (a + b)/2$$

$$\sigma_X^2 = \frac{(b - a)^2}{12}$$

Weibull

$$f_X(x) = \frac{bx^{b-1}}{\bar{\sigma}_0} \exp\left(-\frac{(x)^b}{\bar{\sigma}_0}\right) \quad ; \; (x, b, \bar{\sigma}_0) \geq 0$$

$$\bar{X} = \frac{\Gamma(1 + b^{-1})}{1/(\sqrt[b]{\bar{\sigma}_0})}$$

$$\sigma_X^2 = \frac{\Gamma(1 + 2b^{-1}) - [\Gamma(1 + b^{-1})]^2}{1/[\sqrt[b]{(\bar{\sigma}_0)^2}]}$$

Appendix 2-C: Z-Transform Pairs

$x(n);\ n \ge 0$	$X(z)$	**ROC;** $\|z\| > R$
$\delta(n)$	1	0
1	$\frac{z}{z-1}$	1
n	$\frac{z}{(z-1)^2}$	1
n^2	$\frac{z(z+1)}{(z-1)^3}$	1
a^n	$\frac{z}{z-a}$	$\|a\|$
na^n	$\frac{az}{(z-a)^2}$	$\|a\|$
$\frac{a^n}{n!}$	$e^{a/z}$	0
$(n+1)a^n$	$\frac{z^2}{(z-a)^2}$	$\|a\|$
$\sin n\omega T$	$\frac{z\sin\omega T}{z^2 - 2z\cos\omega T + 1}$	1
$\cos n\omega T$	$\frac{z(z-\cos\omega T)}{z^2 - 2z\cos\omega T + 1}$	1
$a^n \sin n\omega T$	$\frac{az\sin\omega T}{z^2 - 2az\cos\omega T + a^2}$	$\frac{1}{\|a\|}$
$a^n \cos n\omega T$	$\frac{z(z-a^2\cos\omega T)}{z^2 - 2az\cos\omega T + a^2}$	$\frac{1}{\|a\|}$
$\frac{n(n-1)}{2!}$	$\frac{z}{(z-1)^3}$	1
$\frac{n(n-1)(n-2)}{3!}$	$\frac{z}{(z-1)^4}$	1

$x(n);\ n \geq 0$	$X(z)$	**ROC;** $\lvert z \rvert > R$
$\frac{(n+1)(n+2)a^n}{2!}$	$\frac{z^3}{(z-a)^3}$	$\lvert a \rvert$
$\frac{(n+1)(n+2)\ldots(n+m)a^n}{m!}$	$\frac{z^{m+1}}{(z-a)^{m+1}}$	$\lvert a \rvert$

Chapter 3

Modulation Techniques

3.1. Introduction

The basic communication system was discussed in Chapter 1. The message signal (data to be transmitted) is invariably not suitable for transmission in its original format. Hence, direct transmission of the signal in its original format is not effective. Instead, the message is used to modulate a carrier signal which is more suitable for transmission. In this context, modulation is a deliberate and systematic modification by the message signal of some attributes (characteristics) of the carrier signal such as amplitude, phase or frequency. In general, modulation is used to perform a frequency translation of the signal bandwidth; modulation advantages include the following:

1. Translate the signal to a higher frequency band (i.e., shorter wavelength) to increase radiation (emission of the signal into the channel) efficiency and reduce the size and cost of the transmitter.
2. It is much easier to amplify or filter a signal in one frequency band than another. For example, a DC signal my be converted to an AC signal, amplified and then converted back to DC to avoid the technical challenges associated with DC signal amplification.
3. Certain signals (e.g., Amplitude Modulation (AM) and Frequency Modulation (FM) radio signals and television signals) are only allowed to be transmitted within internationally agreed upon and assigned frequency bands.
4. In many cases, it is very desirable to transmit more than one signal over a single channel. In this case, frequency multiplexing is used to perform different translation on each message to a different frequency band.
5. Finally, modulation facilitates changing the signal bandwidth. For example, one can reduce the bandwidth of the signal to reduce the transmission cost at the expense of reducing the signal fidelity. Alternatively, the bandwidth can be increased to ensure greater immunity to noise, but at the expense of increasing the cost of transmission.

Figure 3.1 shows a block diagram of a simple communication system. The message signal $x_m(t)$ is used to modulate a carrier signal $x_c(t)$, and the modulated signal $x(t)$ is transmitted through the channel. The received signal is $y(t)$ contaminated with noise. The output of the demodulator is the signal $y_m(t)$ which is not an exact replica of the signal $x_m(t)$. Nonetheless, much if not all of the noise can be filtered out and the signal $y_m(t)$ can be an excellent estimate of $x_m(t)$.

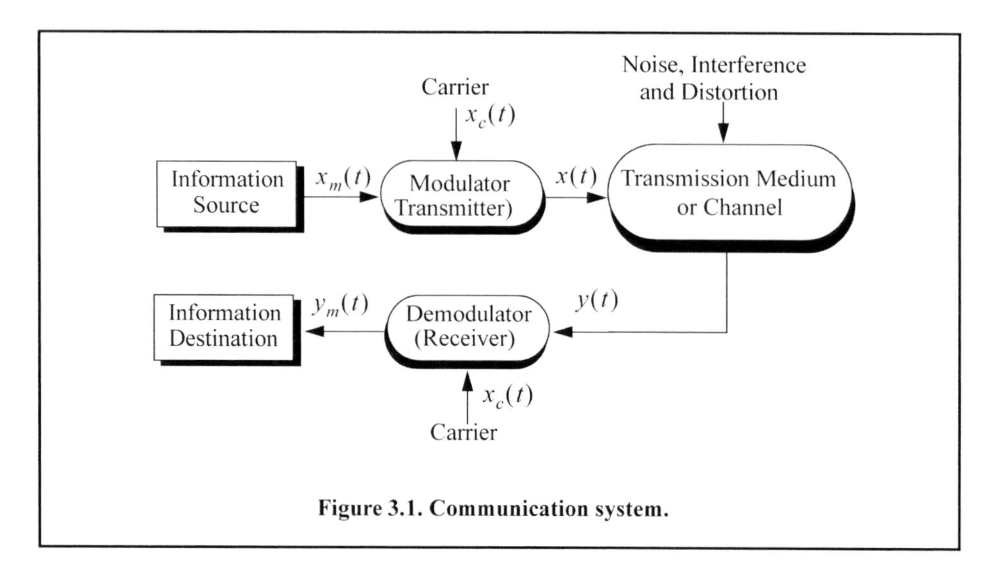

Figure 3.1. Communication system.

Several modulation techniques can be used. The carrier signal can be continuous sine wave or pulsed and the modulation can be periodic, pulsed, or random. In this book, the emphasis is on continuous sine wave carriers. In this case, the carrier signal is given by

$$x_c(t) = a(t)\cos(2\pi f_c t + \phi(t)) . \qquad \text{Eq. (3.1)}$$

where $a(t)$ and $\phi(t)$ are slow time-varying signals, the signal $a(t)$ is the envelope of the signal $x_c(t)$, the frequency f_c is the carrier frequency and $\phi(t)$ is the phase modulation.

The carrier signal can be modulated in amplitude, phase or frequency. Four types of amplitude modulation techniques are discussed in this chapter. They are:

1. Amplitude Modulation Suppressed Carrier (AM-SC)
2. Amplitude Modulation Large Carrier (AM-LC), also known referred to as the double sided large carrier modulation
3. Single Side Band (SSB) amplitude modulation, and
4. Frequency modulation

Phase or frequency modulation techniques are denoted as angle modulation since the phase angle of the modulated carrier signal conveys the message signal. In general, angle modulation is referred to as nonlinear modulation technique since a linear change in phase or frequency occurs as a result of the modulation process. The channel introduces zero mean Gaussian noise to the transmitted signal. For this purpose, the signal $y(t)$ can be written as

$$y(t) = x(t) + n(t) \qquad \text{Eq. (3.2)}$$

where $n(t)$ is the noise signal.

3.2. Amplitude Modulation - Suppressed Carrier

The general equation for an Amplitude Modulation - Suppressed Carrier (AM-SC) signal is given by

$$x(t) = x_m(t)x_c(t). \quad \text{Eq. (3.3)}$$

The message signal, $x_m(t)$, (also known as the modulating signal) is assumed to be a slow time-varying lowpass signal with maximum bandwidth B_m Hertz. The signal $x_c(t)$ is called the carrier signal and $x(t)$ is the modulated signal. The frequency f_c is referred to as the carrier frequency, and it is assumed to be much larger than the maximum frequency content of the message signal, i.e., $f_c \gg B_m$. The general form for the carrier signal is given in Eq. (3.1). In amplitude modulation, it is given by

$$x_c(t) = a_c\cos(2\pi f_c t), \quad \text{Eq. (3.4)}$$

where a_c is a non-time-varying constant and φ is assumed to be zero. Substituting Eq. (3.4) into Eq. (3.3) yields

$$x(t) = a_c x_m(t)\cos(2\pi f_0 t). \quad \text{Eq. (3.5)}$$

The Fourier transform of the modulated signal is computed as,

$$X(f) = FT\{x_m(t)a_c\cos(2\pi f_c t)\}. \quad \text{Eq. (3.6)}$$

It follows that,

$$X(f) = X_m(f) \otimes \left[\frac{a_c}{2}\delta(f-f_c) + \frac{a_c}{2}\delta(f+f_c)\right] \quad \text{Eq. (3.7)}$$

where $\otimes$ is the convolution operator, δ is the delta Dirac function, and $X_m(f)$ is the Fourier transform for the signal $x_m(t)$. Performing the convolution and collecting terms yield

$$X(f) = \frac{a_c}{2}X_m(f-f_c) + \frac{a_c}{2}X_m(f+f_c). \quad \text{Eq. (3.8)}$$

Observation of Eq. (3.8) indicates that amplitude modulation translates the frequency spectrum of the message signal to $\pm f_c$. This type of modulation is often called Double Sideband Suppressed Carrier (DSB-SC) since the amplitude spectrum of the carrier signal is not immediately identified from Eq. (3.8). This is illustrated in Figure 3.2. In this example, the message signal bandwidth is B_m. Note that this amplitude spectrum being symmetric with respect to the origin is an indication that the message signal is a real signal. Thus, using Eq. (3.8) the amplitude spectrum of the modulated signal is also symmetric with respect to the carrier frequency, as illustrated in Figure 3.2. The portion of the spectrum above f_c ($f > |f_c|$) is the upper sideband while the portion below f_c ($f < |f_c|$) is the lower sideband. Hence, the transmission bandwidth, B, is twice that of the message bandwidth, that is

$$B = 2B_m. \quad \text{Eq. (3.9)}$$

Demodulation of DSB-SC signals at the receiver side is very complex since phase reversal can occur when the message signal changes sign from positive to negative or vice versa. Therefore, DSB-SC modulation is not a very popular method of communication. Phase reversal is illustrated in Figure 3.3 using

$$\begin{aligned} x_m(t) &= \cos(2\pi 10t) \\ x_c(t) &= \cos(2\pi 100t) \end{aligned}. \quad \text{Eq. (3.10)}$$

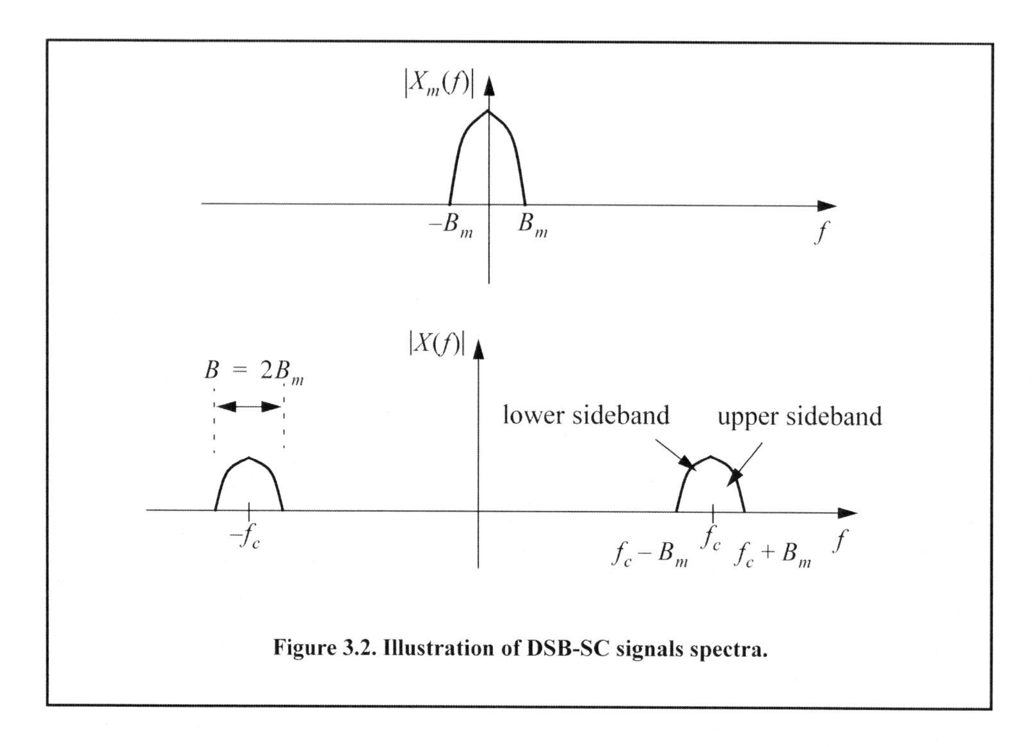

Figure 3.2. Illustration of DSB-SC signals spectra.

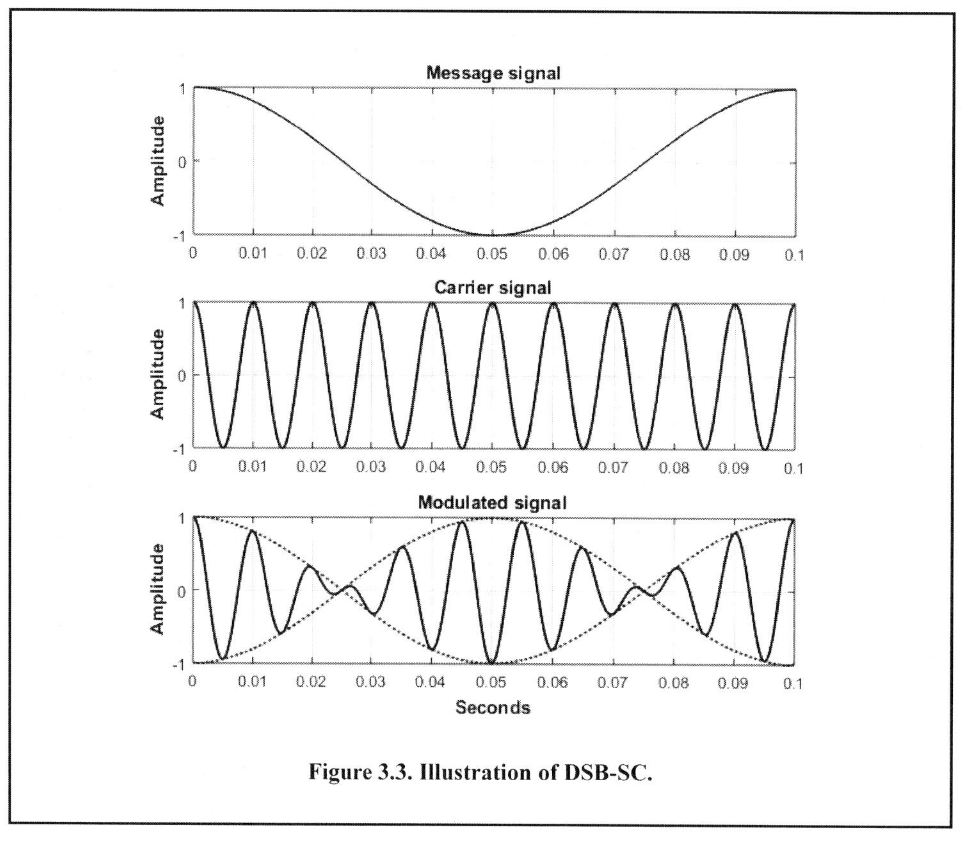

Figure 3.3. Illustration of DSB-SC.

3.3. Double Sideband Amplitude Modulation - Large Carrier

The primary distinguishing characteristic of the Double Sideband - Large Carrier (DSB-LC) amplitude modulation is that the envelope of the modulated signal carrier is identical to the modulating signal (the message). This is accomplished by adding the unmodulated carrier signal to a scaled and frequency translated message signal. More precisely, the modulated signal is now given by

$$x(t) = [a + x_m(t)] \times a_c \cos(2\pi f_c t) \qquad \text{Eq. (3.11)}$$

where, as before, the message signal, $x_m(t)$ is assumed to be a slow time-varying lowpass signal with maximum bandwidth B_m Hertz, $a_c \cos(2\pi f_c t)$ is the carrier signal, and a is the DC bias. Figure 3.4 shows a simple block diagram for an amplitude modulator. Rewrite Eq. (3.11) as

$$x(t) = A_c[1 + m x_m^n(t)] \cos(2\pi f_c t) \qquad \text{Eq. (3.12)}$$

where $A_c = a a_c$, m is the modulation index defined as

$$m = \frac{|Min\{x_m(t)\}|}{a}; \qquad \text{Eq. (3.13)}$$

the term $|Min\{x_m(t)\}|$ indicates the absolute value for the minimum value of $x_m(t)$. Finally, the normalized message signal $x_m^n(t)$ is

$$x_m^n(t) = \frac{x_m(t)}{|Min\{x_m(t)\}|}. \qquad \text{Eq. (3.14)}$$

For ease of notation, we will drop the superscript n and assume that the messuage signal has been normalized, that is

$$|x_m(t)| = |x_m^n(t)| \le 1; \qquad \text{Eq. (3.15)}$$

therefore, Eq. (3.12) can be rewritten as

$$x(t) = A_c[1 + m x_m(t)] \cos(2\pi f_c t). \qquad \text{Eq. (3.16)}$$

Observation of Eq. (3.16) indicates that the carrier signal amplitude has been modified (modulated) by the translated message signal. The Fourier transform of the modulated carrier signal is computed as

$$X(f) = FT\{A_c[1 + m x_m(t)] \cos(2\pi f_c t)\}. \qquad \text{Eq. (3.17)}$$

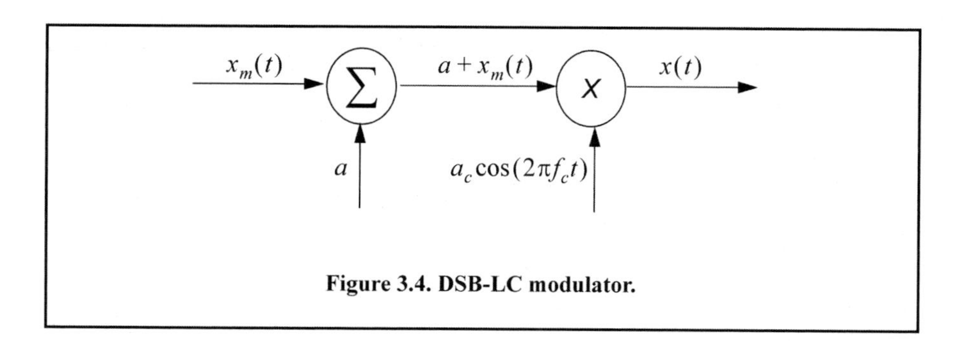

Figure 3.4. DSB-LC modulator.

It follows that

$$X(f) = mX_m(f) \otimes \left[\frac{A_c}{2}\delta(f-f_c) + \frac{A_c}{2}\delta(f+f_c)\right] + \frac{A_c}{2}\delta(f-f_c) + \frac{A_c}{2}\delta(f+f_c) \qquad \text{Eq. (3.18)}$$

where $X_m(f)$ is the FT of the message signal $x_m(t)$, $\otimes$ is the convolution operator and $\delta(\ )$ is the delta Dirac function. Performing the convolution and collecting terms yield

$$X(f) = m\frac{A_c}{2}X_m(f-f_c) + m\frac{A_c}{2}X_m(f+f_c) + \frac{A_c}{2}\delta(f-f_c) + \frac{A_c}{2}\delta(f+f_c)\,. \qquad \text{Eq. (3.19)}$$

Let the amplitude spectrum of the lowpass signal $x_m(t)$ be as illustrated in the upper portion of Figure 3.5. Thus, using Eq. (3.19) the amplitude spectrum of the modulated signal is symmetric with respect to the carrier frequency, as illustrated in Figure 3.5. Observation of Figure 3.5 shows that the carrier signal spectrum is present in the amplitude spectrum of the modulated signal, thus, the reason behind calling this modulation scheme a double sided large carrier modulation.

At the receiver side, the message signal is extracted from the envelope of the signal $y(t)$ (see Figure 3.1). The quality of the message signal estimate extracted from the modulated signal, at the receiver side, depends heavily on the modulation index m. Drawing on the example signals given in Eq. (3.10), Figure 3.6 shows three possibilities. The top plot shows the modulated signal when $m < 1$, under-modulation. The middle plot is known as the 100 percent modulation case, where $m = 1$.

Finally, the bottom plot shows the case of over-modulation where $m > 1$. In this last case, phase reversal of the modulated signal occurs and the envelope may become negative causing envelop distortion; thus, extraction of a clean estimate of the message signal may not be possible.

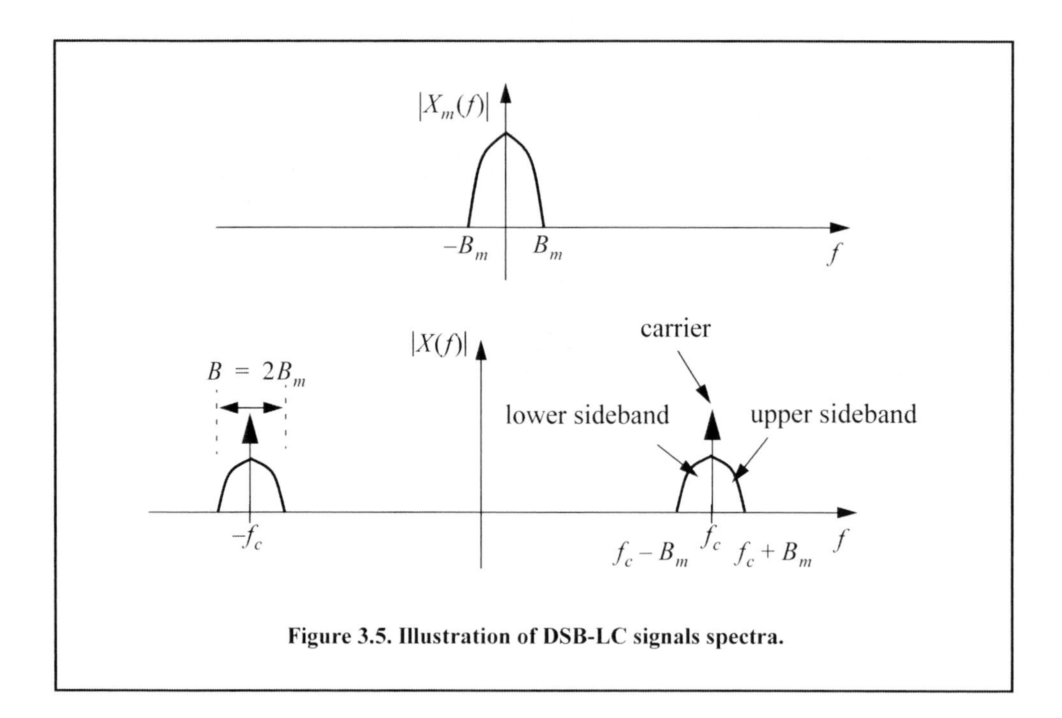

Figure 3.5. Illustration of DSB-LC signals spectra.

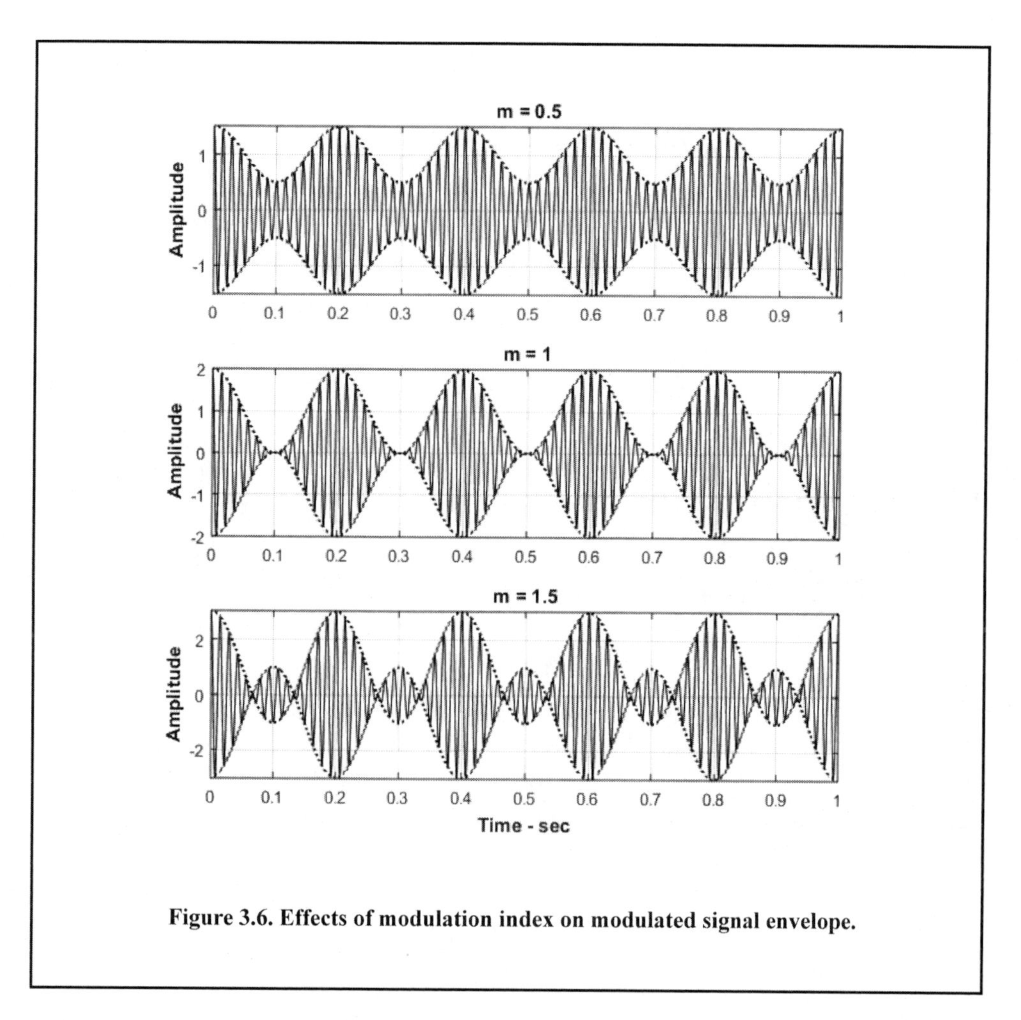

Figure 3.6. Effects of modulation index on modulated signal envelope.

The modulated signal total power, P_t, is easily obtained using time averaging. More precisely,

$$P_t = \langle x^2(t) \rangle = \lim_{T \to \infty} \frac{1}{T} \int_{-T}^{T} x(t)x(t) \ dt . \qquad \text{Eq. (3.20)}$$

But since the modulated signal is periodic with period T, where

$$T = 1/f_c \qquad \text{Eq. (3.21)}$$

it follows that Eq. (3.20) can be written as

$$P_t = \frac{1}{T} \int_{-T/2}^{T/2} x(t)x(t) \ dt . \qquad \text{Eq. (3.22)}$$

Substituting Eq. (3.16) into Eq. (3.22) and collecting terms yield

$$P_t = \frac{1}{T}\int_{-T/2}^{T/2} A_c^2\{1 + 2mx_m(t) + m^2x_m^2(t)\}\{\cos(2\pi f_c t)\}^2 dt. \qquad \text{Eq. (3.23)}$$

Assuming that the message signal does not contain any DC component (i.e., its time average is zero), then Eq. (3.23) reduces to

$$P_t = \frac{A_c^2}{2}\{1 + m^2\langle(t)\rangle\} \quad . \qquad \text{Eq. (3.24)}$$

The modulation efficiency, E_m is defined as

$$E_m = \frac{\frac{A_c^2 m^2}{2}\langle x_m^2(t)\rangle}{\frac{A_c^2}{2}(1 + m^2\langle x_m^2(t)\rangle)} = \frac{m^2\langle x_m^2(t)\rangle}{1 + m^2\langle x_m^2(t)\rangle}. \qquad \text{Eq. (3.25)}$$

The efficiency is maximum when the modulation index is unity. A special case is when $x_m(t)$ is a sine wave, where

$$\langle x_m^2(t)\rangle = \langle[\cos(2\pi f_m t)]^2\rangle = 0.5 ; \qquad \text{Eq. (3.26)}$$

thus, the maximum efficient is *33.3%*. Note that when the modulation index is 0.5 then the efficiency E_m drops to *11%*. Thus, most of the modulated signal power is wasted in the transmission.

Demodulation at the receiver side is performed via simple techniques of envelope detection. An example of an envelope detection circuitry is shown in Figure 3.7. The input signal is $y(t)$ and the output is $y_m(t)$ which is an estimate of the signal $x_m(t)$. As indicated by Figure 3.7, as long as the input signal voltage is greater than $y_m(t)$, then the capacitor is charged through the diode up to the peak of $y(t)$. Alternatively, when the input signal drops below the capacitor voltage, the diode is blocked and the capacitor begins to slowly discharge through the resistor. Therefore, the output signal $y_m(t)$ closely follows the envelope of the input signal $y(t)$.

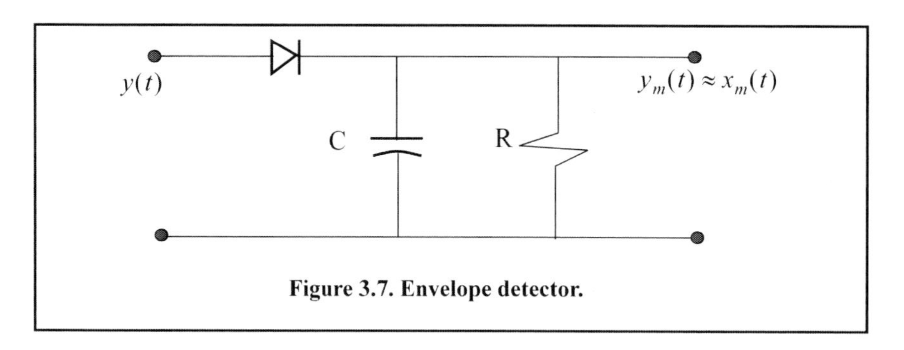

Figure 3.7. Envelope detector.

Example:

Determine the modulation efficiency and the output spectrum for a DSB-LC modulator using a modulation index $m = 0.5$, $A_c = 4V$, *and the input signal is*

$$x_m(t) = 2\cos\left(2\pi 10t - \frac{\pi}{5}\right).$$

Solution:

The first step is to determine the efficiency. The figure below shows a plot of the message signal as a function of normalized time (i.e., t/T) where $T = 1/f_m$. From the figure the minimum value is -2 and it occurs at $tf_m = 0.6$.

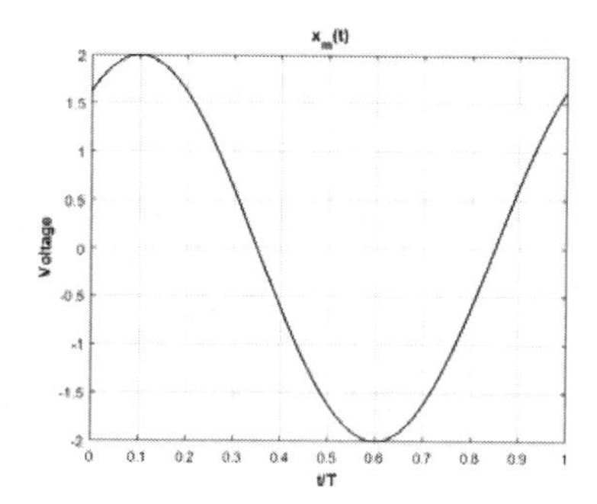

It follows that the normalized message signal is

$$x_m^n(t) = \frac{2}{2}\cos\left(2\pi f_m t - \frac{\pi}{5}\right) = \cos\left(2\pi f_m t - \frac{\pi}{5}\right).$$

The mean square value for $x_m^n(t)$ is 0.5; therefore, the efficiency is

$$E_m = \frac{(0.5^2) \times 0.5)}{1 + (0.5^2) \times 0.5)} = 11.1\,;$$

it follows that

$$x(t) = 4\left\{1 + 0.5\left[\cos\left(2\pi f_m t - \frac{\pi}{5}\right)\right]\right\}\cos(2\pi f_c t)$$

$$x(t) = 4\cos(2\pi f_c t) + \cos\left(2\pi(f_c + f_m)t - \frac{\pi}{5}\right) + \cos\left(2\pi(f_c - f_m)t - \frac{\pi}{5}\right).$$

The amplitude spectrum for $x(t)$ is as shown below. The spectrum is centered around $\pm f_c$ and comprises 6 spectral lines at $-f_c \pm f_c$ and at $f_c \pm f_m$. Note that the amplitude spectrum has symmetry around the carrier frequency.

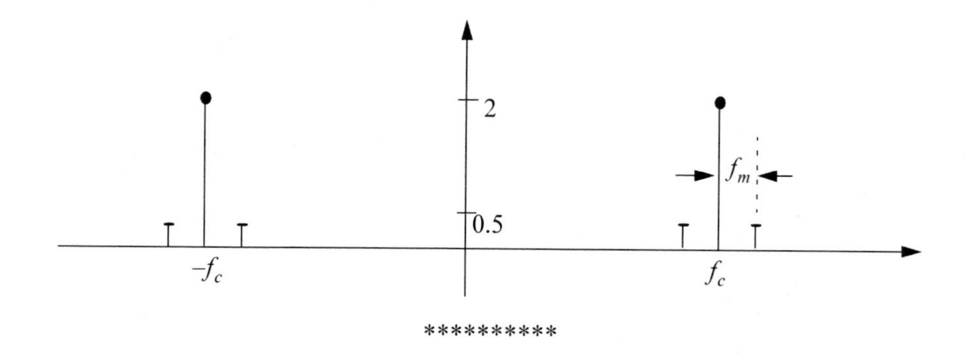

****

3.4. Single Sideband Amplitude Modulation

Both of the DSB-SC and DSB-LC transmit two sidebands with redundant information, since the message signal can be extracted from only one of the two sidebands. Thus, wasting transmitting power and bandwidth occurs. Suppressing one sideband halves the bandwidth requirements and produces Single Sideband (SSB) amplitude modulation. This reduction in transmission bandwidth is accomplished at the expense of increased complexity of the system.

There are several ways through which one can produce an upper sideband, or lower sideband, AM-SSB signal. The following discussion addresses one of those techniques. The generation of an SSB signal is illustrated in Figure 3.8 (upper sideband). The SSB signal is obtained from the message signal spectrum $X_m(f)$. First, the message signal $x_m(t)$ is replaced by its analytic signal $\psi_m(t)$, where (see Chapter 2),

$$\psi_m(t) = x_m(t) + j\hat{x}_m(t) \qquad \text{Eq. (3.27)}$$

$\hat{x}_m(t)$ is the Hilbert transform of the signal $x_m(t)$. The spectrum of the analytic signal is

$$\Psi_m(f) = X_m(f)(1 + \text{sgn}(f)), \qquad \text{Eq. (3.28)}$$

and the function $\text{sgn}(f)$ is

$$\text{sgn}(f) = \begin{Bmatrix} 1 & ;f > 0 \\ 0 & ;f = 0 \\ -1 & ,f < 0 \end{Bmatrix}. \qquad \text{Eq. (3.29)}$$

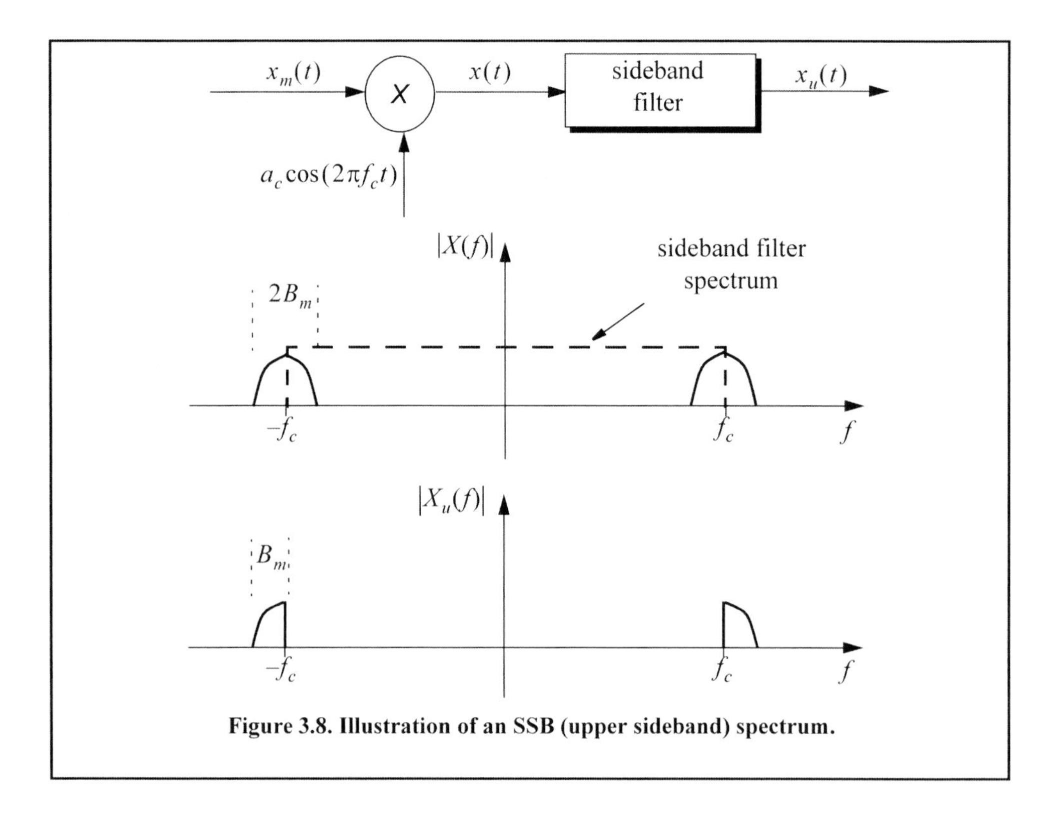

Figure 3.8. Illustration of an SSB (upper sideband) spectrum.

The second step is to map the signal $\psi_m(t)$ into a bandpass analytic signal by frequency translation. More precisely,

$$\psi(t) = \frac{A_c}{2}\psi_m(t)e^{j2\pi f_c t}, \qquad \text{Eq. (3.30)}$$

where f_c is the carrier frequency and $A_c/2$ is a scaling factor. Third, the SSB (upper side-band) signal $x_u(t)$ is computed as

$$x_u(t) = Re\{\psi(t)\} = \frac{A_c}{2}\{x_m(t)\cos(2\pi f_c t) - \hat{x}_m(t)\sin(2\pi f_c t)\}. \qquad \text{Eq. (3.31)}$$

The Lower sideband SSB signal is generated in a similar fashion, except in this case, Eq. (3.27) is replaced by

$$\psi_m(t) = x_m(t) - j\hat{x}_m(t) \qquad \text{Eq. (3.32)}$$

and by following the same analysis, the lower sideband SSB is given by

$$x_l(t) = \frac{A_c}{2}\{x_m(t)\cos(2\pi f_c t) + \hat{x}_m(t)\sin(2\pi f_c t)\}. \qquad \text{Eq. (3.33)}$$

3.5. Frequency Modulation

The discussion presented in this section is restricted to sinusoidal modulating signals. In this case, the general formula for a Frequency Modulated (FM) waveform can be expressed by

$$x(t) = A_c \cos\left(2\pi f_c t + k_f \int_0^t \cos 2\pi f_m u du\right); \qquad \text{Eq. (3.34)}$$

f_c is the carrier frequency, $\cos 2\pi f_m t$ is the modulating message signal, A_c is a constant, and $k_f = 2\pi\Delta f_{peak}$, where Δf_{peak} is the peak frequency deviation. The phase is given by

$$\phi(t) = 2\pi f_c t + 2\pi\Delta f_{peak}\int_0^t \cos 2\pi f_m u du = 2\pi f_c t + \beta \sin 2\pi f_m t \qquad \text{Eq. (3.35)}$$

where β is the FM modulation index given by

$$\beta = \frac{\Delta f_{peak}}{f_m}. \qquad \text{Eq. (3.36)}$$

Consider the FM waveform $x(t)$ given by

$$x(t) = A_c \cos(2\pi f_c t + \beta \sin 2\pi f_m t) \qquad \text{Eq. (3.37)}$$

which can be written as

$$x(t) = A_c Re\{e^{j2\pi f_c t} e^{j\beta \sin 2\pi f_m t}\} \qquad \text{Eq. (2.38)}$$

where $Re\{ \}$ denotes the real part. Since the signal $\exp(j\beta\sin 2\pi f_m t)$ is periodic with period $T = 1/f_m$, it can be expressed using the complex exponential Fourier series as

$$e^{j\beta\sin 2\pi f_m t} = \sum_{n=-\infty}^{\infty} C_n e^{jn2\pi f_m t} \qquad \text{Eq. (3.39)}$$

where the Fourier series coefficients C_n are given by

$$C_n = \frac{1}{2\pi}\int_{-\pi}^{\pi} e^{j\beta\sin 2\pi f_m t}\, e^{-jn2\pi f_m t}\, dt . \qquad \text{Eq. (3.40)}$$

Make the change of variable $u = 2\pi f_m t$, and recognize that the Bessel function of the first kind of order n is

$$J_n(\beta) = \frac{1}{2\pi}\int_{-\pi}^{\pi} e^{j(\beta\sin u - nu)}\, du . \qquad \text{Eq. (3.41)}$$

The Fourier series coefficients are $C_n = J_n(\beta)$, and consequently Eq. (3.39) becomes

$$e^{j\beta\sin 2\pi f_m t} = \sum_{n=-\infty}^{\infty} J_n(\beta) e^{jn2\pi f_m t} \qquad \text{Eq. (3.42)}$$

which is known as the Bessel-Jacobi equation. Figure 3.9 shows a plot of Bessel functions of the first kind for $n = 0, 1, 2, 3$. Table 3.1 shows some values of the Bessel function.

The total power in the signal $x(t)$ is

$$P = \frac{1}{2}A_c^2 \sum_{n=-\infty}^{\infty} |J_n(\beta)|^2 = \frac{1}{2}A_c^2 . \qquad \text{Eq. (3.43)}$$

Substituting Eq. (3.42) into Eq. (3.38) yields

$$x(t) = A_c Re\left\{ e^{j2\pi f_c t} \sum_{n=-\infty}^{\infty} J_n(\beta) e^{jn2\pi f_m t} \right\} . \qquad \text{Eq. (3.44)}$$

Expanding Eq. (3.44) yields

$$x(t) = A_c \sum_{n=-\infty}^{\infty} J_n(\beta)\cos(2\pi f_c + n2\pi f_m)t . \qquad \text{Eq. (3.45)}$$

Finally, since $J_n(\beta) = J_{-n}(\beta)$ for n odd and $J_n(\beta) = -J_{-n}(\beta)$ for n even, one can rewrite Eq. (3.45) as

$$x(t) = A_c\{J_0(\beta)\cos 2\pi f_c t + J_1(\beta)[\cos(2\pi f_c + 2\pi f_m)t - \cos(2\pi f_c - 2\pi f_m)t] + J_2(\beta)[\cos(2\pi f_c + 4\pi f_m)t + \cos(2\pi f_c - 4\pi f_m)t] + J_3(\beta)[\cos(2\pi f_c + 6\pi f_m)t - \cos(2\pi f_c - 6\pi f_m)t] + J_4(\beta)[\cos((2\pi f_c + 8\pi f_m)t + \cos(2\pi f_c - 8\pi f_m)t)] + \dots\} \quad \text{Eq. (3.46)}$$

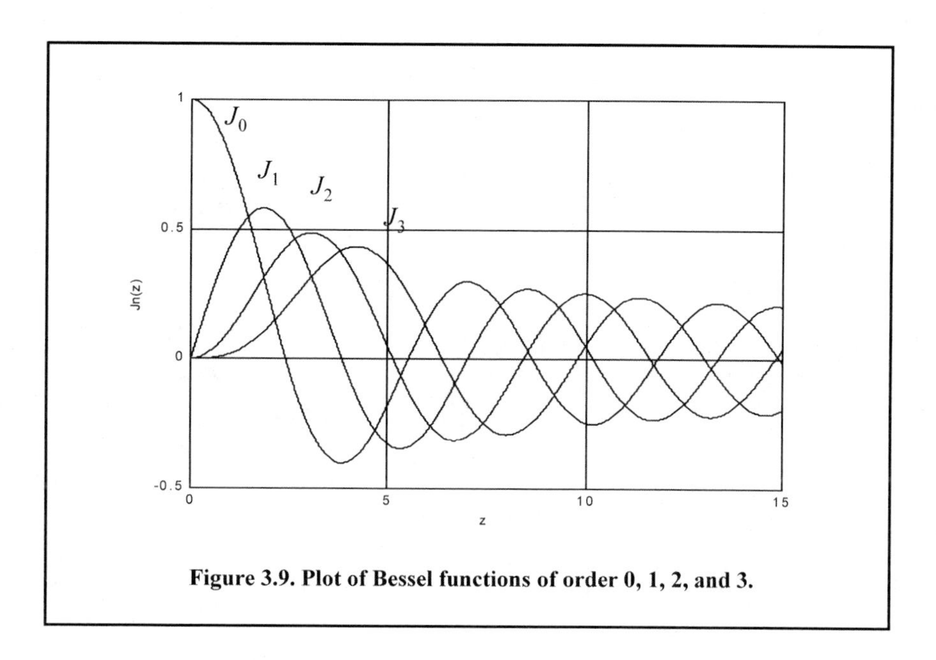

Figure 3.9. Plot of Bessel functions of order 0, 1, 2, and 3.

Table 3.1. Some values of the Bessel function, $J_n(\beta)$.

n	$\beta = 0.1$	$\beta = 0.2$	$\beta = 0.5$	$\beta = 1$	$\beta = 2$	$\beta = 5$
0	0.997	0.990	0.938	0.765	0.224	−0.178
1	0.050	0.100	0.242	0.440	0.577	−0.328
2	0.001	0.005	0.031	0.115	0.353	0.047
3				0.020	0.129	0.365
4				0.002	0.034	0.391
5					0.007	0.261
6					0.001	0.131
7						0.053
8						0.018
9						0.006

which can be rewritten as

$$x(t) = A_c\left\{J_0(\beta)\cos 2\pi f_c t + \sum_{n = even}^{\infty} J_n(\beta)[\cos(2\pi f_c + 2n\pi f_m)t + \right. \qquad \text{Eq. (3.47)}$$

$$\left. \cos(2\pi f_c - 2n\pi f_m)t] + \sum_{q = odd}^{\infty} J_q(\beta)[\cos(2\pi f_c + 2q\pi f_m)t - \cos(2\pi f_c - 2q\pi f_m)t]\right\}$$

The spectrum of $x(t)$ is composed of pairs of spectral lines centered at f_c, as sketched in Figure 3.10. The spacing between adjacent spectral lines is f_m. The central spectral line has an amplitude equal to $A_c J_o(\beta)$, while the amplitude of the *nth* spectral line is $A_c J_n(\beta)$.

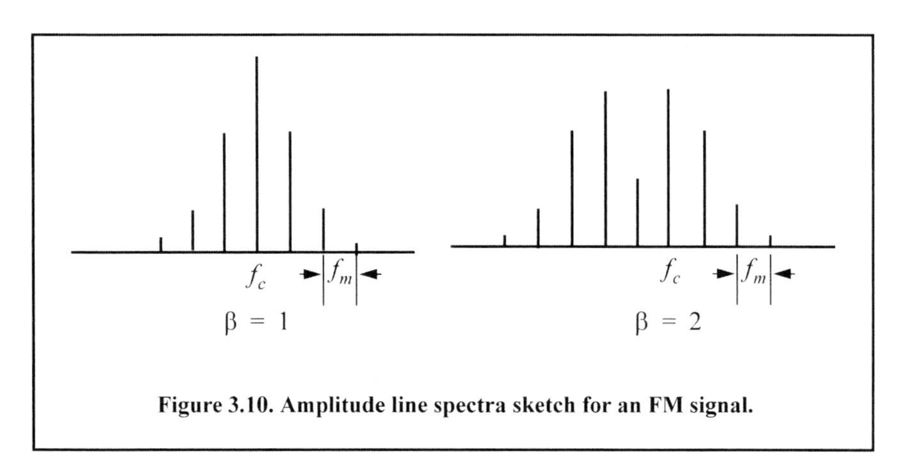

Figure 3.10. Amplitude line spectra sketch for an FM signal.

As indicated by Eq. (3.47) the bandwidth of FM signals is infinite. However, the magnitudes of spectral lines of the higher orders (i.e., large n) are small, and thus the bandwidth can be approximated using Carson's rule as,

$$B \approx 2(\beta + 1)f_m . \qquad \text{Eq. (3.48)}$$

When β is small, only $J_0(\beta)$ and $J_1(\beta)$ have significant values. Thus, we may approximate Eq. (3.47) by

$$x(t) \approx A_c\{J_0(\beta)\cos 2\pi f_c t + J_1(\beta)[\cos(2\pi f_c + 2\pi f_m)t - \cos(2\pi f_c - 2\pi f_m)t]\} . \qquad \text{Eq. (3.49)}$$

Finally, for small β, the Bessel functions can be approximated by

$$\begin{aligned} J_0(\beta) &\approx 1 \\ J_1(\beta) &\approx \beta/2 \end{aligned} . \qquad \text{Eq. (3.50)}$$

Thus, Eq. (3.49) may be approximated by

$$x(t) \approx A_c\left\{\cos 2\pi f_c t + \frac{1}{2}\beta[\cos(2\pi f_c + 2\pi f_m)t - \cos(2\pi f_c - 2\pi f_m)t]\right\} . \qquad \text{Eq. (3.51)}$$

Example:

If the modulation index is $\beta = 0.5$, *give an expression for the signal* $x(t)$.

Solution:

From Bessel function tables we get $J_0(0.5) = 0.9385$ *and* $J_1(0.5) = 0.2423$; *then using Eq. (3.51) we get*

$$x(t) \approx A_c\{(0.9385)\cos 2\pi f_0 t + (0.2423)[\cos(2\pi f_0 + 2\pi f_m)t - \cos(2\pi f_0 - 2\pi f_m)t]\}.$$

****

Example:

Consider an FM transmitter with output signal $x(t) = 100\cos(2000\pi t + \varphi(t))$. *The frequency deviation is* $4Hz$, *and the modulating waveform is* $x_m(t) = 10\cos 16\pi t$. *Determine the FM signal bandwidth. How many spectral lines will pass through a bandpass filter whose bandwidth is* $58Hz$ *centered at* $1000Hz$?

Solution:

The peak frequency deviation is $\Delta f_{peak} = 4 \times 10 = 40Hz$. *It follows that*

$$\beta = \frac{\Delta f_{peak}}{f_m} = \frac{40}{8} = 5.$$

Using Eq. (3.48) we get

$$B \approx 2(\beta + 1)f_m = 2 \times (5 + 1) \times 8 = 96Hz.$$

Only seven spectral lines pass through the bandpass filter as illustrated below.

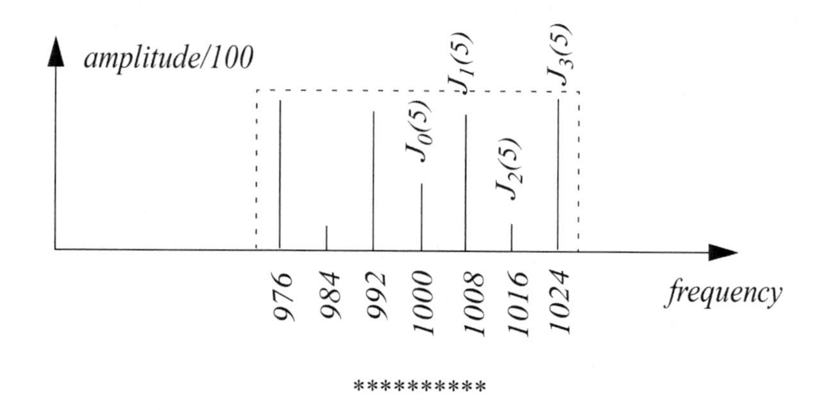

****

3.6. Spectra of a Few Common Radar Signals

The spectrum of a given signal describes the spread of its energy in the frequency domain. An energy signal (finite energy) can be characterized by its Energy Spectrum Density (ESD) function, while a power signal (finite power) is characterized by the Power Spectrum Density (PSD) function. The units of the ESD are Joules/Hertz, and the PSD has units Watts/Hertz. In this section we will discuss the spectra for the most common radar signals and their associated modulation techniques.

3.6.1. Continuous Wave Signal - Unmodulated

Consider the unmodulated Continuous Wave (CW) radar waveform given by

$$x_1(t) = \cos 2\pi f_c t \qquad \text{Eq. (3.52)}$$

where f_c is the radar center frequency. The FT of $x_1(t)$ is

$$X_1(f) = \frac{1}{2}[\delta(f - f_c) + \delta(f + f_c)] \qquad \text{Eq. (3.53)}$$

$\delta(\ )$ is the Dirac delta function. As indicated by the amplitude spectrum shown in Figure 3.11, the signal $x_1(t)$ has infinitesimal bandwidth, located at $\pm f_c$.

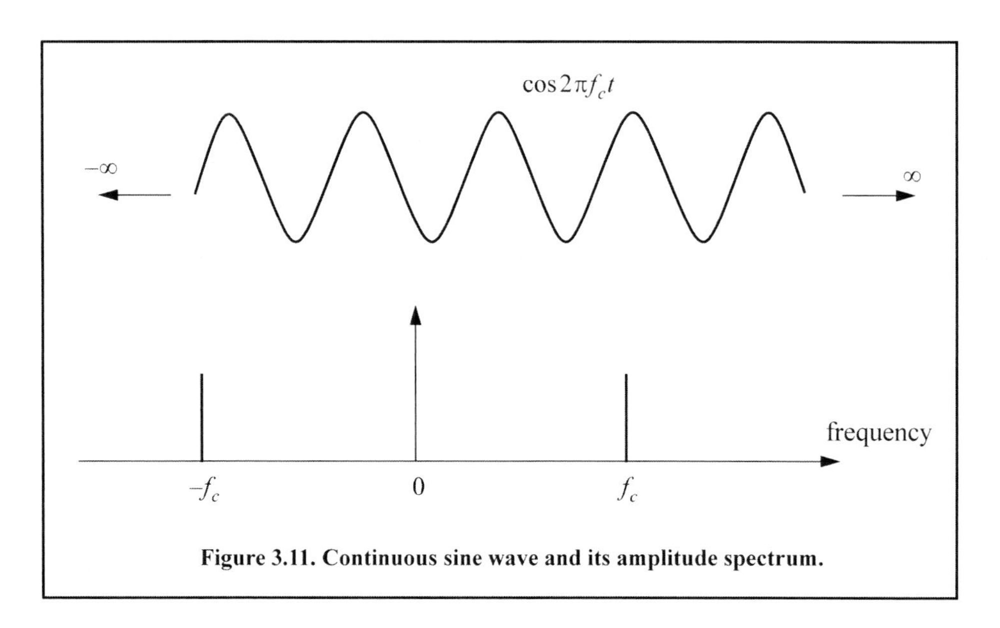

Figure 3.11. Continuous sine wave and its amplitude spectrum.

3.6.2. Finite Duration Pulse Signal - Amplitude Modulation

Consider the time-domain radar signal $x_2(t)$ given by

$$x_2(t) = x_1(t)Rect\left(\frac{t}{\tau_0}\right) = Rect\left(\frac{t}{\tau_0}\right)\cos 2\pi f_c t \qquad \text{Eq. (3.54)}$$

where $\cos 2\pi f_c t$ is the carrier signal, and $Rect$ is the modulating signal defined by

$$Rect\left(\frac{t}{\tau_0}\right) = \left\{\begin{array}{ll} 1 & -\frac{\tau_0}{2} \le t \le \frac{\tau_0}{2} \\ 0 & otherwise \end{array}\right\}. \qquad \text{Eq. (3.55)}$$

The Fourier transform of the $Rect$ function is

$$FT\left\{Rect\left(\frac{t}{\tau_0}\right)\right\} = \tau_0 Sinc(f\tau_0) \qquad \text{Eq. (3.56)}$$

where

$$Sinc(u) = \frac{\sin(\pi u)}{\pi u}. \qquad \text{Eq. (3.57)}$$

It follows that the FT is

$$X_2(f) = X_1(f) \otimes \tau_0 Sinc(f\tau_0) = \frac{1}{2}[\delta(f-f_c) + \delta(f+f_c)] \otimes \tau_0 Sinc(f\tau_0), \qquad \text{Eq. (3.58)}$$

which can be written as

$$X_2(f) = \frac{\tau_0}{2}\{Sinc[(f-f_c)\tau_0] + Sinc[(f+f_c)\tau_0]\}. \qquad \text{Eq. (3.59)}$$

The amplitude spectrum of $x_2(t)$ is shown in Figure 3.12. It is made up of two $Sinc$ functions, as defined in Eq. (3.59), centered at $\pm f_c$.

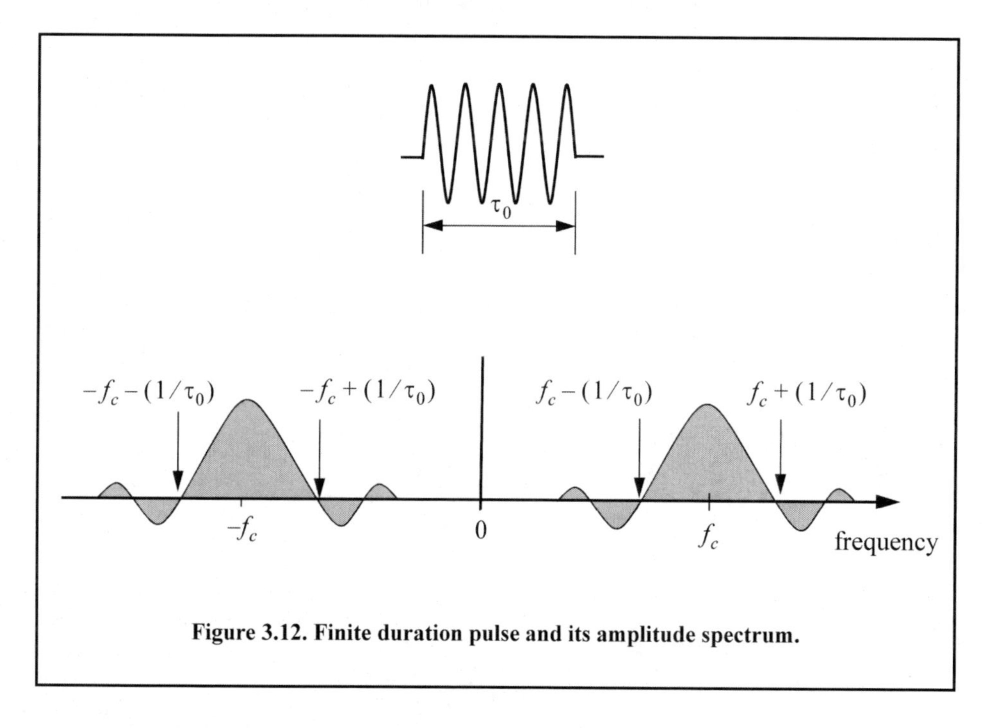

Figure 3.12. Finite duration pulse and its amplitude spectrum.

3.6.3. Periodic Pulse Signal - Amplitude Modulation

In this case, consider the coherent gated CW radar waveform $x_3(t)$ given by

$$x_3(t) = \sum_{n=-\infty}^{\infty} x_1(t) Rect\left(\frac{t-nT}{\tau_0}\right) = \cos 2\pi f_c t \sum_{n=-\infty}^{\infty} Rect\left(\frac{t-nT}{\tau_0}\right). \qquad \text{Eq. (3.60)}$$

The signal $x_3(t)$ is periodic, with period T (recall that $f_r = 1/T$ is the PRF); of course the condition $f_r \ll f_c$ is assumed. The FT of the signal $x_3(t)$ is

$$X_3(f) = X_1(f) \otimes FT\left\{ \sum_{n=-\infty}^{\infty} Rect\left(\frac{t-nT}{\tau_0}\right) \right\} = \quad . \qquad \text{Eq. (3.61)}$$

$$\frac{1}{2}[\delta(f-f_c) + \delta(f+f_c)] \otimes FT\left\{ \sum_{n=-\infty}^{\infty} Rect\left(\frac{t-nT}{\tau_0}\right) \right\}$$

The complex exponential Fourier series of the summation inside Eq. (3.61) is

$$\sum_{n=-\infty}^{\infty} Rect\left(\frac{t-nT}{\tau_0}\right) = \sum_{n=-\infty}^{\infty} X_n e^{j\frac{nt}{T}} \qquad \text{Eq. (3.62)}$$

where the Fourier series coefficients X_n are given by

$$X_n = \frac{1}{T} FT\left\{ Rect\left(\frac{t}{\tau_0}\right) \right\}\Bigg|_{f=\frac{n}{T}} = \frac{\tau_0}{T} Sinc(f\tau_0)\Big|_{f=\frac{n}{T}} = \frac{\tau_0}{T} Sinc\left(\frac{n\tau_0}{T}\right). \qquad \text{Eq. (3.63)}$$

It follows that

$$FT\left\{ \sum_{n=-\infty}^{\infty} X_n e^{j\frac{nt}{T}} \right\} = \left(\frac{\tau_0}{T}\right) \sum_{n=-\infty}^{\infty} Sinc(nf_r\tau_0)\delta(f-nf_r) \qquad \text{Eq. (3.64)}$$

where the relation $f_r = 1/T$ was used. Substituting Eq. (3.64) into Eq. (3.61) yields the FT of $x_3(t)$. That is

$$X_3(f) = \frac{\tau_0}{2T}[\delta(f-f_c) + \delta(f+f_c)] \otimes \sum_{n=-\infty}^{\infty} Sinc(nf_r\tau_0)\delta(f-nf_r). \qquad \text{Eq. (3.65)}$$

The amplitude spectrum of $x_3(t)$ has two parts centered at $\pm f_c$. The spectrum of the summation part is an infinite number of delta functions repeated every f_r, where the *nth* line is modulated in amplitude with the value corresponding to $Sinc(nf_r\tau_0)$. Therefore, the overall spectrum consists of an infinite number of lines separated by f_r and have a $\sin u/u$ envelope that corresponds to X_n as illustrated in Figure 3.13.

3.6.4. Finite Duration Pulse Train Signal - Amplitude Modulation

Define the radar signal $x_4(t)$ as

$$x_4(t) = \cos(2\pi f_c t) \sum_{n=0}^{N-1} Rect\left(\frac{t-nT}{\tau_0}\right) = \cos 2\pi f_c t \times g(t) \qquad \text{Eq. (3.66)}$$

where

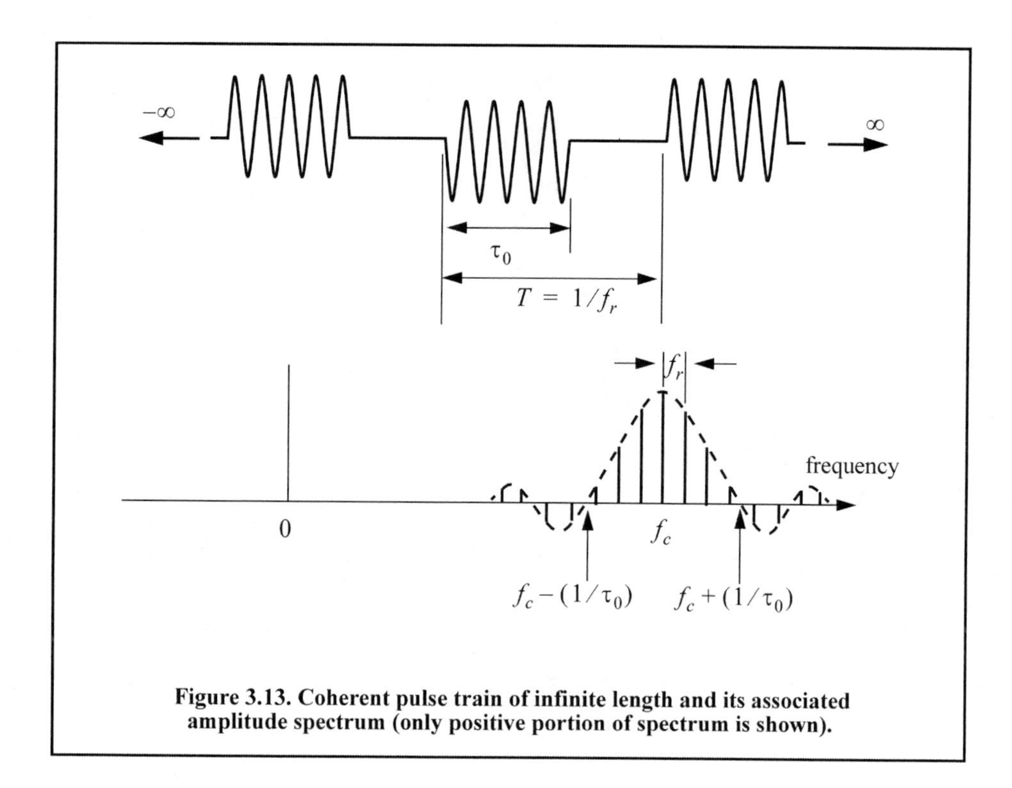

Figure 3.13. Coherent pulse train of infinite length and its associated amplitude spectrum (only positive portion of spectrum is shown).

$$g(t) = \sum_{n=0}^{N-1} Rect\left(\frac{t-nT}{\tau_0}\right). \qquad \text{Eq. (3.67)}$$

The FT of the signal $x_4(t)$ is

$$X_4(f) = \frac{1}{2}G(f) \otimes [\delta(f-f_c) + \delta(f+f_c)] \qquad \text{Eq. (3.68)}$$

where $G(f)$ is the FT of $g(t)$. This means that the amplitude spectrum of the signal $x_4(t)$ is equal to replicas of $G(f)$ centered at $\pm f_c$. Given this conclusion, one can then focus on computing $G(f)$. The signal $g(t)$ can be written as (see top portion of Figure 3.14)

$$g(t) = \sum_{n=-\infty}^{\infty} g_1(t) Rect\left(\frac{t-nT}{\tau_0}\right) \qquad \text{Eq. (3.69)}$$

where

$$g_1(t) = Rect\left(\frac{t}{NT}\right). \qquad \text{Eq. (3.70)}$$

It follows that the FT of Eq. (3.69) can be computed using analysis similar to that which led to Eq. (3.65). More precisely,

$$G(f) = \frac{\tau_0}{T}G_1(f) \otimes \sum_{n=-\infty}^{\infty} Sinc(nf_r\tau_0)\delta(f-nf_r) \qquad \text{Eq. (3.71)}$$

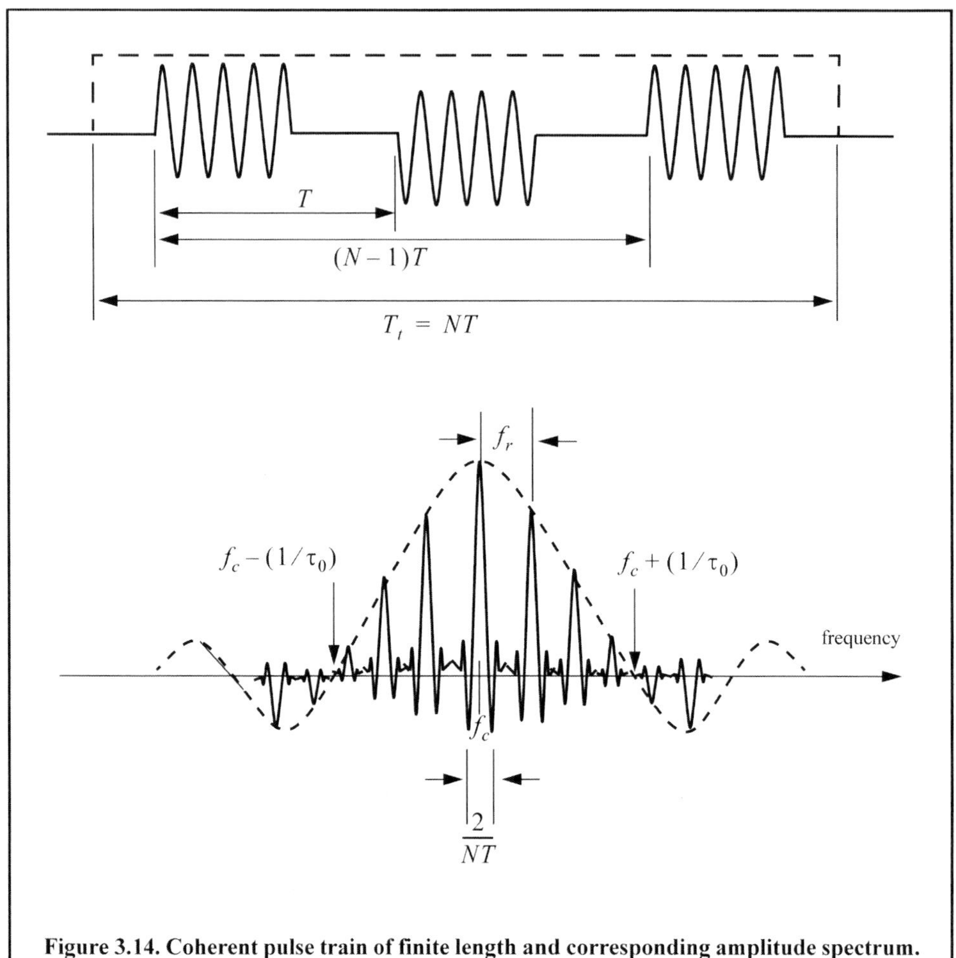

Figure 3.14. Coherent pulse train of finite length and corresponding amplitude spectrum.

and the FT of $g_1(t)$ is

$$G_1(f) = FT\left\{Rect\left(\frac{t}{T_t}\right)\right\} = T_t Sinc(fT_t) . \qquad \text{Eq. (3.72)}$$

Using these results, the FT of $x_4(t)$ can be written as

$$X_4(f) = \left(\frac{T_t\tau_0}{2T}\right)\left(Sinc(fT_t) \otimes \sum_{n=-\infty}^{\infty} Sinc(nf_r\tau_0)\delta(f-nf_r)\right) \otimes [\delta(f-f_c) + \delta(f+f_c)] . \qquad \text{Eq. (3.73)}$$

Therefore, the overall spectrum of $x_4(t)$ consists of a two equal positive and negative portions, centered at $\pm f_c$. Each portion is made up of N $Sinc(fT_t)$ functions repeated every f_r with envelope corresponding to $Sinc(nf_r\tau_0)$. This is illustrated in the lower part of Figure 3.14; only the positive portion of the spectrum is shown.

3.6.5. Linear Frequency Modulation (LFM) Signal

Frequency or phase modulated signals can be used to achieve much wider operating bandwidths. Linear Frequency Modulation (LFM) is very commonly used in modern radar systems. In this case, the frequency is swept linearly across the pulse width, either upward (up-chirp) or downward (down-chirp). Figure 3.15 shows a typical example of an LFM waveform. The pulse width is τ_0, and the bandwidth is B.

The LFM up-chirp instantaneous phase can be expressed by

$$\phi(t) = 2\pi\left(f_c t + \frac{\mu}{2}t^2\right) \qquad -\frac{\tau_0}{2} \le t \le \frac{\tau_0}{2}, \qquad \text{Eq. (3.74)}$$

where $\mu = B/\tau_0$ is the LFM coefficient. Thus, the instantaneous frequency is

$$f(t) = \frac{1}{2\pi}\frac{d}{dt}\phi(t) = f_c + \mu t \qquad -\frac{\tau_0}{2} \le t \le \frac{\tau_0}{2}. \qquad \text{Eq. (3.75)}$$

Similarly, the down-chirp instantaneous phase and frequency are

$$\phi(t) = 2\pi\left(f_c t - \frac{\mu}{2}t^2\right) \qquad -\frac{\tau_0}{2} \le t \le \frac{\tau_0}{2} \qquad \text{Eq. (3.76)}$$

$$f(t) = \frac{1}{2\pi}\frac{d}{dt}\phi(t) = f_c - \mu t \qquad -\frac{\tau_0}{2} \le t \le \frac{\tau_0}{2}. \qquad \text{Eq. (3.77)}$$

A typical LFM waveform can be expressed by

$$x_1(t) = Rect\left(\frac{t}{\tau_0}\right)e^{j2\pi\left(f_c t + \frac{\mu}{2}t^2\right)} \qquad \text{Eq. (3.78)}$$

where $Rect(t/\tau_0)$ denotes a rectangular pulse of width τ_0. Remember that the signal $x_1(t)$ is the analytic signal for the LMF waveform. It follows that

$$x_1(t) = \tilde{x}(t)e^{j2\pi f_c t} \qquad \text{Eq. (3.79)}$$

$$\tilde{x}(t) = Rect\left(\frac{t}{\tau}\right)e^{j\pi\mu t^2}. \qquad \text{Eq. (3.80)}$$

The spectrum of the signal $x_1(t)$ is determined from its complex envelope $\tilde{x}(t)$. The complex exponential term in Eq. (3.80) introduces a frequency shift about the center frequency f_c. Taking the FT of $\tilde{x}(t)$ yields

$$\tilde{X}(f) = \int_{-\infty}^{\infty} Rect\left(\frac{t}{\tau_0}\right)e^{j\pi\mu t^2}\, e^{-j2\pi f t}dt = \int_{-\frac{\tau_0}{2}}^{\frac{\tau_0}{2}} e^{j\pi\mu t^2}\, e^{-j2\pi f t}dt. \qquad \text{Eq. (3.81)}$$

Let $\mu' = \pi\mu = \pi B/\tau_0$, and perform the change of variable

$$\left(z = \sqrt{\frac{2}{\pi}}\left(\sqrt{\mu'}t - \frac{\pi f}{\sqrt{\mu'}}\right)\right) \quad ; \quad \sqrt{\frac{\pi}{2\mu'}}\, dz = dt \quad . \qquad \text{Eq. (3.82)}$$

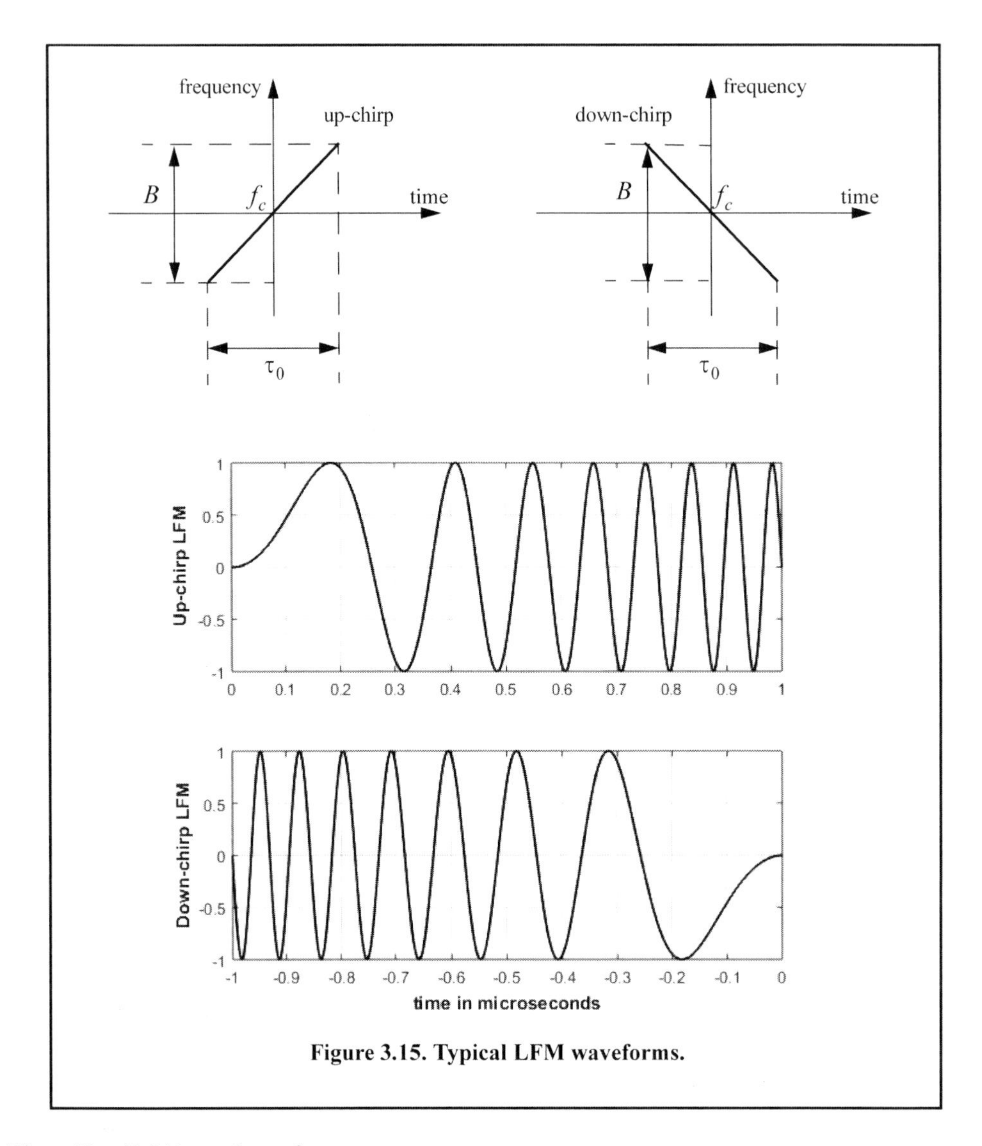

Figure 3.15. Typical LFM waveforms.

Thus, Eq. (3.81) can be written as

$$\tilde{X}(f) = \sqrt{\frac{\pi}{2\mu'}}\ e^{-j(\pi f)^2/\mu'} \int_{-z_1}^{z_2} e^{j\pi z^2/2}\ dz \qquad \text{Eq. (3.83)}$$

$$\tilde{X}(f) = \sqrt{\frac{\pi}{2\mu'}}\ e^{-j(\pi f)^2/\mu'} \left\{ \int_0^{z_2} e^{j\pi z^2/2}\ dz - \int_0^{-z_1} e^{j\pi z^2/2}\ dz \right\} \qquad \text{Eq. (3.84)}$$

$$z_1 = -\sqrt{\frac{2\mu'}{\pi}}\left(\frac{\tau_0}{2} + \frac{\pi f}{\mu'}\right) = \sqrt{\frac{B\tau_0}{2}}\left(1 + \frac{f}{B/2}\right) \qquad \text{Eq. (3.85)}$$

$$z_2 = \sqrt{\frac{\mu'}{\pi}}\left(\frac{\tau_0}{2} - \frac{\omega}{\mu'}\right) = \sqrt{\frac{B\tau_0}{2}}\left(1 - \frac{f}{B/2}\right). \qquad \text{Eq. (3.86)}$$

The Fresnel integrals, denoted by $C(z)$ and $S(z)$, are defined by

$$C(z) = \int_0^z \cos\left(\frac{\pi \upsilon^2}{2}\right) d\upsilon \text{ and } S(z) = \int_0^z \sin\left(\frac{\pi \upsilon^2}{2}\right) d\upsilon. \qquad \text{Eq. (3.87)}$$

Fresnel integrals can be approximated by

$$C(z) \approx \frac{1}{2} + \frac{1}{\pi z}\sin\left(\frac{\pi}{2}z^2\right) \qquad ; z \gg 1 \qquad \text{Eq. (3.88)}$$

$$S(z) \approx \frac{1}{2} - \frac{1}{\pi z}\cos\left(\frac{\pi}{2}z^2\right) \qquad ; z \gg 1. \qquad \text{Eq. (3.89)}$$

Note that $C(-z) = -C(z)$ and $S(-z) = -S(z)$. Figure 3.16 shows a plot of both $C(z)$ and $S(z)$ for $0 \le z \le 4.0$.

Using Eqs. (3.88) and (3.89) into Eq. (3.84) and performing the integration yield

$$\tilde{X}(f) = \sqrt{\frac{\pi}{2\mu'}}\, e^{-j(\pi f)^2/(\mu')}\{[C(z_2) + C(z_1)] + j[S(z_2) + S(z_1)]\}. \qquad \text{Eq. (3.90)}$$

Figure 3.17 shows typical plots for the LFM amplitude spectrum. This square-like spectrum is widely known as the Fresnel spectrum.

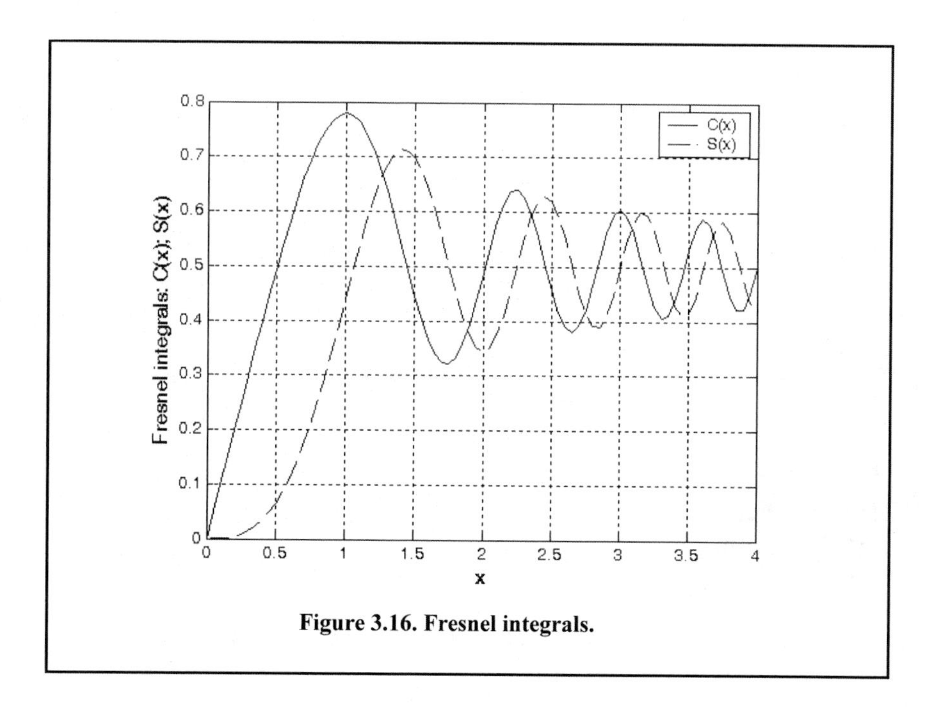

Figure 3.16. Fresnel integrals.

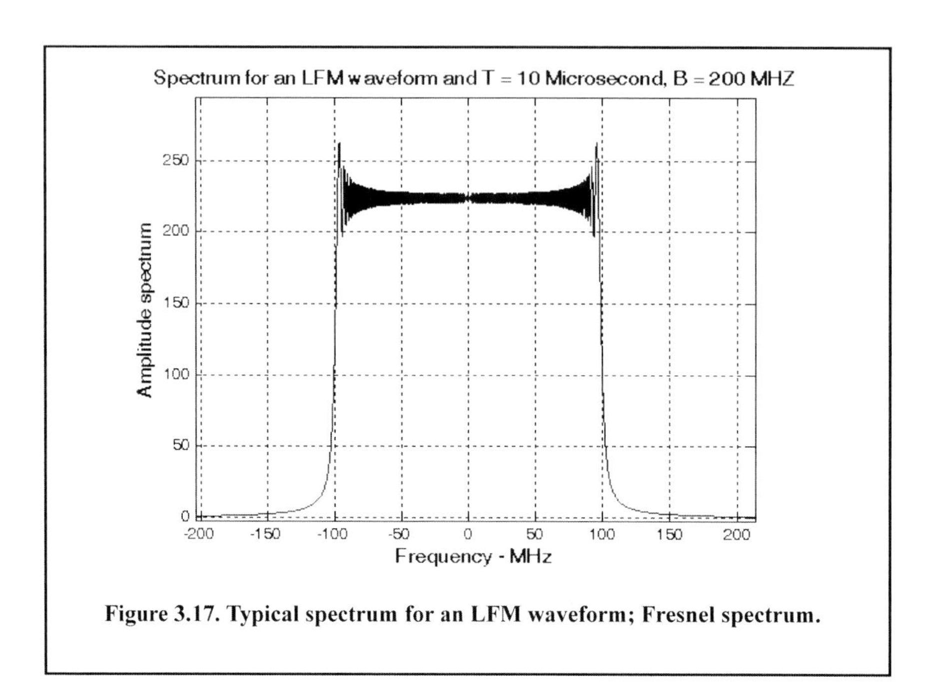

Figure 3.17. Typical spectrum for an LFM waveform; Fresnel spectrum.

Problems

3.1. A modulating signal $x(t)$ is applied to a DSB-SC; the carrier frequency is $f_c = 200Hz$. Sketch the spectral density of the resulting DSB-SC waveform where $x(t) = \cos 100\pi t$.

3.2. A given AM DSB-LC transmitter is tested using a dummy resistive load. With no modulation present, the average output power is 5 KW. But when a 1KHz sinewave input signal is present (with peak amplitude 1Volt), the output power is 7KW. (a) Assume a message signal that has a peak-to-rms ratio equal to 8, and assuming a 100% modulation, what is the resulting average transmitter power? (b) Find the modulation index.

3.3. A 1GHz carrier frequency is modulated by a 10KHz sinewave so that the peak frequency deviation is 100Hz. (a) Find the approximate FM signal bandwidth. (b) Find the bandwidth of the modulating signal. (c) Repeat part (a) with frequency deviation 40 KHz. Hint: you may use Bessel functions to verify your answer.

3.4. A DSB-Lc signal is generated using a 1KHz carrier and $x(t) = \cos 200\pi t$. The modulation index is $m = 0.8$. The lower sideband is attenuated using an ideal lowpass filter. Generate an expression for the resulting SSB signal if it develops 0.58 Watts across a 1Ω resistor.

3.5. An AM modulator operates using the message signal

$$x(t) = -4\cos(30\pi t) - \cos 90\pi t .$$

The modulated carrier signal is $50\cos 400\pi t$, and the modulation index is $m = 0.4$. Determine the normalized message signal and compute the modulator efficiency.

3.6. Two SSB-AM signals are generated. The first is obtained by side-band filtering of the DSB-AM signal $Ax(t)\cos 2\pi f_c t$. The second SSB signal is derived using phase shift to generate $0.5Ax(t)(\cos 2\pi f_c t) \pm 0.5A\hat{x}(t)(\sin 2\pi f_c t)$. Show that the time average powers of both signals are equal.

3.7. A sinusoidal message signal $x(t) = 150(\cos 300\pi t)$ is applied to an FM modulator with frequency deviation equal to 10. Determine the bandwidth of the modulator if a power ratio $P \geq 0.85$ is desired.

3.8. A certain sinewave whose frequency is f_m Hz is used as the modulating signal in bot an AM (DSB-LC) and a FM system. After modulation the peak frequency deviation of the FM signal is set to three times the bandwidth of the AM signal. The magnitudes of the side bands spaced at $\pm f_m$ from the carrier in both systems are equal, and the total average powers are equal. Determine the modulation indices for both systems.

3.9. Prove that

$$\sum_{n=-\infty}^{\infty} J_n(z) = 1 .$$

3.10. Show that $J_{-n}(z) = (-1)^n J_n(z)$. Hint: You may utilize the relation

$$J_n(z) = \frac{1}{\pi}\int_0^{\pi} \cos(z\sin y - ny)dy .$$

3.11. Hyperbolic frequency modulation (HFM) is better than LFM for high radial velocities. The HFM phase is

$$\phi_h(t) = \frac{\omega_0^2}{\mu_h}\ln\left(1 + \frac{\mu_h \alpha t}{\omega_0}\right)$$

where μ_h is an HFM coefficient and α is a constant. (a) Give an expression for the instantaneous frequency of an HFM pulse of duration τ'_h. (b) Show that HFM can be approximated by LFM. Express the LFM coefficient μ_l in terms of μ_h and in terms of B and τ'.

3.12. Prove that the effective duration of a finite pulse train is equal to $(T_t \tau_0)/T$, where τ_0 is the pulse width, T is the PRI, and T_t is as defined in Figure 3.14.

Chapter 4

Radar Systems - Basic Concepts

4.1. Radar Classifications

Radars can be classified as ground based, airborne, spaceborne, or ship based radar systems. They can also be classified into numerous categories based on the specific radar characteristics, such as the frequency band, antenna type, and waveforms utilized. Another classification is concerned with the mission and/or the functionality of the radar. This includes weather, acquisition and search, tracking, track-while-scan, fire control, early warning, over the horizon, terrain following and terrain avoidance radars.

Phased array radars utilize phased array antennas, and are often called multifunction (multimode) radars. A phased array is a composite antenna formed from two or more basic radiators. Array antennas synthesize narrow directive beams that may be steered, mechanically or electronically. Electronic steering is achieved by controlling the phase of the electric current feeding the array elements, and thus the name phased arrays is adopted.

Radars are most often classified by the types of waveforms they use, or by their operating frequency. Considering the waveforms first; radars can be Continuous Wave (CW) or Pulsed Radars (PR). CW radars are those that continuously emit electromagnetic energy, and use separate transmit and receive antennas. Unmodulated CW radars can accurately measure target radial velocity (Doppler shift) and angular position. Target range information cannot be extracted without utilizing some form of modulation.

The primary use of unmodulated CW radars is in target velocity search and track, and in missile guidance. Pulsed radars use a train of pulsed waveforms (mainly with modulation). In this category, radar systems can be classified on the basis of the Pulse Repetition Frequency (PRF), as low PRF, medium PRF, and high PRF radars. Low PRF radars are primarily used for ranging where target velocity (Doppler shift) is not of interest. High PRF radars are mainly used to measure target velocity. Continuous wave as well as pulsed radars can measure both target range and radial velocity by utilizing different modulation schemes.

Table 4.1 has the radar classifications based on the operating frequency. High Frequency (HF) radars utilize the electromagnetic waves' reflection off the ionosphere to detect targets beyond the horizon. Some examples include the United States Over The

Horizon Backscatter (U.S. OTH/B) radar which operates in the frequency range of $5 - 28MHZ$, the U.S. Relocatable Over The Horizon Radar (ROTHR), and the Soviet Woodpecker radar.

Very High Frequency (VHF) and Ultra High Frequency (UHF) bands are used for very long range Early Warning Radars (EWR). Some examples include the U.S. Ballistic Missile Early Warning System (BMEWS) search and track monopulse radar which operates at $245MHz$, the Perimeter and Acquisition Radar (PAR) which is a very long range multifunction phased array radar, and the early warning PAVE PAWS multifunction UHF phased array radar. Due to the very large wavelength and to the sensitivity requirements for very long range measurements, large apertures are needed in such radar systems.

Radars in the L-Band are primarily ground and ship based systems that are used in long range military and air traffic control search operations. Most ground and ship based medium range radars operate in the S-Band. For example, the Airport Surveillance Radar (ASR) used for air traffic control, and the ship based U.S. Navy AEGIS multifunction phased array are S-Band radars. The Airborne Warning And Control System (AWACS) and the National Weather Service Next Generation Doppler Weather Radar (NEXRAD) are also S-Band radars. However, most weather detection radar systems are C-Band radars. Medium range search and fire control military radars and metric instrumentation radars are also C-Band.

The X-Band is used for radar systems where the size of the antenna constitutes a physical limitation; this include most military multimode airborne radars. Radar systems that require fine target detection capabilities and yet cannot tolerate the atmospheric attenuation of higher frequency bands may also be X-Band. The higher frequency bands (Ku, K, and Ka) suffer severe weather and atmospheric attenuation. Therefore, radars utilizing these frequency bands are limited to short range applications, such as the police traffic radars, short range terrain avoidance and terrain following radars. Milli-Meter Wave (MMW) radars are mainly limited to very short range Radio Frequency (RF) seekers and experimental radar systems.

Table. 4.1. Radar frequency bands.

Letter designation	Frequency (GHz)	New band designation (GHz)
HF	0.003 - 0.03	A
VHF	0.03 - 0.3	A < 0.25; B > 0.25
UHF	0.3 - 1.0	B < 0.5; C > 0.5
L-Band	1.0 - 2.0	D
S-Band	2.0 - 4.0	E < 3.0; F > 3.0
C-Band	4.0 - 8.0	G < 6.0; H > 6.0
X-Band	8.0 - 12.5	I < 10.0; J > 10.0
Ku-Band	12.5 - 18.0	J
K-Band	18.0 - 26.5	J < 20.0; K > 20.0
Ka-Band	26.5 - 40.0	K
MMW	Normally > 34.0	L < 60.0; M > 60.0

4.2. Range Measurements

Figure 4.1 shows a simplified pulsed radar block diagram. The time control box generates the synchronization timing signals required throughout the system. A modulated signal is generated and sent to the antenna by the modulator/transmitter block. Switching the antenna between the transmitting and receiving modes is controlled by the duplexer. The duplexer allows one antenna to be used for both transmit and receive. During transmission it directs the radar electromagnetic energy towards the antenna. Alternatively, on reception, it directs the received radar echoes to the receiver. The receiver amplifies the radar returns and prepares them for signal processing. Extraction of target information is performed by the signal processor block. The target's range, R, is computed by measuring the time delay, Δt, that it takes a pulse to travel the two-way path between the radar and the target. Since electromagnetic waves travel at the speed of light, $c = 3 \times 10^8 m/second$, then

$$R = (c\Delta t)/2 \qquad \text{Eq. (4.1)}$$

where R is in meters and Δt is in seconds. The factor of $\frac{1}{2}$ is needed to account for the two-way time delay.

In general, a pulsed radar transmits and receives a train of pulses, as illustrated by Figure 4.2. The Inter Pulse Period (IPP) is T, and the pulse width is τ. The IPP is often referred to as the Pulse Repetition Interval (PRI). The inverse of the PRI is the Pulse Repetition Frequency (PRF), which is denoted by f_r

$$f_r = \frac{1}{PRI} = \frac{1}{T} = PRF. \qquad \text{Eq. (4.2)}$$

During each PRI the radar radiates energy only for τ seconds and listens for target returns for the rest of the PRI. The radar transmitting duty cycle (factor) d_t, is defined as the ratio $d_t = \tau/T$. The radar average transmitted power is then

$$P_{av} = P_t \times d_t, \qquad \text{Eq. (4.3)}$$

where P_t denotes the radar peak transmitted power. The pulse energy is $E_p = P_t\tau = P_{av}T = P_{av}/f_r$.

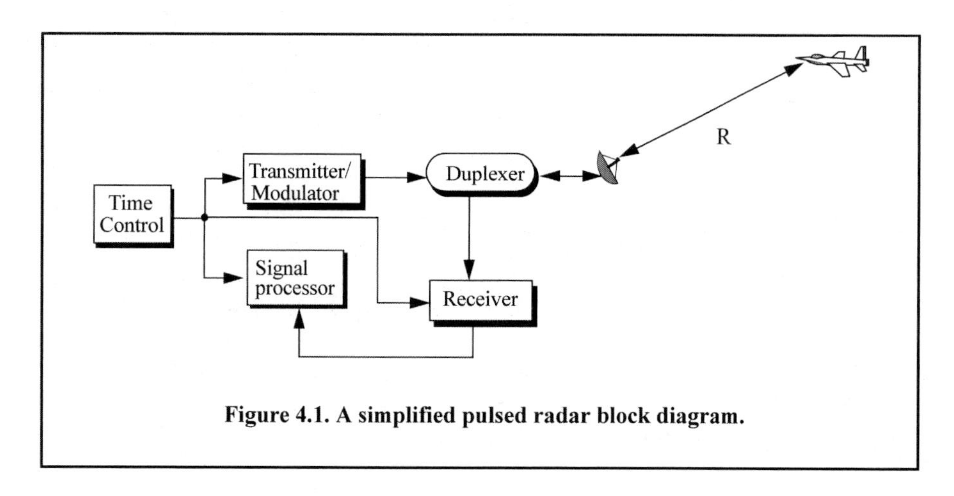

Figure 4.1. A simplified pulsed radar block diagram.

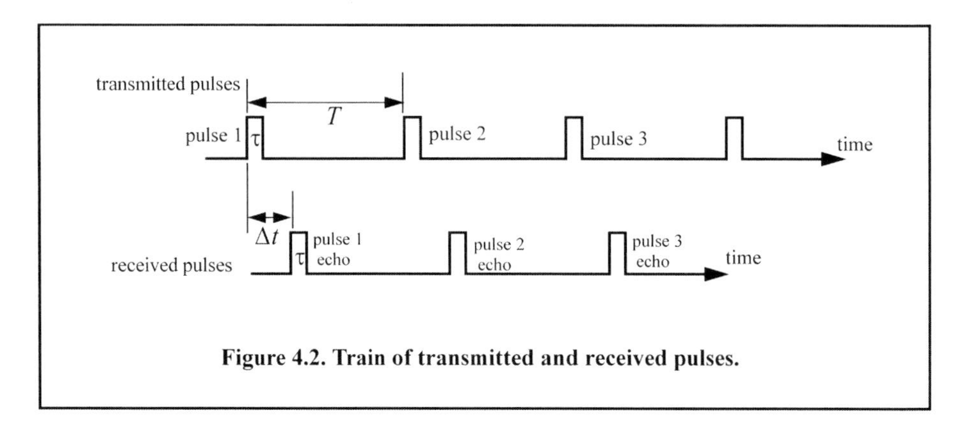

Figure 4.2. Train of transmitted and received pulses.

Example:

A certain airborne pulsed radar has peak power $P_t = 10KW$*, and uses two PRFs,* $f_{r1} = 10KHz$ *and* $f_{r2} = 30KHz$*. What are the required pulse widths for each PRF so that the average transmitted power is constant and is equal to* $1500 Watts$*? Compute the pulse energy in each case.*

Solution:

Since P_{av} *is constant, then both PRFs have the same duty cycle. More precisely,*

$$d_t = \frac{1500}{10 \times 10^3} = 0.15 .$$

The pulse repetition intervals are

$$T_1 = \frac{1}{10 \times 10^3} = 0.1ms \; ; \; T_2 = \frac{1}{30 \times 10^3} = 0.0333ms .$$

It follows that

$$\tau_1 = 0.15 \times T_1 = 15\mu s \; ; \; \tau_2 = 0.15 \times T_2 = 5\mu s$$

and

$$E_{p1} = P_t \tau_1 = 10 \times 10^3 \times 15 \times 10^{-6} = 0.15 Joules$$

$$E_{p2} = P_2 \tau_2 = 10 \times 10^3 \times 5 \times 10^{-6} = 0.05 Joules .$$

****

4.2.1. Unambiguous Range

The range corresponding to the two-way time delay T is known as the radar unambiguous range, R_u. Consider the case shown in Figure 4.3. Echo 1 represents the radar return from a target at range $R_1 = c\Delta t/2$ due to pulse 1. Echo 2 could be interpreted as the return from the same target due to pulse 2, or it may be the return from a far away target at range R_2 due to pulse 1 again. In this case,

$$R_2 = \frac{c\Delta t}{2} \quad or = \frac{c(T+\Delta t)}{2}. \qquad \text{Eq. (4.4)}$$

Clearly, range ambiguity is associated with echo 2. Therefore, once a pulse is transmitted, the radar must wait a sufficient length of time so that returns from targets at maximum range are back before the next pulse is emitted. It follows that the maximum unambiguous range must correspond to half of the PRI,

$$R_u = c\frac{T}{2} = \frac{c}{2f_r}. \qquad \text{Eq. (4.5)}$$

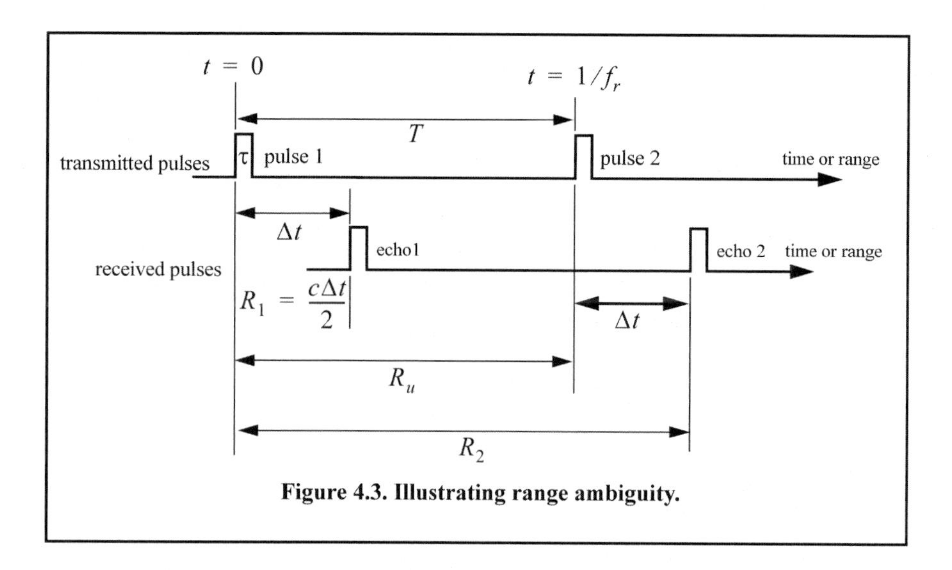

Figure 4.3. Illustrating range ambiguity.

4.2.2. Range Resolution

Range resolution, denoted as ΔR, is a radar metric that describes its ability to detect targets in close proximity to each other as distinct objects. Radar systems are normally designed to operate between a minimum range R_{min} and maximum range R_{max}. The distance between R_{min} and R_{max} is divided into M range bins (gates), each of width ΔR,

$$M = \frac{R_{max} - R_{min}}{\Delta R}. \qquad \text{Eq. (4.6)}$$

Targets separated by at least ΔR will be completely resolved in range, as illustrated in Figure 4.4 and Table 4.2. Targets within the same range bin can be resolved in cross range (azimuth) utilizing signal processing techniques.

Consider two targets located at ranges R_1 and R_2, corresponding to time delays t_1 and t_2, respectively. Denote the difference between those two ranges as ΔR,

$$\Delta R = R_2 - R_1 = c\frac{(t_2 - t_1)}{2} = c\frac{\delta t}{2}. \qquad \text{Eq. (4.7)}$$

Now, let us try to answer the following question: What is the minimum δt such that, target 1 at R_1 and target 2 at R_2 will appear completely resolved in range (different range bins)? In other words, what is the minimum ΔR?

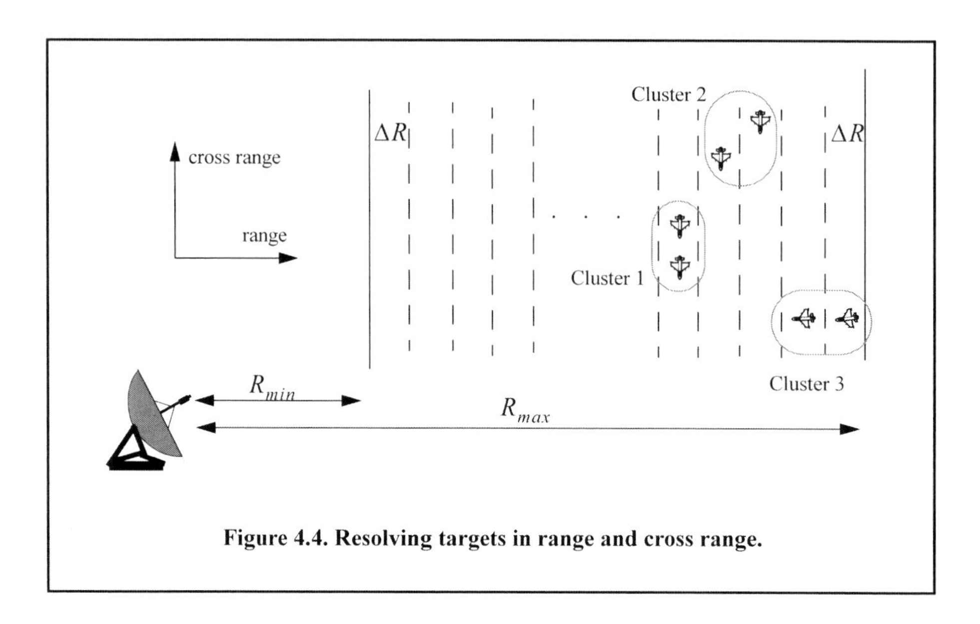

Figure 4.4. Resolving targets in range and cross range.

Table 4.2. Summary of Figure 4.4.

Cluster	**Resolved in range**	**Resolved in cross range**
#1	No	Yes
#2	Yes	Yes
#3	Yes	No

First, assume that the two targets are separated by $c\tau/4$; τ is the pulse width. In this case, when the pulse trailing edge strikes target 2 the leading edge would have traveled back a distance $c\tau$, and the returned pulse would be composed of returns from both targets (i.e., unresolved return), as shown in Figure 4.5a. However, if the two targets are at least $c\tau/2$ apart, then as the pulse trailing edge strikes the first target the leading edge will start to return from target 2, and two distinct returned pulses will be produced, as illustrated by Figure 4.5b. Thus, ΔR should be greater or equal to $c\tau/2$. And since the radar bandwidth B is equal to $1/\tau$, then

$$\Delta R = \frac{c\tau}{2} = \frac{c}{2B}. \qquad \text{Eq. (4.8)}$$

In general, radar users and designers alike seek to minimize ΔR in order to enhance the radar performance. As suggested by Eq. (4.8), in order to achieve fine range resolution one must minimize the pulse width. However, this will reduce the average transmitted power and increase the operating bandwidth. Achieving fine range resolution while maintaining adequate average transmitted power can be accomplished by using pulse compression techniques, which will be discussed in a subsequent chapter.

***Example*:**

Consider a radar system with an unambiguous range of 100 Km, and a bandwidth 0.5 MHz. Compute the required PRF, PRI, ΔR, and τ.

Solution:

$$PRF = \frac{c}{2R_u} = \frac{3 \times 10^8}{2 \times 10^5} = 1500 \ Hz$$

$$PRI = \frac{1}{PRF} = \frac{1}{1500} = 0.6667 \ ms$$

$$\Delta R = \frac{c}{2B} = \frac{3 \times 10^8}{2 \times 0.5 \times 10^6} = 300 \ m \text{ and } \tau = \frac{2\Delta R}{c} = \frac{2 \times 300}{3 \times 10^8} = 2 \ \mu s .$$

****

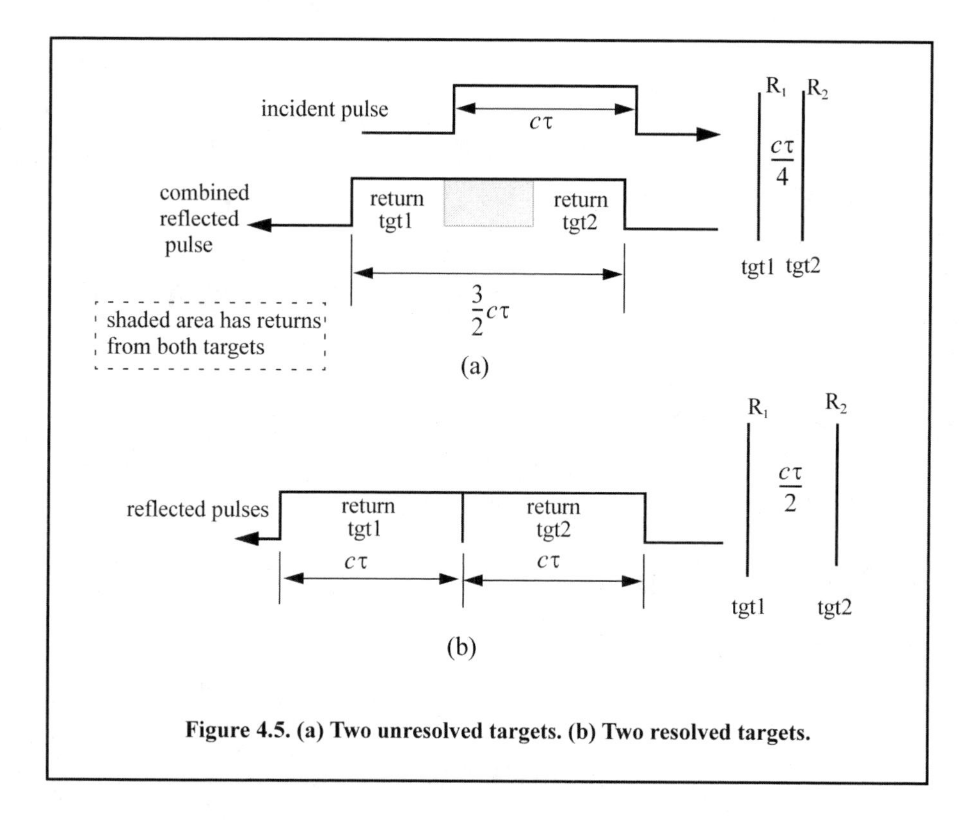

Figure 4.5. (a) Two unresolved targets. (b) Two resolved targets.

4.3. Doppler Frequency

Radars use Doppler frequency to extract target radial velocity (range rate), as well as to distinguish between moving and stationary targets or objects such as clutter. The Doppler phenomenon describes the shift in the center frequency of an incident waveform due to the target motion with respect to the source of radiation. Depending on the direction of the target's motion this frequency shift may be positive or negative. A waveform incident on a target has equiphase wavefronts separated by λ, the wavelength. A closing target will cause the reflected equiphase wavefronts to get closer to each other (smaller wavelength). Alternatively, an opening or receding target (moving away from the radar) will cause the reflected equiphase wavefronts to expand (larger wavelength), as illustrated in Figure 4.6.

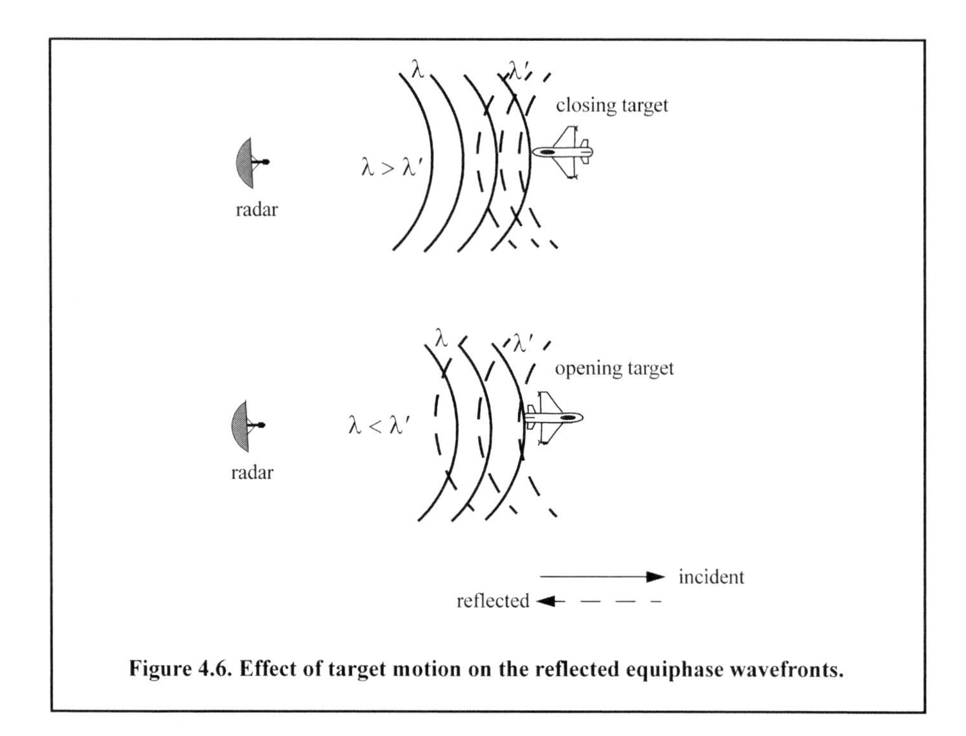

Figure 4.6. Effect of target motion on the reflected equiphase wavefronts.

Consider a pulse of width τ (seconds) incident on a target which is moving towards the radar at velocity v, as shown in Figure 4.7. Define d as the distance (in meters) that the target moves into the pulse during the interval Δt,

$$d = v\Delta t, \qquad \text{Eq. (4.9)}$$

where Δt is equal to the time span between the pulse leading edge striking the target and the trailing edge striking the target. Since the pulse is moving at the speed of light and the trailing edge has moved distance $c\tau - d$, then

$$\Delta t = (c\tau - d)/c. \qquad \text{Eq. (4.10)}$$

Combining Eq. (4.9) and Eq. (4.10) yields,

$$d = \frac{vc}{v+c}\tau. \qquad \text{Eq. (4.11)}$$

Now, in Δt seconds the pulse leading edge has moved in the direction of the radar a distance s,

$$s = c\Delta t. \qquad \text{Eq. (4.12)}$$

Therefore, the reflected pulse width is now τ' seconds, or L meters,

$$L = c\tau' = s - d. \qquad \text{Eq. (4.13)}$$

Substituting Eq. (4.11) and Eq. (4.12) into the right-hand side of Eq. (4.13) yields,

$$c\tau' = c\Delta t - \frac{vc}{v+c}\tau \qquad \text{Eq. (4.14)}$$

$$c\tau' = \frac{c^2}{v+c}\tau - \frac{vc}{v+c}\tau = \frac{c^2 - vc}{v+c}\tau, \qquad \text{Eq. (4.15)}$$

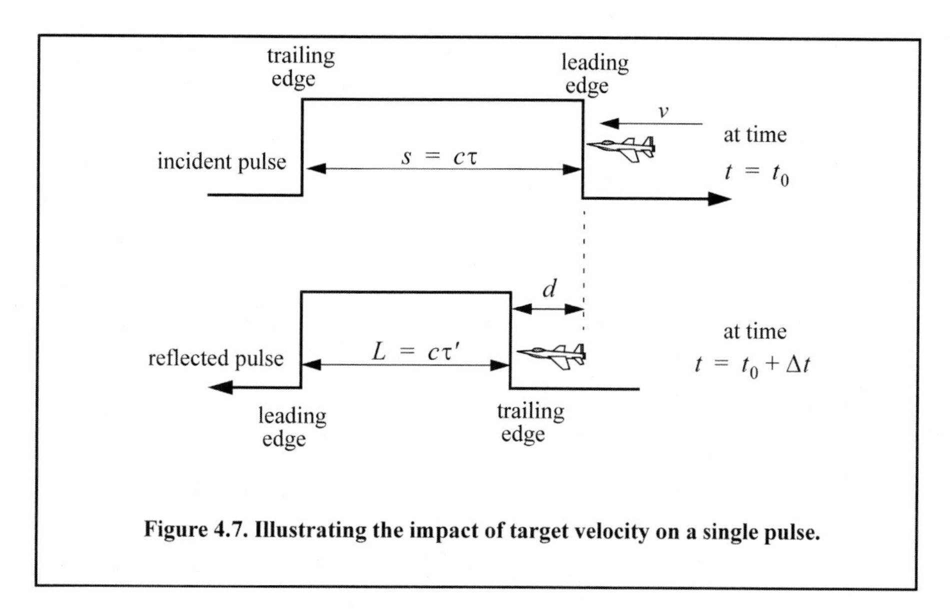

Figure 4.7. Illustrating the impact of target velocity on a single pulse.

finally,

$$\tau' = \frac{c - v}{c + v}\tau . \qquad \text{Eq. (4.16)}$$

In practice, the factor $(c - v)/(c + v)$ is often referred to as the time dilation factor. Notice that if $v = 0$, then $\tau' = \tau$. In a similar fashion, one can compute τ' for an opening target. In this case,

$$\tau' = \frac{v + c}{c - v}\tau . \qquad \text{Eq. (4.17)}$$

To derive an expression for Doppler frequency, consider the illustration shown in Figure 4.8. It takes the leading edge of pulse 2 Δt seconds to travel a distance $(c/f_r)–d$ to strike the target. Over the same time interval, the leading edge of pulse 1 travels the same distance $c\Delta t$. More precisely,

$$d = v\Delta t \qquad \text{Eq. (4.18)}$$

$$\frac{c}{f_r} - d = c\Delta t \qquad \text{Eq. (4.19)}$$

solving for Δt yields

$$\Delta t = \frac{c/f_r}{c + v} \qquad \text{Eq. (4.20)}$$

$$\Delta t = \frac{c/f_r}{c + v} . \qquad \text{Eq. (4.21)}$$

It follows that

$$d = \frac{cv/f_r}{c + v} . \qquad \text{Eq. (4.22)}$$

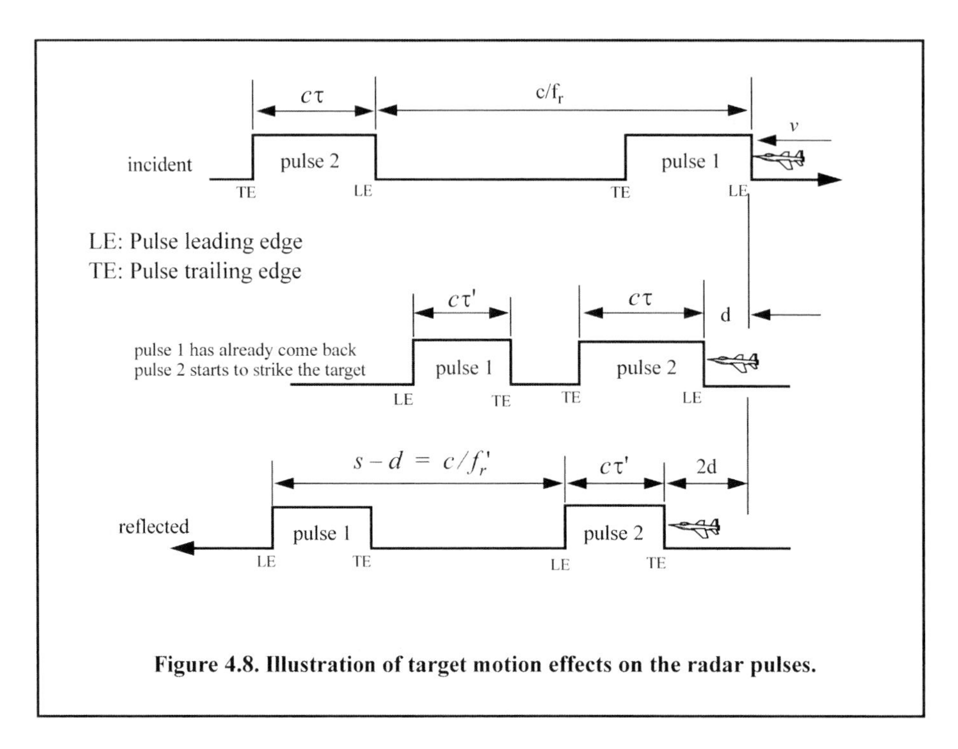

Figure 4.8. Illustration of target motion effects on the radar pulses.

The reflected pulse spacing is now $s - d$ and the new PRF is f_r', where

$$s - d = \frac{c}{f_r'} = c\Delta t - \frac{cv/f_r}{c + v}. \qquad \text{Eq. (4.23)}$$

It follows that the new PRF is related to the original PRF by,

$$f_r' = \frac{c + v}{c - v} f_r. \qquad \text{Eq. (4.24)}$$

However, since the number of cycles does not change, the frequency of the reflected signal will go up by the same factor. Denoting the new frequency by f_c', it follows

$$f_c' = \frac{c + v}{c - v} f_c, \qquad \text{Eq. (4.25)}$$

where f_c is the carrier frequency of the incident signal. The Doppler frequency f_d is defined as the difference $f_c' - f_c$. More precisely,

$$f_d = f_c' - f_c = \frac{c + v}{c - v} f_c - f_c = \frac{2v}{c - v} f_c \qquad \text{Eq. (4.26)}$$

but since $v \ll c$ and $c = \lambda f_c$, then

$$f_d \approx \frac{2v}{c} f_c = \frac{2v}{\lambda}. \qquad \text{Eq. (4.27)}$$

Equation (4.27) indicates that the Doppler shift is proportional to the target velocity, and thus, one can extract f_d from range rate and vice versa.

The result in Eq. (4.27) can also be derived using the following approach: Figure 4.9 shows a closing target with velocity v. Let R_0 refer to the range at time t_0 (time reference); then the range to the target at any time t is

$$R(t) = R_0 - v(t - t_0) . \qquad \text{Eq. (4.28)}$$

The signal received by the radar is then given by

$$x_r(t) = x(t - \phi(t)) , \qquad \text{Eq. (4.29)}$$

where $x(t)$ is the transmitted signal, and

$$\phi(t) = \frac{2}{c}(R_0 - vt + vt_0) . \qquad \text{Eq. (4.30)}$$

Substituting Eq. (4.30) into Eq. (4.29) and collecting terms yield,

$$x_r(t) = x\left(\left(1 + \frac{2v}{c}\right)t - \varphi_0\right) \qquad \text{Eq. (4.31)}$$

the constant phase φ_0 is

$$\varphi_0 = \frac{2R_0}{c} + \frac{2v}{c} t_0 . \qquad \text{Eq. (4.32)}$$

Define the compression or scaling factor γ by

$$\gamma = 1 + \frac{2v}{c} . \qquad \text{Eq. (4.33)}$$

Note that for a receding target the scaling factor is $\gamma = 1 - (2v/c)$. Using Eq. (4.33) we can rewrite Eq.(4.31) as

$$x_r(t) = x(\gamma t - \varphi_0) . \qquad \text{Eq. (4.34)}$$

Equation (4.34) is a time-compressed version of the returned signal from a stationary target ($v = 0$). Hence, based on the scaling property of the Fourier transform, the spectrum of the received signal will be expanded in frequency by a factor of γ.

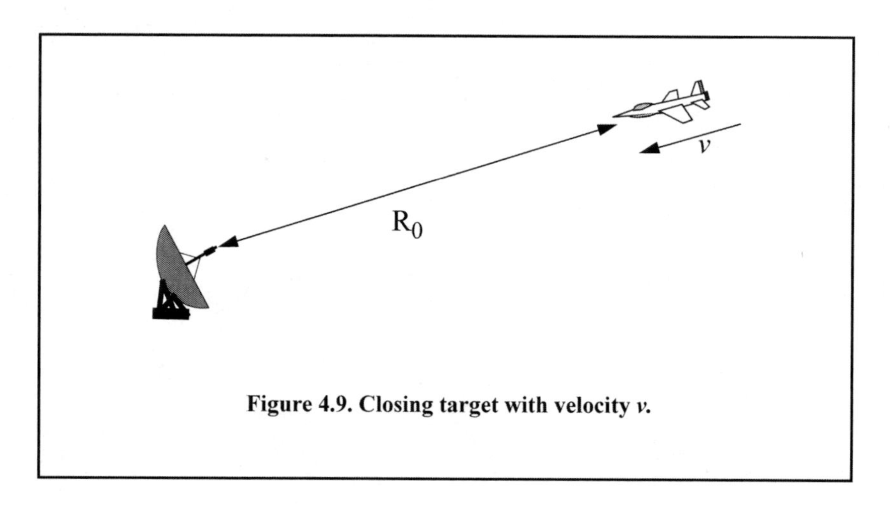

Figure 4.9. Closing target with velocity *v*.

Consider the special case when

$$x(t) = y(t)\cos\omega_c t \tag{Eq. (4.35)}$$

where $\omega_c = 2\pi f_c$ is the radar center frequency in radians. The received signal $x_r(t)$ is then given by

$$x_r(t) = y(\gamma t - \varphi_0)\cos(\gamma\omega_c t - \varphi_0) . \tag{Eq. (4.36)}$$

The Fourier transform of Eq. (4.36) is

$$X_r(\omega) = \frac{1}{2\gamma}\left(Y\left(\frac{\omega}{\gamma} - \omega_c\right) + Y\left(\frac{\omega}{\gamma} + \omega_c\right)\right) \tag{Eq. (4.37)}$$

where for simplicity the effects of the constant phase φ_0 have been ignored in Eq. (4.36). Therefore, the spectrum of the received signal is now centered at $\gamma\omega_c$ instead of ω_c. The difference between the two values corresponds to the amount of Doppler shift incurred due to the target motion,

$$\omega_d = \omega_c - \gamma\omega_c \tag{Eq. (4.38)}$$

where ω_d is the Doppler frequency in radians per second. Substituting the value of γ in Eq. (4.38) and using $\omega = 2\pi f$ yield

$$f_d = \frac{2v}{c} f_c = \frac{2v}{\lambda} \tag{Eq. (4.39)}$$

which is the same as Eq. (4.27). It can be shown that for a receding target the Doppler shift is $f_d = -2v/\lambda$. This is illustrated in Figure 4.10.

In both Eq. (4.39) and Eq. (4.27) the target radial velocity with respect to the radar is equal to v, but this is not always the case. In fact, the amount of Doppler frequency depends on the target velocity component in the direction of the radar (radial velocity). Figure 4.11 shows three targets all having velocity v: Target 1 has zero Doppler shift; target 2 has maximum Doppler frequency as defined in Eq. (4.39). The amount of Doppler frequency due to target 3 is $f_d = 2v\cos\theta/\lambda$, where $v\cos\theta$ is the radial velocity; and θ is the total angle between the radar line of sight and the target.

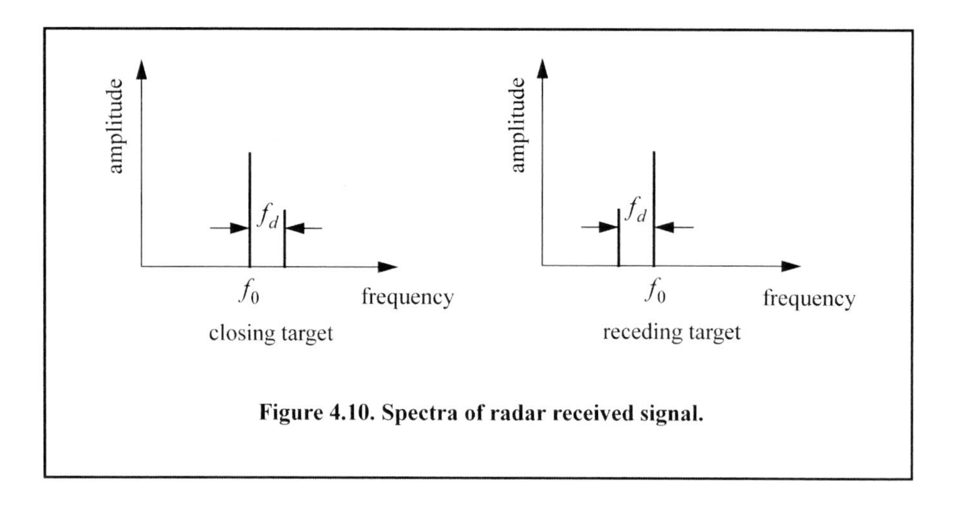

Figure 4.10. Spectra of radar received signal.

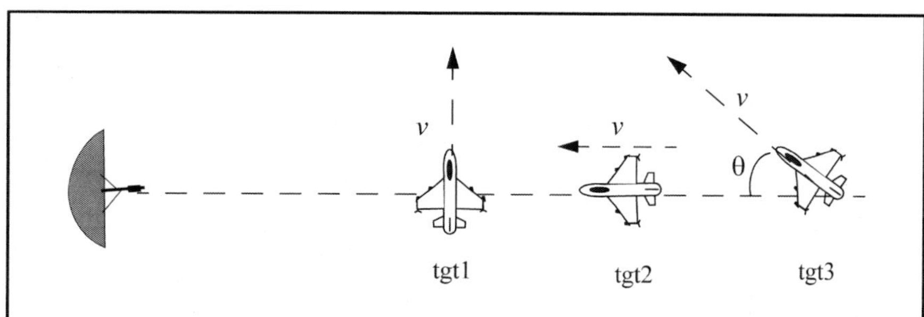

Figure 4.11. Target 1 generates zero Doppler. Target 2 generates maximum Doppler. Target 3 is in-between.

Thus, a more general expression for f_d that accounts for the total angle between the radar and the target is

$$f_d = \frac{2v}{\lambda}\cos\theta \qquad \text{Eq. (4.40)}$$

and for an opening target

$$f_d = \frac{-2v}{\lambda}\cos\theta \qquad \text{Eq. (4.41)}$$

where $\cos\theta = \cos\theta_e \cos\theta_a$. The angles θ_e and θ_a are, respectively, the elevation and azimuth angles, see Figure 4.12.

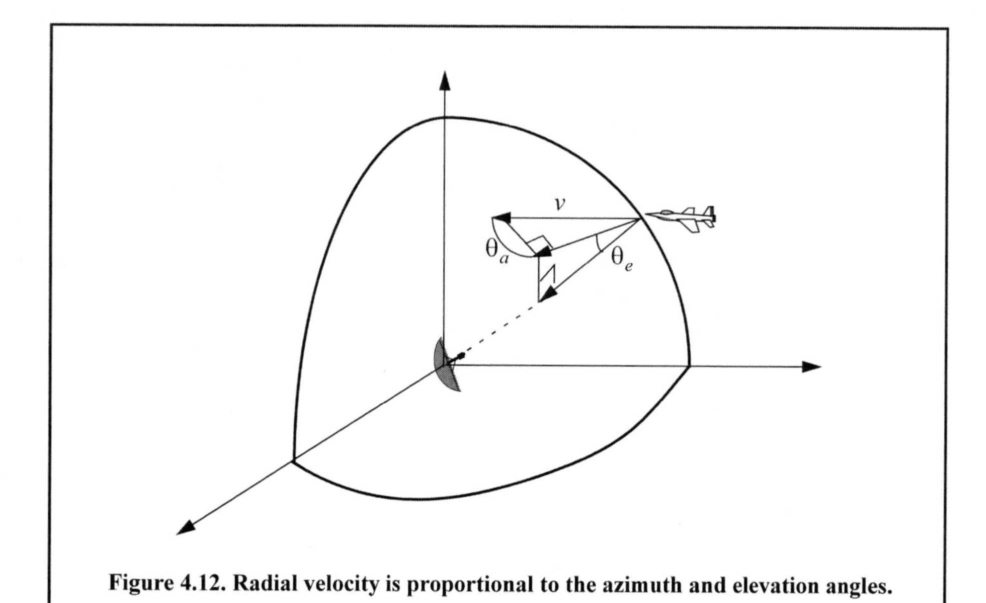

Figure 4.12. Radial velocity is proportional to the azimuth and elevation angles.

Example:

Compute the Doppler frequency measured by the radar shown in the figure below.

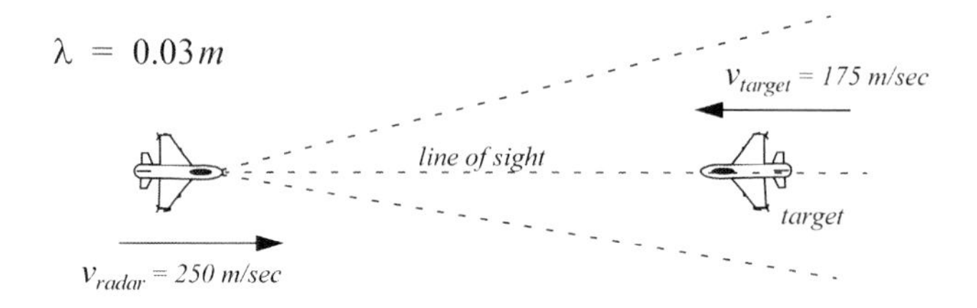

Solution:

The relative radial velocity between the radar and the target is $v_{radar} + v_{target}$. *Thus using Eq. (4.39), we get*

$$f_d = 2\frac{(250 + 175)}{0.03} = 28.3KHz.$$

Similarly, if the target were opening the Doppler frequency is

$$f_d = 2\frac{250 - 175}{0.03} = 5KHz.$$

****

4.4. Coherence

A radar is said to be coherent if the phase of any two transmitted pulses is consistent; i.e., there is a continuity in the signal phase from one pulse to the next. One can view coherence as the radar's ability to maintain an integer multiple of wavelengths between the equiphase wavefront from the end of one pulse to the equiphase wavefront at the beginning of the next pulse. Coherence can be achieved by using a STAble Local Oscillator (STALO). A radar is said to be coherent-on-receive or quasi-coherent if it stores in its memory a record of the phases of all transmitted pulses. In this case, the receiver phase reference is normally the phase of the most recently transmitted pulse.

Coherence also refers to the radar's ability to accurately measure (extract) the received signal phase. Since Doppler represents a frequency shift in the received signal, only coherent or coherent-on-receive radars can extract Doppler information. This is because the instantaneous frequency of a signal is proportional to the time derivative of the signal phase.

4.5. Radar Wave Propagation

In order to accurately predict radar performance, one must account for the earth's shape and atmosphere on the radar signals. These effects include ground reflections from the surface of the earth, diffraction of electromagnetic waves, bending or refraction of radar waves due to the earth's atmosphere, and attenuation or absorption of radar energy by the gases in the atmosphere.

The earth's impact on the radar waves manifests itself by introducing an additional power term in the radar signal. This term is referred to as the *pattern propagation factor* and is symbolically denoted by F_p. The propagation factor can actually introduce constructive as well as destructive interference with the radar signal depending on the radar frequency and the geometry under consideration.

4.5.1. Earth's Atmosphere

The earth's atmosphere comprises several layers, as illustrated in Figure 4.13. The first layer, which extends in altitude to about *30Km*, is known as the troposphere. Electromagnetic waves refract (bend downward) as they travel in the troposphere. The troposphere refractive effect is related to its dielectric constant, which is a function of the pressure, temperature, water vapor, and gaseous content. Additionally, due to gases and water vapor in the atmosphere, radar energy suffers a loss. This loss is known as the atmospheric attenuation. Atmospheric attenuation increases significantly in the presence of rain, fog, dust, and clouds. The region above the troposphere (altitude from *30* to *85Km*) behaves like free space, and thus little refraction occurs in this region. This region is known as the interference zone.

The ionosphere extends from about *85Km* to about *1000Km*. It has very low gas density compared to the troposphere. It contains a significant amount of ionized free electrons. The ionization is primarily caused by the sun's ultraviolet and X-rays. This presence of free electrons in the ionosphere affects electromagnetic wave propagation in different ways. These effects include refraction, absorption, noise emission, and polarization rotation. The degree of degradation depends heavily on the frequency of the incident waves. For example, frequencies lower than about *4* to *6MHz* are completely reflected from the lower region of the ionosphere.

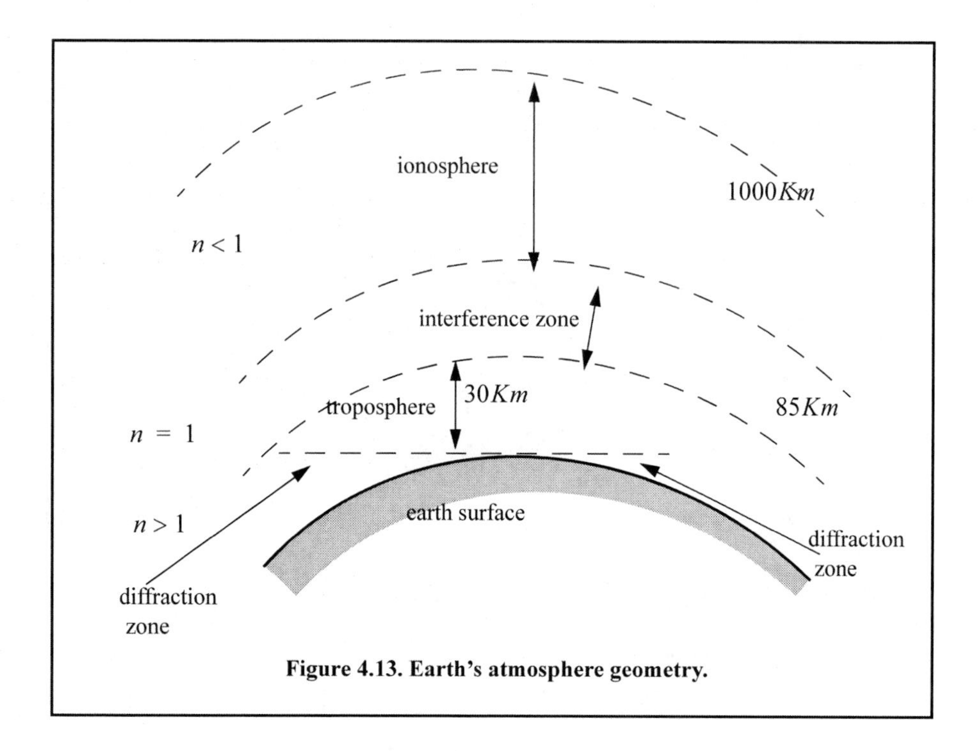

Figure 4.13. Earth's atmosphere geometry.

Frequencies higher than *30MHz* may penetrate the ionosphere with some level of attenuation. In general, as the frequency is increased, most of the ionosphere's effects become less prominent. The region below the horizon, close to the earth's surface, is called the diffraction region. Diffraction is a term used to describe the bending of radar waves. In free space, electromagnetic waves travel in straight lines. However, in the presence of the earth's atmosphere, they bend (refract), as illustrated in Figure 4.14. Refraction is a term used to describe the deviation of radar wave propagation from straight lines. The deviation from straight line propagation is caused by the variation of the index of refraction. The index of refraction is defined as

$$n = c/v \qquad \text{Eq. (4.42)}$$

where c is the velocity of electromagnetic waves in free space and v is the wave group velocity in the medium. In the troposphere, the index of refraction decreases uniformly with altitude, while in the ionosphere the index of refraction is minimum at the level of maximum electron density. Alternatively, the interference zone acts like free space and in it the index of refraction is unity.

In order to effectively study the effects of the atmosphere on the propagation of radar waves, it is necessary to have accurate knowledge of the height variation of the index of refraction in the troposphere and the ionosphere. The index of refraction is a function of the geographic location on the earth, weather, time of day or night, and the season of the year. Therefore, analyzing the atmospheric propagation effects under all parametric conditions becomes an overwhelming task. Typically, this problem is simplified by analyzing atmospheric models that are representative of an average of atmospheric conditions.

In radar applications, one can assume a *well-mixed atmosphere* condition, where the index of refraction decreases in a smooth monotonic fashion with height. The rate of change of the earth's index of refraction n with altitude h is normally referred to as the refractivity gradient, dn/dh. As a result of the negative rate of change in dn/dh, electromagnetic waves travel at slightly higher velocities in the upper troposphere than in the lower part. As a result of this, waves traveling horizontally in the troposphere gradually bend downward. In general, since the rate of change in the refractivity index is very slight, waves do not curve downward appreciably unless they travel very long distances through the atmosphere.

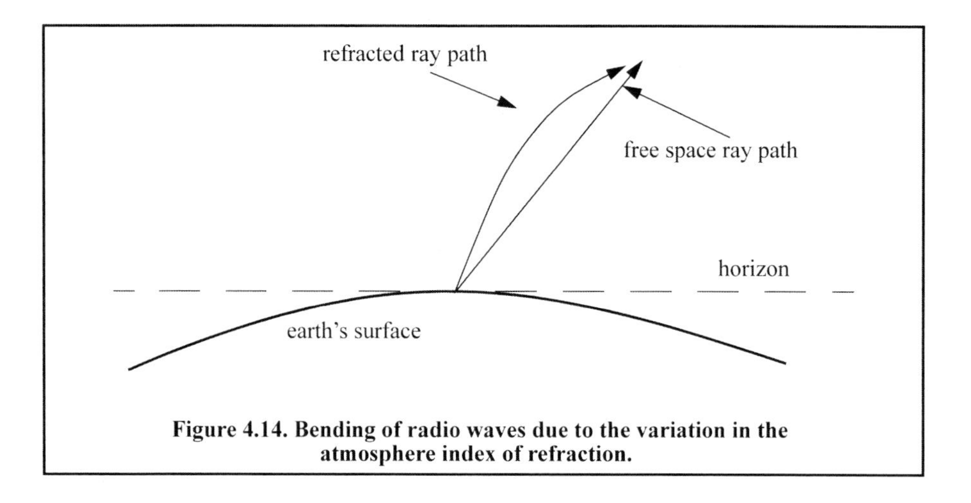

Figure 4.14. Bending of radio waves due to the variation in the atmosphere index of refraction.

Refraction affects radar waves in two different ways depending on height. For targets that have altitudes typically above *100m*, the effect of refraction is illustrated in Figure 4.15. In this case, refraction imposes limitations on the radar's capability to measure target position, and introduces an error in measuring the elevation angle. In a well-mixed atmosphere and very low altitudes (less than *100m*), the refractivity gradient close to the earth's surface is almost constant. However, temperature changes and humidity lapses close to the earth's surface may cause serious changes in the refractivity profile. When the refractivity index becomes large enough, electromagnetic waves bend around the curve of the earth.

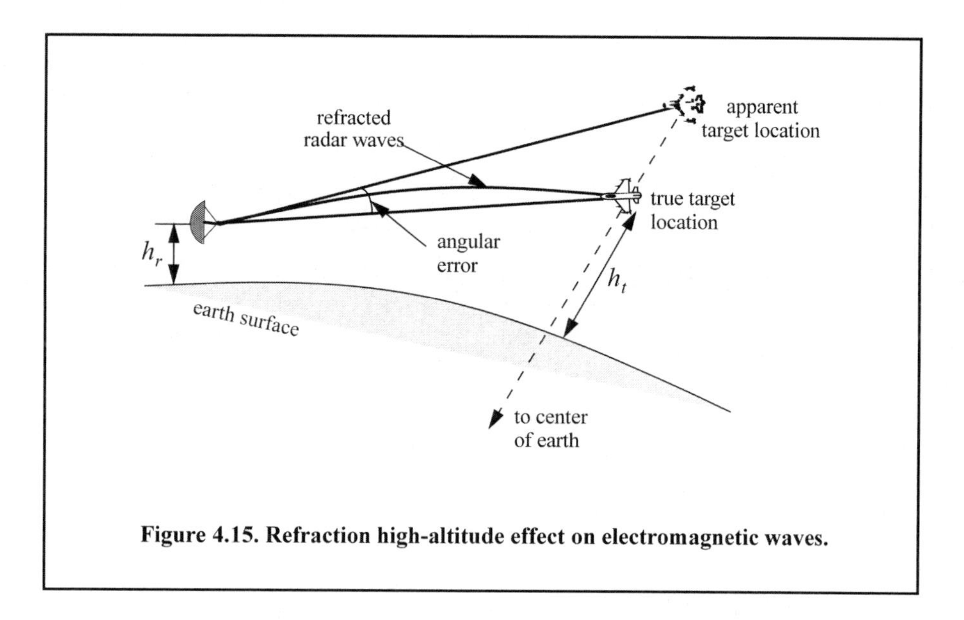

Figure 4.15. Refraction high-altitude effect on electromagnetic waves.

4.6. Atmospheric Models

The amount of bending electromagnetic waves experience due to refraction has a lot to do with the medium propagation index of refraction n, defined in Eq. (4.42). Because the index of refraction is not constant as one rises in altitude, it is necessary to analyze the formulas for the index of refraction as a function of height or altitude. Over the last several decades, this topic has been a subject of study by many scientists and physicists; thus, open source references on the subject are abundant in the literature. However, due to differences in notation used as well as the application being studied, it is rather difficult to sift through all available information in a timely and productive manner, particularly for the non-experts in the field. In this chapter, the subject is analyzed in the context of radar wave propagation in the atmosphere.

4.6.1. Index of Refraction in the Troposphere

As mentioned earlier, the index of refraction is a function of water vapor, air temperature, and air pressure in the medium, which all vary as a function of height. Because the rate of change of the index of refraction as a function of height is so small, it is very common to introduce a new quantity referred to as *refractivity* N, where

$$N = (n-1) \times 1 \times 10^6 . \qquad \text{Eq. (4.43)}$$

Using this notation, refractivity in the troposphere is given by

$$N = \frac{K_1}{T}\left(P + \frac{K_2 P_w}{T}\right) \qquad \text{Eq. (4.44)}$$

where T is the air temperature of the medium in degrees *Kelvin*, P is the total air pressure in *millibars*, P_w is the partial pressure of water vapor in *millibars*, and K_1, K_2 are constants. The first term of Eq. (4.44) (i.e., $(K_1 P)/T$) applies to all frequencies, while the second term (i.e., $(K_1 K_2 P_w)/T^2$) is applicable to radio frequencies only. Experts in the field differ on the exact values for K_1, K_2 based on their relevant applications. However, for most radar applications K_1 can be assumed to be $77.6°$ *Kelvin/millibar* and K_2 is $4810°$ *Kelvin*. Therefore, Eq. (4.44) can now be written as,

$$N = \frac{77.6}{T}\left(P + \frac{4810 P_w}{T}\right) . \qquad \text{Eq. (4.45)}$$

The lowest values of N occur in dry areas where both P and P_w are low. In the United States, the surface value of N, denoted by N_0, varies between *285* and *345* in the winter, and from *275* to *385* in the summer. Note that Eq. (4.45) is valid for heights up to $h \le 50Km$.

If the values for T, P, and P_w are known everywhere and at all times, then N can be computed everywhere. However, knowing these variables everywhere and at all times is a very daunting task. Therefore, approximations are made for N, where the assumption that pressure and water vapor tend to decrease with height in a well-mixed atmosphere is taken into consideration. On average, the refractivity will decrease exponentially from N_0 in accordance with the following relation,

$$N = N_0 e^{-c_e \cdot h} \qquad \text{Eq. (4.46)}$$

where h is the altitude in *Km* and c_e is a constant (in Km^{-1}) related to refractivity by

$$c_e = -\frac{\left.\left(\frac{d}{dh}N\right)\right|_{h=0}}{N_0} . \qquad \text{Eq. (4.47)}$$

In general, c_e can be computed from Eq. (4.46) using two different altitudes.

4.6.2. Index of Refraction in the Ionosphere

Unlike the troposphere, refraction in the ionosphere occurs because of the high electron density (ionization) inside the ionosphere and not due to water vapor or other variables. The average electron density as a function of height is given by the Chapman function as

$$\rho_e = \rho_{max} \cdot e^{\frac{1 - z - e^{-z}}{2}} \qquad \text{Eq. (4.48)}$$

where ρ_e is the electron density in electrons per cubic meters, ρ_{max} is the maximum electron density along the propagation path, and z is the normalized altitude or normalized height. The normalized height is given by

$$z = \frac{(h - h_m)}{H} \qquad \text{Eq. (4.49)}$$

where h_m is the height of maximum electron density and the height scale H is given by

$$H = \frac{(kT)}{mg} \qquad \text{Eq. (4.50)}$$

where k is Boltzmann's constant, T is the temperature in degrees *Kelvin*, m is the mean molecular mass of an air particle, and g is the gravitational constant.

Electrons in the ionosphere travel in spiral paths along the earth's magnetic field lines at an angular rate ϖ_p given by

$$\varpi_p^2 = \frac{\rho_e Q}{m\varepsilon_0} \qquad \text{Eq. (4.51)}$$

where Q is the charge of an electron ($1.6022 \times 10^{-19} Columbs$) and ε_0 is the permittivity of free space ($8.8542 \times 10^{-12} Columbs/m$). The index of refraction is given by

$$n = \sqrt{1 - \left(\frac{\varpi_p}{\omega}\right)^2} \qquad \text{Eq. (4.52)}$$

where $\omega = 2\pi f$ is the radar wave frequency in radians per second and f is the frequency in Hertz. Substituting Eq. (4.51) into Eq. (4.52) and collecting terms yields

$$n = \sqrt{1 - \frac{80.6\rho_e}{f^2}} \approx 1 - \frac{40.3\rho_e}{f^2}. \qquad \text{Eq. (4.53)}$$

Note that Eq. (4.53) is valid for $h > 50Km$ and the refractivity is given by

$$N \approx -\frac{40.3\rho_e \times 10^6}{f^2}. \qquad \text{Eq. (4.54)}$$

Figure 4.16 shows a plot for the total range error incurred versus radar range due to refraction at $f = 9.5GHz$ for a few elevation angles, denoted by the greek letter β.

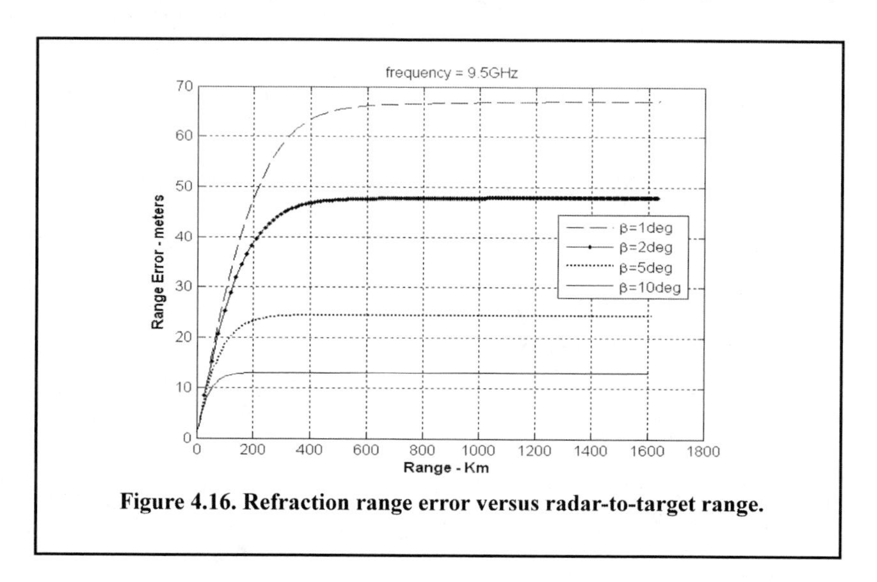

Figure 4.16. Refraction range error versus radar-to-target range.

4.6.3. Four-Third Earth Model

A very common way of dealing with refraction is to replace the actual earth with an imaginary earth whose effective radius is $r_e = kr_0$, where r_0 is the actual earth radius, and k is

$$k = \frac{1}{1 + r_0(dn/dh)}. \qquad \text{Eq. (4.55)}$$

When the refractivity gradient is assumed to be constant with altitude and is equal to 39×10^{-9} per meter, then $k = 4/3$. Using an effective earth radius $r_e = (4/3)r_0$ produces what is known as the *four-third earth model.* In general, choosing

$$r_e = r_0(1 + 6.37 \times 10^{-3}(dn/dh)) \qquad \text{Eq. (4.56)}$$

produces a propagation model where waves travel in straight lines. Selecting the correct value for k depends heavily on the region's meteorological conditions. At low altitudes (typically less than *10Km*) when using the *4/3* earth model, one can assume that radar waves (beams) travel in straight lines and do not refract. This is illustrated in Figure 4.17.

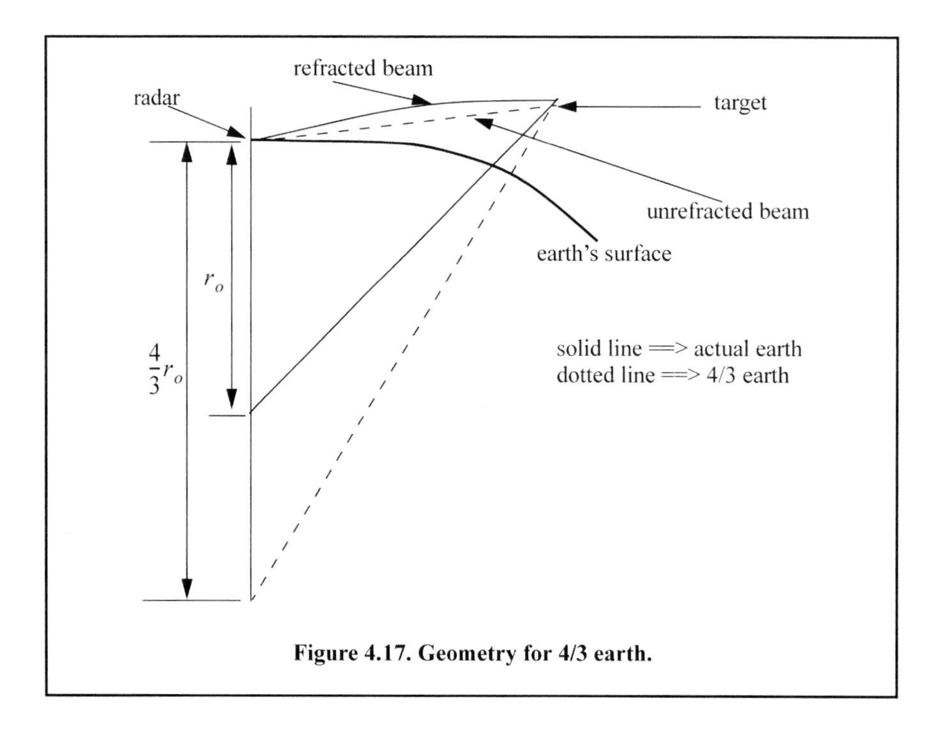

Figure 4.17. Geometry for 4/3 earth.

4.6.4. Target Height Equation

Using ray tracing (geometric optics), an integral-relating range-to-target height with the elevation angle as a parameter can be derived and calculated. However, such computations are complex and numerically intensive. Thus, in practice, radar systems deal with refraction in two different ways, depending on height. For altitudes higher than *3Km*, actual target heights are estimated from look-up tables or from charts of target height versus range for different elevation angles.

Blake[1] derived the *height-finding equation* for the 4/3 earth (see Figure 4.18) as

$$h = h_r + 6076R\sin\theta + 0.6625R^2(\cos\theta)^2 \qquad \text{Eq. (4.57)}$$

where h and h_r are in feet and R is nautical miles. The distance to the horizon for a radar located at height h_r can be calculated with the help of Figure 4.19. For the right-angle triangle OBA we get

$$r_h = \sqrt{(r_0 + h_r)^2 - r_0^2} \qquad \text{Eq. (4.58)}$$

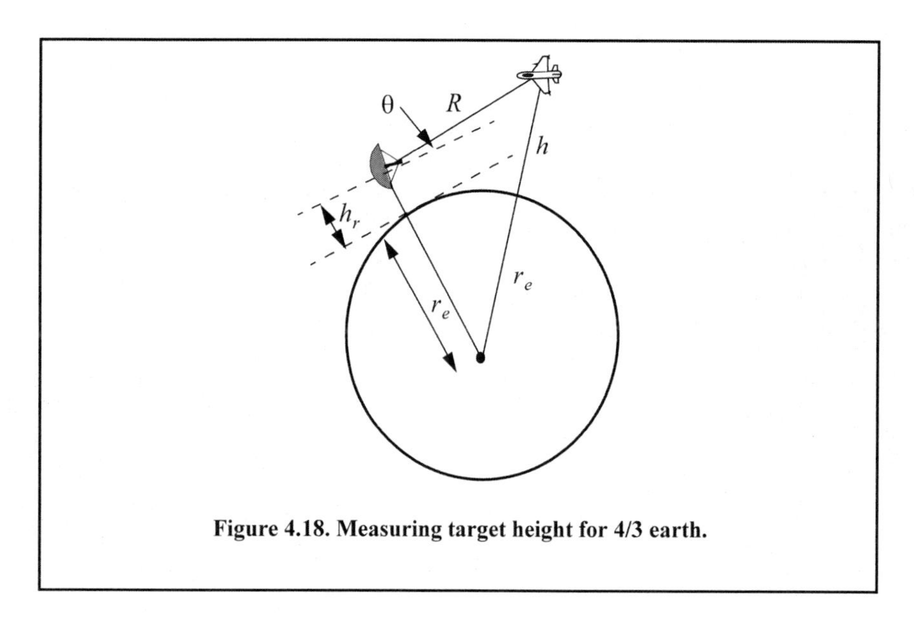

Figure 4.18. Measuring target height for 4/3 earth.

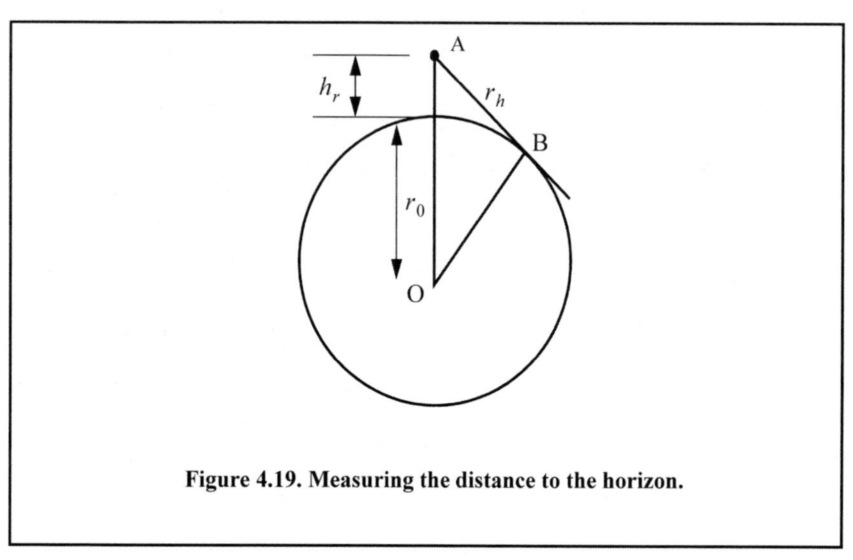

Figure 4.19. Measuring the distance to the horizon.

1. Blake, L. V., *Radar Range-Performance Analysis*, Artech House, 1986.

where r_h is the distance to the horizon. By expanding Eq. (4.58) and collecting terms, one can derive the expression for the distance to the horizon as

$$r_h^2 = 2r_0h_r + h_r^2 . \qquad \text{Eq. (4.59)}$$

Finally, since $r_0 \gg h_r$ Eq. (4.59) is approximated by

$$r_h \approx \sqrt{2r_0h_r}, \qquad \text{Eq. (4.60)}$$

and when refraction is accounted for, Eq. (4.60) becomes

$$r_h \approx \sqrt{2r_eh_r}. \qquad \text{Eq. (4.61)}$$

4.7. Ground Reflection

When radar waves are reflected from the earth's surface, they suffer a loss in amplitude and a change in phase. Three factors that contribute to these changes that are the overall ground reflection coefficient are the reflection coefficient for a flat surface, the divergence factor due to earth's curvature, and the surface roughness.

4.7.1. Smooth Surface Reflection Coefficient

The smooth surface reflection coefficient depends on the frequency, on the surface dielectric coefficient, and on the radar grazing angle. The vertical polarization and the horizontal polarization reflection coefficients are

$$\Gamma_v = \frac{\varepsilon \sin\psi_g - \sqrt{\varepsilon - (\cos\psi_g)^2}}{\varepsilon \sin\psi_g + \sqrt{\varepsilon - (\cos\psi_g)^2}} \qquad \text{Eq. (4.62)}$$

$$\Gamma_h = \frac{\sin\psi_g - \sqrt{\varepsilon - (\cos\psi_g)^2}}{\sin\psi_g + \sqrt{\varepsilon - (\cos\psi_g)^2}} \qquad \text{Eq. (4.63)}$$

where ψ_g is the grazing angle (incident angle) and ε is the complex dielectric constant of the surface, and are given by

$$\varepsilon = \varepsilon' - j\varepsilon'' = \varepsilon' - j60\lambda\sigma \qquad \text{Eq. (4.64)}$$

where λ is the wavelength and σ the medium conductivity in mhos/meter.

Typical values of ε' and ε'' can be found tabulated in the literature. Note that when $\psi_g = 90°$ one gets

$$\Gamma_h = \frac{1 - \sqrt{\varepsilon}}{1 + \sqrt{\varepsilon}} = -\frac{\varepsilon - \sqrt{\varepsilon}}{\varepsilon + \sqrt{\varepsilon}} = -\Gamma_v \qquad \text{Eq. (4.65)}$$

while when the grazing angle is very small ($\psi_g \approx 0$), one has

$$\Gamma_h = -1 = \Gamma_v \qquad \text{Eq. (4.66)}$$

Figure 4.20 shows typical plots for Γ_h and Γ_v for seawater at $28°C$ where $\varepsilon' = 65$ and $\varepsilon'' = 30.7$ at the X-band.

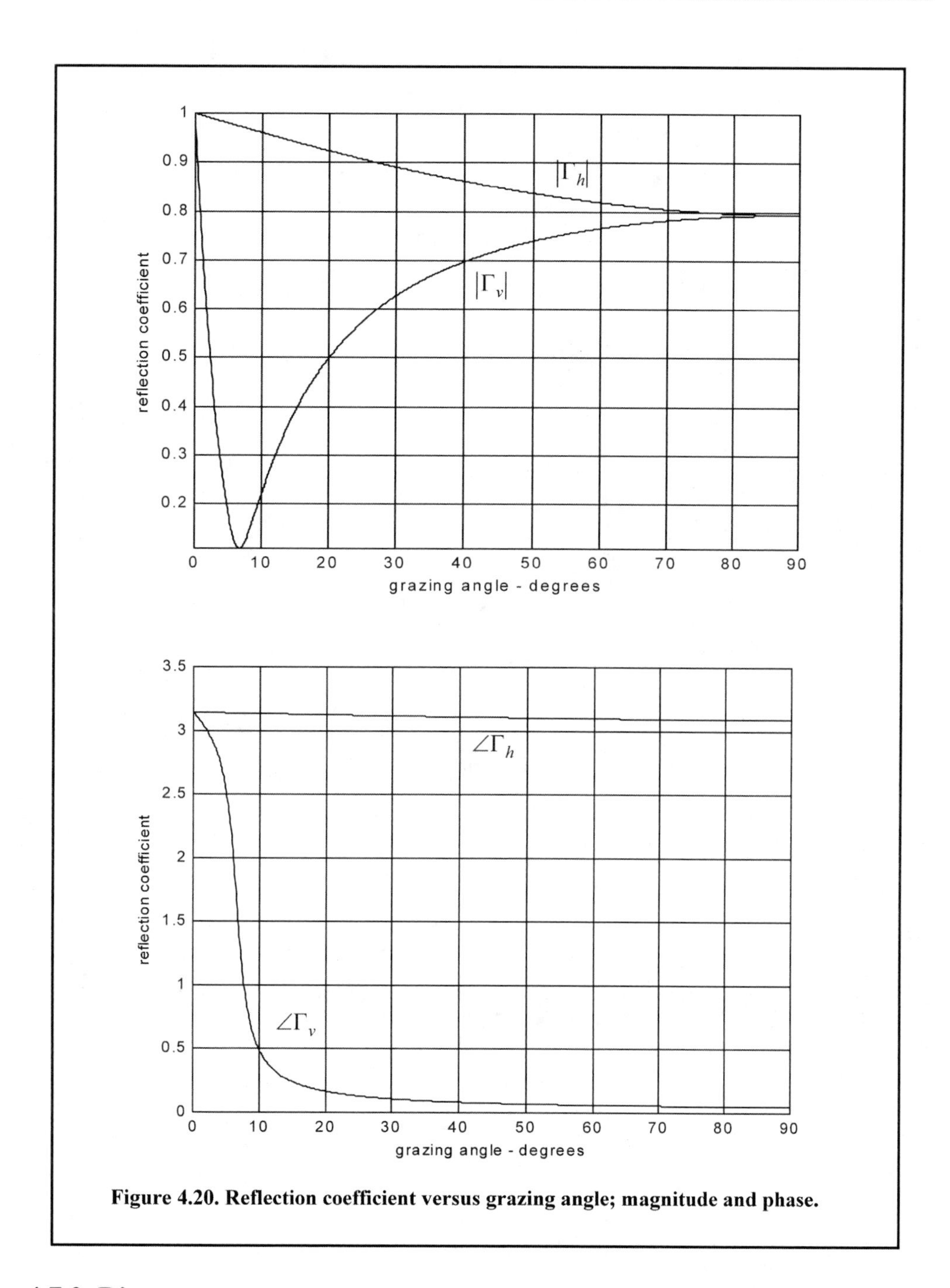

Figure 4.20. Reflection coefficient versus grazing angle; magnitude and phase.

4.7.2. Divergence

The overall reflection coefficient is also affected by the round earth divergence factor, D. When an electromagnetic wave is incident on a round earth surface, the reflected wave diverges because of the earth's curvature. This is illustrated in Figure 4.21. Due to divergence, the reflected energy is defocused, and the radar power density is reduced. The divergence factor can be derived using geometrical considerations.

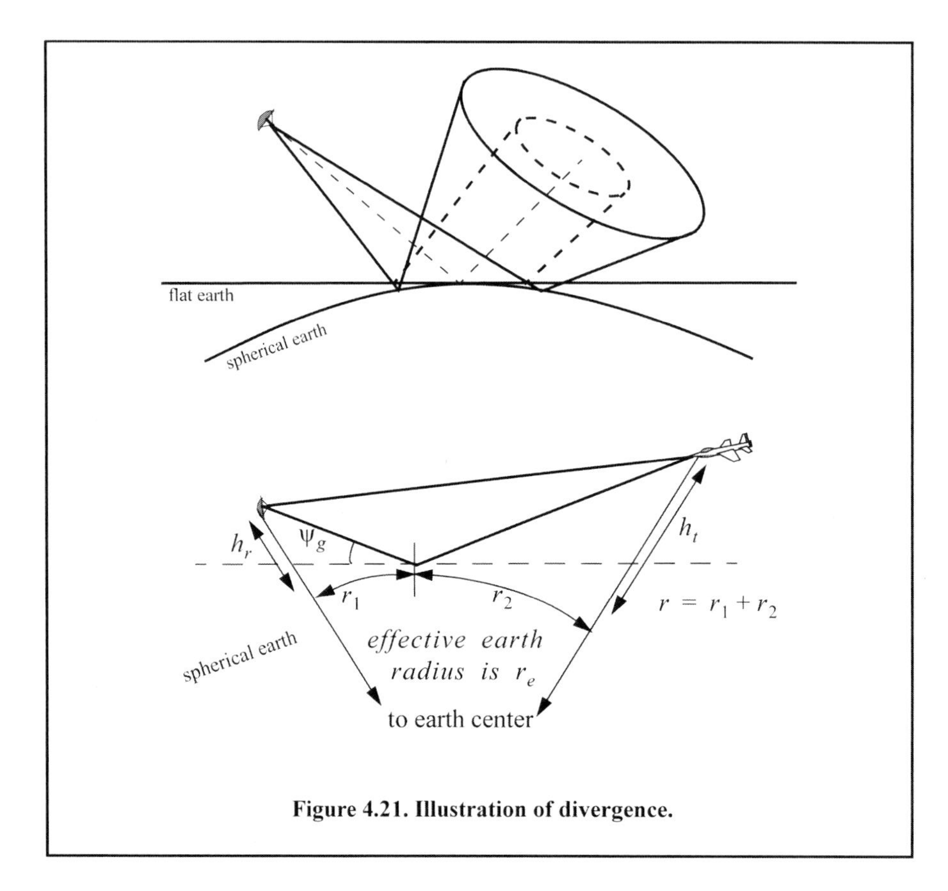

Figure 4.21. Illustration of divergence.

The divergence factor can be expressed as

$$D = \sqrt{\frac{r_e\ r\ \sin\psi_g}{[(2r_1r_2/\cos\psi_g) + r_e r\sin\psi_g](1 + h_r/r_e)(1 + h_t/r_e)}} \qquad \text{Eq. (4.67)}$$

where all the parameters in Eq. (4.67) are defined in Figure 4.21. Since the grazing ψ_g is always small when the divergence D is very large, the following approximation is adequate in most radar cases of interest,

$$D \approx \frac{1}{\sqrt{1 + \dfrac{4r_1r_2}{r_e r\sin 2\psi_g}}}\,. \qquad \text{Eq. (4.68)}$$

4.7.3. Rough Surface Reflection

In addition to divergence, surface roughness also affects the reflection coefficient. Surface roughness is given by

$$S_r = e^{-2\left(\frac{2\pi h_{rms}\sin\psi_g}{\lambda}\right)^2} \qquad \text{Eq. (4.69)}$$

where h_{rms} is the rms surface height irregularity. Another form for the rough surface reflection coefficient that is more consistent with experimental results is given by

$$S_r = e^{-z} I_0(z) \qquad \text{Eq. (4.70)}$$

$$z = 2\left(\frac{2\pi h_{rms} \sin\psi_g}{\lambda}\right)^2 \qquad \text{Eq. (4.71)}$$

where I_0 is the modified Bessel function of order zero. Figure 4.22 shows a plot of the rough surface reflection coefficient versus $f_{MHz} h_{rms} \sin\psi_g$. The solid line uses Eq. (4.69) while the dashed line uses Eq. (4.70).

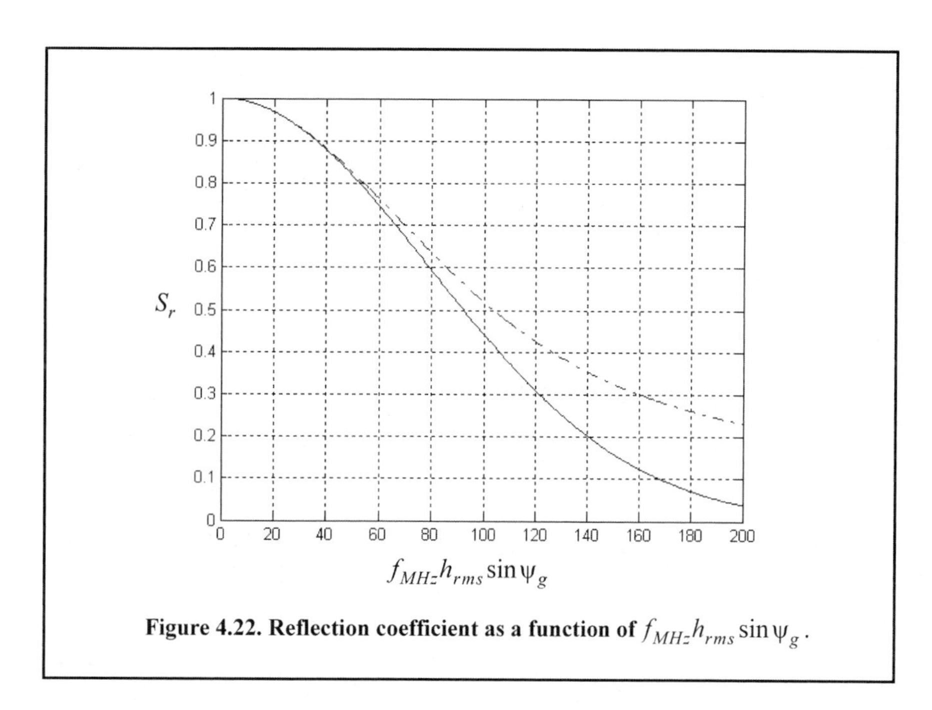

Figure 4.22. Reflection coefficient as a function of $f_{MHz} h_{rms} \sin\psi_g$.

4.7.4. Total Reflection Coefficient

In general, rays reflected from rough surfaces undergo changes in phase and amplitude, which results in the diffused (non-coherent) portion of the reflected signal. Combining the effects of smooth surface reflection coefficient, divergence, and the rough surface reflection coefficient, one can express the total reflection coefficient Γ_t as

$$\Gamma_t = \Gamma_{(h,\,v)} D S_r, \qquad \text{Eq. (4.72)}$$

where $\Gamma_{(h,\,v)}$ is the horizontal or vertical smooth surface reflection coefficient, D is divergence, and S_r is the rough surface reflection coefficient.

4.8. Multipath Effects

As mentioned earlier, the propagation factor describes the constructive/destructive interference of the electromagnetic waves diffracted from the earth's surface (which can be either flat or curved). In this section, we will derive the propagation factor due to flat earth.

Consider the geometry shown in Figure 4.23. The radar is located at height h_r. The target is at range R, and is located at a height h_t. The grazing angle is ψ_g. The radar energy emanating from its antenna will reach the target via two paths: the "direct path" AB and the "indirect path" ACB. The lengths of the paths AB and ACB are normally very close to one another, and thus the difference between the two paths is very small. Denote the direct path as R_d, the indirect path as R_i, and the difference as $\Delta R = R_i - R_d$. It follows that the phase difference between the two paths is given by

$$\Delta\Phi = (2\pi\Delta R)/\lambda \qquad \text{Eq. (4.73)}$$

where λ is the radar wavelength.

The indirect signal amplitude arriving at the target is less than the signal amplitude arriving via the direct path. This is because the antenna gain in the direction of the indirect path is less than that along the direct path, and because the signal reflected from the earth's surface at point C is modified in amplitude and phase in accordance with the earth's reflection coefficient, Γ. The earth reflection coefficient is given by

$$\Gamma = \rho e^{j\varphi} \qquad \text{Eq. (4.74)}$$

where ρ is less than unity and φ describes the phase shift induced on the indirect path signal due to surface roughness.

The direct signal (in volts) arriving at the target via the direct path can be written as

$$E_d = e^{j\omega_0 t} e^{j\frac{2\pi}{\lambda}R_d} \qquad \text{Eq. (4.75)}$$

where the time harmonic term $\exp(j\omega_0 t)$ represents the signal's time dependency and the exponential term $\exp(j(2\pi/\lambda)R_d)$ represents the signal spatial phase. The indirect signal at the target is

$$E_i = \rho e^{j\varphi} e^{j\omega_0 t} e^{j\frac{2\pi}{\lambda}R_i} \qquad \text{Eq. (4.76)}$$

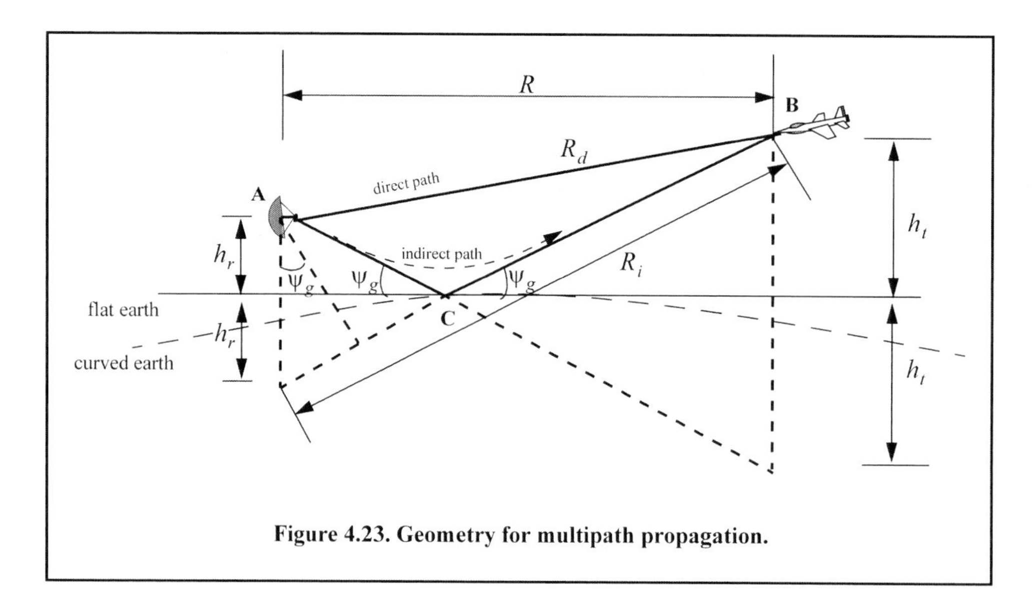

Figure 4.23. Geometry for multipath propagation.

where $\rho\exp(j\varphi)$ is the surface reflection coefficient. Therefore, the overall signal arriving at the target is

$$E = E_d + E_i = e^{j\omega_0 t} e^{j\frac{2\pi}{\lambda}R_d}\left(1 + \rho e^{j\left(\varphi + \frac{2\pi}{\lambda}(R_i - R_d)\right)}\right). \qquad \text{Eq. (4.77)}$$

Due to reflections from the earth's surface, the overall signal strength is then modified at the target by the ratio of the signal strength in the presence of earth to the signal strength at the target in free space. It follows that the propagation factor is computed as

$$F_p = \left|\frac{E_d}{E_d + E_i}\right| = \left|1 + \rho e^{j\varphi} e^{j\Delta\Phi}\right|, \qquad \text{Eq. (4.78)}$$

which can be rewritten as

$$F_p = \left|1 + \rho e^{j\alpha}\right| \qquad \text{Eq. (4.79)}$$

where $\alpha = \Delta\Phi + \varphi$. Using Euler's identity ($e^{j\alpha} = \cos\alpha + j\sin\alpha$), Eq. (4.79) can be written as

$$F_p = \sqrt{1 + \rho^2 + 2\rho\cos\alpha}. \qquad \text{Eq. (4.80)}$$

It follows that the signal power at the target is modified by the factor F_p^2. By using reciprocity, the signal power at the radar is computed by multiplying the radar equation by the factor F_p^4.

The propagation factor for free space and no multipath is $F_p = 1$. Denote the radar detection range in free space (i.e., $F_p = 1$) as R_0. It follows that the detection range in the presence of the atmosphere and multipath interference is

$$R = \frac{R_0 F_p}{(L_a)^{1/4}} \qquad \text{Eq. (4.81)}$$

where L_a is the two-way atmospheric loss at range R. Atmospheric attenuation will be discussed in a later section. Thus, for the purpose of illustrating the effect of multipath interference on the propagation factor, assume that $L_a = 1$. In this case, Eq. (4.81) is modified to

$$R = R_0 F_p. \qquad \text{Eq. (4.82)}$$

Using the geometry of Figure 4.23, the direct and indirect paths are

$$R_d = \sqrt{R^2 + (h_t - h_r)^2} \qquad \text{Eq. (4.83)}$$

$$R_i = \sqrt{R^2 + (h_t + h_r)^2}. \qquad \text{Eq. (4.84)}$$

Eqs. (4.83) and (4.84) are approximated using the truncated binomial series expansion as

$$R_d \approx R + \frac{(h_t - h_r)^2}{2R} \qquad \text{Eq. (4.85)}$$

$$R_i \approx R + \frac{(h_t + h_r)^2}{2R}. \qquad \text{Eq. (4.86)}$$

This approximation is valid for low grazing angles, where $R \gg h_t, h_r$. It follows that

$$\Delta R = R_i - R_d \approx \frac{2h_t h_r}{R}. \qquad \text{Eq. (4.87)}$$

Substituting Eq. (4.87) into Eq. (4.73) yields the phase difference due to multipath propagation between the two signals (direct and indirect) arriving at the target. More precisely,

$$\Delta\Phi = \frac{2\pi}{\lambda}\Delta R \approx \frac{4\pi h_t h_r}{\lambda R}. \qquad \text{Eq. (4.88)}$$

At this point, assume a smooth surface with reflection coefficient $\Gamma = -1$. This assumption means that waves reflected from the surface suffer no amplitude loss, and that the induced surface phase shift is equal to $180°$. Using Eq. (4.73) and Eq. (4.80) along with these assumptions yields

$$F_p^2 = 2 - 2\cos\Delta\Phi = 4(\sin(\Delta\Phi/2))^2. \qquad \text{Eq. (4.89)}$$

Substituting Eq. (4.88) into Eq. (4.89) yields

$$F_p^2 = 4\left(\sin\frac{2\pi h_t h_r}{\lambda R}\right)^2. \qquad \text{Eq. (4.90)}$$

Using reciprocity, the expression for the propagation factor at the radar is then given by

$$F_p^4 = 16\left(\sin\frac{2\pi h_t h_r}{\lambda R}\right)^4. \qquad \text{Eq. (4.91)}$$

Since the sine function varies between 0 and 1, the signal power at the radar will then vary between 0 and 16. Therefore, the fourth power relation between signal power and the target range results in varying the target range from 0 to twice the actual range in free space. In addition to that, the field strength at the radar will now have holes that correspond to the nulls of the propagation factor.

The nulls of the propagation factor occur when the sine is equal to zero. More precisely,

$$\frac{2h_r h_t}{\lambda R} = n \qquad \text{Eq. (4.92)}$$

where $n = \{0, 1, 2, \ldots\}$. The maxima occur at

$$\frac{4h_r h_t}{\lambda R} = n + 1. \qquad \text{Eq. (4.93)}$$

The target heights that produce nulls in the propagation factor are $\{h_t = n(\lambda R/2h_r); n = 0, 1, 2, \ldots\}$, and the peaks are produced from target heights $\{h_t = n(\lambda R/4h_r); n = 1, 2, \ldots\}$. Therefore, due to the presence of surface reflections, the antenna elevation coverage is transformed into a lobed pattern structure as illustrated by Figure 4.24. Note that, due to the presence of surface reflections, the antenna elevation coverage is transformed into a lobed pattern structure. The lobe widths are directly proportional to λ, and inversely proportional to h_r. A target located at a maxima will be detected at twice its free space range. Alternatively, at other angles, the detection range will be less than that in free space. A target located at a maxima will be detected at twice its free space range. Alternatively, at other angles, the detection range will be less than that in free space.

Figure 4.25 presents a plot for the propagation factor loss versus range using $f = 3GHz$; $h_r = 30.48m$; and $h_t = 60.96m$. In this case, the target reference range is at $R_o = 185.2Km$. Divergence effects are not included; neither is the reflection coefficient. More precisely, $D = \Gamma_t = 1$. Figure 4.26 shows the relative signal level with and without multipath losses. Note that multipath losses affect the signal level by introducing numerous nulls in the signal level. These nulls will typically cause the radar to lose track of targets passing through such nulls.

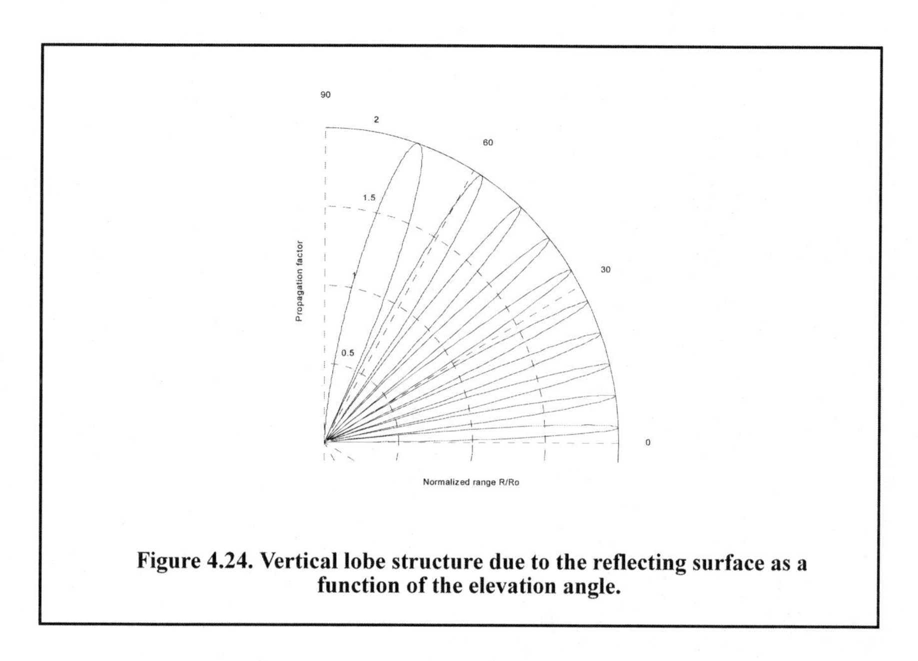

Figure 4.24. Vertical lobe structure due to the reflecting surface as a function of the elevation angle.

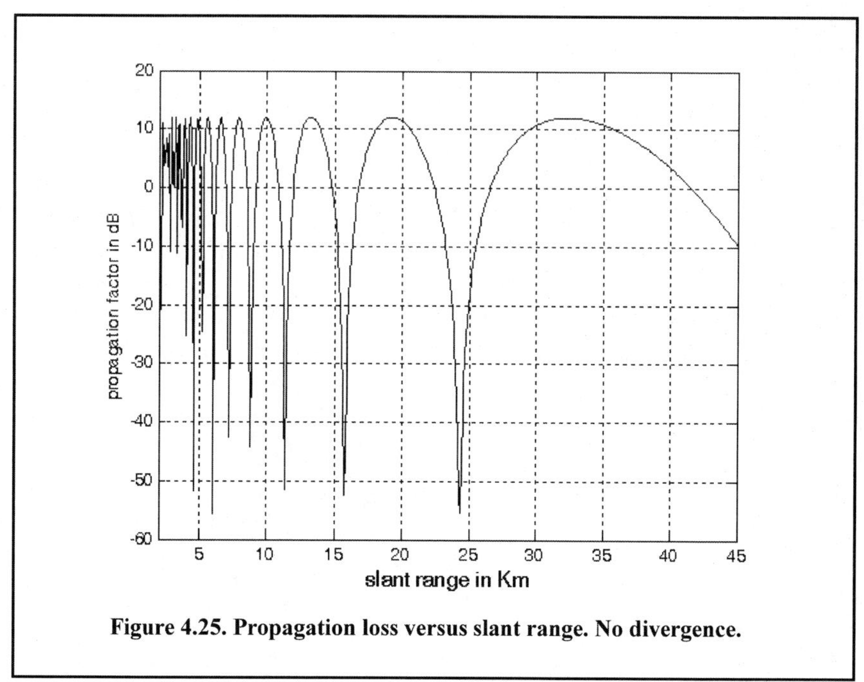

Figure 4.25. Propagation loss versus slant range. No divergence.

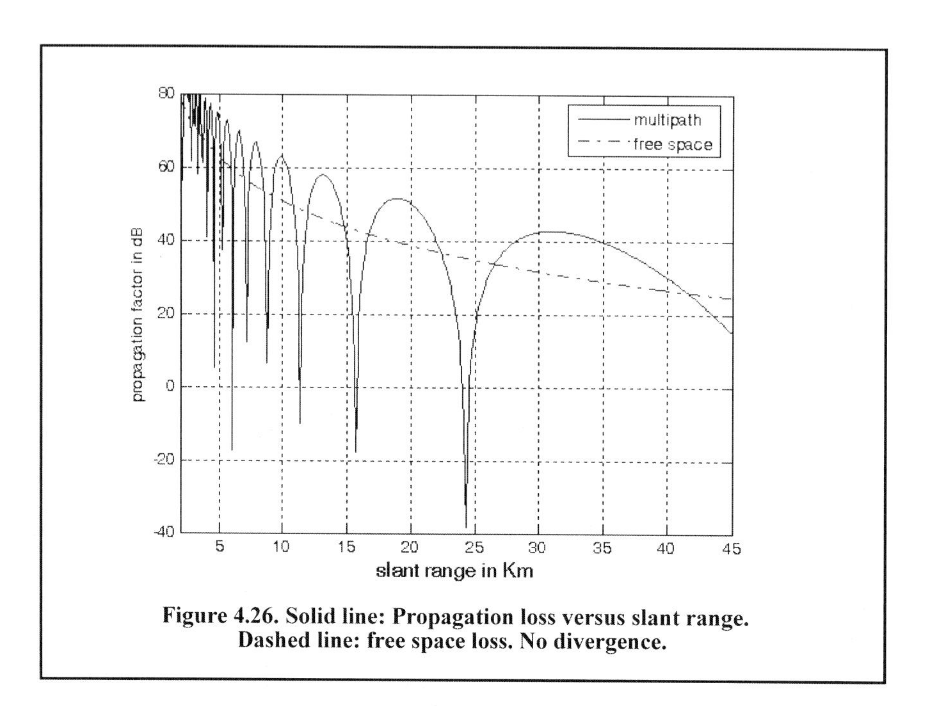

Figure 4.26. Solid line: Propagation loss versus slant range. Dashed line: free space loss. No divergence.

4.9. Atmospheric Attenuation

Radar electromagnetic waves travel in free space without suffering any energy loss. However, due to gases (mainly oxygen) and water vapor present along the radar wave propagation path, a loss in radar energy occurs. This loss is known as atmospheric attenuation. Most of this lost radar energy is normally absorbed by gases and water vapor and transformed into heat, while a small portion of this lost energy is used in molecular transformation of the atmosphere particles.

The two-way atmospheric attenuation over a range R can be expressed as

$$L_{atmosphere} = e^{-2\alpha R} \qquad \text{Eq. (4.94)}$$

where α is the one-way attenuation coefficient. Water vapor attenuation peaks at about $22.3GHz$, while attenuation due to oxygen peaks at about 60 and $118GHz$. Atmospheric attenuation is severe for frequencies higher than $35GHz$. This is the reason why ground based radars rarely use frequencies higher than $35GHz$. Atmospheric attenuation is a function range, frequency, and elevation angle. Figure 4.27 shows a typical two-way atmospheric attenuation plot versus range at $300MHz$, with the elevation angle as a parameter. Figure 4.28 is similar to Figure 4.27, except it is for $3GHz$.

4.10. Attenuation Due to Precipitation

Radar waves propagating through rain precipitation suffer loss in signal power. This power loss is due to absorption by and scattering from the rain droplets. Clearly, heavier rain rate will result in more absorption and scattering, thus leading to more power loss. Attenuation due to rain is also a function of frequency or radar wavelength.

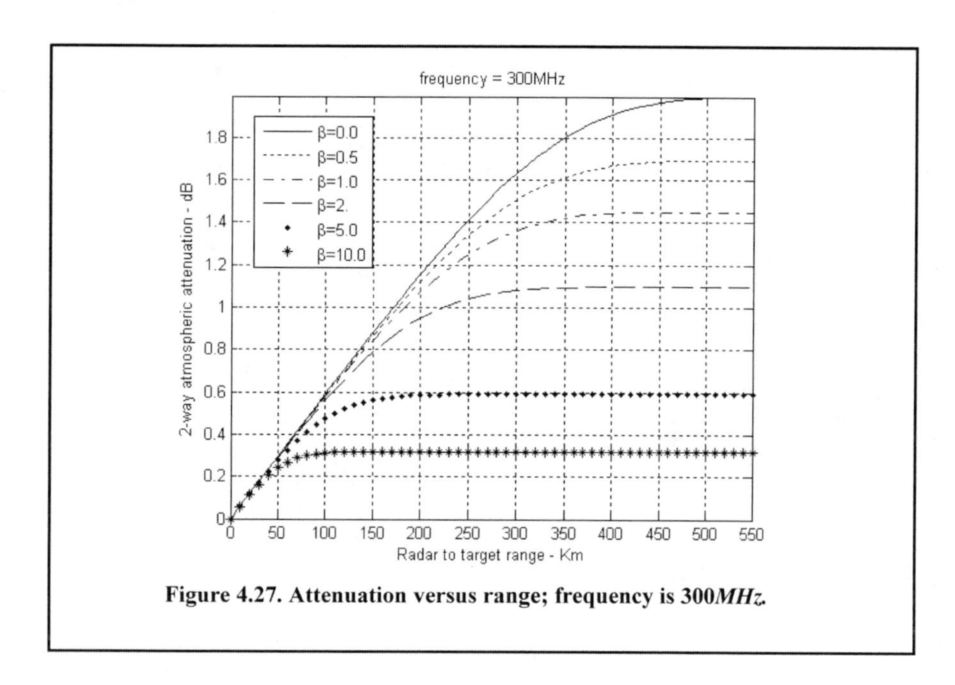

Figure 4.27. Attenuation versus range; frequency is 300*MHz*.

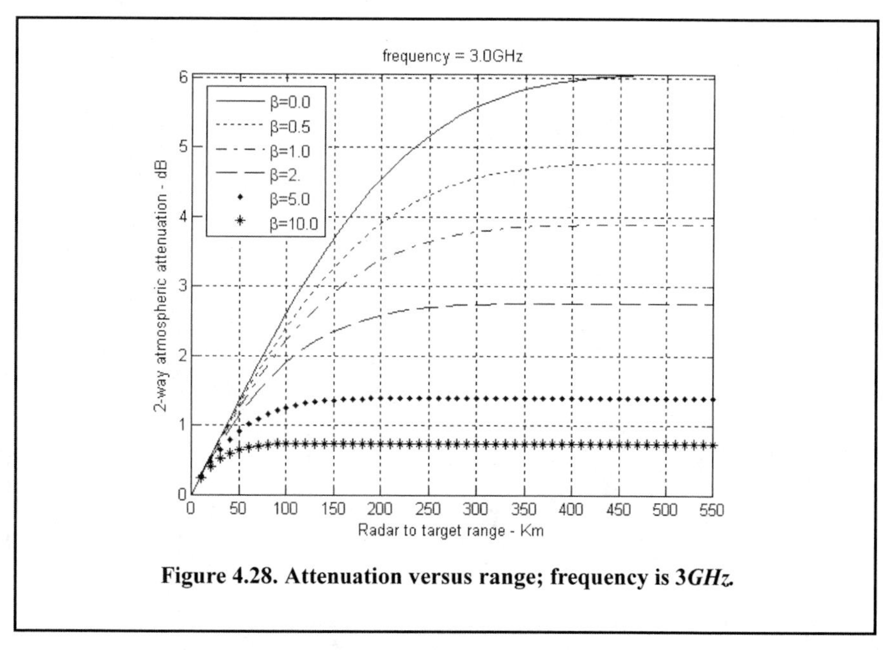

Figure 4.28. Attenuation versus range; frequency is 3*GHz*.

For example, the one-way attenuation, measured in dB/Km, due to rain precipitation is given by

$$A_r = \begin{cases} 3.43 \times 10^{-4} r^{0.97} & \lambda = 10cm \\ 1.8 \times 10^{-3} r^{1.05} & \lambda = 5cm \\ 1.0 \times 10^{-2} r^{1.21} & \lambda = 3.2cm \end{cases} \qquad \text{Eq. (4.95)}$$

where r is the rainfall rate in *mm/hr*. A more general formula for this attenuation is given by

$$A_r = K_A f^{\alpha} r \qquad \text{Eq. (4.96)}$$

where f is the frequency in *GHz*, K_A and α are constants yet to be defined. Almost all open literature sources do not agree on specific values for these two constants, where α varies from about *2.39* to *3.84* while K_A varies from 1.21×10^{-5} to 8.33×10^{-6}. This author recommends using $K_A = 0.0002$ and $\alpha = 2.25$. It follows that

$$A_r = 0.0002\ f^{2.25} r\ dB/Km . \qquad \text{Eq. (4.97)}$$

Figure 4.29 illustrates the behavior of rain attenuation as a function of frequency. Clearly, and as one would expect, as the wavelength becomes smaller, the rain attenuation becomes more dominant. The one-way attenuation in dB/Km due to snow precipitation has been reported in the literature as one of the following two formulas

$$A_s = \frac{0.035r^2}{\lambda^4} + \frac{0.0022r}{\lambda} \qquad \text{Eq. (4.98)}$$

$$A_s = \frac{0.00349r^{1.6}}{\lambda^4} + \frac{0.00224r}{\lambda} \qquad \text{Eq. (4.99)}$$

where r is the snow fall rate in *millimeters* of water content per hour and λ is the radar wavelength in *centimeters*. Both of Eqs. (4.98) and (4.99) give fairly accurate results with Eq. (4.98) having the edge. Figure 4.30 illustrates the behavior of snow attenuation as a function of frequency. Clearly, and as one would expect, as the wavelength becomes smaller, the snow attenuation becomes more dominant.

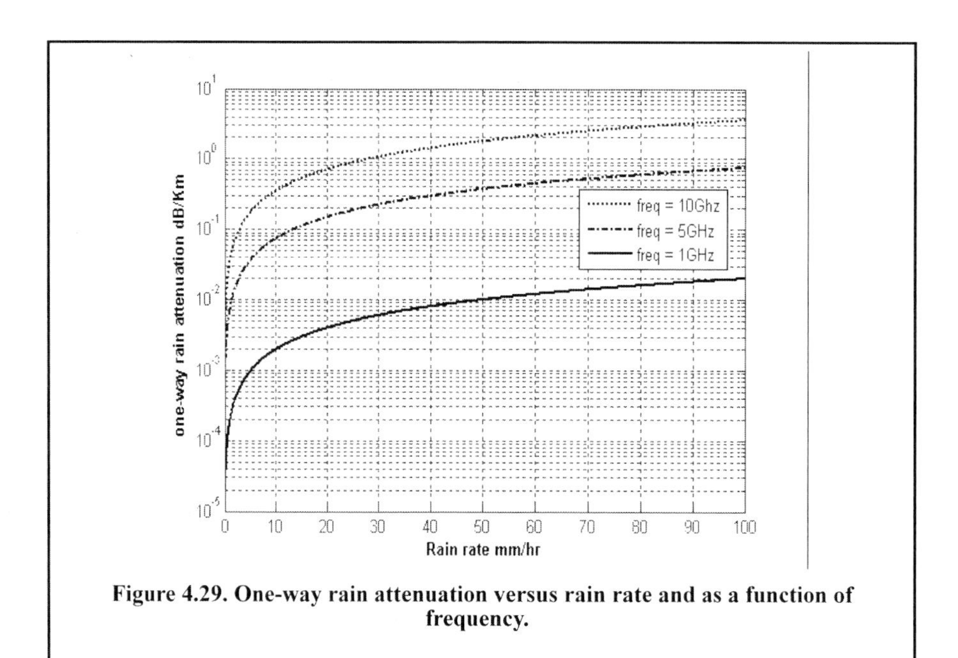

Figure 4.29. One-way rain attenuation versus rain rate and as a function of frequency.

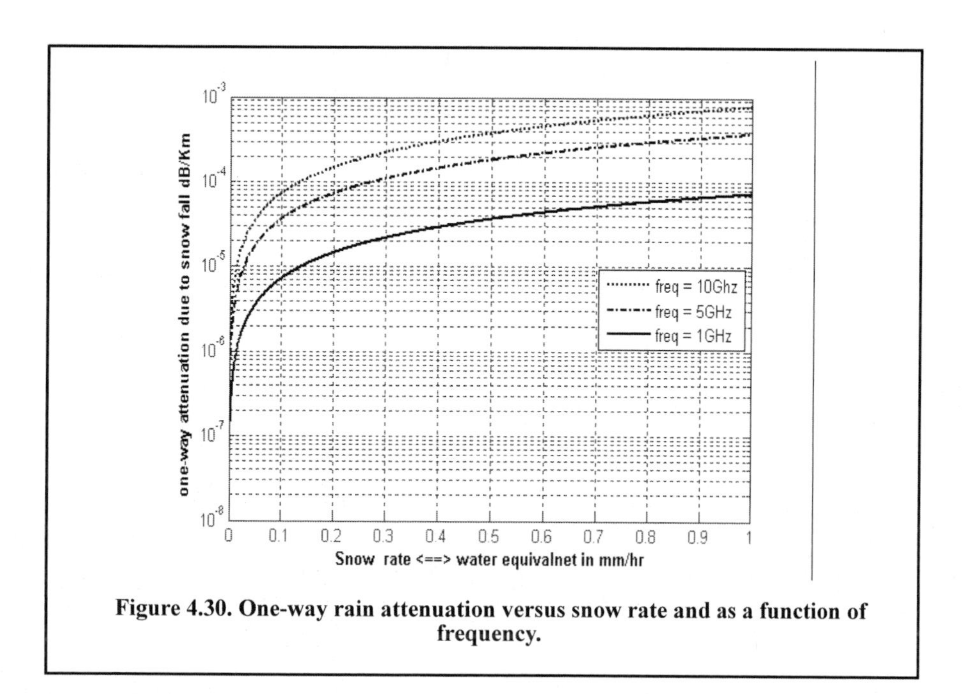

Figure 4.30. One-way rain attenuation versus snow rate and as a function of frequency.

Problems

4.1. (a) Calculate the maximum unambiguous range for a pulsed radar with PRF of $200Hz$ and $750Hz$. (b) What are the corresponding PRIs?

4.2. A certain pulsed radar uses pulse width $\tau = 1\mu s$. Compute the corresponding range resolution.

4.3. An X-band radar uses PRF of $3KHz$. Compute the unambiguous range and the required bandwidth so that the range resolution is $30m$. What is the duty cycle?

4.4. Compute the Doppler shift associated with a closing target with velocity 100, 200, and 350 meters per second. In each case, compute the time dilation factor. Assume that $\lambda = 0.3m$.

4.5. Compute the round-trip delays, minimum PRIs, and corresponding PRFs for targets located 30*Km*, 80*Km*, and 150*Km* away from the radar.

4.6. Assume an S-band radar. What are the Doppler frequencies for the following target range rates: 50m/s; 200m/s; and 250m/s?

4.7. Repeat the previous problem for an X-Band radar (9.5*GHz*).

4.8. A certain L-band radar has center frequency $1.5GHz$, and PRF $f_r = 10KHz$. What is the maximum Doppler shift that can be measured by this radar?

4.9. Starting with Eq. (4.28), derive an expression for the Doppler shift associated with a receding target.

4.10. In reference to Figure 4.12, compute the Doppler frequency for $v = 150m/s$, $\theta_a = 30°$, and $\theta_e = 15°$. Assume that $\lambda = 0.1m$.

4.11. A pulsed radar system has a range resolution of $30cm$. Assuming sinusoid pulses at $45KHz$, determine the pulse width and the corresponding bandwidth.

4.12. (a) Develop an expression for the minimum PRF of a pulsed radar. (b) Compute $f_{r_{min}}$ for a closing target whose velocity is $400m/s$. (c) What is the unambiguous range? Assume that $\lambda = 0.2m$.

4.13. A certain target has the following characteristics: its range away from the radar given in its corresponding x-, y-, and z- components is $\{25Km, 32Km, 12Km\}$. The target velocity vector is $v_z = v_y = 0$, and $v_x = -250m/s$. Compute the composite target range and range rate. If the radar's operating frequency is $9GHz$, what is the corresponding Doppler frequency?

4.14. Compute the Doppler shift associated with a closing target with velocity 100, 200, and 350 meters per second. In each case compute the time dilation factor. Assume that $\lambda = 0.3m$.

4.15. An L-Band pulsed radar is designed to have an unambiguous range of $100Km$ and range resolution $\Delta R \le 100m$. The maximum resolvable Doppler frequency corresponds to $v_{target} \le 350m/\text{sec}$. Compute the maximum required pulse width, the PRF, and the average transmitted power if $P_t = 500W$.

4.16. Reproduce Figure 4.20 using $f = 8GHz$ and (a) $\varepsilon' = 2.8$ and $\varepsilon'' = 0.032$ (dry soil); (b) $\varepsilon' = 47$ and $\varepsilon'' = 19$ (sea water at $0°C$); (c) $\varepsilon' = 50.3$ and $\varepsilon'' = 18$ (lake water at $0°C$).

4.17. A radar at altitude $h_r = 10m$ and a target at altitude $h_t = 300m$, and assuming a spherical earth, calculate r_1, r_2, and ψ_g.

4.18. In the previous problem, assuming that you may be able to use the small grazing angle approximation: (a) calculate the ratio of the direct to the indirect signal strengths at the target. (b) If the target is closing on the radar with velocity $v = 300m/s$, calculate the Doppler shift along the direct and indirect paths. Assume $\lambda = 3cm$.

4.19. Derive an asymptotic form for Γ_h and Γ_v when the grazing angle is very small.

4.20. Using the law of cosines, derive Eq. (4.68) from Eq. (4.67).

4.21. In reference to Figure 4.23 assume a radar height of $h_r = 100m$ and a target height of $h_t = 500m$. The range is $R = 20Km$. (a) Calculate the lengths of the direct and indirect paths. (b) Calculate how long it will take a pulse to reach the target via the direct and indirect paths.

4.22. Calculate the range to the horizon corresponding to a radar at $5Km$ and $10Km$ of altitude. Assume 4/3 earth.

4.23. Using Eq. (4.57), determine h when $h_r = 15m$ and $R = 35Km$.

4.24. An exponential expression for the index of refraction is given by

$$n = 1 + 315 \times 10^{-6}\exp(-0.136h)$$

where the altitude h is in *Km*. Calculate the index of refraction for a well-mixed atmosphere at 10% and 50% of the troposphere.

4.25. The atmospheric attenuation can be included in the radar equation as another loss term. Consider an X-band radar whose detection range at $20Km$ includes a $0.25dB/Km$ atmospheric loss. Calculate the corresponding detection range with no atmospheric attenuation.

Chapter 5

The Radar Equation

5.1. The Radar Range Equation

Consider a radar with an omni-directional (isotropic) antenna (one that radiates energy equally in all directions). Since isotropic antennas have spherical radiation patterns, one can define the peak power density (power per unit area) at any point in space away from the radar as

$$P_D = \frac{Peak\ transmitted\ power}{area\ of\ a\ sphere} \quad \frac{Watts}{m^2}. \qquad \text{Eq. (5.1)}$$

The power density, in $Watts/m^2$, at range R away from the radar (assuming a lossless propagation medium) is

$$P_D = \frac{P_t}{(4\pi R^2)} \qquad \text{Eq. (5.2)}$$

where P_t is the peak transmitted power and $4\pi R^2$ is the surface area of a sphere of radius R. Radar systems utilize directional antennas in order to increase the power density in a certain direction. Directional antennas are usually characterized by the antenna gain G and the antenna effective aperture A_e. They are related by

$$G = \frac{(4\pi A_e)}{\lambda^2} \qquad \text{Eq. (5.3)}$$

where λ is the radar operating wavelength. The relationship between the antenna's effective aperture A_e and the physical aperture A is

$$A_e = \rho A \qquad \text{Eq. (5.4)}$$
$$0 \le \rho \le 1$$

where ρ is referred to as the aperture efficiency, and good antennas require $\rho \to 1$. In this book, unless otherwise noted, A and A_e are used interchangeably to refer to the antenna's aperture, and it will be assumed that antennas have the same gain in the transmitting and receiving modes. In practice, $\rho \approx 0.7$ is widely accepted. The power density at a distance R away from a radar using a directive antenna of gain G is then given by

$$P_D = \frac{P_t G}{4\pi R^2}. \qquad \text{Eq. (5.5)}$$

When the radar radiated energy impinges upon a target, the induced surface currents on that target radiate electromagnetic energy in all directions. The amount of the radiated energy is proportional to the target size, orientation, physical shape, and material, which are all lumped together in one target-specific parameter called the Radar Cross Section (RCS) denoted symbolically by the Greek letter σ.

The radar cross section is defined as the ratio of the power reflected back to the radar to the power density incident on the target,

$$\sigma = \frac{P_r}{P_D} m^2, \qquad \text{Eq. (5.6)}$$

where P_r is the power reflected from the target. Thus, the total power delivered to the radar signal processor by its antenna is

$$P_{Dr} = \frac{P_t G \sigma}{(4\pi R^2)^2} A_e. \qquad \text{Eq. (5.7)}$$

Substituting the value of A_e from Eq. (5.3) into Eq. (5.7) yields

$$P_{Dr} = \frac{P_t G^2 \lambda^2 \sigma}{(4\pi)^3 R^4}. \qquad \text{Eq. (5.8)}$$

Let S_{min} denote the minimum detectable signal power by the radar. It follows that the maximum radar range R_{max} is

$$R_{max} = \left(\frac{P_t G^2 \lambda^2 \sigma}{(4\pi)^3 S_{min}}\right)^{1/4}. \qquad \text{Eq. (5.9)}$$

Eq. (5.9) suggests that in order to double the radar maximum range, one must increase the peak transmitted power P_t sixteen times; or equivalently, one must increase the effective aperture four times.

In practical situations the returned signals received by the radar will be corrupted with noise, which introduces unwanted voltages at all radar frequencies. Noise is random in nature and can be characterized by its Power Spectral Density (PSD) function. The noise power N is a function of the radar operating bandwidth, B. More precisely,

$$N = Noise\ PSD \times B. \qquad \text{Eq. (5.10)}$$

The receiver input noise power is

$$N_i = kT_s B \qquad \text{Eq. (5.11)}$$

where $k = 1.38 \times 10^{-23}$ *Joule/degree Kelvin* is Boltzmann's constant, and T_s is the total effective system noise temperature in degrees *Kelvin*. It is always desirable that the minimum detectable signal (S_{min}) be greater than the noise power. The fidelity of a radar receiver is normally described by a figure of merit referred to as the noise figure, F. The noise figure is defined as

$$F = \frac{(SNR)_i}{(SNR)_o} = \frac{S_i/N_i}{S_o/N_o} \qquad \text{Eq. (5.12)}$$

where $(SNR)_i$ and $(SNR)_o$ are, respectively, the Signal to Noise Ratios (SNR) at the input and output of the receiver. The input signal power is S_i, and the input noise power immediately at the antenna terminal is N_i. The values S_o and N_o are, respectively, the output signal and noise powers.

The receiver effective noise temperature excluding the antenna is

$$T_e = T_o(F-1). \qquad \text{Eq. (5.13)}$$

It follows that the total effective system noise temperature T_s is given by

$$T_s = T_e + T_a = T_0(F-1) + T_a = T_oF - T_o + T_a \qquad \text{Eq. (5.14)}$$

where T_a is the antenna temperature. In many radar applications it is desirable to set the antenna temperature T_a to T_0 and thus, Eq. (5.14) is reduced to

$$T_s = T_oF. \qquad \text{Eq. (5.15)}$$

Using Eq. (5.15) in Eq. (5.11) and substituting the result into Eq. (5.12) yields

$$S_i = kT_oBF(SNR)_o. \qquad \text{Eq. (5.16)}$$

Thus, the minimum detectable signal power can be written as

$$S_{min} = kT_oBF(SNR)_{o_{min}}. \qquad \text{Eq. (5.17)}$$

The radar detection threshold is set equal to $(SNR)_{o_{min}}$, the minimum output SNR. Substituting Eq. (5.17) in Eq. (5.9) gives

$$R_{max} = \left(\frac{P_tG^2\lambda^2\sigma}{(4\pi)^3kT_oBF(SNR)_{o_{min}}}\right)^{1/4} \qquad \text{Eq. (5.18)}$$

or equivalently,

$$(SNR)_{o_{min}} = \frac{P_tG^2\lambda^2\sigma}{(4\pi)^3kT_oBFR_{max}^4}. \qquad \text{Eq. (5.19)}$$

In general, radar losses denoted by L reduce the overall SNR, and hence

$$(SNR)_o = \frac{P_tG^2\lambda^2\sigma}{(4\pi)^3kT_oBFLR^4}. \qquad \text{Eq. (5.20)}$$

Although Eq. (5.20) is widely known and used as the Radar Range Equation, it is not quite correct unless the antenna temperature is equal to $290K$. In real-world cases, the antenna temperature may vary from a few degrees *Kelvin* to several thousand degrees. However, the actual error will be small if the radar receiver noise figure is large. In order to accurately account for the radar antenna temperature, one must use Eq. (5.15) in Eq. (5.20). Thus, the radar equation is now given by

$$(SNR)_o = \frac{P_tG^2\lambda^2\sigma}{(4\pi)^3kT_sBLR^4}. \qquad \text{Eq. (5.21)}$$

Figure 5.1 shows typical plots for SNR versus detection range for three different RCS values. In this figure, the following parameters were used: Peak power $P_t = 1.5MW$, operating frequency $f_0 = 5.6GHz$, antenna gain $G = 45dB$, radar losses $L = 6dB$, noise figure $F = 3dB$. The radar bandwidth is $B = 5MHz$. The radar minimum and maximum detection ranges are $R_{min} = 25Km$ and $R_{max} = 165Km$.

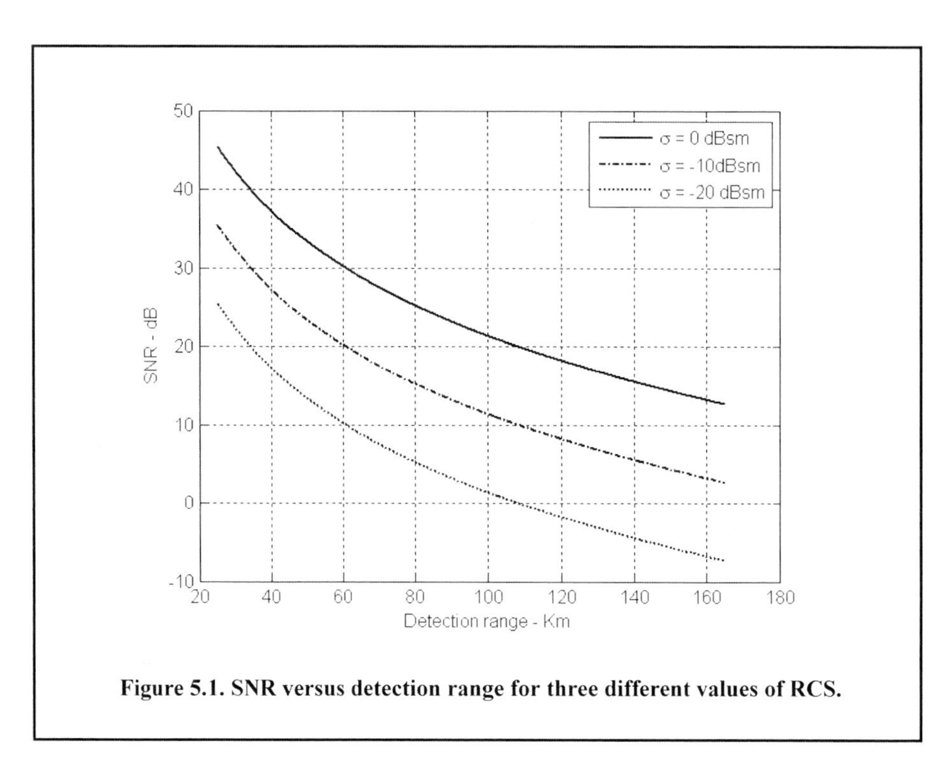

Figure 5.1. SNR versus detection range for three different values of RCS.

Example:

Assume a certain C-band radar with the following parameters: Peak power $P_t = 1.5MW$, *operating frequency* $f_0 = 5.6GHz$, *antenna gain* $G = 45dB$, *effective temperature* $T_o = 290K$, *pulse width* $\tau = 0.2\mu\sec$. *The radar threshold is* $(SNR)_{min} = 20dB$. *Assume target cross section* $\sigma = 0.1m^2$. *Compute the maximum range.*

Solution:

The radar bandwidth is

$$B = \frac{1}{\tau} = \frac{1}{0.2 \times 10^{-6}} = 5MHz.$$

The wavelength is

$$\lambda = \frac{c}{f_0} = \frac{3 \times 10^8}{5.6 \times 10^9} = 0.054m.$$

From Eq. (5.18) one gets

$$(R^4)_{dB} = (P_t + G^2 + \lambda^2 + \sigma - (4\pi)^3 - kT_oB - F - (SNR)_{o_{min}})_{dB}$$

where, before summing, the dB calculations are carried out for each of the individual parameters on the right-hand side. One can now construct the following table with all parameters computed in dB

P_t	λ^2	G^2	kT_eB	$(4\pi)^3$	F	$(SNR)_{o_{min}}$	σ
61.761	−25.421	$90dB$	−136.988	32.976	$3dB$	$20dB$	−10

It follows that

$$R^4 = 61.761 + 90 - 25.352 - 10 - 32.976 + 136.987 - 3 - 20 = 197.420dB$$

$$R^4 = 10^{(197.420/10)} = 55.208 \times 10^{18} m^4$$

$$R = \sqrt[4]{55.208 \times 10^{18}} = 86.199Km.$$

Thus, the maximum detection range is $86.2Km$.

****

5.2. Low PRF Radar Equation

Consider a pulsed radar with pulse width τ, PRI T, and peak transmitted power P_t. The average transmitted power is $P_{av} = P_t d_t$, where $d_t = \tau/T$ is the transmission duty factor. One can define the receiving duty factor d_r as

$$d_r = (T-\tau)/T = 1-\tau f_r. \quad \text{Eq. (5.22)}$$

Thus, for low PRF radars ($T \gg \tau$) the receiving duty factor is $d_r \approx 1$.

Define the "time on target" T_i (the time that a target is illuminated by the beam) as

$$T_i = n_p/f_r \Rightarrow n_p = T_i f_r \quad \text{Eq. (5.23)}$$

where n_p is the total number of pulses that strike the target, and f_r is the radar PRF. Assuming low PRF, the single pulse radar equation is given by

$$(SNR)_1 = \frac{P_t G^2 \lambda^2 \sigma}{(4\pi)^3 R^4 k T_o BFL}, \quad \text{Eq. (5.24)}$$

and for n_p coherently integrated pulses we get

$$(SNR)_{n_p} = \frac{P_t G^2 \lambda^2 \sigma\ n_p}{(4\pi)^3 R^4 k T_o BFL}. \quad \text{Eq. (5.25)}$$

By using Eq. (5.23) and using $B = 1/\tau$, the low PRF radar equation can be written as

$$(SNR)_{n_p} = \frac{P_t G^2 \lambda^2 \sigma T_i f_r \tau}{(4\pi)^3 R^4 k T_o FL}. \quad \text{Eq. (5.26)}$$

5.3. High PRF Radar Equation

In this case, the transmitted signal is a periodic train of pulses, with pulse width of τ and period T. This pulse train can be represented using an exponential Fourier series, where the central power spectrum line (DC component) for this series contains most of the signal's power. Its value is $(\tau/T)^2$, and it is equal to the square of the transmit duty factor. Thus, the single pulse radar equation for a high PRF radar is

$$SNR = \frac{P_t G^2 \lambda^2 \sigma d_t^2}{(4\pi)^3 R^4 k T_o B F L d_r} \qquad \text{Eq. (5.27)}$$

where, in this case, one can no longer ignore the receive duty factor, since its value is comparable to the transmit duty factor. In fact, $d_r \approx d_t = \tau f_r$. Additionally, the operating radar bandwidth is now matched to the radar integration time (time-on-target), $B = 1/T_i$. It follows that

$$SNR = \frac{P_t \tau f_r T_i G^2 \lambda^2 \sigma}{(4\pi)^3 R^4 k T_o F L} \qquad \text{Eq. (5.28)}$$

and finally,

$$SNR = \frac{P_{av} T_i G^2 \lambda^2 \sigma}{(4\pi)^3 R^4 k T_o F L} \qquad \text{Eq. (5.29)}$$

where P_{av} was substituted for $P_t \tau f_r$. Note that the product $P_{av} T_i$ is a "kind of energy" product, which indicates that high PRF radars can enhance detection performance by using relatively low power and longer integration time.

Example:

Compute the single pulse SNR for a high PRF radar with the following parameters: peak power $P_t = 100KW$, antenna gain $G = 20dB$, operating frequency $f_0 = 5.6GHz$, losses $L = 8dB$, noise figure $F = 5dB$, dwell interval $T_i = 2s$, duty factor $d_t = 0.3$. The range of interest is $R = 50Km$. Assume target RCS $\sigma = 0.01m^2$.

Solution:

From Eq. (5.29) we have

$$(SNR)_{dB} = (P_{av} + G^2 + \lambda^2 + \sigma + T_i - (4\pi)^3 - R^4 - kT_o - F - L)_{dB}$$

The following table gives all parameters in dB:

P_{av}	λ^2	T_i	kT_0	$(4\pi)^3$	R^4	σ
44.771	−25.421	3.01	−203.977	32.976	187.959	−20

$SNR = 44.771 + 40 - 25.421 - 20 + 3.01 - 32.976 + 203.977 - 187.959 - 5 - 8 = 12.4dB$.

****

5.4. Surveillance Radar Equation

The primary job for surveillance radars is to continuously scan a specified volume of space searching for targets of interest. Once detection is established, target information such as range, angular position, and possibly target velocity are extracted by the radar signal and data processors. Depending on the radar design and antenna, different search patterns can be adopted. A two-dimensional (2-D) fan beam search pattern is shown in Figure 5.2a. In this case, the beamwidth is wide enough in elevation to cover the desired search volume along that coordinate; however, it has to be steered in azimuth. Figure 5.2b shows a stacked beam search pattern; here the beam has to be steered in azimuth and elevation.

Search volumes are normally specified by a search solid angle Ω in steradians, as illustrated in Figure 5.3. Define the radar search volume extent for both azimuth and elevation as Θ_A and Θ_E. Consequently, the search volume is computed as

$$\Omega = (\Theta_A \Theta_E)/(57.296)^2 \ steradians \qquad \text{Eq. (5.30)}$$

where both Θ_A and Θ_E are given in degrees. The radar antenna $3dB$ beamwidth can be expressed in terms of its azimuth and elevation beamwidths θ_a and θ_e, respectively. It follows that the antenna solid angle coverage is $\theta_a\theta_e$ and, thus, the number of antenna beam positions n_B required to cover a solid angle Ω is

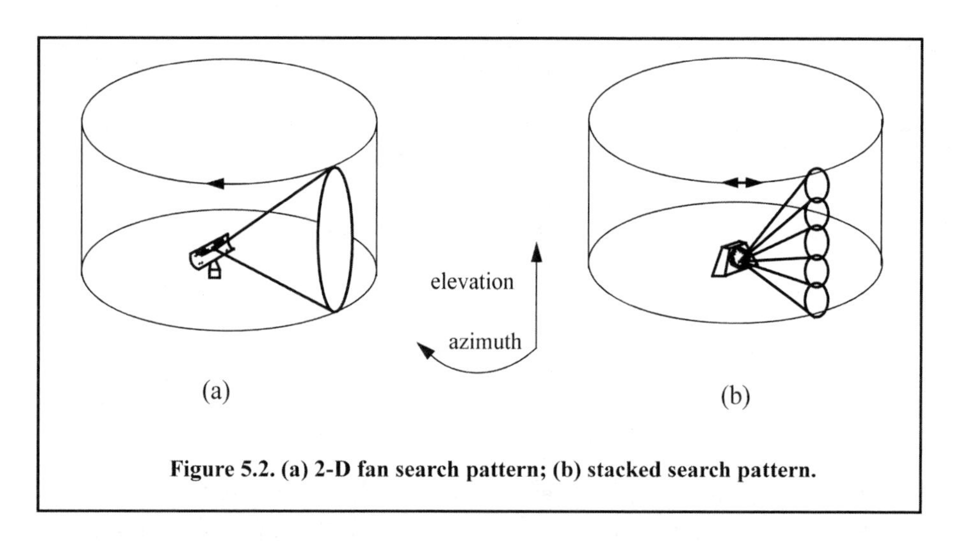

Figure 5.2. (a) 2-D fan search pattern; (b) stacked search pattern.

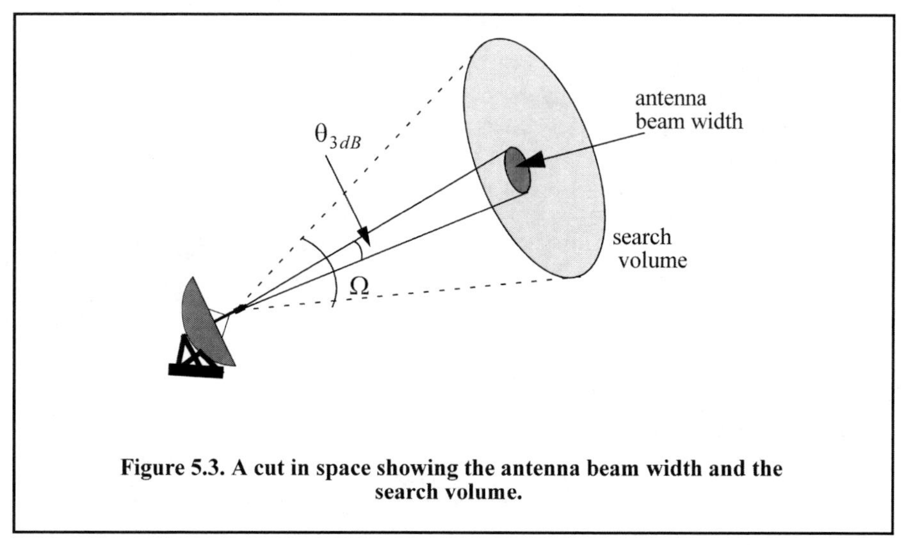

Figure 5.3. A cut in space showing the antenna beam width and the search volume.

$$n_B = \frac{\Omega}{(\theta_a\theta_e)/(57.296)^2}. \qquad \text{Eq. (5.31)}$$

In order to develop the search radar equation, start with Eq. (5.20), which is repeated here for convenience, as Eq. (5.32):

$$SNR = \frac{P_t G^2 \lambda^2 \sigma}{(4\pi)^3 k T_o B F L R^4}. \qquad \text{Eq. (5.32)}$$

Using the relations $\tau = 1/B$ and $P_t = P_{av}T/\tau$, where T is the PRI and τ is the pulse width, yields

$$SNR = \frac{T}{\tau} \frac{P_{av} G^2 \lambda^2 \sigma \tau}{(4\pi)^3 k T_o F L R^4}. \qquad \text{Eq. (5.33)}$$

Define the time it takes the radar to scan a volume defined by the solid angle Ω as the scan time T_{sc}. The time on target can then be expressed in terms of T_{sc} as

$$T_i = \frac{T_{sc}}{n_B} = \frac{T_{sc}}{\Omega}\theta_a\theta_e. \qquad \text{Eq. (5.34)}$$

Assume that during a single scan only one pulse per beam per PRI illuminates the target. It follows that $T_i = T$ and, thus, Eq. (5.33) can be written as

$$SNR = \frac{P_{av} G^2 \lambda^2 \sigma}{(4\pi)^3 k T_o F L R^4} \frac{T_{sc}}{\Omega}\theta_a\theta_e. \qquad \text{Eq. (5.35)}$$

Substituting Eqs. (5.3) into Eq. (5.35) and collecting terms yields the search radar equation (based on a single pulse per beam per PRI) as

$$SNR = \frac{P_{av} A_e \sigma}{4\pi k T_o F L R^4} \frac{T_{sc}}{\Omega}. \qquad \text{Eq. (5.36)}$$

The quantity $P_{av}A_e$ in Eq. (5.36) is known as the power aperture product. In practice, the power aperture product is widely used to categorize the radar's ability to fulfill its search mission. Normally, a power aperture product is computed to meet a predetermined SNR and radar cross section for a given search volume defined by Ω. Figure 5.4 shows plots of the power aperture product versus range and of the average power versus aperture area for three RCS choices. In this case, the following radar parameters were used:

σ	T_{sc}	$\theta_e = \theta_a$	R	$nf \times loss$	snr
$0.1\ m^2$	$2.5\,\text{sec}$	$2°$	$250Km$	$13dB$	$15dB$

Example:

Compute the power aperture product for an X-band radar with the following parameters: signal-to-noise ratio $SNR = 15dB$; losses $L = 8dB$; search volume $\Omega = 2°$; scan time $T_{sc} = 2.5s$; noise figure $F = 5dB$. Assume a $-10dBsm$ target cross section, and range $R = 250Km$. Also, compute the peak transmitted power corresponding to 30% duty factor, if the antenna gain is 45dB. Assume a circular aperture (see Problem 5.7).

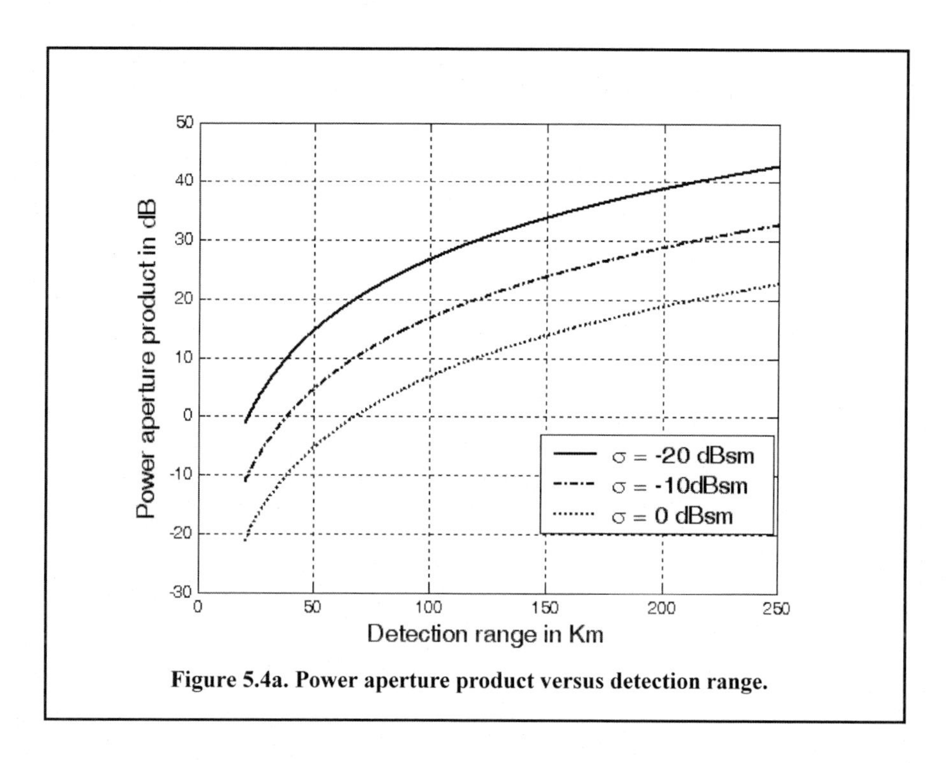

Figure 5.4a. Power aperture product versus detection range.

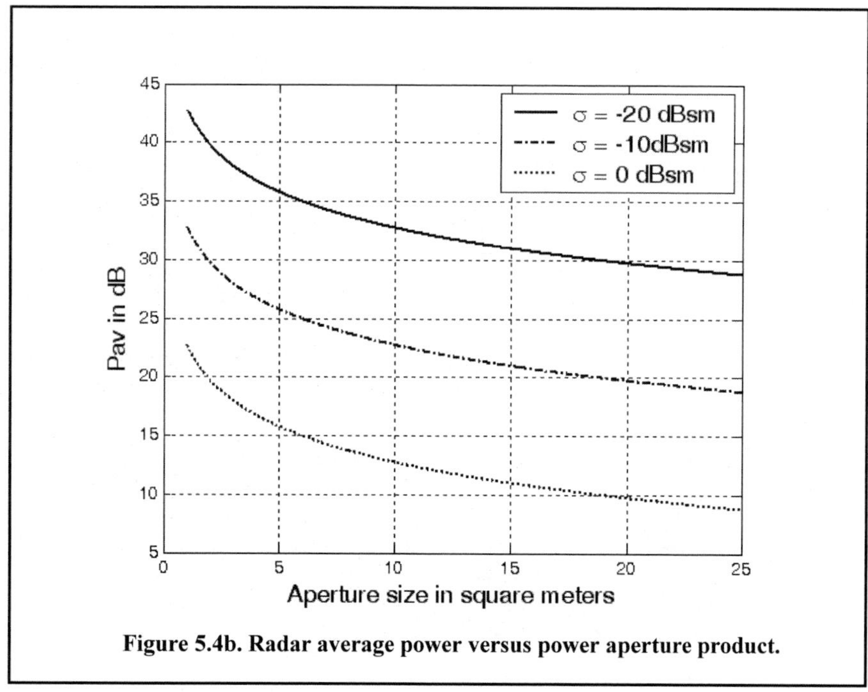

Figure 5.4b. Radar average power versus power aperture product.

Solution:

The angular coverage is $2°$ *in both azimuth and elevation. It follows that the solid angle coverage is*

$$\Omega = \frac{2 \times 2}{(57.23)^2} = -29.132dB.$$

The factor $360/2\pi = 57.23$ *converts angles into solid angles. From Problem 5.7, we get*

$$(SNR)_{dB} = (P_{av} + A + \sigma + T_{sc} - 16 - R^4 - kT_o - L - F - \Omega)_{dB}.$$

σ	T_{sc}	16	R^4	kT_o
$-10dB$	$3.979dB$	$12.041dB$	$215.918dB$	$-203.977dB$

It follows that

$$15 = P_{av} + A - 10 + 3.979 - 12.041 - 215.918 + 203.977 - 5 - 8 + 29.133.$$

Then the power aperture product is

$$P_{av} + A = 28.87dB \;.$$

Now, assume the radar wavelength to be $\lambda = 0.03m$ *; then*

$$A = \frac{G\lambda^2}{4\pi} = 3.550dB$$

$$P_{av} = -A + 28.87 = 25.32dB$$

$$P_{av} = 10^{32.532} = 340.408W$$

$$P_t = \frac{P_{av}}{d_t} = \frac{340.408}{0.3} = 1.135KW.$$

****

5.5. Radar Equation with Jamming

Any deliberate electronic effort intended to disturb normal radar operations is usually referred to as an Electronic Countermeasure (ECM). This includes chaff, radar decoys, radar RCS alterations (e.g., radio frequency absorbing materials), and of course, radar jamming. Jammers can be categorized into two general types: (1) barrage jammers and (2) deceptive jammers (repeaters). When strong jamming is present, detection capability is determined by receiver signal-to-noise plus interference ratio rather than SNR. In fact, in most cases, detection is established based on the signal-to-interference ratio alone.

Barrage jammers attempt to increase the noise level across the entire radar operating bandwidth, consequently lowering the receiver SNR, and, in turn, making it difficult to detect the desired targets. This is the reason why barrage jammers are often called maskers (since they mask the target returns). Barrage jammers can be deployed in the main beam or in the sidelobes of the radar antenna. If a barrage jammer is located in the radar main beam, it can take advantage of the antenna maximum gain to amplify the broadcasted noise signal. Alternatively, sidelobe barrage jammers must either use more power or operate at a much shorter range than main-beam jammers. Main-beam barrage jammers can be deployed either onboard the attacking vehicle or act as an escort to the target. Sidelobe jammers are often deployed to interfere with a specific radar, and since they do not stay close to the target, they have a wide variety of standoff deployment options.

Repeater jammers carry receiving devices onboard in order to analyze the radar's transmission, and then send back false target-like signals in order to confuse the radar. There are two common types of repeater jammers: spot noise repeaters and deceptive repeaters. The spot noise repeater measures the transmitted radar signal bandwidth and then jams only a specific range of frequencies. The deceptive repeater sends back altered signals that make the target appear in some false position (ghosts). These ghosts may appear at different ranges or angles than the actual target. Furthermore, there may be several ghosts created by a single jammer. By not having to jam the entire radar bandwidth, repeater jammers are able to make more efficient use of their jamming power. Radar frequency agility may be the only way possible to defeat spot noise repeaters.

In general, a jammer is characterized by its operating bandwidth B_J and Effective Radiated Power (ERP), which is proportional to the jammer transmitter power P_J. More precisely,

$$ERP = (P_J G_J)/L_J \qquad \text{Eq. (5.37)}$$

where G_J is the jammer antenna gain and L_J is the total jammer losses. The effect of a jammer on a radar is measured by the Signal-to-Jammer ratio (S/J).

5.5.1. Self-Screening Jammers (SSJ)

Self-screening jammers (SSJ), also known as self-protecting jammers and as main-beam jammers, are a class of ECM systems carried on the platform they are protecting. Escort jammers (carried on platforms that accompany the attacking vehicles) can also be treated as SSJs if they appear at the same range as that of the target(s).

Assume a radar with an antenna gain G, wavelength λ, aperture A_r, bandwidth B_r, receiver losses L, and peak power P_t. The single pulse power received by the radar from a target of RCS σ, at range R, is

$$S = \frac{P_t G^2 \lambda^2 \sigma \tau}{(4\pi)^3 R^4 L} \qquad \text{Eq. (5.38)}$$

where τ is the radar pulse width. The power received by the radar from an SSJ jammer at the same range is

$$J = \frac{P_J G_J}{4\pi R^2} \frac{A_r}{B_J L_J} \qquad \text{Eq. (5.39)}$$

where P_J, G_J, B_J, L_J are, respectively, the jammer's peak power, antenna gain, operating bandwidth, and losses. Using the relation

$$A_r = \lambda^2 G / 4\pi , \qquad \text{Eq. (5.40)}$$

Eq. (5.39) can be written as

$$J = \frac{P_J G_J}{4\pi R^2} \frac{\lambda^2 G}{4\pi} \frac{1}{B_J L_J} . \qquad \text{Eq. (5.41)}$$

Note that for jammers to be effective, they require $B_J > B_r$. This is needed in order to compensate for the fact that the jammer bandwidth is usually larger than the operating bandwidth of the radar. Jammers are normally designed to operate against a wide variety of radar systems with different bandwidths.

Substituting Eq. (5.37) into Eq. (5.41) yields

$$J = ERP\frac{\lambda^2 G}{(4\pi)^2 R^2}\frac{1}{B_J}. \qquad \text{Eq. (5.42)}$$

Thus, the S/J ratio for an SSJ case is obtained from Eqs. (5.38) and (5.42) as,

$$\frac{S}{J} = \frac{P_t \tau G \sigma B_J}{(ERP)(4\pi)R^2 L}, \qquad \text{Eq. (5.43)}$$

and when pulse compression is used, with time-bandwidth-product G_{PC}, then Eq. (5.43) can be written as

$$\frac{S}{J} = \frac{P_t G \sigma B_J G_{PC}}{(ERP)(4\pi)R^2 B_r L}. \qquad \text{Eq. (5.44)}$$

The jamming power reaches the radar on a one-way transmission basis, whereas the target echoes involve two-way transmission. Thus, the jamming power is generally greater than the target signal power. In other words, the ratio S/J is less than unity. However, as the target becomes closer to the radar, there will be a certain range such that the ratio S/J is equal to unity. This range is known as the cross-over range. The range window where the ratio S/J is sufficiently larger than unity is denoted as the detection range. The range window where the ratio S/J is larger than unity is denoted as the detection range as illustrated in Figure 5.5. In order to compute the cross-over range R_{co}, set S/J to unity in Eq. (5.44) and solve for range. It follows that

$$(R_{CO})_{SSJ} = \left(\frac{P_t G \sigma B_J}{4\pi B_r L(ERP)}\right)^{1/2}. \qquad \text{Eq. (5.45)}$$

5.5.2. Stand-Off Jammers (SOJ)

Stand-off jammers (SOJ) emit ECM signals from long ranges that are beyond the defense's lethal capability. The power received by the radar from an SOJ jammer at range R_J is

$$J = \frac{P_J G_J}{4\pi R_J^2}\frac{\lambda^2 G'}{4\pi}\frac{1}{B_J L_J} = \frac{ERP}{4\pi R_J^2}\frac{\lambda^2 G'}{4\pi}\frac{1}{B_J} \qquad \text{Eq. (5.46)}$$

where all terms in Eq. (5.46) are the same as those for the SSJ case except for G'. The gain term G' represents the radar antenna gain in the direction of the jammer and is normally considered to be the sidelobe gain.

The SOJ radar equation is then computed as

$$\frac{S}{J} = \frac{P_t \tau G^2 R_J^2 \sigma B_J}{4\pi(ERP)G' R^4 L} \qquad \text{Eq. (5.47)}$$

and when pulse compression is used, with time-bandwidth-product G_{PC}, then Eq. (5.47) can be written as

$$\frac{S}{J} = \frac{P_t G^2 R_J^2 \sigma B_J G_{PC}}{4\pi (ERP) G' R^4 B_r L}. \qquad \text{Eq. (5.48)}$$

Again, the cross-over range is that corresponding to $S = J$; it is given by

$$(R_{CO})_{SOJ} = \left(\frac{P_t G^2 R_J^2 \sigma B_J G_{PC}}{4\pi (ERP) G' B_r L} \right)^{1/4}. \qquad \text{Eq. (5.49)}$$

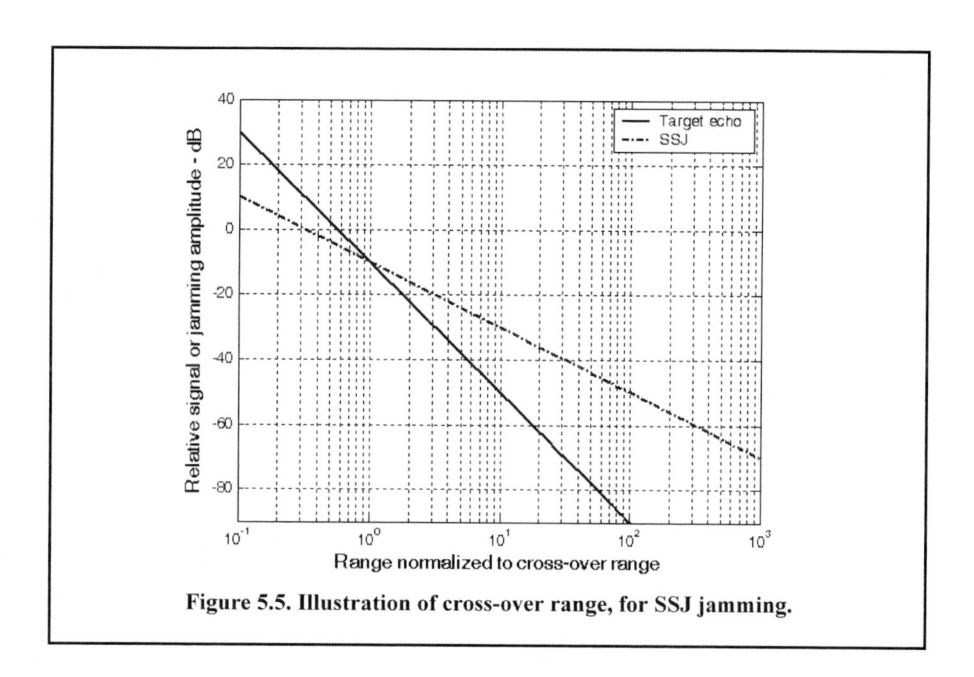

Figure 5.5. Illustration of cross-over range, for SSJ jamming.

5.6. Bistatic Radar Equation

Radar systems that use the same antenna for both transmitting and receiving are called monostatic radars. Bistatic radars use transmit and receive antennas that are placed at different locations. Under this definition CW radars, although they use separate transmit and receive antennas, are not considered bistatic radars unless the distance between the two antennas is considerable. Figure 5.6 shows the geometry associated with bistatic radars. The angle, β, is called the bistatic angle. A synchronization link between the transmitter and receiver is necessary in order to maximize the receiver's knowledge of the transmitted signal so that it can extract maximum target information.

The synchronization link may provide the receiver with the following information: (1) the transmitted frequency in order to compute the Doppler shift, and (2) the transmit time or phase reference in order to measure the total scattered path $(R_t + R_r)$. Frequency and phase reference synchronization can be maintained through line-of-sight communications between the transmitter and receiver. However, if this is not possible, the receiver may use a stable reference oscillator for synchronization.

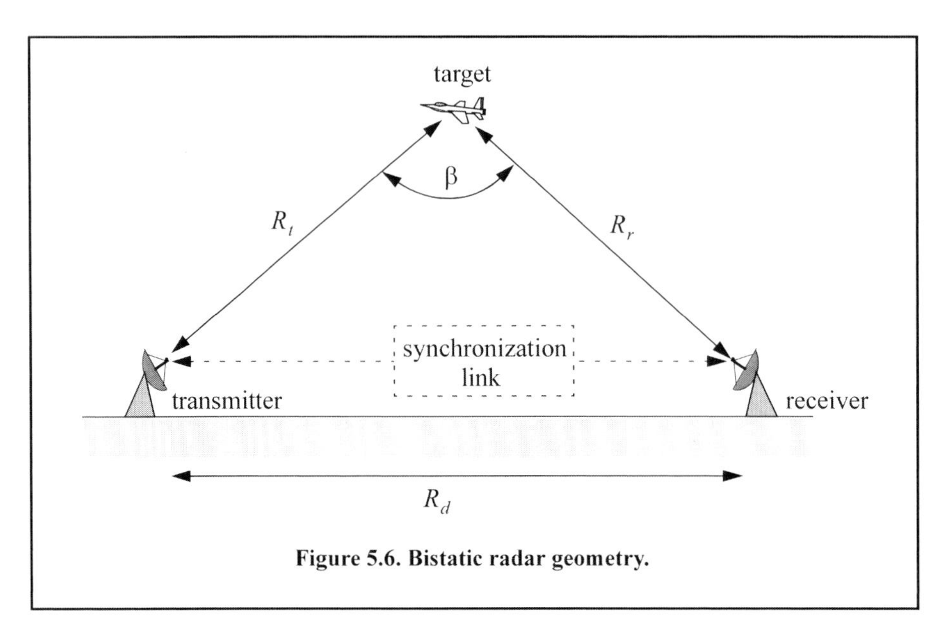

Figure 5.6. Bistatic radar geometry.

One major distinction between monostatic and bistatic radar operations has to do with the measured bistatic target RCS, denoted by σ_B. In the case of a small bistatic angle, the bistatic RCS is similar to the monostatic RCS; but, as the bistatic angle approaches $180°$, the bistatic RCS becomes very large and can be approximated by

$$\sigma_{B_{max}} \approx (4\pi A_t^2)/\lambda^2 \qquad \text{Eq. (5.50)}$$

where λ is the wavelength and A_t is the target projected area. The bistatic radar equation can be derived in a similar fashion to the monostatic radar equation. Referring to Figure 5.5, the power density at the target is

$$P_D = (P_t G_t)/(4\pi R_t^2) \qquad \text{Eq. (5.51)}$$

where P_t is the peak transmitted power, G_t is the gain of the transmitting antenna, and R_t is the range from the radar transmitter to the target.

The effective power scattered off a target with bistatic RCS σ_B is

$$P' = P_D \sigma_B \qquad \text{Eq. (5.52)}$$

and the power density at the receiver antenna is

$$P_{refl} = \frac{P'}{4\pi R_r^2} = \frac{P_D \sigma_B}{4\pi R_r^2}. \qquad \text{Eq. (5.53)}$$

R_r is the range from the target to the receiver. Substituting Eq. (5.51) into Eq. (5.53) yields

$$P_{refl} = \frac{P_t G_t \sigma_B}{(4\pi)^2 R_t^2 R_r^2}. \qquad \text{Eq. (5.54)}$$

The total power delivered by an antenna with aperture A_e is

$$P_{Dr} = \frac{P_t G_t \sigma_B A_e}{(4\pi)^2 R_t^2 R_r^2}.$$ Eq. (5.55)

Substituting $(G_r \lambda^2 / 4\pi)$ for A_e yields

$$P_{Dr} = \frac{P_t G_t G_r \lambda^2 \sigma_B}{(4\pi)^3 R_t^2 R_r^2}$$ Eq. (5.56)

where G_r is the gain of the receiver antenna. Finally, when transmitter and receiver losses, L_t and L_r, are taken into consideration, the bistatic radar equation can be written as

$$P_{Dr} = \frac{P_t G_t G_r \lambda^2 \sigma_B}{(4\pi)^3 R_t^2 R_r^2 L_t L_r}.$$ Eq. (5.57)

5.7. Radar Cross Section (RCS)

The radar cross section σ is a function of the target size, shape, and material and can be used by the radar as a means of discrimination. The value of a target cross section σ that is used in the radar equation is the equivalent cross section of a sphere that will produce the same echo at the radar. Radar RCS usually fluctuates over a period of time (RCS scintillation) as a function of frequency and the target orientation with respect to the radar (aspect angle). Electromagnetic waves are normally diffracted or scattered when incident on an object; and in order to accurately determine a target cross section, one must solve Maxwell's equations, with the proper set of boundary conditions. Obtaining a closed form solution of Maxwell's equations for a scattered field from a complex object is not an easy task; thus numerical solutions are often implemented using high-speed computers. However, Maxwell's equations can be solved to determine the RCS for simple shaped objects such as a sphere, cone, paraboloid, and corner reflector.

In practice, radar designers use simple shaped targets (such as spheres) of known cross sections to experimentally calibrate the radar. In the early 1900s Mie developed an exact solution for the energy backscattered from a conductive sphere. The normalized exact sphere RCS is a Mie series given by

$$\frac{\sigma}{\pi r^2} = \left(\frac{j}{kr}\right) \sum_{n=1}^{\infty} (-1)^n (2n+1) \left[\left(\frac{kr J_{n-1}(kr) - n J_n(kr)}{kr H_{n-1}^{(1)}(kr) - n H_n^{(1)}(kr)} \right) - \left(\frac{J_n(kr)}{H_n^{(1)}(kr)} \right) \right]$$ Eq. (5.58)

where r is the radius of the sphere, $k = 2\pi/\lambda$, λ is the wavelength, J_n is the spherical Bessel of the first kind of order n, and $H_n^{(1)}$ is the Hankel function of order n, and is given by

$$H_n^{(1)}(kr) = J_n(kr) + jY_n(kr)$$ Eq. (5.59)

where Y_n is the spherical Bessel function of the second kind of order n. Figure 5.7 shows a plot for the normalized RCS of a sphere as a function of its circumference in wavelength units. Three regions are identified. First is the optical region (corresponds to a large sphere). In this case,

$$\sigma \approx \pi r^2 \qquad r \gg \lambda.$$ Eq. (5.60)

Second is the Rayleigh region (small sphere). In this case,

$$\sigma \approx 9\pi r^2 (kr)^4 \qquad r \ll \lambda . \tag{Eq. (5.61)}$$

The region between the optical and Rayleigh regions is oscillatory in nature and is called the Mie or resonance region. Table 5.1 shows RCS formulas, as a function of the wavelength, for some simple conductive objects.

Target cross sections of complex or extended targets such as aircrafts, ships, and missiles are complicated and difficult to obtain. In such cases, the best radar RCS estimates are those obtained experimentally. However, experimental RCS measurements may not always be possible. In such cases, estimates of the target physical shape and dimensions are used to compute RCS estimate(s) using computer simulations. Depending on the accuracy of the algorithms used and on the target complexity, dedicated computers may be required to run for hours or longer to compute a certain target RCS signature.

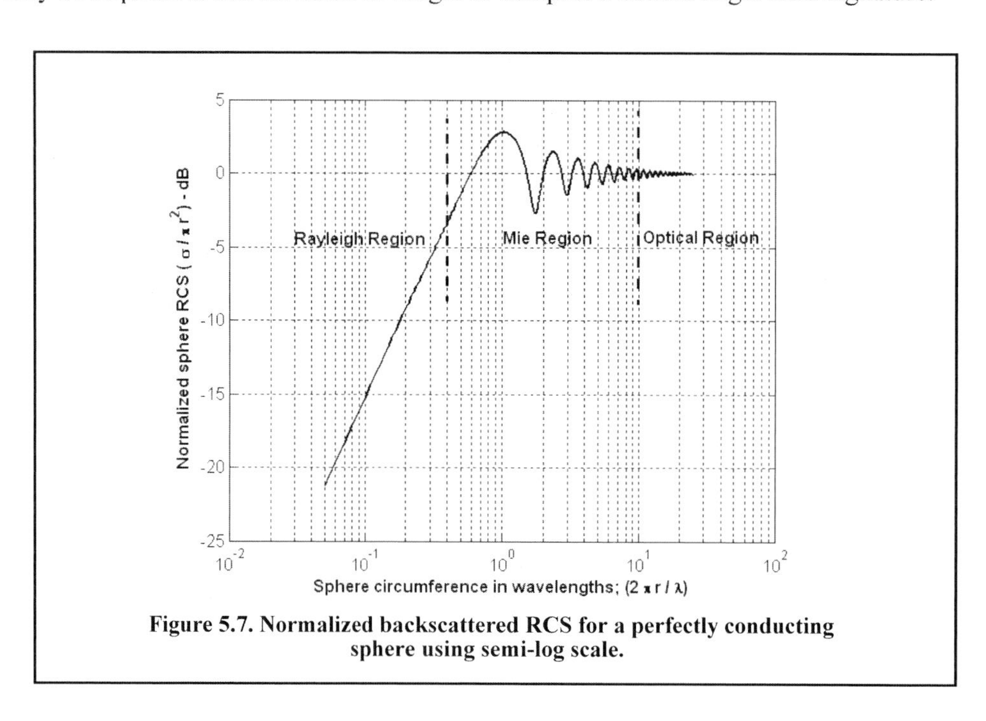

Figure 5.7. Normalized backscattered RCS for a perfectly conducting sphere using semi-log scale.

Table 5.1. Known RCS for some simple conductive objects.

Object	Aspect	RCS
large sphere	any	πr^2 ; r is radius of the sphere
cone	axial	$(\lambda^2/16\pi)(\tan\theta)^4$; θ is cone half angle
paraboloid	axial	πr^2 ; r is apex radius of curvature
triangular corner reflector	any axis	$4\pi r^4/3\lambda^2$; r is edge length
dipole	broadside	$0.88\lambda^2$

5.8. Target Scintillation.

Since a target cross section is very sensitive to aspect angle it will, unless the target is stationary, change (fluctuate) over a period of time. Swerling has calculated the detection probability densities for different types of target fluctuations known as Swerling I through Swerling IV. Targets that do not have any fluctuations are normally referred to as Swerling 0 or Swerling V targets. The concept of Swerling targets is illustrated in Figure 5.8.

In Swerling I, the radar received pulses have constant amplitude throughout an entire scan, but are uncorrelated from scan to scan (slow fluctuation). The probability density function (*pdf*) corresponding to Swerling I targets is

$$f_{\Sigma}(\sigma) = \frac{1}{\sigma_{av}} \exp\left(-\frac{\sigma}{\sigma_{av}}\right) \qquad \sigma \geq 0 \qquad \text{Eq. (5.62)}$$

where σ_{av} denotes the average RCS over all target fluctuation. Swerling II target fluctuation is more rapid than Swerling I, but the measurements are pulse to pulse uncorrelated. The *pdf* is also given by Eq. (5.62). Swerlings I and II apply to targets consisting of many independent fluctuating point scatterers of approximately equal physical dimensions.

Swerlings III and IV have the same *pdf*, and it is given by

$$f_{\Sigma}(\sigma) = \frac{4\sigma}{\sigma_{av}^2} \exp\left(-\frac{2\sigma}{\sigma_{av}}\right). \qquad \text{Eq. (5.63)}$$

Figure 5.9 shows a typical plot of the *pdf*s for Swerling cases. The fluctuations in Swerling III are similar to Swerling I; while in Swerling IV they are similar to Swerling II fluctuations. Table 5.2 gives a summary of Swerling type, target cross section fluctuations. Swerlings II and IV are more applicable to targets that can be represented by one dominant scatterer and many other small reflectors.

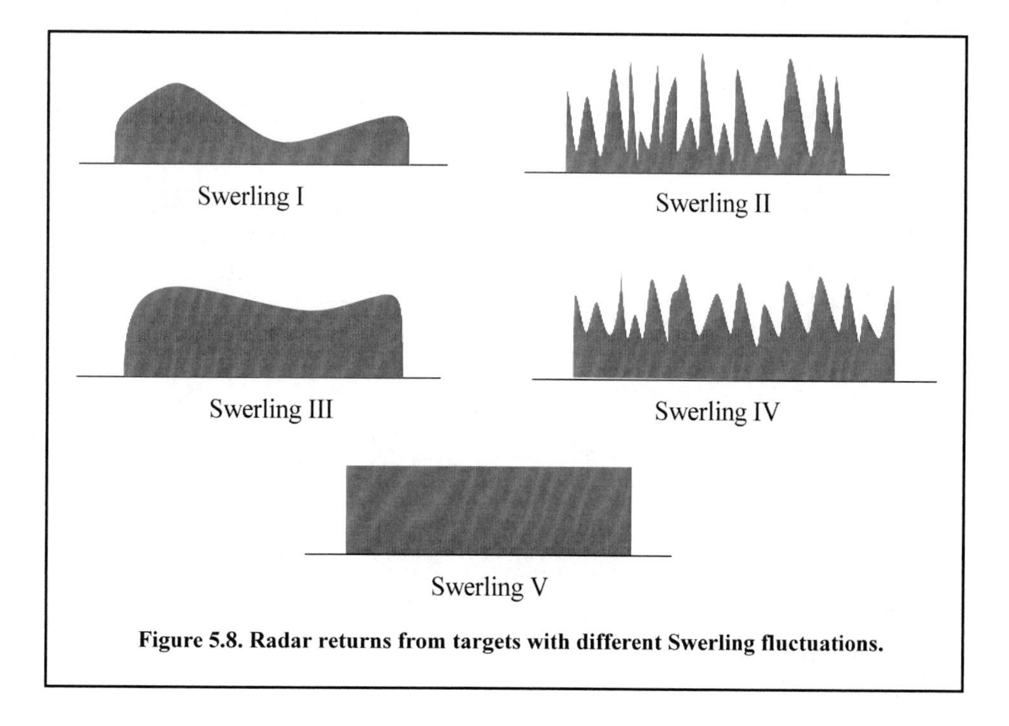

Figure 5.8. Radar returns from targets with different Swerling fluctuations.

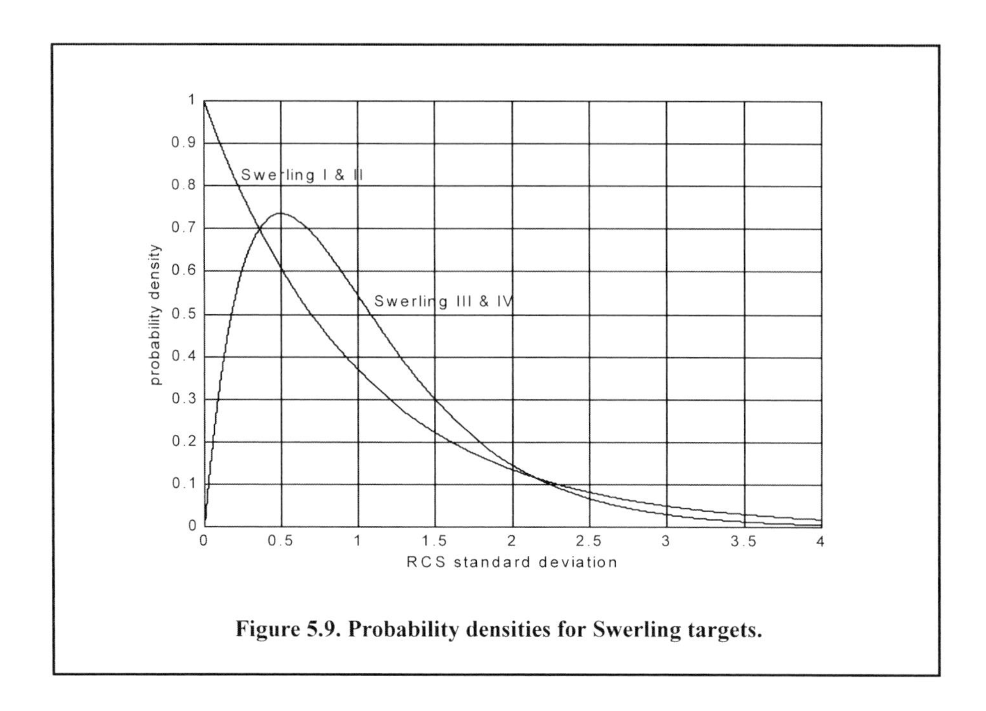

Figure 5.9. Probability densities for Swerling targets.

Table 5.2. Swerling models I through IV.

Swerling type	Representation	Fluctuation
Swerling I	Many small equal scatterers	scan-to-scan
Swerling II	Many small equal scatterers	pulse-to-pulse
Swerling III	One dominant plus many small scatterers	scan-to-scan
Swerling IV	One dominant plus many small scatterers	pulse-to-pulse

5.9. Noise Figure

Any signal other than the target returns in the radar receiver is considered to be noise. This includes interfering signals from outside the radar and thermal noise generated within the receiver itself. Thermal noise (thermal agitation of electrons) and shot noise (variation in carrier density of a semiconductor) are the two main internal noise sources within a radar receiver. The power spectral density of thermal noise is given by

$$S_n(\omega) = \frac{|\omega|h}{\pi\left[\exp\left(\frac{|\omega|h}{2\pi kT}\right) - 1\right]} \qquad \text{Eq. (5.64)}$$

where $|\omega|$ is the absolute value of the frequency in radians per second, T is the temperature of the conducting medium in degrees *Kelvin*, k is Boltzman's constant, and h is Planck's constant ($h = 6.625 \times 10^{-34}\ Joule\ s$). When the condition $|\omega| \ll 2\pi kT/h$ is true, it can be shown that Eq. (5.64) is approximated by

$$S_n(\omega) \approx 2kT. \qquad \text{Eq. (5.65)}$$

This approximation is widely accepted, since, in practice, radar systems operate at frequencies less than $100GHz$; and, for example, if $T = 290K$, then $2\pi kT/h \approx 6000GHz$. The mean-square noise voltage (noise power) generated across a $1ohm$ resistance is then

$$\langle n^2 \rangle = \frac{1}{2\pi} \int_{-2\pi B}^{2\pi B} 2kT \quad d\omega = 4kTB \tag{Eq. (5.66)}$$

where B is the system bandwidth in Hertz.

Any electrical system containing thermal noise and having input resistance R_{in} can be replaced by an equivalent noiseless system with a series combination of a noise equivalent voltage source and a noiseless input resistor R_{in} added at its input. This is illustrated in Figure 5.10. The amount of noise power that can physically be extracted from $\langle n^2 \rangle$ is one fourth the value computed in Eq. (5.66). Consider a noisy system with power gain A_P, as shown in Figure 5.11. The noise figure is defined by

$$F_{dB} = 10 \log \frac{total\ noise\ power\ out}{noise\ power\ out\ due\ to\ R_{in}\ alone} . \tag{Eq. (5.67)}$$

More precisely,

$$F_{dB} = 10 \log \frac{N_o}{N_i \ A_p} \tag{Eq. (5.68)}$$

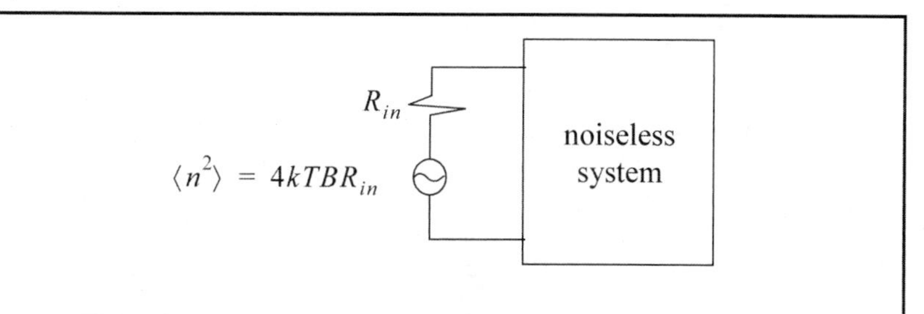

Figure 5.10. Noiseless system with an input noise voltage source.

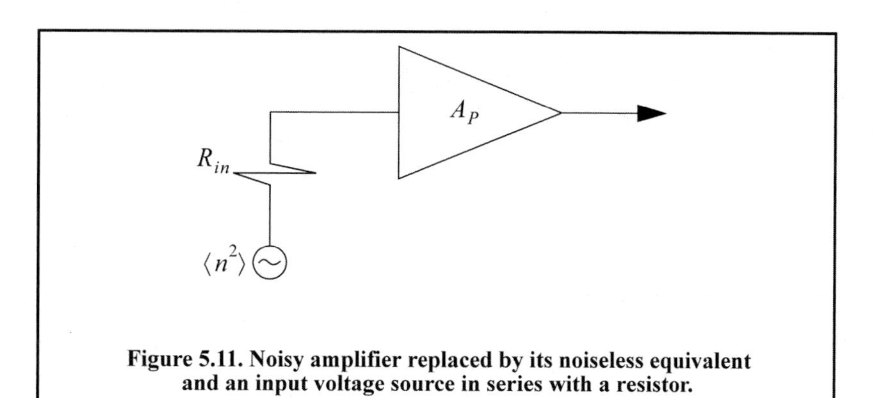

Figure 5.11. Noisy amplifier replaced by its noiseless equivalent and an input voltage source in series with a resistor.

where N_o and N_i are, respectively, the noise power at the output and input of the system. If we define the input and output signal power by S_i and S_o, respectively, then the power gain is

$$A_P = \frac{S_o}{S_i}. \quad \text{Eq. (5.69)}$$

It follows that

$$F_{dB} = 10\log\left(\frac{S_i/Ni}{S_o/N_o}\right) = \left(\frac{S_i}{N_i}\right)_{dB} - \left(\frac{S_o}{N_o}\right)_{dB} \quad \text{Eq. (5.70)}$$

where

$$\left(\frac{S_i}{N_i}\right)_{dB} > \left(\frac{S_o}{N_o}\right)_{dB}. \quad \text{Eq. (5.71)}$$

Thus, the noise figure is the loss in the signal-to-noise ratio due to the added thermal noise of the amplifier $((SNR)_o = (SNR)_i - F \ \ in \ \ dB)$.

One can also express the noise figure in terms of the system's effective temperature T_e. Consider the amplifier shown in Figure 5.11, and let its effective temperature be T_e. Assume the input noise temperature is T_o. Thus, the input noise power is

$$N_i = kT_oB \quad \text{Eq. (5.72)}$$

and the output noise power is

$$N_o = kT_oB \ A_p + kT_eB \ A_p \quad \text{Eq. (5.73)}$$

where the first term on the right-hand side of Eq. (5.73) corresponds to the input noise, and the latter term is due to thermal noise generated inside the system. It follows that the noise figure can be expressed as

$$F = \frac{(SNR)_i}{(SNR)_o} = \frac{S_i}{kT_oB} kBA_p \frac{T_o + T_e}{S_o} = 1 + \frac{T_e}{T_o}. \quad \text{Eq. (5.74)}$$

Equivalently, we can write

$$T_e = (F-1)T_o. \quad \text{Eq. (5.75)}$$

Example:

An amplifier has a 4dB noise figure; the bandwidth is $B = 500KHz$. *Calculate the input signal power that yields a unity SNR at the output. Assume* $T_o = 290K$ *and an input resistance of 1ohm.*

Solution:

The input noise power is

$$kT_oB = 1.38 \times 10^{-23} \times 290 \times 500 \times 10^3 = 2.0 \times 10^{-15} W.$$

Assuming a voltage signal, then the input noise mean squared voltage is

$$\langle n_i^2 \rangle = kT_oB = 2.0 \times 10^{-15} \ v^2$$

$$F = 10^{0.4} = 2.51 .$$

From the noise figure definition we get

$$\frac{S_i}{N_i} = F\left(\frac{S_o}{N_o}\right) = F$$

$$\langle s_i^2 \rangle = F\langle n_i^2 \rangle = 2.51 \times 2.0 \times 10^{-15} = 5.02 \times 10^{-15}\ v^2 .$$

Finally,

$$\sqrt{\langle s_i^2 \rangle} = 70.852nv .$$

****

Consider a cascaded system as in Figure 5.12. Network 1 is defined by noise figure F_1, power gain G_1, bandwidth B, and temperature T_{e1}. Similarly, network 2 is defined by F_2, G_2, B, and T_{e2}. Assume the input noise has temperature T_0. The output signal power is

$$S_o = S_i G_1 G_2 . \qquad \text{Eq. (5.76)}$$

The input and output noise powers are, respectively, given by

$$N_i = kT_oB \qquad \text{Eq. (5.77)}$$

$$N_o = kT_0BG_1G_2 + kT_{e1}BG_1G_2 + kT_{e2}BG_2 \qquad \text{Eq. (5.78)}$$

where the three terms on the right-hand side of Eq. (5.78), respectively, correspond to the input noise power, thermal noise generated inside network 1, and thermal noise generated inside network 2.

Now, use the relation $T_e = (F-1)T_0$ along with Eq. (5.76) and Eq. (5.77) to express the overall output noise power as

$$N_o = F_1N_iG_1G_2 + (F_2-1)N_iG_2 . \qquad \text{Eq. (5.79)}$$

It follows that the overall noise figure for the cascaded system is

$$F = \frac{(S_i/N_i)}{(S_o/N_o)} = F_1 + \frac{F_2-1}{G_1} . \qquad \text{Eq. (5.80)}$$

In general, for an n-stage system we get

$$F = F_1 + \frac{F_2-1}{G_1} + \frac{F_3-1}{G_1G_2} + \ldots + \frac{F_n-1}{G_1G_2G_3 \cdot \cdot \cdot G_{n-1}} . \qquad \text{Eq. (5.81)}$$

Also, the n-stage system effective temperatures can be computed as

$$T_e = T_{e1} + \frac{T_{e2}}{G_1} + \frac{T_{e3}}{G_1G_2} + \ldots + \frac{T_{en}}{G_1G_2G_3 \cdot \cdot \cdot G_{n-1}} . \qquad \text{Eq. (5.82)}$$

As suggested by Eq. (5.81) and Eq. (5.82), the overall noise figure is mainly dominated by the first stage. Thus, radar receivers employ low-noise power amplifiers in the first stage in order to minimize the overall receiver noise figure. However, for radar systems that are built for low RCS operations, every stage should be included in the analysis.

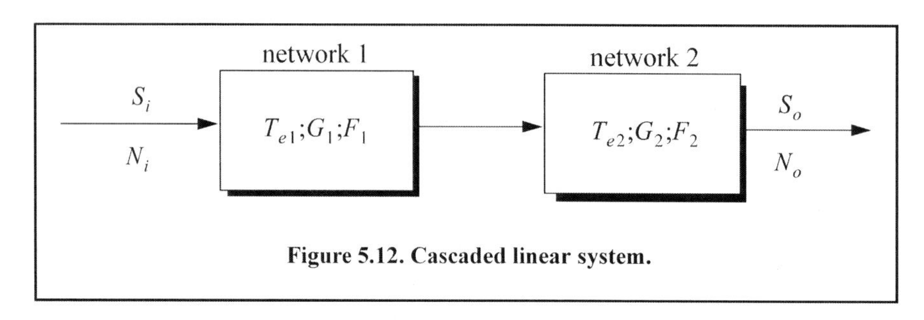

Figure 5.12. Cascaded linear system.

Example:

A radar receiver consists of an antenna with cable loss $L = 1dB = F_1$, an RF amplifier with $F_2 = 6dB$, and gain $G_2 = 20dB$, followed by a mixer whose noise figure is $F_3 = 10dB$ and conversion loss $L = 8dB$, and finally, an integrated circuit IF amplifier with $F_4 = 6dB$ and gain $G_4 = 60dB$. Find the overall noise figure.

Solution:

From Eq. (5.81)

$$F = F_1 + \frac{F_2 - 1}{G_1} + \frac{F_3 - 1}{G_1 G_2} + \frac{F_4 - 1}{G_1 G_2 G_3} \cdot$$

G_1	G_2	G_3	G_4	F_1	F_2	F_3	F_4
$-1dB$	$20dB$	$-8dB$	$60dB$	$1dB$	$6dB$	$10dB$	$6dB$
0.7943	100	0.1585	10^6	1.2589	3.9811	10	3.9811

It follows that

$$F = 1.2589 + \frac{3.9811 - 1}{0.7943} + \frac{10 - 1}{100 \times 0.7943} + \frac{3.9811 - 1}{0.158 \times 100 \times 0.7943} = 5.3629$$

$$F = 10\log(5.3628) = 7.294dB .$$

****

5.10. Radar Losses

As indicated by the radar equation, the receiver SNR is inversely proportional to the radar losses. Hence, any increase in radar losses causes a drop in the SNR, thus decreasing the probability of detection, since it is a function of the SNR. Often, the principal difference between a good radar design and a poor radar design is the radar losses. Radar losses include ohmic (resistance) losses and statistical losses. In this section we will briefly summarize radar losses.

5.5.1. Transmit and Receive Losses

Transmit and receive losses occur between the radar transmitter and antenna input port, and between the antenna output port and the receiver front end, respectively. Such losses are often called plumbing losses. Typically, plumbing losses are on the order of *1* to *2* dBs.

5.5.2. Antenna Pattern Loss and Scan Loss

So far, when we used the radar equation we assumed maximum antenna gain. This is true only if the target is located along the antenna's boresight axis. However, as the radar scans across a target, the antenna gain in the direction of the target is less than maximum, as defined by the antenna's radiation pattern. The loss in the SNR due to not having maximum antenna gain on the target at all times is called the antenna pattern (shape) loss. Once an antenna has been selected for a given radar, the amount of antenna pattern loss can be mathematically computed.

For example, consider a $\sin x/x$ antenna radiation pattern as shown in Figure 5.13. It follows that the average antenna gain over an angular region of $\pm\theta/2$ about the boresight axis is

$$G_{av} \approx 1 - \left(\frac{\pi r}{\lambda}\right)^2 \frac{\theta^2}{36} \qquad \text{Eq. (5.83)}$$

where r is the aperture radius and λ is the wavelength. The proof of Eq. (5.83) is left as an exercise. In practice, Gaussian antenna patterns are often adopted. In this case, if θ_{3dB} denotes the antenna 3dB beam width, then the antenna gain can be approximated by

$$G(\theta) = \exp\left(-\frac{2.776\theta^2}{\theta_{3dB}^2}\right). \qquad \text{Eq. (5.84)}$$

If the antenna scanning rate is so fast that the gain on receive is not the same as on transmit, additional scan loss has to be calculated and added to the beam shape loss. Scan loss can be computed in a similar fashion to beam shape loss. Phased array radars are often prime candidates for both beam shape and scan losses.

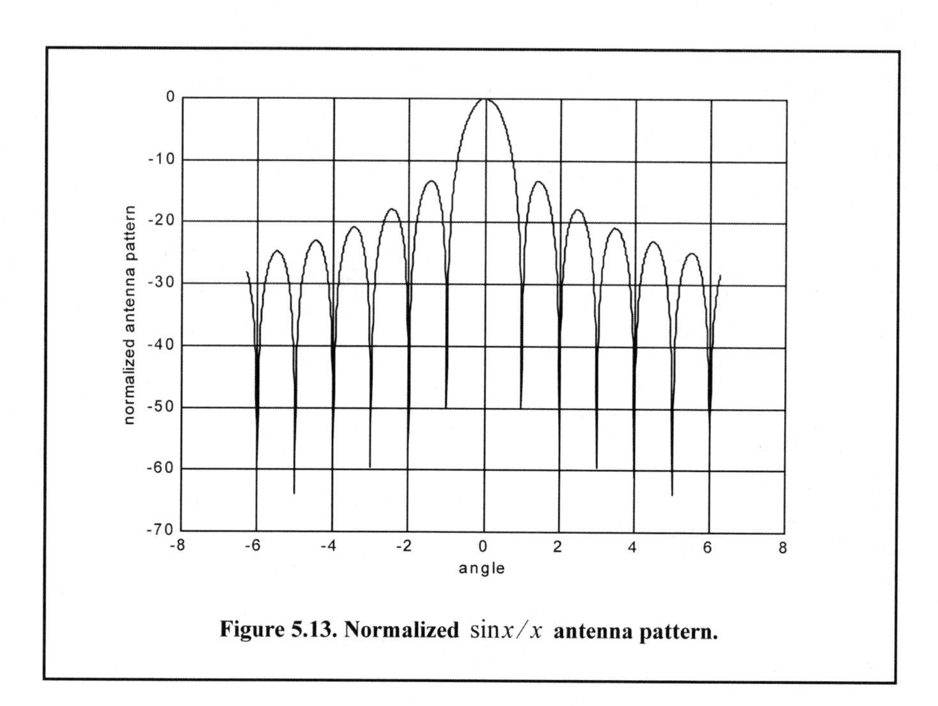

Figure 5.13. Normalized $\sin x/x$ antenna pattern.

5.5.3. Atmospheric Loss

Detailed discussion of atmospheric loss and propagation effects is in Chapter 7. Atmospheric attenuation is a function of the radar operating frequency, target range, and elevation angle. Atmospheric attenuation can be as high as a few dBs.

5.5.4. Collapsing Loss

When the number of integrated returned noise pulses is larger than the target returned pulses, a drop in the SNR occurs. This is called collapsing loss. The collapsing loss factor is defined as

$$\rho_c = \frac{n+m}{n} \qquad \text{Eq. (5.85)}$$

where n is the number of pulses containing both signal and noise, while m is the number of pulses containing noise only. Radars detect targets in azimuth, range, and Doppler. When target returns are displayed in one coordinate, such as range, noise sources from azimuth cells adjacent to the actual target return converge in the target vicinity and cause a drop in the SNR. This is illustrated in Figure 5.14.

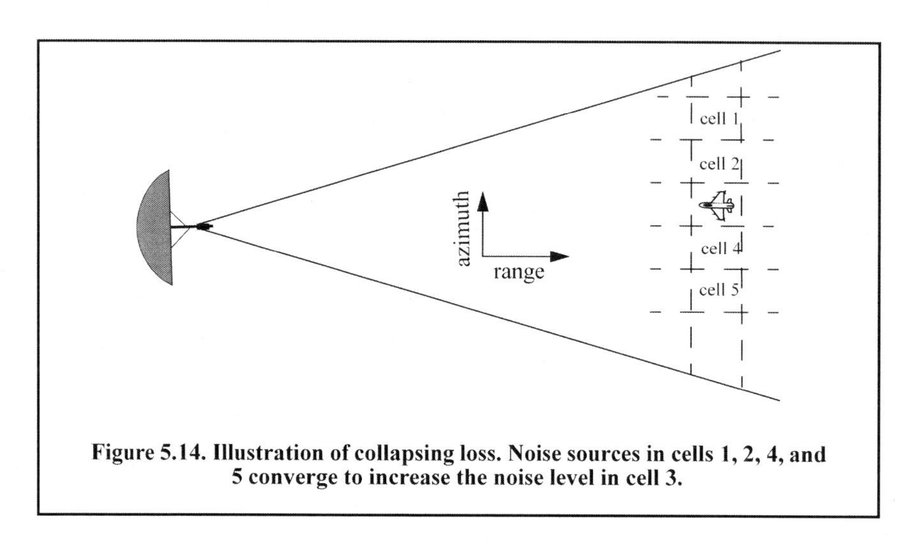

Figure 5.14. Illustration of collapsing loss. Noise sources in cells 1, 2, 4, and 5 converge to increase the noise level in cell 3.

5.5.5. Processing Losses

Detector Approximation

The output voltage signal of a radar receiver that utilizes a linear detector is

$$v(t) = (v_I^2(t) + v_Q^2(t))^{1/2}, \qquad \text{Eq. (5.86)}$$

where (v_I, v_Q) are the in-phase and quadrature components. For a radar using a square law detector, we have $v^2(t) = v_I^2(t) + v_Q^2(t)$.

Since in real hardware the operations of squares and square roots are time consuming, many algorithms have been developed for detector approximation. This approximation results in a loss of the signal power, typically *0.5* to *1* dB.

Constant False Alarm Rate (CFAR) Loss

In many cases the radar detection threshold is constantly adjusted as a function of the receiver noise level in order to maintain a constant false alarm rate. For this purpose, Constant False Alarm Rate (CFAR) processors are utilized in order to keep the number of false alarms under control in a changing and unknown background of interference. CFAR processing can cause a loss in the SNR level on the order of *1* dB.

Three different types of CFAR processors are primarily used. They are adaptive threshold CFAR, nonparametric CFAR, and nonlinear receiver techniques. Adaptive CFAR assumes that the interference distribution is known and approximates the unknown parameters associated with these distributions. Nonparametric CFAR processors tend to accommodate unknown interference distributions. Nonlinear receiver techniques attempt to normalize the root mean square amplitude of the interference.

Quantization Loss

Finite word length (number of bits) and quantization noise cause an increase in the noise power density at the output of the Analog to Digital (A/D) converter. The A/D noise level is $q^2/12$, where q is the quantization level.

Range Gate Straddle

The radar receiver is normally mechanized as a series of contiguous range gates (bins). Each range bin is implemented as an integrator matched to the transmitted pulse width. Since the radar receiver acts as a filter that smears (smooths) the received target echoes. The smoothed target return envelope is normally straddled to cover more than one range gate.

Typically, three gates are affected; they are called the early, on, and late gates. If a point target is located exactly at the center of a range gate, then the early and late samples are equal. However, as the target starts to move into the next gate, the late sample becomes larger while the early sample gets smaller. In any case, the amplitudes of all three samples should always roughly add up to the same value. Figure 5.15 illustrates the concept of range straddling. The envelope of the smoothed target echo is likely to be Gaussian shape. In practice, triangular shaped envelopes may be easier and faster to implement. Since the target is likely to fall anywhere between two adjacent range bins, a loss in the SNR occurs (per range gate). More specifically, a target's returned energy is split between three range bins. Typically, straddle loss of about *2* to *3* dBs is not unusual.

Doppler Filter Straddle

Doppler filter straddle is similar to range gate straddle. However, in this case the Doppler filter spectrum is spread (widened) due to weighting functions. Weighting functions are normally used to reduce the side lobe levels. Since the target Doppler frequency can fall anywhere between two Doppler filters, signal loss occurs. This is illustrated in Figure 5.16, where due to weighting, the cross-over frequency f_{co} is smaller than the filter cut-off frequency f_c which normally corresponds to the 3dB power point.

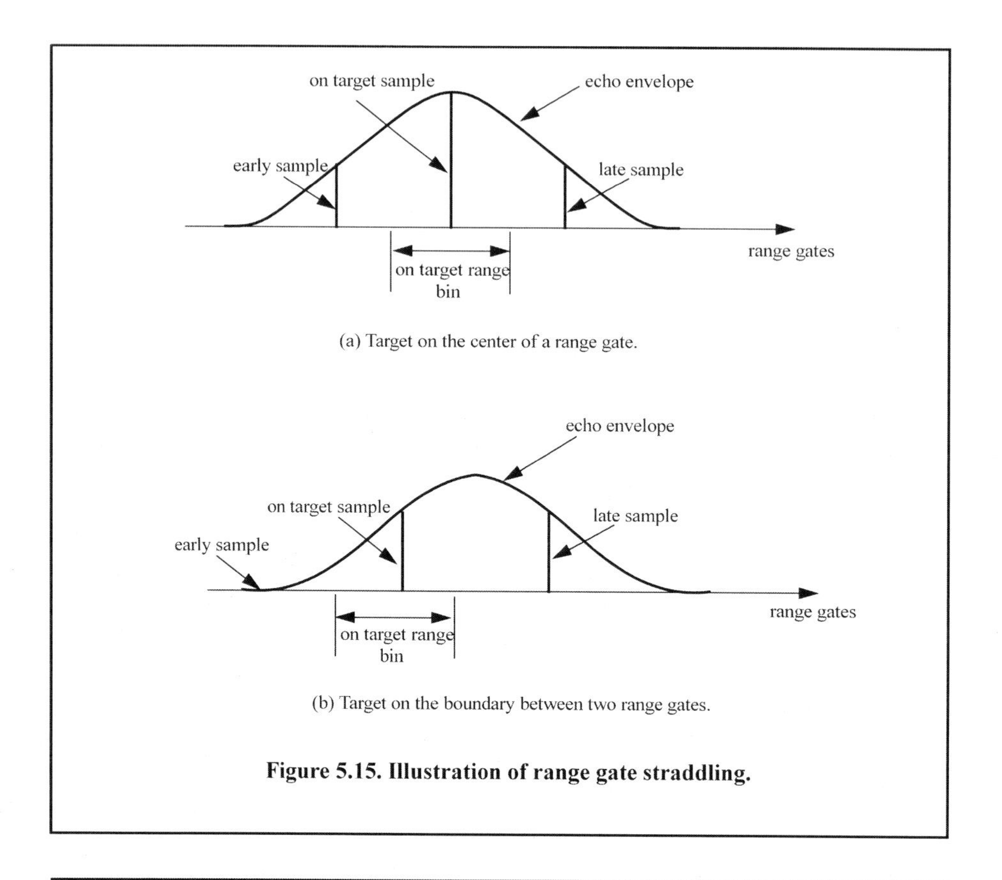

Figure 5.15. Illustration of range gate straddling.

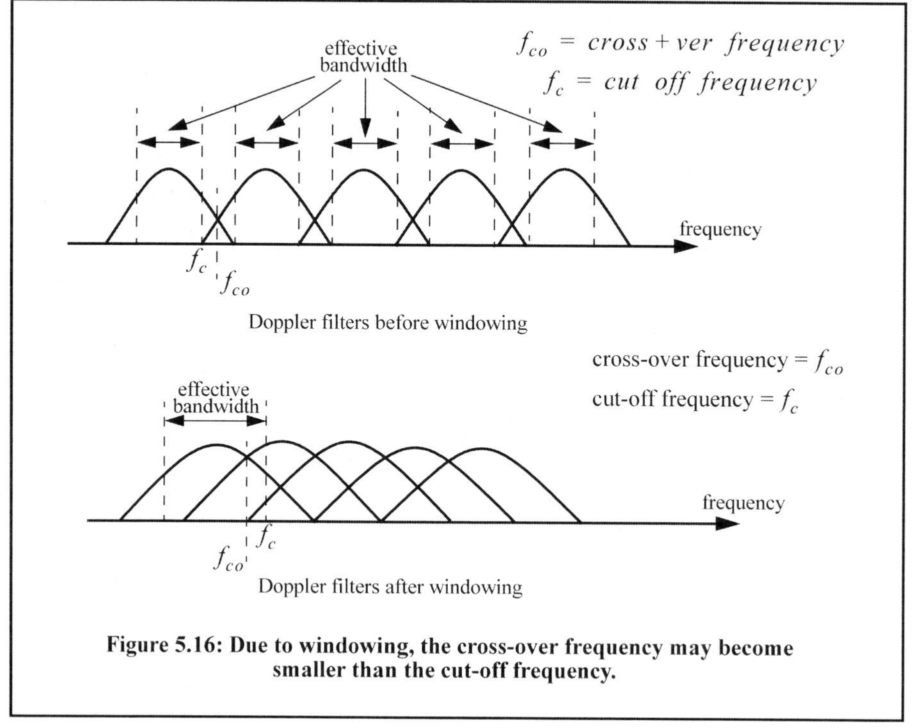

Figure 5.16: Due to windowing, the cross-over frequency may become smaller than the cut-off frequency.

Example:

Consider the smoothed target echo voltage shown below. Assume 1Ω *resistance. Find the power loss due to range gate straddling over the interval* $\{0, \tau\}$.

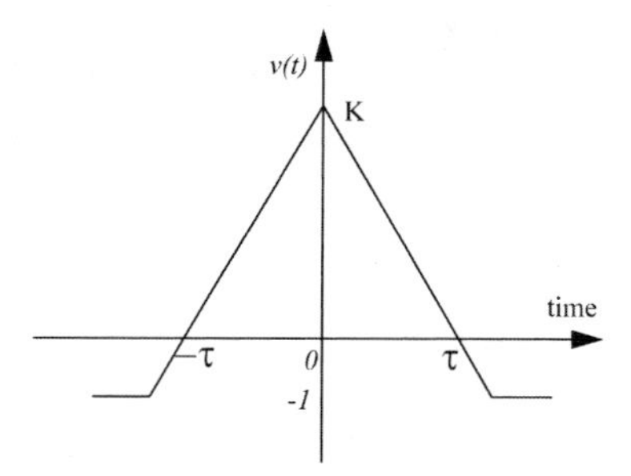

Solution:

The smoothed voltage can be written as

$$v(t) = \left\{ \begin{array}{ll} K + \left(\frac{K+1}{\tau}\right)t & ;t < 0 \\ K - \left(\frac{K+1}{\tau}\right)t & ;t \geq 0 \end{array} \right\}.$$

The power loss due to straddle over the interval $\{0, \tau\}$ *is*

$$L_s = \frac{v^2}{K^2} = 1 - 2\left(\frac{K+1}{K\tau}\right)t + \left(\frac{K+1}{K\tau}\right)^2 t^2.$$

The average power loss is then

$$\bar{L}_s = \frac{2}{\tau}\int_0^{\tau/2} \left(1 - 2\left(\frac{K+1}{K\tau}\right)t + \left(\frac{K+1}{K\tau}\right)^2 t^2\right) dt = 1 - \frac{K+1}{2K} + \frac{(K+1)^2}{12K^2}$$

and, for example, if $K = 15$, *then* $\bar{L}_s = 2.5dB$.

****

5.6. Radar Reference Range

Many radar design issues can be derived or computed based on the radar reference range R_{ref} which is often provided by the radar end user. It simply describes that range at which a certain SNR value, referred to as SNR_{ref}, has to be achieved using a specific reference pulsewidth τ_{ref} for a pre-determined target cross section, σ_{ref}. The target is assumed to be on the radar line of sight, as illustrated in Figure 5.17. The radar equation at the reference range is

$$R_{ref} = \left(\frac{P_t G^2 \lambda^2 \sigma_{ref} \tau_{ref}}{(4\pi)^3 k T_e F L (SNR)_{ref}}\right)^{1/4}. \quad \text{Eq. (5.87)}$$

The radar equation at any other detection range for any other combination of SNR, RCS, and pulsewidth can be given as

$$R = R_{ref}\left(\frac{\tau}{\tau_{ref}}\ \frac{\sigma}{\sigma_{ref}}\ \frac{SNR_{ref}}{SNR}\ \frac{1}{L_p}\right)^{1/4} \qquad \text{Eq. (5.88)}$$

where the additional loss term L_p is introduced to account for the possibility that the non-reference target may not be on the radar line of sight, and to account for other losses associated with the specific scenario. Other forms of Eq. (5.88) can be in terms of the SNR. More precisely,

$$SNR = SNR_{ref}\ \frac{\tau}{\tau_{ref}}\ \frac{1}{L_p}\ \frac{\sigma}{\sigma_{ref}}\left(\frac{R_{ref}}{R}\right)^4 . \qquad \text{Eq. (5.89)}$$

As an example, consider the radar with the following parameters: $\sigma_{ref} = 0.1m^2$, $R_{ref} = 86Km$, and $SNR_{ref} = 20dB$. The reference pulsewidth is $\tau_{ref} = 0.1\mu\sec$. Using Eq. (5.89) we compute the SNR at $R = 120Km$ for a target whose RCS is $\sigma = 0.2m^2$. Assume $L_p = 2dB$ to be equal to $(SNR)_{120Km} = 15.2dB$.

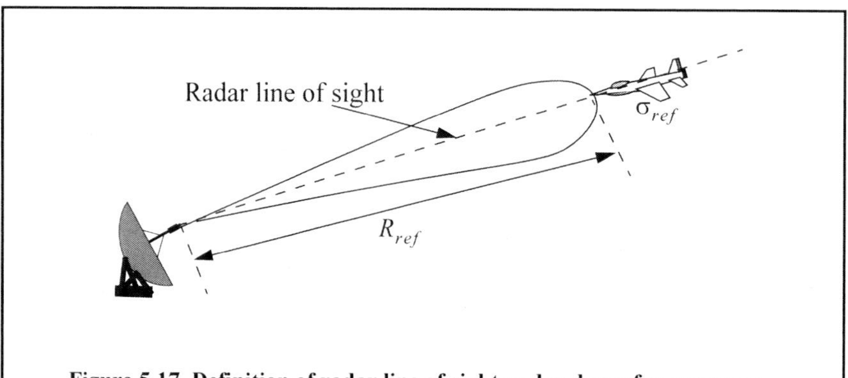

Figure 5.17. Definition of radar line of sight and radar reference range.

5.7. Pulse Integration

When a target is located within the radar beam during a single scan, it may reflect several pulses. By adding the returns from all pulses returned by a given target during a single scan, the radar sensitivity (SNR) can be increased. The number of returned pulses depends on the antenna scan rate and the radar PRF. More precisely, the number of pulses returned from a given target is given by

$$n_P = \frac{\theta_a T_{sc} f_r}{2\pi} \qquad \text{Eq. (5.90)}$$

where θ_a is the azimuth antenna beamwidth, T_{sc} is the scan time, and f_r is the radar PRF. The number of reflected pulses may also be expressed as

$$n_P = \frac{\theta_a f_r}{\dot{\theta}_{scan}} \qquad \text{Eq. (5.91)}$$

where $\dot{\theta}_{scan}$ is the antenna scan rate in degrees per second. Note that when using Eq. (5.90), θ_a is expressed in radians, while when using Eq. (5.91) it is expressed in degrees. As an example, consider a radar with an azimuth antenna beamwidth $\theta_a = 3°$, antenna scan rate $\dot{\theta}_{scan} = 45°/\sec$ (antenna scan time, $T_{sc} = 8 seconds$), and a PRF $f_r = 300Hz$. Using either Eqs. (5.90) or (5.91) yields $n_P = 20$ pulses.

The process of adding radar returns from many pulses is called radar pulse integration. Pulse integration can be performed on the quadrature components prior to the envelope detector. This is called coherent integration or pre-detection integration. Coherent integration preserves the phase relationship between the received pulses. Thus a build up in the signal amplitude is achieved. Alternatively, pulse integration performed after the envelope detector (where the phase relation is destroyed) is called non-coherent or post-detection integration.

Radar designers should exercise caution when utilizing pulse integration for the following reasons. First, during a scan a given target will not always be located at the center of the radar beam (i.e., have maximum gain). In fact, during a scan a given target will first enter the antenna beam at the 3-dB point, reach maximum gain, and finally leave the beam at the 3-dB point again. Thus, the returns do not have the same amplitude even though the target RCS may be constant and all other factors which may introduce signal loss remain the same. This is illustrated in Figure 5.18, and is normally referred to as antenna beamshape loss.

Other factors that may introduce further variation to the amplitude of the returned pulses include target RCS and propagation path fluctuations. Additionally, when the radar employs fast scan rates, an additional loss term is introduced due to the motion of the beam between transmission and reception. This is referred to as scan loss. A distinction should be made between scan loss due to a rotating antenna and the term scan loss that is normally associated with phased array antennas.

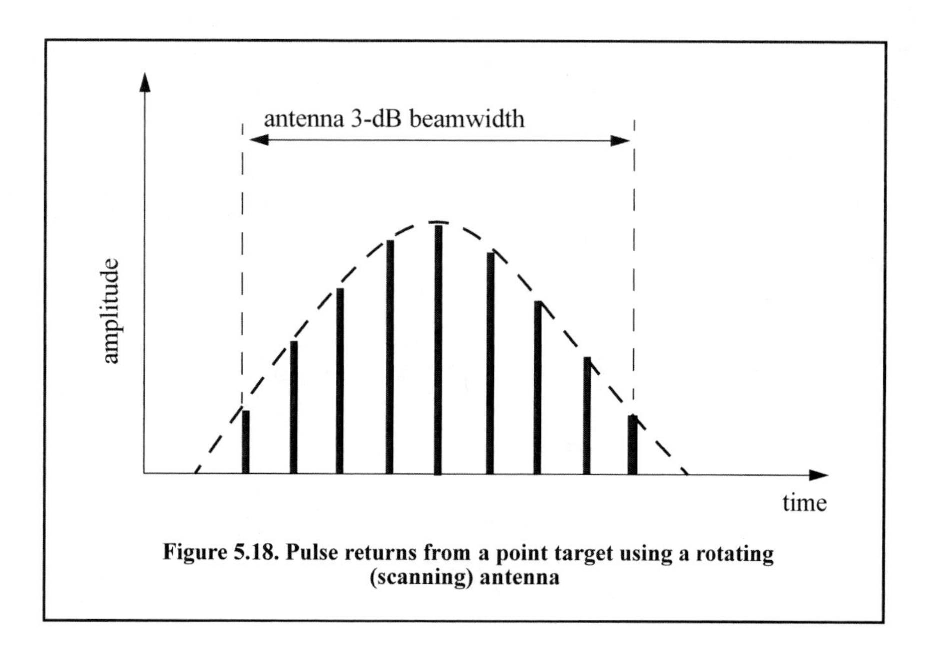

Figure 5.18. Pulse returns from a point target using a rotating (scanning) antenna

Finally, since coherent integration utilizes the phase information from all integrated pulses, it is critical that any phase variation between all integrated pulses be known with a great level of confidence. Consequently, target dynamics (such as target range, range rate, tumble rate, RCS fluctuation, etc.) must be estimated or computed accurately so that coherent integration can be meaningful. In fact, if a radar coherently integrates pulses from targets without proper knowledge of the target dynamics, it suffers a loss in SNR rather than the expected SNR build up. Knowledge of target dynamics is not as critical when employing non-coherent integration; nonetheless, target range rate must be estimated so that only the returns from a given target within a specific range bin are integrated. In other words, one must avoid range walk (i.e., avoid having a target cross between adjacent range bins during a single scan).

A comprehensive analysis of pulse integration should take into account issues such as the probability of detection P_D, probability of false alarm P_{fa}, the target statistical fluctuation model, and the noise or interference statistical models. However, in this section an overview of pulse integration is introduced in the context of radar measurements as it applies to the radar equation. The basic conclusions presented in this section concerning pulse integration are still valid. When a more comprehensive analysis of pulse integration is presented, however, the exact implementation, the mathematical formulation, and /or the numerical values used will vary.

5.7.1. Coherent Integration

In coherent integration, when a perfect integrator is used (100% efficiency) to integrate n_P pulses, the SNR is improved by the same factor. Otherwise, integration loss occurs, which is always the case for non-coherent integration. Coherent integration loss occurs when the integration process is not optimum. This could be due to target fluctuation, instability in the radar local oscillator, or propagation path changes. Denote the single pulse SNR required to produce a given probability of detection as $(SNR)_1$. The SNR resulting from coherently integrating n_P pulses is then given by

$$SNR)_{CI} = n_P(SNR)_1. \qquad \text{Eq. (5.92)}$$

Coherent integration cannot be applied over a large number of pulses, particularly if the target RCS is varying rapidly. If the target radial velocity is known and no acceleration is assumed, the maximum coherent integration time is limited to

$$t_{CI} = \sqrt{\lambda/2a_r} \qquad \text{Eq. (5.93)}$$

where λ is the wavelength and a_r is the target radial acceleration. Coherent integration time can be extended if the target radial acceleration can be compensated for by the radar.

5.7.2. Non-Coherent Integration

Non-coherent integration is often implemented after the envelope detector, also known as the quadratic detector. Non-coherent integration is less efficient than coherent integration. Actually, the non-coherent integration gain is always smaller than the number of non-coherently integrated pulses. This loss in integration is referred to as post detection or square law detector loss. Marcum and Swerling showed that this loss is somewhere between $\sqrt{n_P}$ and n_P. DiFranco and Rubin presented an approximation of this loss as

$$L_{NCI} = 10\log(\sqrt{n_p}) - 5.5 \ dB . \qquad \text{Eq. (5.94)}$$

Note that as n_P becomes very large, the integration loss approaches $\sqrt{n_P}$.

The subject of integration loss is treated in great levels of detail in the literature. Different authors use different approximations for the integration loss associated with non-coherent integration. However, all these different approximations yield very comparable results. Therefore, in the opinion of these authors the use of one formula or another to approximate integration loss becomes somewhat subjective. In this book, the integration loss approximation reported by Barton and used by Curry will be adopted. In this case, the non-coherent integration loss which can be used in the radar equation is

$$L_{NCI} = \frac{1+(SNR)_1}{(SNR)_1} . \qquad \text{Eq. (5.95)}$$

It follows that the SNR when n_P pulses are integrated non-coherently is

$$(SNR)_{NCI} = \frac{n_P(SNR)_1}{L_{NCI}} = n_P(SNR)_1 \times \frac{(SNR)_1}{1+(SNR)_1} . \qquad \text{Eq. (5.96)}$$

5.5.3. Detection Range with Pulse Integration

The process of determining the radar sensitivity or equivalently the maximum detection range when pulse integration is used is as follows. First, decide whether to use coherent or non-coherent integration. Keep in mind the issues discussed in the beginning of this section when deciding whether to use coherent or non-coherent integration.

Second, determine the minimum $(SNR)_{CI}$ or $(SNR)_{NCI}$ required for adequate detection and track. Typically, for ground based surveillance radars that can be on the order of 13 to 15 dB. The third step is to determine how many pulses should be integrated. The choice of n_P is affected by the radar scan rate, the radar PRF, the azimuth antenna beamwidth, and of course the target dynamics (remember that range walk should be avoided or compensated for, so that proper integration is feasible). Once n_P and the required SNR are known one can compute the single pulse SNR (i.e., the reduction in SNR). For this purpose use Eq. (5.92) in the case of coherent integration. In the non-coherent integration case, Curry[1] presents an attractive formula for this calculation, as follows

$$(SNR)_1 = \frac{(SNR)_{NCI}}{2n_P} + \sqrt{\frac{(SNR)^2_{NCI}}{4n_P^2} + \frac{(SNR)_{NCI}}{n_P}} . \qquad \text{Eq. (5.97)}$$

Finally, use $(SNR)_1$ from Eq. (5.97) in the radar equation to calculate the radar detection range. Observe that due to the integration reduction in SNR the radar detection range is now larger than that for the single pulse when the same SNR value is used. This is illustrated using the following mini design case study.

1. Curry, G. R., *Radar System Performance Modeling*, Artech House, Norwood, MA, 2001

5.8. Design Exercises

Design Problem #1 - Problem Statement:

Design a ground based radar that is capable of detecting aircraft and missiles at 10 Km and 2 Km altitudes, respectively. The maximum detection range for either target type is 60 Km. Assume that an aircraft average RCS is 6 dBsm, and that a missile average RCS is $0.01m^2$ (i.e., -10 dBsm). The radar azimuth and elevation search extents are respectively $\Theta_A = 360°$ and $\Theta_E = 10°$. The required scan rate is 2 seconds, and the range resolution is 150 meters. Assume a noise figure F = 8 dB, and total receiver noise L = 10 dB. Use a fan beam with azimuth beamwidth less than 3 degrees. The SNR is equal to15 dB.

Design Problem #1 - A Design

The range resolution requirement is $\Delta R = 150m$; thus by using

$$\Delta R = \frac{c\tau}{2}, \qquad \text{Eq. (5.98)}$$

we calculate the required pulsewidth to be $\tau = 1\mu sec$, or equivalently require the bandwidth $B = 1MHz$. The statement of the problem lends itself to radar sizing in terms of power aperture product. For this purpose, one must first compute the maximum search volume at the detection range that satisfies the design requirements. The radar search volume is

$$\Omega = \frac{\Theta_A \Theta_E}{(57.296)^2} = \frac{360 \times 10}{(57.296)^2} = 1.097 \ steradians. \qquad \text{Eq. (5.99)}$$

At this point, the designer is ready to use the radar search equation to compute the power aperture product. Use the parameters in Table 5.3 as inputs to the radar equation.

Table 5.3. Radar parameters.

Parameter	Units	Value
sensitivity SNR	*dB*	*15*
scan time	*seconds*	*2*
missile radar cross section	*dBsm*	*-10*
aircraft radar cross section	*dBsm*	*6*
missile detection range	*Km*	*60*
aircraft detection range	*Km*	*60*
effective temperature	*Kelvin*	*290*
noise figure	*dB*	*8*
radar losses	*dB*	*10*
search volume azimuth extent	*degrees*	*360*
search volume elevation extent	*degrees*	*10*

Figure 5.19 shows a plot of the power aperture versus detection range. Since the detection range is 60 Km, then we can read from this figure the corresponding power aperture product for both the missile and aircraft cases. They are,

$$\begin{aligned} PAP_{missile} &= 38.53dB \\ PAP_{aircraft} &= 22.53dB \end{aligned} . \qquad \text{Eq. (5.100)}$$

Choosing the more stressing case as the design baseline (i.e., selecting the power-aperture-product resulting from the missile analysis), yields

$$P_{av} \times A_e = 10^{3.853} = 7128.53 \Rightarrow A_e = \frac{7128.53}{Pav} . \qquad \text{Eq. (5.101)}$$

Choose $A_e = 1.75m^2$ to calculate the average power as,

$$P_{av} = \frac{7128.53}{1.75} = 4.073KW \qquad \text{Eq. (5.102)}$$

and assuming an aperture efficiency of $\rho = 0.8$, this yields the physical aperture area. More precisely,

$$A = \frac{A_e}{\rho} = \frac{1.75}{0.8} = 2.1875m^2 . \qquad \text{Eq. (5.103)}$$

Use $f_0 = 2.0GHz$ as the radar operating frequency. Then by using $A_e = 1.75m^2$ we calculate $G = 29.9dB$. Now one must determine the antenna azimuth beamwidth. The antenna gain is related to the antenna 3-dB beamwidth by

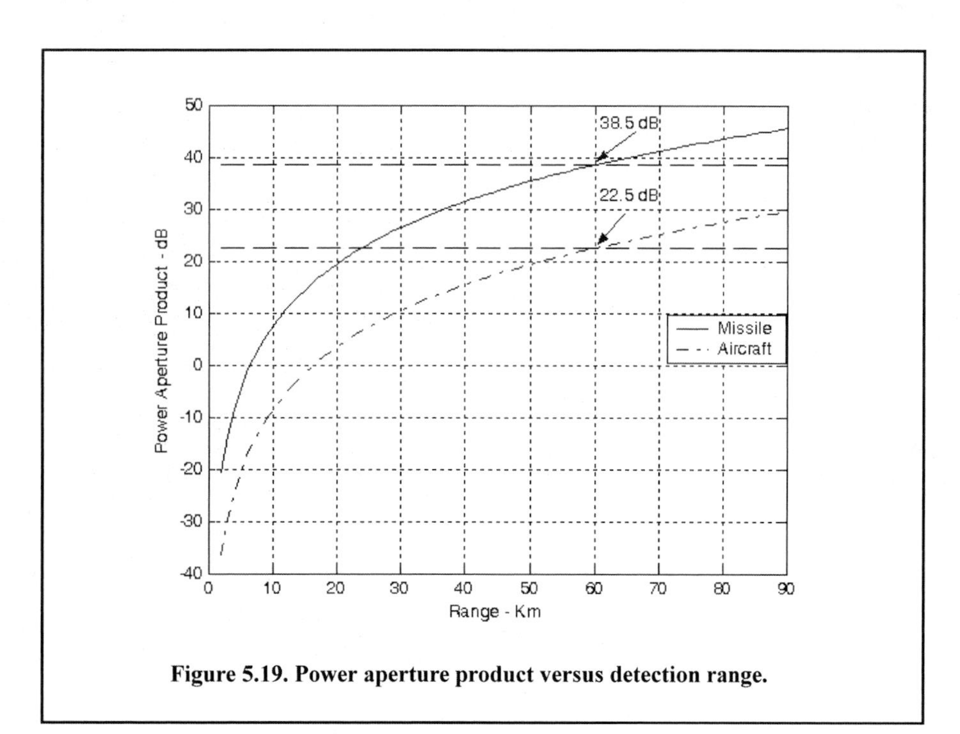

Figure 5.19. Power aperture product versus detection range.

$$G = k\frac{4\pi}{\theta_e\theta_a} \qquad \text{Eq. (5.104)}$$

where $k \le 1$ and depends on the physical aperture shape; the angles θ_e and θ_a are the antenna's elevation and azimuth beamwidths, respectively, in radians. An excellent approximation of Eq. (5.104) is

$$G \approx \frac{26000}{\theta_e\theta_a}. \qquad \text{Eq. (5.105)}$$

Assume a fan beam with $\theta_e = \Theta_E = 15°$. It follows that

$$\theta_a = \frac{26000}{\theta_e G} = \frac{26000}{10 \times 977.38} = 2.66° \Rightarrow \theta_a = 46.43 mrad. \qquad \text{Eq. (5.106)}$$

Design Problem #2 - Problem Statement

A MMW radar has the following specifications:

Center frequency $f = 94GHz$

Pulsewidth $\tau = 50 \times 10^{-9} seconds$

Peak power $P_t = 4W$

Azimuth coverage $\Delta\alpha = \pm 120°$

Pulse repetition frequency $PRF = 10KHz$

Noise figure $F = 7dB$

Antenna diameter $D = 12in$

Antenna gain $G = 47dB$

Radar cross section of target is $\sigma = 20m^2$

System losses $L = 10dB$

Radar scan time $T_{sc} = 3 seconds$.

Compute the following:

The wavelength λ

Range resolution ΔR

Bandwidth B

Antenna half power beamwidth

Antenna scan rate

Time on target

The range that corresponds to 10 dB SNR

Plot the SNR as a function of range.

Compute the number of pulses on the target that can be used for integration and the corresponding new detection range when pulse integration is used, assuming that the SNR stays unchanged.

Design Problem #2 - A Design

The wavelength λ is

$$\lambda = \frac{c}{f} = \frac{3 \times 10^8}{94 \times 10^9} = 0.00319m. \qquad \text{Eq. (5.107)}$$

The range resolution ΔR is

$$\Delta R = \frac{c\tau}{2} = \frac{(3 \times 10^8)(50 \times 10^{-9})}{2} = 7.5m. \qquad \text{Eq. (5.108)}$$

Radar operating bandwidth B is

$$B = \frac{1}{\tau} = \frac{1}{50 \times 10^{-9}} = 20MHz. \qquad \text{Eq. (5.109)}$$

The antenna 3-dB beamwidth is

$$\theta_{3dB} = 1.25\frac{\lambda}{D} = 0.7499°. \qquad \text{Eq. (5.110)}$$

Time on target is

$$T_i = \frac{\theta_{3dB}}{\dot{\theta}_{scan}} = \frac{0.7499°}{80°/\sec} = 9.38m\sec \quad . \qquad \text{Eq. (5.111)}$$

It follows that the number of pulses available for integration is

$$n_P = \frac{\theta_{3dB}}{\dot{\theta}_{scan}} f_r = 9.38 \times 10^{-3} \times 10 \times 10^3 \Rightarrow 94 \ \ pulses. \qquad \text{Eq. (5.112)}$$

Coherent Integration Case

From the radar we compute $R_{ref} = 2.245Km$. The SNR improvement due to coherently integrating 94 pulses is

$$I = 10 \times \log 94 = 19.73dB. \qquad \text{Eq. (5.113)}$$

However, since it is requested that the SNR remains at 10 dB, we can calculate the new detection range as

$$R_{CI}\big|_{n_P = 94} = 2.245 \times (94)^{1/4} = 6.99Km. \qquad \text{Eq. (5.114)}$$

Using the radar equation with $R = 6.99Km$ yields $SNR = -9.68dB$.

This means that using 94 pulses integrated coherently at 6.99 Km where each pulse has a SNR of -9.68 dB provides the same detection criteria as using a single pulse with $SNR = 10dB$ and $R = 2.245Km$. This is illustrated in Figure 5.20. Clearly, one can see the improvement of the detection range due to integration. For example, the detection range prior to integration when the $SNR = 5dB$ is about 3.5 Km. However, after the 94 pulse coherent integration, the same SNR value corresponds to a detection range of about 11 Km.

Non-coherent Integration Case

In this case, the integrated SNR is $(SNR)_{NCI} = 10dB$ and $n_P = 94$; it follows that

$$(SNR)_1 = \frac{10}{2 \times 94} + \sqrt{\frac{(10)^2}{4 \times 94^2} + \frac{10}{94}} = 0.38366 \Rightarrow -4.16dB . \qquad \text{Eq. (5.115)}$$

Therefore, the single pulse SNR when 94 pulses are integrated non-coherently is equal to $(SNR)_1 = -4.16dB$.

The integration loss L_{NCI} is

$$L_{NCI} = \frac{1 + 0.38366}{0.38366} = 3.6065 \Rightarrow 5.571dB . \qquad \text{Eq. (5.116)}$$

Therefore, the net non-coherent integration gain is

$$10 \times \log(94) - 5.571 = 14.16dB \Rightarrow 26.06422 , \qquad \text{Eq. (5.117)}$$

and consequently, the maximum detection range is

$$R_{NCI}\Big|_{n_p = 94} = 2.245 \times (26.06422)^{1/4} = 5.073Km . \qquad \text{Eq. (5.118)}$$

This means that using 94 pulses integrated non-coherently at 5.073 Km where the single pulse has $(SNR)_1 = -4.16dB$ provides the same detection criterion as using a single pulse with $(SNR)_1 = 10dB$ at $R = 2.245Km$. This is illustrated in Figure 5.21.

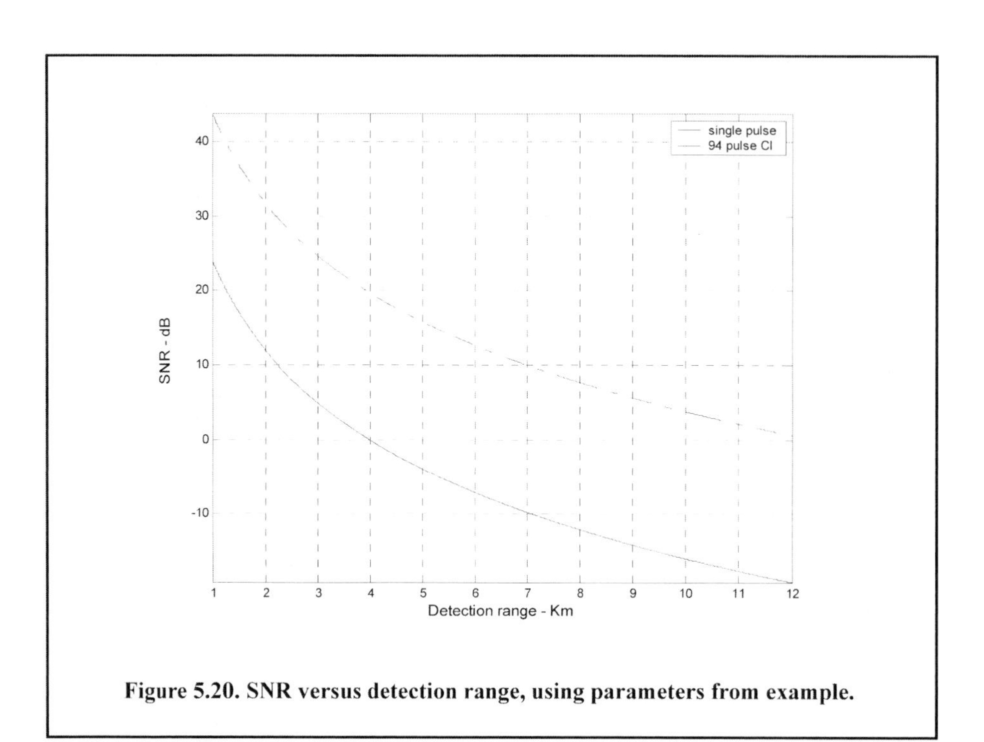

Figure 5.20. SNR versus detection range, using parameters from example.

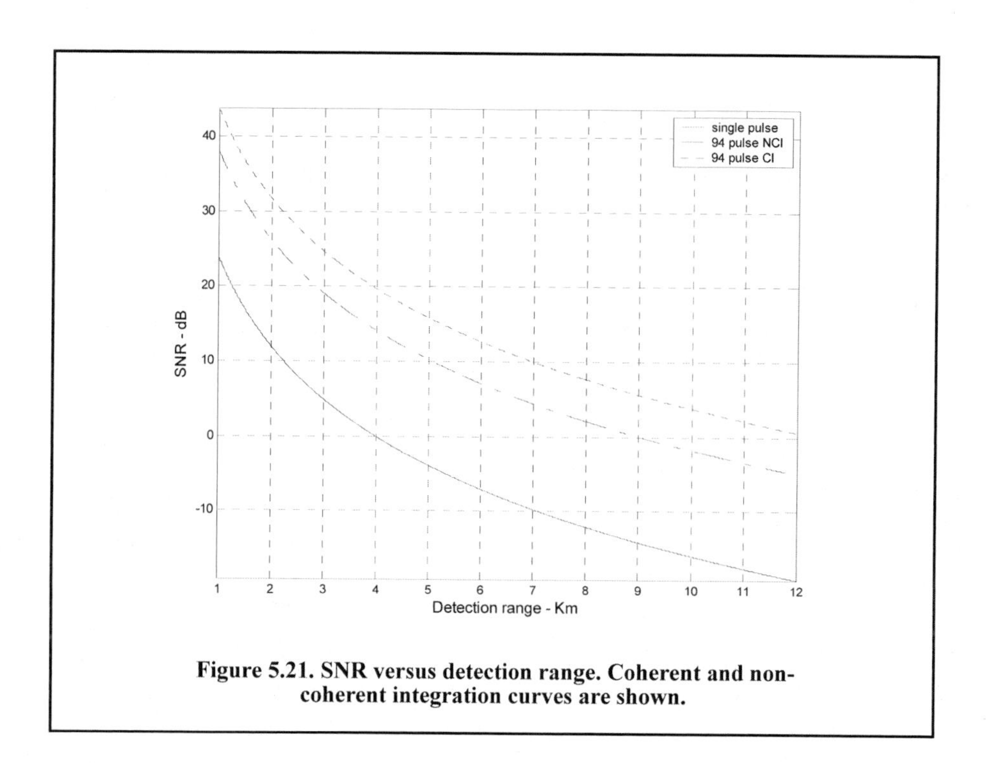

Figure 5.21. SNR versus detection range. Coherent and non-coherent integration curves are shown.

Problems

5.1. Compute the aperture size for an X-band antenna at $f_0 = 9GHz$. Assume antenna gain $G = 10, 20, 30dB$.

5.2. For the radar defined in Problem 4 (see Chapter 4), assume a duty cycle of 30% and peak power of $5KW$. Compute the average power and the amount of radiated energy during the first $20ms$.

5.3. An L-band radar ($1500MHz$) uses an antenna whose gain is $G = 30dB$. Compute the aperture size. If the radar duty cycle is $d_t = 0.2$ and the average power is $25KW$, compute the power density at range $R = 50Km$.

5.4. For the radar described in Problem 5.2, assume the minimum detectable signal is $5dBm$. Compute the radar maximum range for $\sigma = 1.0, 10.0, 20.0m^2$.

5.5. Consider an L-band radar with the following specifications: operating frequency $f_0 = 1500MHz$, bandwidth $B = 5MHz$, and antenna gain $G = 5000$. Compute the peak power, the pulse width, and the minimum detectable signal for this radar. Assume target RCS $\sigma = 10m^2$, the single pulse SNR is $15.4dB$, noise figure $F = 5dB$, temperature $T_0 = 290K$, and maximum range $R_{max} = 150Km$.

5.6. Repeat the example in Section 5.1 with $P_t = 1MW$, $G = 40dB$, and $\sigma = 0.5m^2$.

5.7. Starting with Eq. (3.56) and assuming a circular aperture show that Eq. (5.36) can be written as

$$SNR = \frac{P_{av}A_e\sigma}{16kT_oFLR^4}\frac{T_{sc}}{\Omega}.$$

Hint: Use $A = (\pi D^2)/4$, where D is the diameter.

5.8. Show that the DC component is the dominant spectral line for high PRF waveforms.

5.9. Repeat the example in Section 5.3 with $L = 5dB$, $F = 10dB$, $T = 500K$, $T_i = 1.5s$, $d_t = 0.25$, and $R = 75Km$.

5.10. Consider a low PRF C-band radar operating at $f_0 = 5000MHz$. The antenna has a circular aperture with radius $2m$. The peak power is $P_t = 1MW$ and the pulse width is $\tau = 2\mu s$. The PRF is $f_r = 250Hz$, and the effective temperature is $T_0 = 600K$. Assume radar losses $L = 15dB$ and target RCS $\sigma = 10m^2$. (a) Calculate the radar's unambiguous range; (b) calculate the range R_0 that corresponds to $SNR = 0dB$; (c) calculate the SNR at $R = 0.75R_0$.

5.11. Let the maximum unambiguous range for a low PRF radar be R_{max}. (a) Calculate the SNR at $(1/2)R_{max}$ and $(3/4)R_{max}$. (b) If a target with $\sigma = 10m^2$ exists at $R = (1/2)R_{max}$, what should the target RCS be at $R = (3/4)R_{max}$ so that the radar has the same signal strength from both targets?

5.12. A Millie-Meter Wave (MMW) radar has the following specifications: operating frequency $f_0 = 94GHz$, PRF $f_r = 15KHz$, pulse width $\tau = 0.05ms$, peak power $P_t = 10W$, noise figure $F = 5dB$, circular antenna with diameter $D = 0.254m$, antenna gain $G = 30dB$, target RCS $\sigma = 1m^2$, system losses $L = 8dB$, radar scan time $T_{sc} = 3s$, radar angular coverage $200°$, and atmospheric attenuation $3dB/Km$. Compute the following: (a) wavelength λ; (b) range resolution ΔR; (c) bandwidth B; (d) the SNR as a function of range; (e) the range for which $SNR = 15dB$; (f) antenna beam width; (g) antenna scan rate; (h) time on target; (i) the effective maximum range when atmospheric attenuation is considered.

5.13. Repeat the second example in Section 5.4 with $\Omega = 4°$, $\sigma = 1m^2$, and $R = 400Km$.

5.14. Compute (as a function of B_J/B) the cross-over range for the radar in Problem 2.11. Assume $P_J = 100W$, $G_J = 10dB$, and $L_J = 2dB$.

5.15. Compute (as a function of B_J/B) the cross-over range for the radar in Problem 2.11. Assume $P_J = 200W$, $G_J = 15dB$, and $L_J = 2dB$. Assume $G' = 12dB$ and $R_J = 25Km$.

5.16. Using Figure 5.6 derive an expression for R_r. Assume 100% synchronization between the transmitter and receiver.

5.17. A certain radar is subject to interference from an SSJ jammer. Assume the following parameters: radar peak power $P_t = 55KW$, radar antenna gain $G = 30dB$, radar pulse width $\tau = 2\mu s$, radar losses $L = 10dB$, jammer power $P_J = 150W$, jammer antenna gain $G_J = 12dB$, jammer bandwidth $B_J = 50MHz$, and jammer losses $L_J = 1dB$. Compute the cross-over range for a $5m^2$ target.

5.18. A certain radar has losses of $6dB$ and a receiver noise figure of $8dB$. It has the requirement to detect targets within a search sector that is 360 degrees in azimuth and from 5 to 65 degrees in elevation. It must cover the search sector in 2 seconds. The RCS of the targets of interest is $5dBsm$, and the radar requires $20dB$ of signal-to-noise ratio to declare a detection. The required detection range of the radar is $75Km$. What is the average power aperture that the radar must have to satisfy the above search requirements

5.19. A radar with antenna gain G is subject to a repeater jammer whose antenna gain is G_J. The repeater illuminates the radar with three fourths of the incident power on the jammer. (a) Find an expression for the ratio between the power received by the jammer and the power received by the radar; (b) what is this ratio when $G = G_J = 200$ and $R/\lambda = 10^5$?

5.20. An X-band airborne radar transmitter and an air-to-air missile receiver act as a bistatic radar system. The transmitter guides the missile toward its target by continuously illuminating the target with a CW signal. The transmitter has the following specifications: peak power $P_t = 4KW$; antenna gain $G_t = 25dB$; operating frequency $f_0 = 9.5GHz$. The missile receiver has the following characteristics: aperture $A_r = 0.01m^2$; bandwidth $B = 750Hz$; noise figure $F = 7dB$; and losses $L_r = 2dB$. Assume that the bistatic RCS is $\sigma_B = 3m^2$. Assume $R_r = 35Km$; $R_t = 17Km$. Compute the SNR at the missile.

5.21. Repeat the previous problem when there is $0.1dB/Km$ atmospheric attenuation.

5.22. Consider an antenna with a $\sin x/x$ pattern. Let $x = (\pi r \sin\theta)/\lambda$, where r is the antenna radius, λ is the wavelength, and θ is the off-boresight angle. Derive Eq. (5.83). Hint: Assume small x, and expand $\sin x/x$ as an infinite series.

5.23. Compute the amount of antenna pattern loss for a phased array antenna whose two-way pattern is approximated by $f(y) = [\exp(-2\ln 2(y/\theta_{3dB})^2)]^4$ where θ_{3dB} is the $3dB$ beam width. Assume circular symmetry.

5.24. A certain radar has a range gate size of $30m$. Due to range gate straddle, the envelope of a received pulse can be approximated by a triangular spread over three range bins. A target is detected in range bin 90. You need to find the exact target position with respect to the center of the range cell. (a) Develop an algorithm to determine the position of a target with respect to the center of the cell; (b) assuming that the early, on, and late measurements are, respectively, equal to $4/6$, $5/6$, and $1/6$, compute the exact target position.

5.25. Compute the amount of Doppler filter straddle loss for the filter defined by

$$H(f) = \frac{1}{1 + a^2 f^2} .$$

Assume half-power frequency $f_{3dB} = 500Hz$ and cross-over frequency $f_c = 350Hz$.

5.26. A radar has the following parameters: Peak power $P_t = 65KW$; total losses $L = 5dB$; operating frequency $f_o = 8GHz$; PRF $f_r = 4KHz$; duty cycle $d_t = 0.3$; circular antenna with diameter $D = 1m$; effective aperture is 0.7 of physical aperture; noise figure $F = 8dB$. (a) Derive the various parameters needed in the radar equation; (b) What is the unambiguous range? (c) Plot the SNR versus range ($1Km$ to the radar unambiguous range) for a $5dBsm$ target. (d) If the minimum SNR required for detection is $14dB$, what is the detection range for a $6dBsm$ target? What is the detection range if the SNR threshold requirement is raised to $18dB$?

5.27. A radar has the following parameters: Peak power $P_t = 50KW$; total losses $L = 5dB$; operating frequency $f_o = 5.6GHz$; noise figure $F = 10dB$ pulse width $\tau = 10\mu s$; PRF $f_r = 2KHz$; antenna beamwidth $\theta_{az} = 1°$ and $\theta_{el} = 5°$. (a) What is the antenna gain? (b) What is the effective aperture if the aperture efficiency is 60%? (c) Given a 14 dB threshold detection, what is the detection range for a target whose RCS is $\sigma = 1m^2$?

5.28. A monostatic radar has the following parameters: Transmit power $100Kw$, transmit losses $2dB$, operating Frequency $7GHz$, PRF $2000Hz$, pulse width 10μsec, antenna beamwidth 2° Az X 4° El, receive losses $3dB$, and receiver noise figure $12dB$. Assume that the radar uses pulses that employ $10MHz$ of linear frequency modulation and uses a processor that is matched to the transmitted pulse. (a) What is the antenna gain? (b) What is the effective aperture if the aperture efficiency is 50%? (c) What is the effective radiated power of the radar, in dBm? (d) Given a detection threshold of $13dB$, what is the detection range for a target with a radar cross section of $6dBsm$?

5.29. The following table is constructed from a radar cross section measurement experiment. Calculate the mean and standard deviation of the radar cross section.

Number of samples	RCS, m^2
2	55
6	67
12	73
16	90
20	98
24	110
26	117
19	126
13	133
8	139
5	144
3	150

5.30. A certain radar has losses of $5dB$ and a receiver noise figure of $10dB$. This radar has a detection coverage requirement that extends over 3/4 of a hemisphere and must complete it in 3 seconds. The base line target RCS is $6dBsm$ and the minimum SNR is $15dB$. The radar detection range is less than $80Km$. What is the average power aperture product for this radar so that it can satisfy its mission?

5.31. A radar generates $100KW$ of power and has $1dB$ of loss between the power tube and the antenna. The radar is monostatic with a single antenna that has a gain of $38dB$. The radar is operating at $5GHz$. What is the power at the receiver antenna output for the following targets:

(a) A $1m^2$ RCS target at a range of $30Km$. (b) A $10dBsm$ target at a range of 50 Km.

Assume that the total radar losses of $1dB$.

5.32. A source with equivalent temperature $T_o = 290K$ is followed by three amplifiers with specifications shown in the table below.

Amplifier	**F, dB**	**G, dB**	T_e
1	You must compute	12	350
2	10	22	
3	15	35	

(a) Compute the noise figure for the three cascaded amplifiers. (b) Compute the effective temperature for the three cascaded amplifiers. (c) Compute the overall system noise figure.

5.33. A radar has the following receiver components. They are arranged in the order shown below

Receiver Stages			
Stage #	**Component**	**Gain, dB**	**Noise Figure, dB**
1	Waveguide	-2	2
2	RF Amp	28	5
3	1st Mixer	-3	15
4	IF Amp	100	30

(a) What is the receiver noise figure through the RF amp and referenced to the input of the waveguide (the first component after the antenna)? (b) What is the noise figure of the receiver through the IF amp and referenced to the input of the RF amp? (c) What is the effective noise temperature of the receiver through the IF amp and referenced to the input of the waveguide? (d) Suppose you want to determine how internal noise and sky noise contribute to noise power at various points in the receiver. Specifically, how does the noise power at the output of each component varies as a function of the effective noise temperature of the antenna, T_{ant}, and noise bandwidth, B? Derive four equations that will allow us to easily perform the computations. All of your equations should be of the form $P = B(K_1 T_{ant} + K_2)$ where K_1 and K_2 are constants. Provide a table with the four sets of values for K_1 and K_2.

Part II

Frequency Modulation Radar Systems

Chapter 6

Continuous Wave and Pulsed Radars

6.1. Continuous Wave (CW) Radar

Continuous Wave (CW) radars utilize CW waveforms, which may be considered to be a pure sinewave of the form $\cos 2\pi f_0 t$. Note that in the earlier chapters we used the notation f_c instead of f_0; to this end and for the rest of this book, we will use both notations interchangeably. Spectra of the radar echoes from stationary targets and clutter will be concentrated at f_0. The center frequency for the echoes from a moving target will be shifted by f_d, the Doppler frequency. Thus by measuring this frequency difference CW radars can very accurately extract target radial velocity. Because of the continuous nature of CW emission, range measurement is not possible without some modifications to the radar operations and waveforms.

6.1.1. Functional Block Diagram

In order to ensure a continuous radar energy emission, two antennas are used in CW radars, one for transmission and one for reception. Figure 6.1 shows a simplified CW radar block diagram. The appropriate values of the signal frequency at different locations are noted on the diagram. The individual Narrow Band Filters (NBF) must be as narrow as possible in bandwidth in order to allow accurate Doppler measurements and minimize the amount of noise power. In theory, the bandwidth of a CW radar is infinitesimal (since it corresponds to an infinite duration continuous sinewave). However, systems with infinitesimal bandwidths cannot physically exist, and thus the bandwidth of CW radars is assumed to correspond to that of a gated CW waveform.

The NBF bank can be implemented using a fast Fourier transform (FFT). If the Doppler filter bank is implemented using an FFT of size N_{FFT}, and if the individual NBF bandwidth is Δf, then the effective radar Doppler bandwidth is $(N_{FFT}\Delta f)/2$. The reason for the one-half factor is to account for both negative and positive Doppler shifts. Since range is computed from the radar echoes by measuring a two-way time delay, then single frequency CW radars cannot measure target range. In order for CW radars to be able to measure target range, the transmit and receive waveforms must have some sort of timing marks. By comparing the timing marks at transmit and receive, CW radars can extract target range. The timing mark can be implemented by modulating the transmit waveform, and one commonly used technique is Linear Frequency Modulation (LFM).

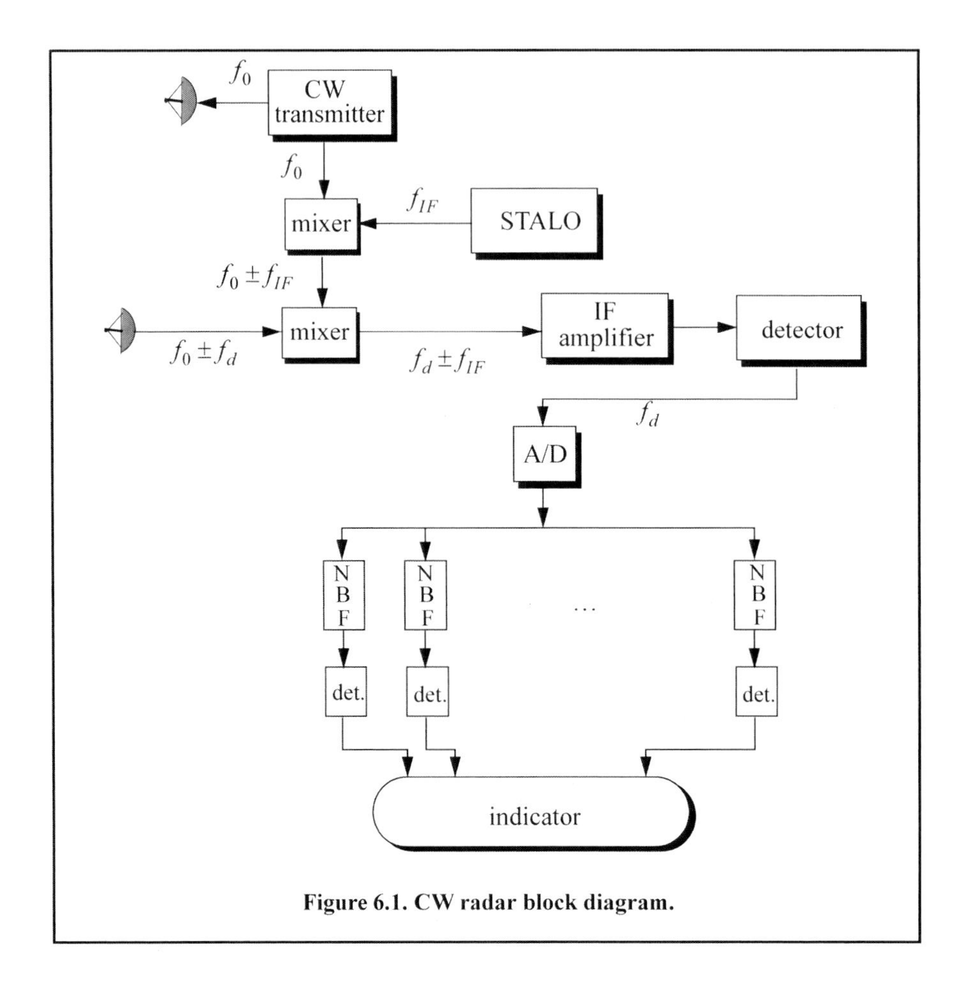

Figure 6.1. CW radar block diagram.

6.1.2. Linear FM (LFM) CW Radar

In Chapter 3 the general formula for an FM waveform was introduced; for convenience, it is repeated here as Eq. (6.1)

$$x(t) = A\cos\left(2\pi f_0 t + k_f\int_0^t \cos 2\pi f_m u du\right), \qquad \text{Eq. (6.1)}$$

where f_0 is the radar operating frequency, $\cos 2\pi f_m t$ is the modulating signal, A is a constant, and $k_f = 2\pi\Delta f_{peak}$, where Δf_{peak} is the peak frequency deviation. Let $x_r(t)$ be the received radar signal from a target at range R. It follows that

$$x_r(t) = A_r\cos(2\pi f_0(t-\Delta t) + \beta\sin 2\pi f_m(t-\Delta t)) \qquad \text{Eq. (6.2)}$$

where the delay Δt is

$$\Delta t = (2R)/c \qquad \text{Eq. (6.3)}$$

and c is the speed of light. CW radar receivers utilize phase detectors in order to extract target range from the instantaneous frequency, as illustrated by Fig. 6.2. A good measurement of the phase detector output implies a good measurement of Δt, and hence range.

CW radars may use LFM waveforms so that both range and Doppler information can be measured. In practical CW radars, the LFM waveform cannot be continually changed in one direction, and thus periodicity in the modulation is normally utilized. Figure 6.3 shows a sketch of a triangular LFM waveform. The modulation does not need to be triangular; it may be sinusoidal, saw-tooth, or some other form. The dashed line in Fig. 6.3 represents the return waveform from a stationary target at range R. The beat frequency f_b is also sketched in Fig. 6.3. It is defined as the difference (due to heterodyning) between the transmitted and received signals. The time delay Δt is a measure of target range, as defined in Eq. (6.3).

In practice, the modulating frequency f_m is selected such that

$$f_m = \frac{1}{2t_0}. \qquad \text{Eq. (6.4)}$$

The rate of frequency change $\dot{f}$, is

$$\dot{f} = \frac{\Delta f}{t_0} = \frac{\Delta f}{(1/2f_m)} = 2f_m\Delta f \qquad \text{Eq. (6.5)}$$

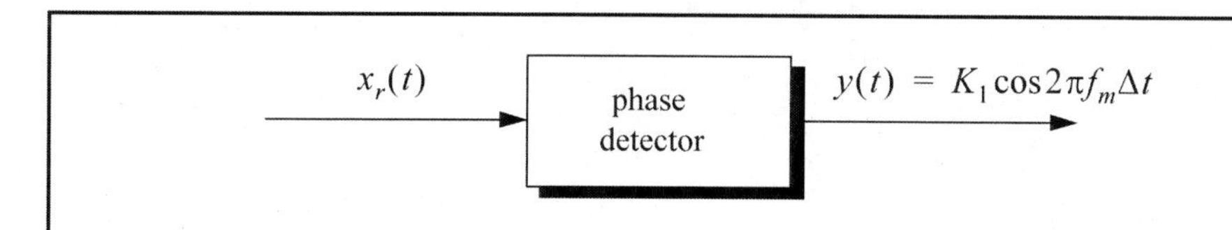

Figure 6.2. Extracting range from an FM signal return. K_1 is a constant.

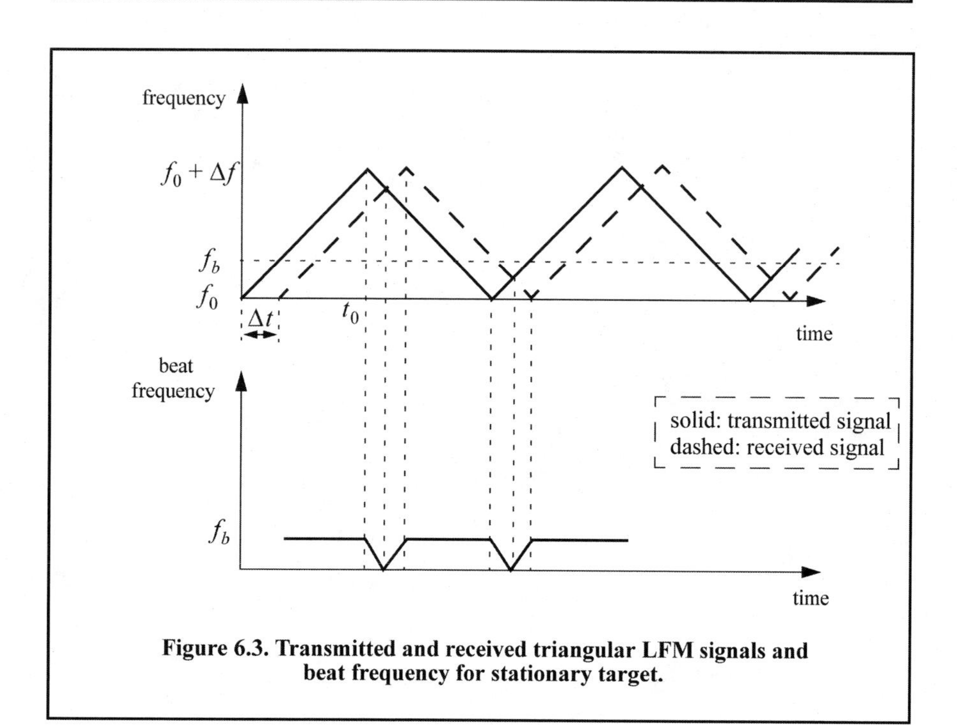

Figure 6.3. Transmitted and received triangular LFM signals and beat frequency for stationary target.

where Δf is the peak frequency deviation. The beat frequency f_b is given by

$$f_b = \Delta t\dot{f} = \frac{2R}{c}\dot{f}. \qquad \text{Eq. (6.6)}$$

Eq. (6.6) can be rewritten as

$$\dot{f} = \frac{c}{2R}f_b. \qquad \text{Eq. (6.7)}$$

Equating Eqs. (6.5) and (6.7) and solving for f_b yield,

$$f_b = (4Rf_m\Delta f)/c. \qquad \text{Eq. (6.8)}$$

Now consider the case when Doppler is present (i.e., non-stationary target). The corresponding triangular LFM transmitted and received waveforms are sketched in Fig. 6.4, along with the corresponding beat frequency. The beat frequency is defined as

$$f_b = f_{received} - f_{transmitted}. \qquad \text{Eq. (6.9)}$$

When the target is not stationary, the received signal will contain a Doppler shift term in addition to the frequency shift due to the time delay Δt. In this case, the Doppler shift term subtracts from the beat frequency during the positive portion of the slope. Alternatively, the two terms will add during the negative portion of the slope. Denote the beat frequency during the positive (up) and negative (down) portions of the slope respectively as f_{bu} and f_{bd}. Then we have

$$f_{bu} = \frac{2R}{c}\dot{f} - \frac{2\dot{R}}{\lambda} \qquad \text{Eq. (6.10)}$$

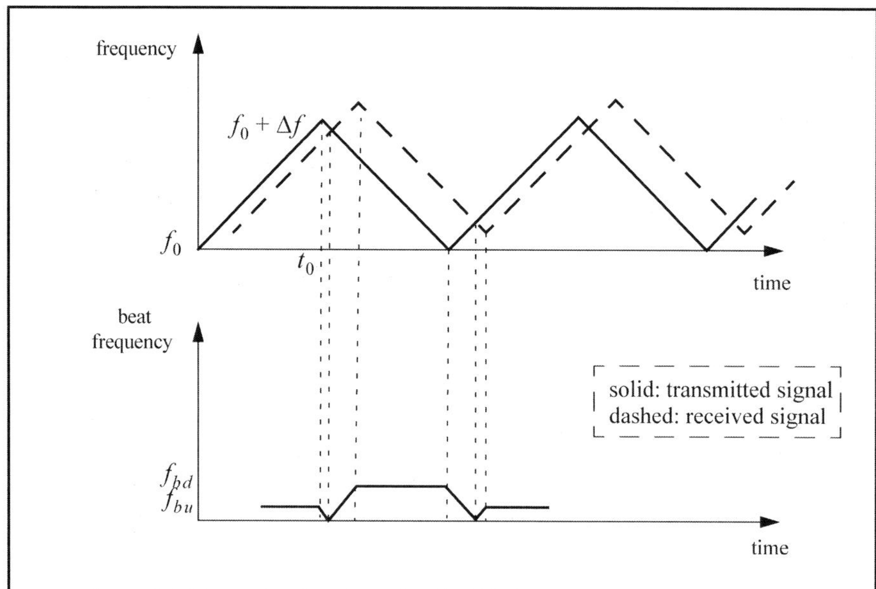

Figure 6.4. Transmitted and received LFM signals and beat frequency, for a moving target.

where $\dot{R}$ is the range rate or the target radial velocity as seen by the radar. The first term of the right-hand side of Eq. (6.10) is due to the range delay defined by Eq. (6.8), while the second term is due to the target Doppler. Similarly, we have

$$f_{bd} = \frac{2R}{c}\dot{f} + \frac{2\dot{R}}{\lambda}. \qquad \text{Eq. (6.11)}$$

Range is computed by adding Eq. (6.10) and Eq. (6.11). More precisely,

$$R = \frac{c}{4\dot{f}}(f_{bu} + f_{bd}). \qquad \text{Eq. (6.12)}$$

The range rate is computed by subtracting Eq. (6.10) from Eq. (6.11),

$$\dot{R} = \frac{\lambda}{4}(f_{bd} - f_{bu}). \qquad \text{Eq. (6.13)}$$

As indicated by Eq. (6.12) and Eq. (6.13), CW radars utilizing triangular LFM can extract both range and range rate information. In practice, the maximum time delay Δt_{max} is normally selected as

$$\Delta t_{max} = 0.1 t_0. \qquad \text{Eq. (6.14)}$$

Thus, the maximum range is given by

$$R_{max} = \frac{0.1 c t_0}{2} = \frac{0.1 c}{4 f_m} \qquad \text{Eq. (6.15)}$$

and the maximum unambiguous range will correspond to a shift equal to $2t_0$.

6.1.3. Multiple Frequency CW Radar

CW radars do not have to use LFM waveforms in order to obtain good range measurements. Multiple frequency schemes allow CW radars to compute very adequate range measurements, without using frequency modulation. In order to illustrate this concept, first consider a CW radar with the following waveform

$$x(t) = A \sin 2\pi f_0 t. \qquad \text{Eq. (6.16)}$$

The received signal from a target at range R is

$$x_r(t) = A_r \sin(2\pi f_0 t - \varphi) \qquad \text{Eq. (6.17)}$$

where the phase φ is equal to

$$\varphi = 2\pi f_0 \frac{2R}{c}. \qquad \text{Eq. (6.18)}$$

Solving for R we obtain

$$R = \frac{c\varphi}{4\pi f_0} = \frac{\lambda}{4\pi}\varphi. \qquad \text{Eq. (6.19)}$$

Clearly, the maximum unambiguous range occurs when φ is maximum, i.e., $\varphi = 2\pi$. Therefore, even for relatively large radar wavelengths, R is limited to impractical small values.

Next, consider a radar with two CW signals, denoted by $x_1(t)$ and $x_2(t)$. More precisely,

$$x_1(t) = A_1 \sin 2\pi f_1 t \qquad \text{Eq. (6.20)}$$

$$x_2(t) = A_2 \sin 2\pi f_2 t . \qquad \text{Eq. (6.21)}$$

The received signals from a moving target are

$$x_{1r}(t) = A_{r1} \sin(2\pi f_1 t - \varphi_1) \qquad \text{Eq. (6.22)}$$

and

$$x_{2r}(t) = A_{r2} \sin(2\pi f_2 t - \varphi_2) \qquad \text{Eq. (6.23)}$$

where $\varphi_1 = (4\pi f_1 R)/c$ and $\varphi_2 = (4\pi f_2 R)/c$. After heterodyning (mixing) with the carrier frequency, the phase difference between the two received signals is

$$\varphi_2 - \varphi_1 = \Delta\varphi = \frac{4\pi R}{c}(f_2 - f_1) = \frac{4\pi R}{c}\Delta f . \qquad \text{Eq. (6.24)}$$

Again R is maximum when $\Delta\varphi = 2\pi$; it follows that the maximum unambiguous range is now

$$R = \frac{c}{2\Delta f} . \qquad \text{Eq. (6.25)}$$

Since $\Delta f \ll c$, the range in Eq. (6.25) is much greater than that computed by Eq. (6.19).

6.1.4. Effect of FM Transmitter Noise on the Receiver

An approximation for a narrow band FM signal with sinusoidal modulation is

$$x(t) \approx A\left\{ \cos 2\pi f_0 t + \frac{1}{2}\beta[\cos(2\pi f_0 + 2\pi f_m)t - \cos(2\pi f_0 - 2\pi f_m)t] \right\} . \qquad \text{Eq. (6.26)}$$

The ratio of the transmitter noise power in the single sideband to the carrier power is then

$$\frac{P_{xx}}{P_t} = \frac{J_1^2(\beta)}{J_0^2(\beta)} \approx \left(\frac{1}{2}\beta\right)^2 = \frac{\Delta f_{peak}^2}{4 f_m^2} . \qquad \text{Eq. (6.27)}$$

In order to consider the noise power at the output of the Intermediate Frequency (IF) mixer in the receiver, assume the time reference to be when the signal strikes the target. Then, at the transmit time, the signal is

$$x(t) = A\cos\left[2\pi f_0\left(t - \frac{\Delta t}{2}\right) + \beta \sin 2\pi f_m\left(t - \frac{\Delta t}{2}\right)\right] \qquad \text{Eq. (6.28)}$$

where $\Delta t = 2R/c$ is the two-way propagation time, and A is a constant. Assuming a coherently derived Local Oscillator (LO), then its signal at the reception time can be expressed as

$$s_{LO}(t) = B\cos\left[2\pi f_L\left(t + \frac{\Delta t}{2}\right) + \beta \sin 2\pi f_m\left(t + \frac{\Delta t}{2}\right)\right] \qquad \text{Eq. (6.29)}$$

where B is a constant. The IF voltage signal is then equal to the sum of the following two signals

$$s_{IF1}(t) = D\cos\left[2\pi f_0\left(t-\frac{\Delta t}{2}\right) + \beta\sin 2\pi f_m\left(t-\frac{\Delta t}{2}\right) + 2\pi f_L\left(t+\frac{\Delta t}{2}\right) + \left(\beta\sin 2\pi f_m\left(t+\frac{\Delta t}{2}\right)\right)\right] \quad \text{Eq. (6.30)}$$

$$s_{IF2}(t) = D\cos\left[2\pi f_i t + \psi_1 + \beta\left\{\sin 2\pi f_m\left(t+\frac{\Delta t}{2}\right) - \sin 2\pi f_m\left(t-\frac{\Delta t}{2}\right)\right\}\right] \quad \text{Eq. (6.31)}$$

where D is a constant and

$$f_i = f_L - f_0$$
$$\phi_1 = \frac{\Delta t}{2}(2\pi f_L - 2\pi f_0) \quad \text{Eq. (6.32)}$$

Finally by using the identity $2\cos\theta\sin\vartheta = \sin(\theta+\vartheta) - \sin(\theta-\vartheta)$ we get the output of the IF filter as (mix down)

$$s_{IF2}(t) = s_{IF}(t) = D\cos\left[2\pi f_i t + \phi_1 + \left(2\beta\sin 2\pi f_m\frac{\Delta t}{2}\right)\cos 2\pi f_m t\right] \quad \text{Eq. (6.33)}$$

and by applying the identity $\cos(\theta+\vartheta) = \cos\theta\cos\vartheta - \sin\theta\sin\vartheta$, we obtain

$$s_{IF}(t) = D\left[\cos(2\pi f_i t + \phi_1)\cos\left(\left\{\left(2\beta\sin 2\pi f_m\frac{\Delta t}{2}\right)\cos 2\pi f_m t\right\}\right. - \sin(2\pi f_i t + \phi_1)\sin\left\{\left(2\beta\sin 2\pi f_m\frac{\Delta t}{2}\right)\cos 2\pi f_m t\right\}\right)\right] \quad \text{Eq. (6.34)}$$

However,

$$\cos\left\{\left(2\beta\sin 2\pi f_m\frac{\Delta t}{2}\right)\cos 2\pi f_m t\right\} = J_0(2\beta\sin\pi f_m\Delta t) \ , + 2J_2(2\beta\sin\pi f_m\Delta t)\sin 4\pi f_m t + \ldots \quad \text{Eq. (6.35)}$$

and

$$\sin\left\{\left(2\beta\sin 2\pi f_m\frac{\Delta t}{2}\right)\cos 2\pi f_m t\right\} = 2J_1(2\beta\sin\pi f_m\Delta t)\cos 2\pi f_m t + 2J_3(2\beta\sin\pi f_m\Delta t)\cos 6\pi f_m t + \ldots \quad \text{Eq. (6.36)}$$

Substituting Eq. (6.35) and Eq. (6.36) into Eq. (6.34), collecting terms and using trigonometric identities, we get

$$s_{IF}(t) = D[J_0(2\beta\sin\pi f_m\Delta t)\cos(2\pi f_i t + \psi_1) - J_1(2\beta\sin\pi f_m\Delta t)\{\sin((2\pi f_i + 2\pi f_m)t + \psi_1) + \sin((2\pi f_i - 2\pi f_m)t + \psi_1)\} + \ldots] \quad \text{Eq. (6.37)}$$

The peak voltage in either of the first side lobes of an FM signal mixed with itself is

$$V_{P1} = J_1(2\beta \sin\pi f_m \Delta t) \quad \text{Eq. (6.38)}$$

and the peak voltage at IF is

$$V_{P_{IF}} = J_0(2\beta \sin\pi f_m \Delta t) \,. \quad \text{Eq. (6.39)}$$

Using the small argument approximation (i.e., $\beta \le 0.5$) we get

$$J_1(2\beta \sin\pi f_m \Delta t) \approx \beta \sin\pi f_m \Delta t \quad \text{Eq. (6.40)}$$

$$J_0(2\beta \sin\pi f_m \Delta t) \approx 1 \,. \quad \text{Eq. (6.41)}$$

Then the power ratio of the single sideband at IF is

$$\frac{P_{xx}}{P_t} = (\beta \sin\pi f_m \Delta t)^2 = \left(\frac{\Delta f_{peak}}{f_m} \sin\pi f_m \Delta t\right)^2 \,. \quad \text{Eq. (6.42)}$$

This is an expression for the power in the sideband after mixing the received signals, that is, the noise power ratio of the receiver and coherent LO due to the FM noise and on the transmitter and a coherent LO. For small angle θ, we have $\sin(\theta \approx \theta)$. It follows that

$$\frac{P_{xx}}{P_t} \approx (\pi \Delta f_{peak} \Delta t)^2 \,. \quad \text{Eq. (6.43)}$$

For low FM side lobes, Eq. (6.43) is also an expression for the ratio of clutter power in the side lobe to the total clutter power received by the radar. Therefore, the clutter power ratio in the FM single sideband can be expressed as

$$P_{sc} = (\pi \Delta f_{peak} \Delta t)^2 P_C \quad \text{Eq. (6.44)}$$

where P_C is the total clutter power, and P_{sc} is the clutter power in a single sideband. Solving for Δf^2 we get

$$\Delta f^2 = \frac{P_{sc}}{P_C}\left(\frac{1}{\pi \Delta t}\right)^2 \,. \quad \text{Eq. (6.45)}$$

Substituting Eq. (6.45) into Eq. (6.27) yields

$$\frac{P_{xx}}{P_t} = \frac{1}{4}\frac{P_{sc}}{P_c}\left(\frac{1}{\pi \Delta t f_m}\right)^2 \,. \quad \text{Eq. (6.46)}$$

Equation (6.46) can be used to determine the FM side lobe power requirements for the transmitter, by letting the power in the received FM side lobe P_s be below the receiver noise power level by (typically) $10dB$. It follows that

$$\frac{P_{xx}}{P_t} = \frac{1}{4}\left(\frac{kTBF}{10P_c}\right)\left(\frac{1}{\pi \Delta t f_m}\right)^2 \quad \text{Eq. (6.47)}$$

where B is the receiver detection (final) bandwidth, and F is the noise figure.

Example:

Compute the power ratio in Eq. (6.47) when $B = 800Hz$, $P_C = -45dB$, $f_m = 1KHz$, *and* $R = 1785m$. *Assume* $T = 300 degreesK$.

Solution:

The time delay Δt *is*

$$\Delta t = \frac{2R}{c} = \frac{2 \times 1785}{3 \times 10^8} = -49.2dB$$

$$(P_s)_{dB} = (kTB + F)_{db} - 10$$

The following table has the rest of the parameters used in Eq. (6.47).

$1/4$	P_s	$1/\pi^2$	$1/\Delta t^2$	$1/f_m^2$	$1/P_C$
$-6dB$	$-148dB$	$-10dB$	$98dB$	$-60dB$	$45dB$

It follows that the power ratio is

$$\frac{P_{xx}}{P_t} = -81dB.$$

****

6.2. Pulsed Radar

Pulsed radars transmit and receive a train of amplitude modulated pulses. Range is extracted from the two-way time delay between a transmitted and received pulse. Doppler measurements can be made in two ways. If accurate range measurements are available between consecutive pulses, then Doppler frequency can be extracted from the range rate $\dot{R} = \Delta R/\Delta t$. This approach works fine as long as the range is not changing drastically over the interval Δt. Otherwise, pulsed radars utilize a Doppler filter bank.

Pulsed radar waveforms can be completely defined by the following:

1. Carrier frequency, which may vary depending on the design requirements and radar mission
2. Pulse width, which is closely related to the bandwidth and defines the range resolution
3. Modulation
4. Pulse repetition frequency (PRF). Different modulation techniques are usually utilized to enhance the radar performance, or to add more capabilities to the radar that otherwise would not have been possible. The PRF must be chosen to avoid Doppler and range ambiguities as well as maximize the average transmitted power.

Radar systems employ low, medium, and high PRF schemes. Low PRF waveforms can provide accurate, long, unambiguous range measurements, but exert severe Doppler ambiguities. Medium PRF waveforms must resolve both range and Doppler ambiguities; however, they provide adequate average transmitted power as compared to low PRFs. High PRF waveforms can provide superior average transmitted power and excellent clutter rejection capabilities. Alternatively, high PRF waveforms are extremely ambiguous in range. Radar systems utilizing high PRFs are often called Pulsed Doppler Radars (PDR). Range and Doppler ambiguities for different PRFs are summarized in Table 6.1.

Table 6.1. PRF Ambiguities.

PRF	Range Ambiguous	Doppler Ambiguous
Low PRF	No	Yes
Medium PRF	Yes	Yes
High PRF	Yes	No

Distinction of a certain PRF as low, medium, or high PRF is almost arbitrary and depends on the radar mode of operations. For example, a $3KHz$ PRF is considered low if the maximum detection range is less than $30Km$. However, the same PRF would be considered medium if the maximum detection range is well beyond $30Km$.

Figure 6.5 shows a simplified pulsed radar block diagram. The range gates can be implemented as filters that open and close at time intervals that correspond to the detection range. The width of such an interval corresponds to the desired range resolution. The radar receiver is often implemented as a series of contiguous (in time) range gates, where the width of each gate is achieved through pulse compression. The clutter rejection can be implemented using MTI or other forms of clutter rejection techniques. The NBF bank is normally implemented using an FFT, where bandwidth of the individual filters corresponds to the FFT frequency resolution.

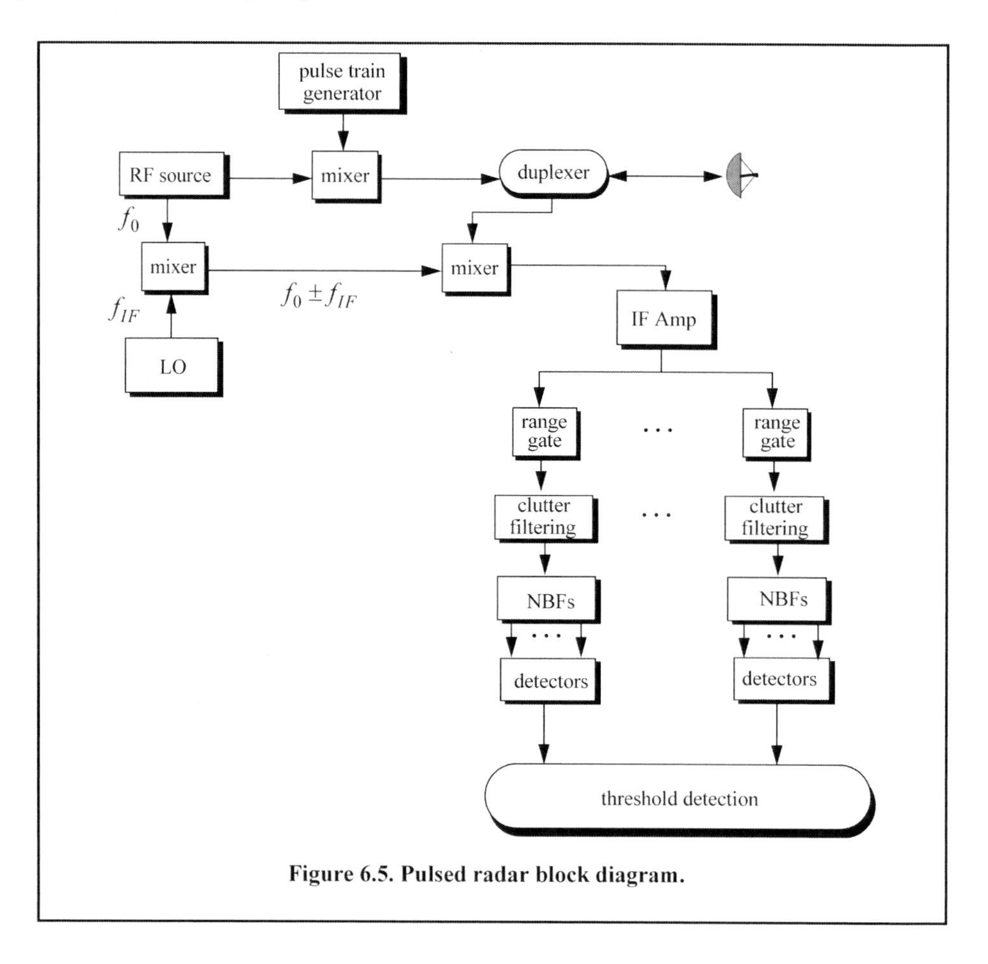

Figure 6.5. Pulsed radar block diagram.

In ground-based radars, the amount of clutter in the radar receiver depends heavily on the radar-to-target geometry. The amount of clutter is considerably higher when the radar beam has to face toward the ground. Radars employing high PRFs have to deal with an increased amount of clutter due to folding in range. Clutter introduces additional difficulties for airborne radars when detecting ground targets and other targets flying at low altitudes. This is illustrated in Fig. 6.6. Returns from ground clutter emanate from ranges equal to the radar altitude to those which exceed the slant range along the mainbeam, with considerable clutter returns in the sidelobes and mainbeam. The presence of such large amounts of clutter interferes with radar detection capabilities and makes it extremely difficult to detect targets in the look-down mode. This difficulty in detecting ground or low-altitude targets has led to the development of pulse Doppler radars where other targets, kinematics such as Doppler effects, are exploited to enhance detection.

Pulse Doppler radars utilize high PRFs to increase the average transmitted power and rely on the target's Doppler frequency for detection. The increase in the average transmitted power leads to an improved SNR, which helps the detection process. However, using high PRFs compromises the radar's ability to detect long-range targets because of range ambiguities associated with high PRF applications.

Pulse Doppler radars (or high PRF radars) have to deal with the additional increase in clutter power due to clutter folding. This has led to the development of a special class of airborne Moving Target Indicator (MTI) filters, often referred to as AMTI. Techniques such as using specialized Doppler filters to reject clutter are very effective and are often employed by pulse Doppler radars. Pulse Doppler radars can measure target Doppler frequency (or its range rate) fairly accurately and use the fact that ground clutter typically possesses limited Doppler shift when compared with moving targets to separate the two returns. This is illustrated in Fig. 6.7.

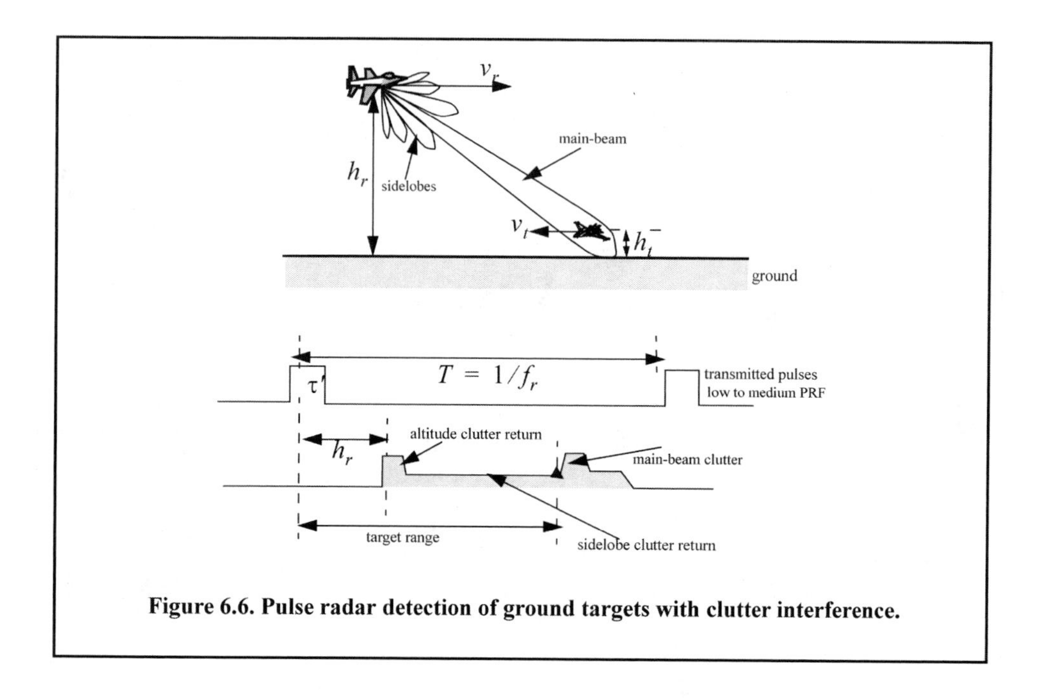

Figure 6.6. Pulse radar detection of ground targets with clutter interference.

Clutter filtering (i.e., AMTI) is used to remove both main-beam and altitude clutter returns, and fast-moving target detection is done effectively by exploiting its Doppler frequency. In many modern pulse Doppler radars, the limiting factor in detecting slow-moving targets is not clutter but rather another source of noise, referred to as phase noise, generated from the receiver local oscillator instabilities.

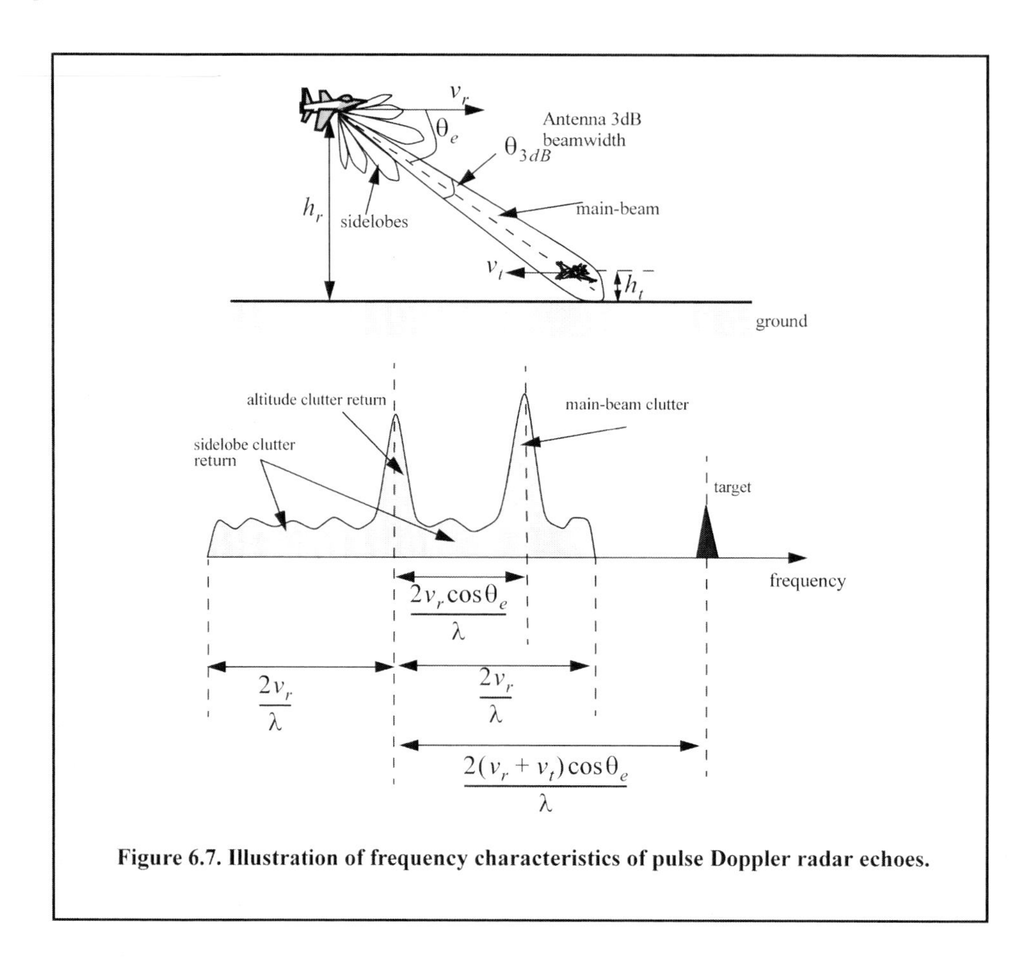

Figure 6.7. Illustration of frequency characteristics of pulse Doppler radar echoes.

6.2.1. Pulse Doppler Radar Signal Processing

The main idea behind pulse Doppler radar signal processing is to divide the footprint (the intersection of the antenna $3dB$ beamwidth with the ground) into resolution cells that constitute a range Doppler map, MAP. The sides of this map are range and Doppler, as illustrated in Fig. 6.8. Fine range resolution, ΔR, is accomplished in real time by utilizing range gating and pulse compression. Frequency (Doppler) resolution is obtained from the coherent processing interval.

To further illustrate this concept, consider the case where N_a is the number of azimuth (Doppler) cells, and N_r is the number of range bins. Hence, the MAP is of size $N_a \times N_r$, where the columns refer to range bins and the rows refer to azimuth cells. For each transmitted pulse within the dwell, the echoes from consecutive range bins are recorded sequentially in the first row of MAP. Once the first row is completely filled (i.e., returns from all range bins have been received), all data (in all rows) are shifted downward one row before the next pulse is transmitted. Thus, one row of MAP is generated for every transmitted pulse. Consequently, for the current observation interval, returns from the first transmitted pulse will be located in the bottom row of MAP, and returns from the last transmitted pulse will be in the top row of MAP.

Fine range resolution is achieved using the matched filter. Clutter rejection (filtering) is performed on each range bin (i.e., rows in the MAP). Then all samples from one dwell within each range bin are processed using an FFT to resolve targets in Doppler. It follows that a peak in a given resolution cell corresponds to a specific target detection at that range and Doppler frequency. Selection of the proper size FFT and its associated parameters were discussed in an earlier chapter.

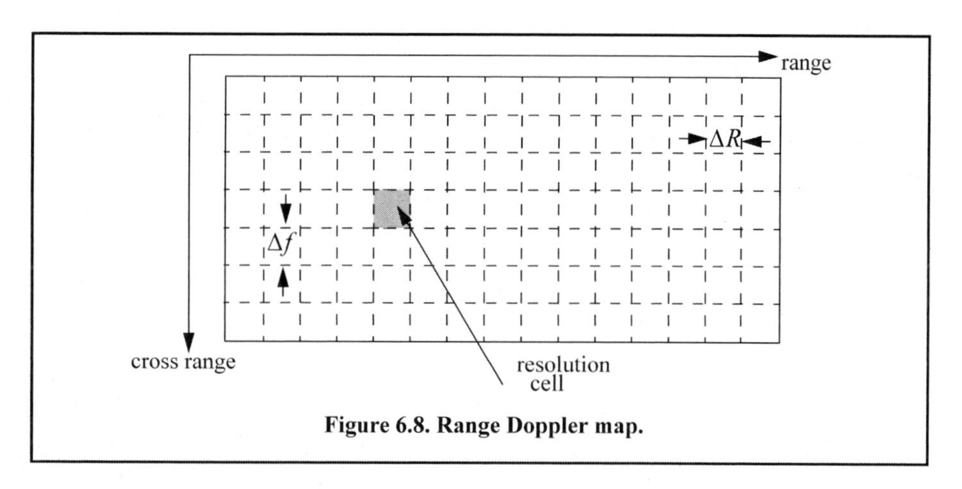

Figure 6.8. Range Doppler map.

6.2.2. Resolving Range Ambiguity

Pulse Doppler radars exhibit severe range ambiguities because they use high PRF pulse streams. In order to resolve these ambiguities, pulse Doppler radars utilize multiple high PRFs (PRF staggering) within each processing interval (dwell). For this purpose, consider a pulse Doppler radar that uses two PRFs, f_{r1} and f_{r2}, on transmit to resolve range ambiguity, as shown in Fig. 6.9. Denote R_{u1} and R_{u2} as the unambiguous ranges for the two PRFs, respectively. Normally, these unambiguous ranges are relatively small and are short of the desired radar unambiguous range R_u (where $R_u \gg R_{u1}, R_{u2}$). Denote the radar desired PRF that corresponds to R_u as f_{rd}.

The choice of f_{r1} and f_{r2} is such that they are relatively prime with respect to one another. One choice is to select $f_{r1} = Nf_{rd}$ and $f_{r2} = (N+1)f_{rd}$ for some integer N. Within one period of the desired PRI ($T_d = 1/f_{rd}$), the two PRFs f_{r1} and f_{r2} coincide only at one location, which is the true unambiguous target position. The time delay T_d establishes the desired unambiguous range. The time delays t_1 and t_2 correspond to the time between the transmit of a pulse on each PRF and receipt of a target return due to the same pulse. Let M_1 be the number of PRF1 intervals between transmit of a pulse and receipt of the true target return. The quantity M_2 is similar to M_1 except it is for PRF2. It follows that over the interval 0 to T_d, the only possible results are $M_1 = M_2 = M$ or $M_1 + 1 = M_2$. The radar needs only to measure t_1 and t_2. First, consider the case when $t_1 < t_2$. In this case,

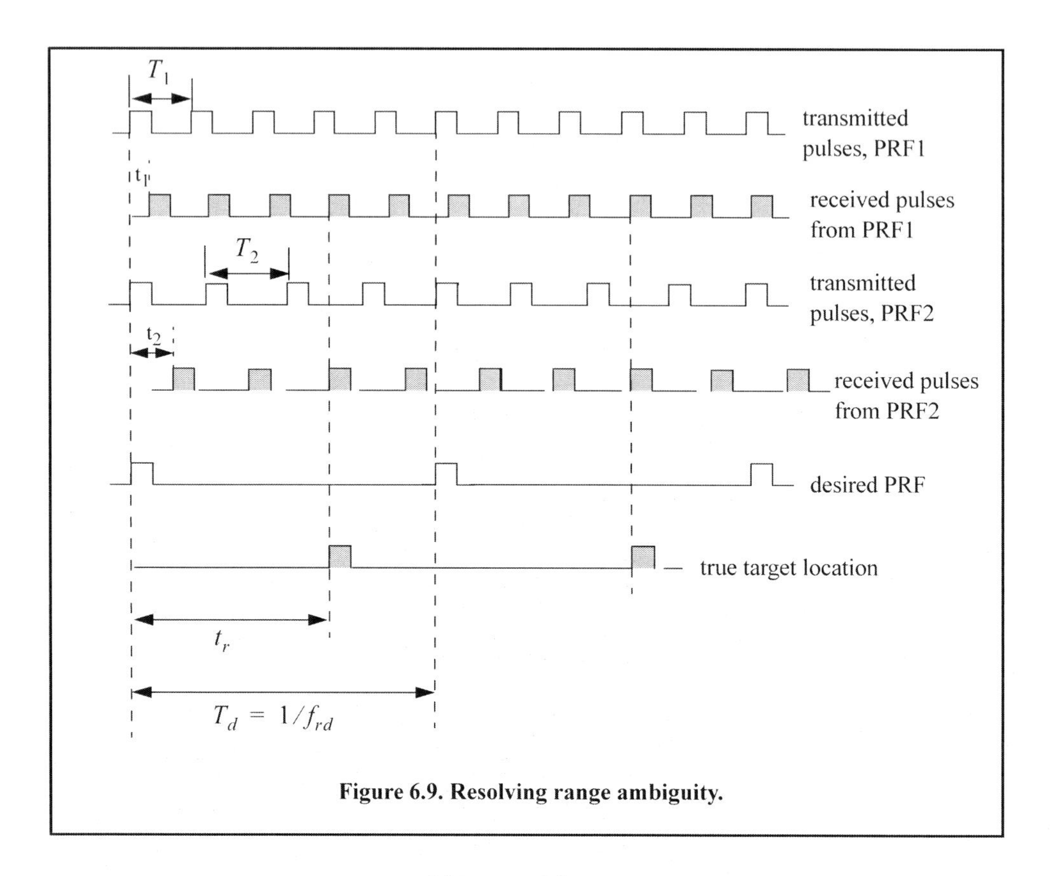

Figure 6.9. Resolving range ambiguity.

$$t_1 + \frac{M}{f_{r1}} = t_2 + \frac{M}{f_{r2}} \tag{Eq. (6.48)}$$

for which we get

$$M = \frac{t_2 - t_1}{T_1 - T_2} \tag{Eq. (6.49)}$$

where $T_1 = 1/f_{r1}$ and $T_2 = 1/f_{r2}$. It follows that the round-trip time to the true target location is

$$t_r = MT_1 + t_1$$
$$t_r = MT_2 + t_2 \tag{Eq. (6.50)}$$

and the true target range is

$$R = ct_r/2 . \tag{Eq. (6.51)}$$

Now, if $t_1 > t_2$, then

$$t_1 + \frac{M}{f_{r1}} = t_2 + \frac{M+1}{f_{r2}} . \tag{Eq. (6.52)}$$

Solving for M we get

$$M = \frac{(t_2 - t_1) + T_2}{T_1 - T_2} \tag{Eq. (6.53)}$$

and the round-trip time to the true target location is

$$t_{r1} = MT_1 + t_1 ,\tag{6.54}$$

and in this case, the true target range is

$$R = (ct_{r1})/2 .\tag{6.55}$$

Finally, if $t_1 = t_2$, then the target is in the first ambiguity. It follows that

$$t_{r2} = t_1 = t_2\tag{6.56}$$

and

$$R = ct_{r2}/2 .\tag{6.57}$$

Since a pulse cannot be received while the following pulse is being transmitted, these times correspond to blind ranges. This problem can be resolved by using a third PRF. In this case, once an integer N is selected, then in order to guarantee that the three PRFs are relatively prime with respect to one another, we may choose $f_{r1} = N(N+1)f_{rd}$, $f_{r2} = N(N+2)f_{rd}$, and $f_{r3} = (N+1)(N+2)f_{rd}$.

6.2.3. Resolving Doppler Ambiguity

In the case where the pulse Doppler radar is utilizing medium PRFs, it will be ambiguous in both range and Doppler. Resolving range ambiguities was discussed in the previous section. In this section, Doppler ambiguity is addressed. Remember that the line spectrum of a train of pulses has $\sin x/x$ envelope, and the line spectra are separated by the PRF, f_r, as illustrated in Fig. 6.10. The Doppler filter bank is capable of resolving target Doppler as long as the anticipated Doppler shift is less than one half the bandwidth of the individual filters (i.e., one half the width of an FFT bin). Thus, pulsed radars are designed such that

$$f_r = 2f_{dmax} = (2v_{rmax})/\lambda\tag{6.58}$$

where f_{dmax} is the maximum anticipated target Doppler frequency, v_{rmax} is the maximum anticipated target radial velocity, and λ is the radar wavelength.

If the Doppler frequency of the target is high enough to make an adjacent spectral line move inside the Doppler band of interest, the radar can be Doppler ambiguous. Therefore, in order to avoid Doppler ambiguities, radar systems require high PRF rates when detecting high-speed targets. When a long-range radar is required to detect a high-speed target, it may not be possible to be both range and Doppler unambiguous. This problem can be resolved by using multiple PRFs. Multiple PRF schemes can be incorporated sequentially within each dwell interval (scan or integration frame), or the radar can use a single PRF in one scan and resolve ambiguity in the next. The latter technique, however, may have problems due to changing target dynamics from one scan to the next.

The Doppler ambiguity problem is analogous to that of range ambiguity. Therefore, the same methodology can be used to resolve Doppler ambiguity. In this case, we measure the Doppler frequencies f_{d1} and f_{d2} instead of t_1 and t_2. If $f_{d1} > f_{d2}$, then we have

$$M = \frac{(f_{d2} - f_{d1}) + f_{r2}}{f_{r1} - f_{r2}} .\tag{6.59}$$

And if $f_{d1} < f_{d2}$,

$$M = \frac{f_{d2} - f_{d1}}{f_{r1} - f_{r2}} \qquad \text{Eq. (6.60)}$$

and the true Doppler is

$$f_d = Mf_{r1} + f_{d1} \qquad ; f_d = Mf_{r2} + f_{d2}. \qquad \text{Eq. (6.61)}$$

Finally, if $f_{d1} = f_{d2}$, then

$$f_d = f_{d1} = f_{d2}. \qquad \text{Eq. (6.62)}$$

Again, blind Dopplers can occur, which can be resolved using a third PRF.

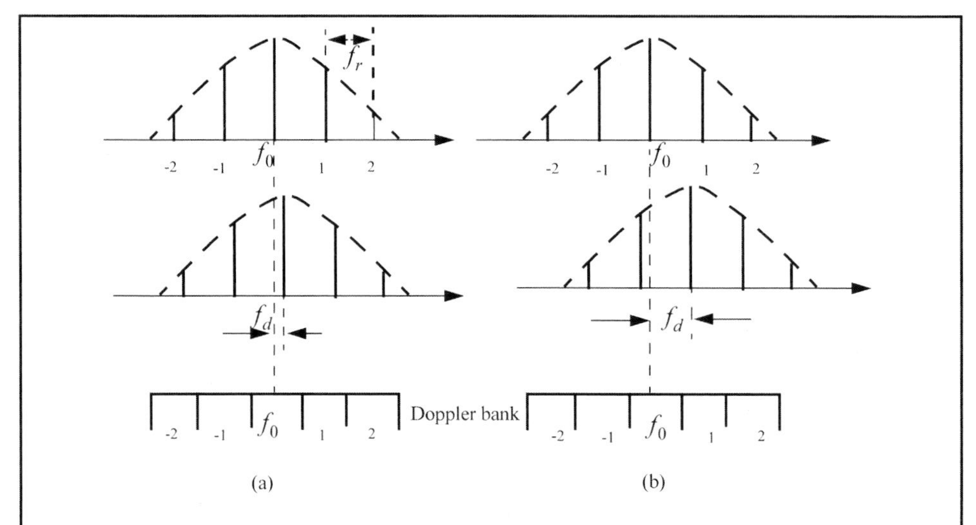

Figure 6.10. Spectra of transmitted and received waveforms, and Doppler bank. (a) Doppler is resolved. (b) Spectral lines have moved into the next Doppler filter. This results in an ambiguous Doppler measurement.

Example:

A certain radar uses two PRFs to resolve range ambiguities. The desired unambiguous range is $R_u = 100Km$. Choose $N = 59$. Compute f_{r1}, f_{r2}, R_{u1}, and R_{u2}.

Solution:

First let us compute the desired PRF, f_{rd}

$$f_{rd} = \frac{c}{2R_u} = \frac{3 \times 10^8}{200 \times 10^3} = 1.5KHz.$$

It follows that

$$f_{r1} = Nf_{rd} = (59)(1500) = 88.5KHz$$

$$f_{r2} = (N+1)f_{rd} = (59+1)(1500) = 90KHz$$

$$R_{u1} = \frac{c}{2f_{r1}} = \frac{3 \times 10^8}{2 \times 88.5 \times 10^3} = 1.695Km$$

$$R_{u2} = \frac{c}{2f_{r2}} = \frac{3 \times 10^8}{2 \times 90 \times 10^3} = 1.667Km\,.$$

****

Example:

Consider a radar with three PRFs; $f_{r1} = 15KHz$, $f_{r2} = 18KHz$, *and* $f_{r3} = 21KHz$. *Assume* $f_0 = 9GHz$. *Calculate the frequency position of each PRF for a target whose velocity is* $550m/s$. *Calculate* f_d *(Doppler frequency) for another target appearing at* $8KHz$, $2KHz$, *and* $17KHz$ *for each PRF.*

Solution:

The Doppler frequency is

$$f_d = 2\frac{vf_0}{c} = \frac{2 \times 550 \times 9 \times 10^9}{3 \times 10^8} = 33KHz\,.$$

Then by using Eq. (6.61) $n_i f_{ri} + f_{di} = f_d$ *where* $i = 1, 2, 3$, *we can write*

$$n_1 f_{r1} + f_{d1} = 15n_1 + f_{d1} = 33$$

$$n_2 f_{r2} + f_{d2} = 18n_2 + f_{d2} = 33$$

$$n_3 f_{r3} + f_{d3} = 21n_3 + f_{d3} = 33\,.$$

We will show here how to compute n_1, *and leave the computations of* n_2 *and* n_3 *to the reader. First, if we choose* $n_1 = 0$, *that means* $f_{d1} = 33KHz$, *which cannot be true since* f_{d1} *cannot be greater than* f_{r1}. *Choosing* $n_1 = 1$ *is also invalid since* $f_{d1} = 18KHz$ *cannot be true either. Finally, if we choose* $n_1 = 2$, *we get* $f_{d1} = 3KHz$, *which is an acceptable value. It follows that the minimum* n_1, n_2, n_3 *that may satisfy the above three relations are* $n_1 = 2$, $n_2 = 1$, *and* $n_3 = 1$. *Thus, the apparent Doppler frequencies are* $f_{d1} = 3KHz$, $f_{d2} = 15KHz$, *and* $f_{d3} = 12KHz$, *as seen below.*

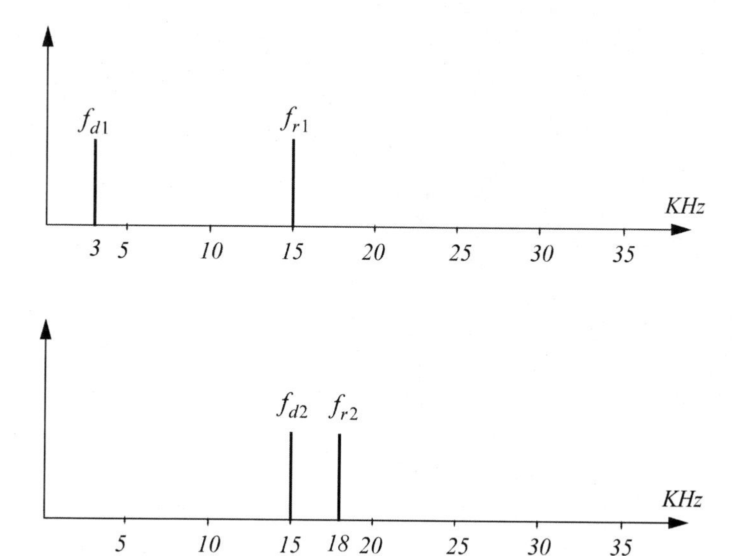

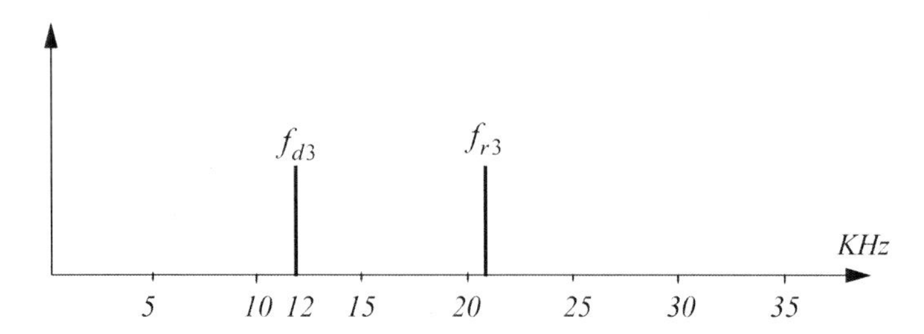

Now for the second part of the problem. Again, by using Eq. (6.61) we have

$$n_1 f_{r1} + f_{d1} = f_d = 15n_1 + 8 \; ; \; n_2 f_{r2} + f_{d2} = f_d = 18n_2 + 2 \; ; \; and$$

$$n_3 f_{r3} + f_{d3} = f_d = 21n_3 + 17 \, .$$

We can now solve for the smallest integers n_1, n_2, n_3 *that satisfy the above three relations. See the table below. Thus,* $n_1 = 2 = n_2$, *and* $n_3 = 1$, *and the true target Doppler is* $f_d = 38KHz$. *It follows that*

$$v_r = 38000 \times \frac{0.0333}{2} = 632.7 \frac{m}{\sec} \, .$$

n	0	1	2	3	4
f_d from f_{r1}	8	23	38	53	68
f_d from f_{r2}	2	20	38	56	
f_d from f_{r3}	17	38	39		

$$v_r = 38000 \times \frac{0.0333}{2} = 632.7 \frac{m}{\sec} \, .$$

****

Problems

6.1. Prove that $\sum_{n=-\infty}^{\infty} J_n(z) = 1$.

6.2. Show that $J_{-n}(z) = (-1)^n J_n(z)$. Hint: you may utilize the relation

$$J_n(z) = \frac{1}{\pi} \int_0^{\pi} \cos(z \sin y - ny) dy \, .$$

6.3. In a multiple frequency CW radar, the transmitted waveform consists of two continuous sinewaves of frequencies $f_1 = 105KHz$ and $f_2 = 115KHz$. Compute the maximum unambiguous detection range.

6.4. In Chapter 4 we developed an expression for the Doppler shift associated with a CW radar (i.e., $f_d = \pm 2v/\lambda$, where the plus sign is used for closing targets and the negative sign is used for receding targets). CW radars can use the system shown below to determine whether the target is closing or receding. Assuming that the emitted signal is $A\cos\omega_0 t$, and the received signal is $kA\cos((\omega_0 \pm \omega_d)t + \varphi)$, show that the direction of the target can be determined by checking the phase shift difference in the outputs $y_1(t)$ and $y_2(t)$.

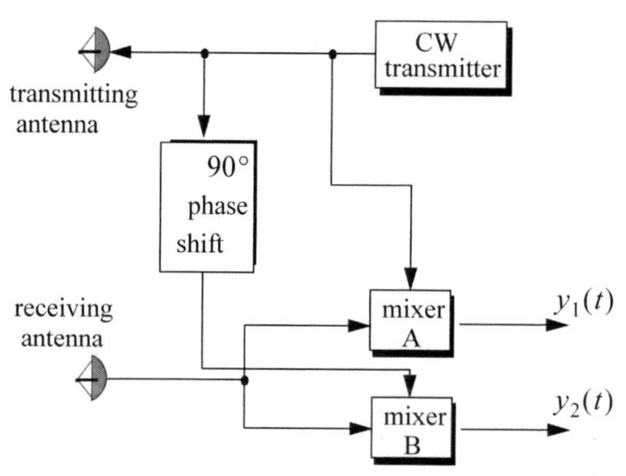

6.5. Consider a radar system using linear frequency modulation. Compute the range that corresponds to $\dot{f} = 20, 10MHz$. Assume a beat frequency $f_b = 1200Hz$.

6.6. A certain radar using linear frequency modulation has a modulation frequency $f_m = 300Hz$, and frequency sweep $\Delta f = 50MHz$. Calculate the average beat frequency differences that correspond to range increments of 10 and 15 meters.

6.7. A CW radar uses linear frequency modulation to determine both range and range rate. The radar wavelength is $\lambda = 3cm$, and the frequency sweep is $\Delta f = 200KHz$. Let $t_0 = 20ms$. (a) Calculate the mean Doppler shift. (b) Compute f_{bu} and f_{bd} corresponding to a target at range $R = 350Km$ which is approaching the radar with radial velocity of $250m/s$.

6.8. Repeat example found in Section 6.1.4, using $B = 1.5KHz$, $P_C = -40dB$, $f_m = 900Hz$, and $R = 2Km$. Assume $T = 375 degreesK$.

6.9. Consider a medium PRF radar on board an aircraft moving at a speed of 350 m/s with PRFs $f_{r1} = 10KHz$, $f_{r2} = 15KHz$, and $f_{r3} = 20KHz$; the radar operating frequency is $9.5GHz$. Calculate the frequency position of a nose-on target with a speed of 300 m/s. Also calculate the closing rate of a target appearing at 6, 5, and $18KHz$ away from the center-line of PRF 10, 15 and $20KHz$, respectively.

6.10. Repeat problem 6.9 when the target is $15°$ off the radar line of sight.

6.11. A certain radar operates at two PRFs, f_{r1} and f_{r2}, where $T_{r1} = (1/f_{r1}) = T/5$ and $T_{r2} = (1/f_{r2}) = T/6$. Show that this multiple PRF scheme will give the same range ambiguity as that of a single PRF with PRI T.

6.12. Consider an X-band radar with wavelength $\lambda = 3cm$, and bandwidth $B = 10MHz$. The radar uses two PRFs, $f_{r1} = 50KHz$ and $f_{r2} = 55.55KHz$. A target is detected at range bin 46 for f_{r1} and at bin 12 for f_{r2}. Determine the actual target range.

6.13. A certain radar uses two PRFs to resolve range ambiguities. The desired unambiguous range is $R_u = 150Km$. Select a reasonable value for N. Compute the corresponding f_{r1}, f_{r2}, R_{u1}, and R_{u2}.

6.14. A certain radar uses three PRFs to resolve range ambiguities. The desired unambiguous range is $R_u = 250Km$. Select $N = 43$. Compute the corresponding f_{r1}, f_{r2}, f_{r3}, R_{u1}, R_{u2} and R_{u3}.

Chapter 7

The Matched Filter Receiver and the Ambiguity Function

7.1. The Matched Filter Receiver

The unique characteristic of the matched filter is that it produces the maximum achievable instantaneous SNR at its output when a signal plus noise (Gaussian noise is assumed in the analysis presented in this book) are present at its input. Maximizing the SNR is key in all radar applications, as was described in Chapter 5 in the context of the radar equation.

It is important to use a radar receiver which can be modeled as an LTI system that maximizes the signal's SNR at its output. For this purpose, the basic radar receiver of interest is often referred to as the *matched filter receiver*. The matched filter is an optimum filter in the sense of SNR because the SNR at its output is maximized at some delay t_0 that corresponds to the true target range R_0 (i.e., $t_0 = (2R_0)/c$). Figure 7.1 shows a simplified block diagram for the radar receiver of interest.

In order to derive the general expression for the transfer function and the impulse response of this optimum filter, adopt the following notation: $h(t)$ is the optimum filter impulse response, $H(f)$ is the optimum filter transfer function, $x(t)$ is the input signal, $X(f)$ is the FT of the input signal, $x_o(t)$ is the output signal, $X_o(f)$ is the FT of the output signal, $n_i(t)$ is the input noise signal, $N_i(f)$ is the input noise PSD (not necessarily white), $n_o(t)$ is the out noise signal, and $N_o(f)$ is the output noise PSD. As one would expect, the impulse response of this optimum filter will take on distinct forms depending on the noise characteristics, i.e., white versus non-white noise.

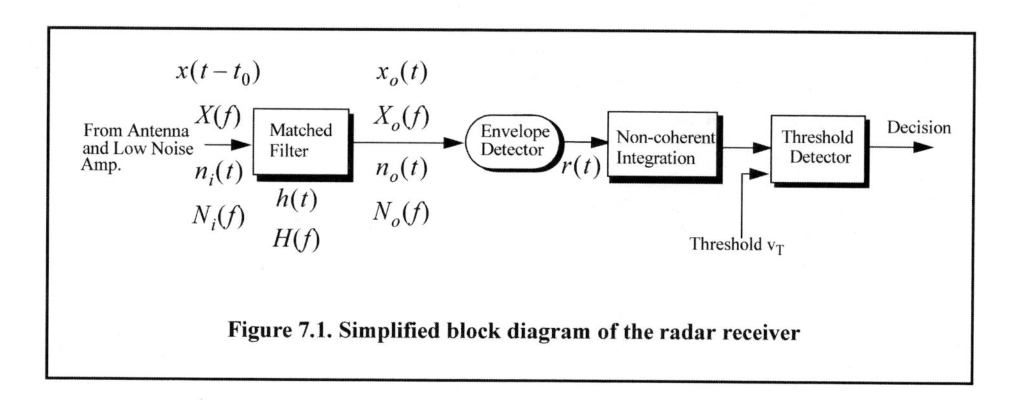

Figure 7.1. Simplified block diagram of the radar receiver

The optimum filter input or received signal (the words *input* and *received* will be used interchangeably in this book) can then be represented by

$$x_i(t) = x(t - t_0) + n_i(t) \quad \text{Eq. (7.1)}$$

where t_0 is an unknown time delay proportional to the target range. The optimum filter output signal is

$$y(t) = x_o(t - t_0) + n_o(t) \quad \text{Eq. (7.2)}$$

where

$$n_o(t) = n_i(t) \otimes h(t) \quad \text{Eq. (7.3)}$$

$$x_o(t) = x(t - t_0) \otimes h(t) . \quad \text{Eq. (7.4)}$$

The operator ($\otimes$) indicates convolution. The FT of Eq. (7.4) is

$$X_o(f) = X(f)H(f)e^{j2\pi f t_0} . \quad \text{Eq. (7.5)}$$

Integrating the right-hand side of Eq. (7.5) over all possible frequencies yields the signal output at time t_0, as

$$x_o(t_0) = \int_{-\infty}^{\infty} X(f)H(f)e^{j2\pi f t_0}\, df . \quad \text{Eq. (7.6)}$$

From Parseval's theorem the modulus square of Eq. (7.6) is the total signal energy, E_x. The total noise power at the output of the filter is calculated using Parseval's theorem as

$$N_o = \int_{-\infty}^{\infty} N_i(f)|H(f)|^2\, df . \quad \text{Eq. (7.7)}$$

Since the output signal power at time t_0 is equal to the modulus square of Eq. (7.6), then the instantaneous SNR at time t_0 is

$$SNR(t_0) = \frac{\left| \int_{-\infty}^{\infty} X(f)H(f)e^{j2\pi f t_0}\, df \right|^2}{\int_{-\infty}^{\infty} N_i(f)|H(f)|^2\, df} = \frac{E_x}{\int_{-\infty}^{\infty} N_i(f)|H(f)|^2\, df} . \quad \text{Eq. (7.8)}$$

Equation (7.8) is the general form of the optimum SNR at the output of the matched filter. Of course, when the noise is white, a simpler formula will result. Remember Schwarz's inequality, which has the form

$$\frac{\left| \int_{-\infty}^{\infty} X_1(f)X_2(f)\, df \right|^2}{\int_{-\infty}^{\infty} |X_1(f)|^2\, df} \le \int_{-\infty}^{\infty} |X_2(f)|^2\, df . \quad \text{Eq. (7.9)}$$

The equal sign in Eq. (7.9) applies when $X_1(f) = KX_2{}^*(f)$ for some arbitrary constant K. Apply Schwarz's inequality to Eq. (7.8) with the following assumptions

$$X_1(f) = H(f)\sqrt{N_i(f)} \qquad \text{Eq. (7.10)}$$

$$X_2(f) = (X(f)e^{j2\pi f t_0})/\sqrt{N_i(f)} \,. \qquad \text{Eq. (7.11)}$$

It follows that the SNR is maximized when

$$H(f) = K\ \{X^*(f)e^{-j2\pi f t_0}\}/N_i(f) \,. \qquad \text{Eq. (7.12)}$$

An alternative way of writing Eq. (7.12) is

$$X(f)H(f)e^{j2\pi f t_0} = \{K|X(f)|^2\}/N_i(f) \,. \qquad \text{Eq. (7.13)}$$

The optimum filter impulse response is computed using inverse FT integral

$$h(t) = \int_{-\infty}^{\infty} K\ \frac{X^*(f)e^{-j2\pi f t_0}}{N_i(f)}\ e^{j2\pi f t}\ df \,. \qquad \text{Eq. (7.14)}$$

7.1.1. White Noise Case

A special case of great interest to radar systems is when the input noise is band-limited white noise with PSD given by

$$N_i(f) = \eta_0/2 \,. \qquad \text{Eq. (7.15)}$$

η_0 is a constant. The transfer function for this optimum filter is then given by

$$H(f) = X^*(f)e^{-j2\pi f t_0} \qquad \text{Eq. (7.16)}$$

where the constant K was set equal to $\eta_0/2$. It follows that

$$h(t) = \int_{-\infty}^{\infty} [X^*(f)e^{-j2\pi f t_0}]\ e^{j2\pi f t}\ df \qquad \text{Eq. (7.17)}$$

which can be written as

$$h(t) = x^*(t_0 - t) \,. \qquad \text{Eq. (7.18)}$$

Observation of Eq. (7.18) indicates that the impulse response of the optimum filter is matched to the input signal, and thus, the term *matched filter* is used for this special case. Under these conditions, the maximum instantaneous SNR at the output is

$$SNR(t_0) = \frac{\left| \int_{-\infty}^{\infty} X(f)H(f)e^{j2\pi f t_0}\ df \right|^2}{\left(\frac{\eta_0}{2}\right)} \,. \qquad \text{Eq. (7.19)}$$

Again, from Parseval's theorem the numerator in Eq. (7.19) is equal to the input signal energy, E_x; consequently one can write the output peak instantaneous SNR as

$$SNR(t_0) = \frac{2E_x}{\eta_0}. \qquad \text{Eq. (7.20)}$$

Note that Eq. (7.20) is unitless since the units for η_0 are in watts per Hertz (or joules). Finally, one can draw the conclusion that the peak instantaneous SNR depends only on the signal energy and input noise power, and is independent of the waveform utilized by the radar.

As indicated by Eq. (7.18), the impulse response $h(t)$ may not be causal if the value for t_0 is less than the signal duration. Thus, an additional time delay term $\tau_0 \geq T$ is added to ensure causality, where T is the signal duration. Thus, a realizable matched filter response is given by

$$h(t) = \begin{cases} x^*(\tau_0 + t_0 - t) & ; t > 0,\ \tau_0 \geq T \\ 0 & ; t < 0 \end{cases}. \qquad \text{Eq. (7.21)}$$

The transfer function for this causal filter is

$$H(f) = \int_{-\infty}^{\infty} x^*(\tau_0 + t_0 - t) e^{-j2\pi ft} dt = \int_{\infty}^{-\infty} x^*(t + \tau_0 + t_0) e^{j2\pi ft} dt = X^*(f) e^{-j2\pi f(\tau_0 + t_0)} . \qquad \text{Eq. (7.22)}$$

Substituting the right-hand side of Eq. (7.22) into Eq. (7.6) yields

$$x_o(\tau_0) = \int_{-\infty}^{\infty} X(f) X^*(f) e^{-j2\pi f(\tau_0 + t_0)} e^{j2\pi f t_0}\, df = \int_{-\infty}^{\infty} |X(f)|^2 e^{-j2\pi f \tau_0}\, df, \qquad \text{Eq. (7.23)}$$

which has a maximum value when τ_0. This result leads to the following conclusion: The peak value of the matched filter output is obtained by sampling its output at times equal to the filter delay after the start of the input signal, and the minimum value for τ_0 is equal to the signal duration T.

Example:

Compute the maximum instantaneous SNR at the output of a linear filter whose impulse response is matched to the signal $x(t) = \exp(-t^2/2T)$.

Solution:

The signal energy is

$$E_x = \int_{-\infty}^{\infty} |x(t)|^2 dt = \int_{-\infty}^{\infty} e^{(-t^2)/T} dt = \sqrt{\pi T}\ joules .$$

It follows that the maximum instantaneous SNR is

$$SNR = \frac{\sqrt{\pi T}}{\frac{\eta_0}{2}} = \frac{2\sqrt{\pi T}}{\eta_0}$$

where $\eta_0/2$ *is the input noise power spectrum density.*

****

7.1.2. The Replica

Again, consider a radar system that uses a finite duration energy signal $x(t)$, and assume that a matched filter receiver is utilized. From Eq. (7.1), the input signal can be written as,

$$x_i(t) = x(t - t_0) + n_i(t) . \qquad \text{Eq. (7.24)}$$

The matched filter output $y(t)$ can be expressed by the convolution integral between the filter's impulse response and $x_i(t)$:

$$y(t) = \int_{-\infty}^{\infty} x_i(u)h(t-u)du . \qquad \text{Eq. (7.25)}$$

Substituting Eq. (7.21) into Eq. (7.25) yields

$$y(t) = \int_{-\infty}^{\infty} x_i(u)x^*(t - \tau_0 - t_0 + u)du = \bar{R}_{x_ix}(t - T_0) \qquad \text{Eq. (7.26)}$$

where $T_0 = \tau_0 + t_0$ and $\bar{R}_{x_ix}(t - T_0)$ is a cross-correlation between $x_i(t)$ and $x(T_0 - t)$. Therefore, the matched filter output can be computed from the cross-correlation between the radar received signal and a delayed replica of the transmitted waveform. If the input signal is the same as the transmitted signal, the output of the matched filter would be the autocorrelation function of the received (or transmitted) signal. In practice, replicas of the transmitted waveforms are normally computed and stored in memory for use by the radar signal processor when needed.

7.2. General Formula for the Output of the Matched Filter

Two cases are analyzed; the first is when a stationary target is present. The second case is concerned with a moving target whose velocity is constant. Assume the range to the target is

$$R(t) = R_0 - v(t - t_0) \qquad \text{Eq. (7.27)}$$

where v is the target radial velocity (i.e., the target velocity component on the radar line of sight). The initial detection range R_0 is given by

$$t_0 = \frac{2R_0}{c} \qquad \text{Eq. (7.28)}$$

where c is the speed of light and t_0 is the round trip delay it takes a certain radar pulse to travel from the radar to the target at range R_0 and back.

The general expression for the radar bandpass signal is

$$x(t) = x_I(t)\cos 2\pi f_0 t - x_Q(t)\sin 2\pi f_0 t \qquad \text{Eq. (7.29)}$$

which can be written using its pre-envelope (analytic signal) as

$$x(t) = Re\{\psi(t)\} = Re\{\tilde{x}(t)e^{j2\pi f_0 t}\} \qquad \text{Eq. (7.30)}$$

where $Re\{\ \}$ indicates "the real part of." Again, $\tilde{x}(t)$ is the complex envelope.

7.2.1. Stationary Target Case

In this case, the received radar return is given by

$$x_i(t) = x\left(t - \frac{2R_0}{c}\right) = x(t - t_0) = Re\{\tilde{x}(t - t_0)e^{j2\pi f_0(t - t_0)}\}. \qquad \text{Eq. (7.31)}$$

It follows that the received (or input) analytic signal is,

$$\psi_i(t) = \{\tilde{x}(t - t_0)e^{-j2\pi f_0 t_0}\}e^{j2\pi f_0 t} \qquad \text{Eq. (7.32)}$$

and by inspection the received (or input) complex envelope is,

$$\tilde{x}_i(t) = \tilde{x}(t - t_0)e^{-j2\pi f_0 t_0}. \qquad \text{Eq. (7.33)}$$

Observation of Eq. (7.33) clearly indicates that the received complex envelope is more than just a delayed version of the transmitted complex envelope. It actually contains an additional phase shift φ_0 which represents the phase corresponding to the two-way optical length for the target range. That is,

$$\varphi_0 = -2\pi f_0 t_0 = -2\pi f_0 2\frac{R_0}{c} = -\frac{2\pi}{\lambda}2R_0 \qquad \text{Eq. (7.34)}$$

where λ is the radar wavelength and is equal to c/f_0. Since a very small change in range can produce significant change in this phase term, this phase is often treated as a random variable with uniform probability density function over the interval $\{0, 2\pi\}$. Furthermore, the radar signal processor will first attempt to remove (correct for) this phase term through a process known as phase unwrapping.

Substituting Eq. (7.33) into Eq. (7.25) provides the output of the matched filter. It is given by

$$y(t) = \int_{-\infty}^{\infty} \tilde{x}_i(u)h(t - u)du \qquad \text{Eq. (7.35)}$$

where the impulse response $h(t)$ is in Eq. (7.18). It follows that

$$y(t) = \int_{-\infty}^{\infty} \tilde{x}(u - t_0)e^{-j2\pi f_0 t_0}\tilde{x}^*(t - t_0 + u)du. \qquad \text{Eq. (7.36)}$$

Make the following change of variables:

$$z = u - t_0 \Rightarrow dz = du. \qquad \text{Eq. (7.37)}$$

Therefore, the output of the matched filter when a stationary target is present is computed from Eq. (7.36) as

$$y(t) = e^{-j2\pi f_0 t_0}\int_{-\infty}^{\infty} \tilde{x}(z)\tilde{x}^*(t - z)dz = e^{-j2\pi f_0 t_0}\bar{R}_x(t). \qquad \text{Eq. (7.38)}$$

$\bar{R}_x(t)$ is the autocorrelation function for the signal $\tilde{x}(t)$ (i.e., the transmitted waveform).

7.2.2. Moving Target Case

In this case, the received signal is not only delayed in time by t_0, but also has a Doppler frequency shift f_d corresponding to the target velocity, where

$$f_d = 2vf_0/c = 2v/(\lambda) \ . \qquad \text{Eq. (7.39)}$$

The pre-envelope of the received signal can be written as

$$\psi_i(t) = \psi\left(t - \frac{2R(t)}{c}\right) = \tilde{x}\left(t - \frac{2R(t)}{c}\right)e^{j2\pi f_0\left(t - \frac{2R(t)}{c}\right)} . \qquad \text{Eq. (7.40)}$$

Substituting Eq. (7.27) into Eq. (7.40) yields

$$\psi_i(t) = \tilde{x}\left(t - \frac{2R_0}{c} + \frac{2vt}{c} - \frac{2vt_0}{c}\right)e^{j2\pi f_0\left(t - \frac{2R_0}{c} + \frac{2vt}{c} - \frac{2vt_0}{c}\right)} . \qquad \text{Eq. (7.41)}$$

Collecting terms yields

$$\psi_i(t) = \tilde{x}\left(t\left(1 + \frac{2v}{c}\right) - t_0\left(1 + \frac{2v}{c}\right)\right)e^{j2\pi f_0\left(t - \frac{2R_0}{c} + \frac{2vt}{c} - \frac{2vt_0}{c}\right)} . \qquad \text{Eq. (7.42)}$$

Define the scaling factor γ as

$$\gamma = 1 + \frac{2v}{c}, \qquad \text{Eq. (7.43)}$$

then Eq. (7.42) can be written as

$$\psi_i(t) = \tilde{x}(\gamma(t - t_0))e^{j2\pi f_0\left(t - \frac{2R_0}{c} + \frac{2vt}{c} - \frac{2vt_0}{c}\right)} . \qquad \text{Eq. (7.44)}$$

Since $c \gg v$, the following approximation can be used

$$\tilde{x}(\gamma(t - t_0)) \approx \tilde{x}(t - t_0) . \qquad \text{Eq. (7.45)}$$

It follows that Eq. (7.44) can now be rewritten as

$$\psi_i(t) = \tilde{x}(t - t_0)e^{j2\pi f_0 t}e^{-j2\pi f_0\frac{2R_0}{c}}e^{j2\pi f_0\frac{2vt}{c}}e^{-j2\pi f_0\frac{2vt_0}{c}} . \qquad \text{Eq. (7.46)}$$

Recognizing that $f_d = (2vf_0)/c$ and $t_0 = (2R_0)/c$, the received pre-envelope signal is

$$\psi_i(t) = \tilde{x}(t - t_0)e^{j2\pi f_0 t}e^{-j2\pi f_0 t_0}e^{j2\pi f_d t}e^{-j2\pi f_d t_0} = \tilde{x}(t - t_0)e^{j2\pi(f_0 + f_d)(t - t_0)} \qquad \text{Eq. (7.47)}$$

which can be rewritten as

$$\psi_i(t) = \{\tilde{x}(t - t_0)e^{j2\pi f_d t}e^{-j2\pi(f_0 + f_d)t_0}\}e^{j2\pi f_0 t} . \qquad \text{Eq. (7.48)}$$

Then by inspection the complex envelope of the received signal is

$$\tilde{x}_i(t) = \tilde{x}(t - t_0)e^{j2\pi f_d t}e^{-j2\pi(f_0 + f_d)t_0} . \qquad \text{Eq. (7.49)}$$

Finally, it is concluded that the complex envelope of the received signal when the target is moving at a constant velocity v is a delayed (by t_0) version of the complex envelope signal of the stationary target case except that:

1. An additional phase shift term corresponding to the target's Doppler frequency is present,
2. The phase shift term $(-2\pi f_d t_0)$ is present.

The output of the matched filter was defined in Eq. (7.25). Substituting Eq. (7.49) into Eq. (7.25) yields

$$y(t) = \int_{-\infty}^{\infty} \tilde{x}(u-t_0) e^{j2\pi f_d u} e^{-j2\pi(f_0+f_d)t_0} \tilde{x}^*(t-t_0+u)\, du \,. \qquad \text{Eq. (7.50)}$$

Applying the change of variables given in Eq. (7.37) and collecting terms provide

$$y(t) = e^{-j2\pi f_0 t_0} \int_{-\infty}^{\infty} \tilde{x}(z)\tilde{x}^*(t-z) e^{j2\pi f_d z} e^{j2\pi f_d t_0} e^{-j2\pi f_d t_0}\, dz \,. \qquad \text{Eq. (7.51)}$$

Observation of Eq. (7.51) shows that the output is a function of both t and f_d. Thus, it is more appropriate to rewrite the output of the matched filter as a two-dimensional function of both variables. That is,

$$y(t;f_d) = e^{-j2\pi f_0 t_0} \int_{-\infty}^{\infty} \tilde{x}(z)\tilde{x}^*(t-z) e^{j2\pi f_d z}\, dz \,. \qquad \text{Eq. (7.52)}$$

It is customary but not necessary to set $t_0 = 0$. Note that if the causal impulse response is used, then the same analysis will hold true. However, in this case, the phase term is equal to $\exp(-j2\pi f_0 T_0)$, instead of $\exp(-j2\pi f_0 t_0)$, where $T_0 = \tau_0 + t_0$.

7.3. Ambiguity Function

The radar ambiguity function represents the modulus of the matched filter output, and it describes the interference caused by the range and/or Doppler shift of a target when compared to a reference target of equal RCS. The ambiguity function evaluated at $(\tau, f_d) = (0, 0)$ is equal to the matched filter output that is perfectly matched to the signal reflected from the target of interest. In other words, returns from the nominal target are located at the origin of the ambiguity function. Thus, the ambiguity function at nonzero τ and f_d represents returns from some range and Doppler different from those for the nominal target.

Define the ambiguity function as the modulus square of the output of the matched filter (given in Eq. (7.52)). More precisely,

$$|\chi(\tau, f_d)|^2 = |y(t;f_d)|^2 = \left| \int_{-\infty}^{\infty} \tilde{x}(t)\tilde{x}^*(t-\tau) e^{j2\pi f_d t} dt \right|^2 . \qquad \text{Eq. (7.53)}$$

The radar ambiguity function is normally used by radar designers as a means of studying different waveforms. It can provide insight about how different radar waveforms may be suitable for the various radar applications. It is also used to determine the range and Doppler resolutions for a specific radar waveform. The three-dimensional (3-D) plot of the ambiguity function versus frequency and time delay is called the radar ambiguity diagram.

Denote E_x as the energy of the signal $\tilde{x}(t)$,

$$E_x = \int_{-\infty}^{\infty} |\tilde{x}(t)|^2 dt . \qquad \text{Eq. (7.54)}$$

The following list includes the properties for the radar ambiguity function:

1. The maximum value for the ambiguity function occurs at $(\tau, f_d) = (0, 0)$ and is equal to $4E_x^2$,

$$max\{|\chi(\tau;f_d)|^2\} = |\chi(0;0)|^2 = (2E_x)^2 \qquad \text{Eq. (7.55)}$$

$$|\chi(\tau;f_d)|^2 \le |\chi(0;0)|^2 . \qquad \text{Eq. (7.56)}$$

2. The ambiguity function is symmetric,

$$|\chi(\tau;f_d)|^2 = |\chi(-\tau;-f_d)|^2 . \qquad \text{Eq. (7.57)}$$

3. The total volume under the ambiguity function is constant,

$$\iint |\chi(\tau;f_d)|^2 \ d\tau \ df_d = (2E_x)^2 . \qquad \text{Eq. (7.58)}$$

4. If the function $X(f)$ is the Fourier transform of the signal $x(t)$, then by using Parseval's theorem we get

$$|\chi(\tau;f_d)|^2 = \left|\int X^*(f) X(f - f_d) e^{-j2\pi f\tau} df\right|^2 . \qquad \text{Eq. (7.59)}$$

5. Suppose that $|\chi(\tau;f_d)|^2$ is the ambiguity function for the signal $\tilde{x}(t)$. Adding a quadratic phase modulation term to $\tilde{x}(t)$ yields

$$\tilde{x}_1(t) = \tilde{x}(t) e^{j\pi\mu t^2} \qquad \text{Eq. (7.60)}$$

where μ is a constant. It follows that the ambiguity function for the signal $\tilde{x}_1(t)$ is given by

$$|\chi_1(\tau;f_d)|^2 = |\chi(\tau;(f_d + \mu\tau))|^2 . \qquad \text{Eq. (7.61)}$$

7.4. Examples of the Ambiguity Function

The ideal radar ambiguity function is represented by a spike of infinitesimal width that peaks at the origin and is zero everywhere else, as illustrated in Figure 7.2. An ideal ambiguity function provides perfect resolution between neighboring targets regardless of how close they may be to each other. Unfortunately, an ideal ambiguity function cannot physically exist because the ambiguity function must have a finite peak value equal to $(2E_x)^2$ and a finite volume also equal to $(2E_x)^2$. Clearly, the ideal ambiguity function cannot meet those conditions.

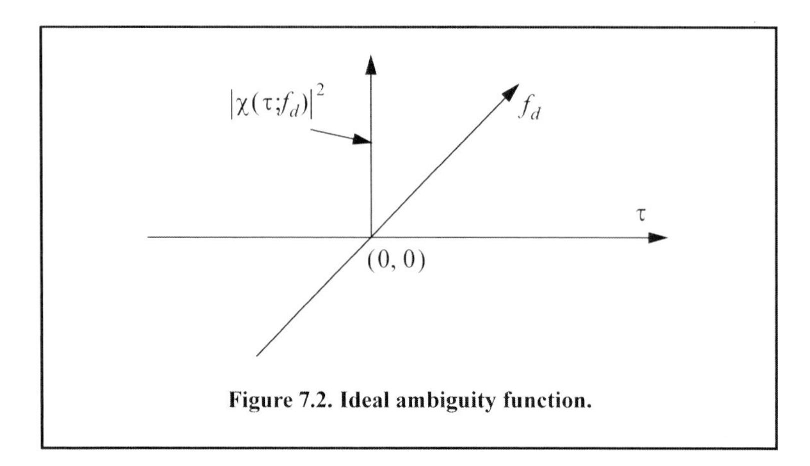

Figure 7.2. Ideal ambiguity function.

7.4.1. Single Pulse Ambiguity Function

The complex envelope of a single pulse is $\tilde{x}(t)$ defined by

$$\tilde{x}(t) = \frac{1}{\sqrt{\tau_0}} Rect\left(\frac{t}{\tau_0}\right). \qquad \text{Eq. (7.62)}$$

In this case, the output of the matched filter is

$$\chi(\tau;f_d) = \int_{-\infty}^{\infty} \tilde{x}(t)\tilde{x}^*(t-\tau)e^{j2\pi f_d t}\,dt. \qquad \text{Eq. (7.63)}$$

Substituting Eq. (7.62) into Eq. (7.63) and performing the integration yield

$$|\chi(\tau;f_d)|^2 = \left|\left(1-\frac{|\tau|}{\tau_0}\right)\frac{\sin(\pi f_d(\tau_0-|\tau|))}{\pi f_d(\tau_0-|\tau|)}\right|^2 \qquad |\tau| \le \tau_0. \qquad \text{Eq. (7.64)}$$

Figures 7.3 a and b show 3-D and contour plots of single pulse ambiguity functions. The ambiguity function cut along the time-delay axis τ is obtained by setting $f_d = 0$. More precisely,

$$|\chi(\tau;0)| = \left(1-\frac{|\tau|}{\tau_0}\right)^2 \qquad |\tau| \le \tau_0. \qquad \text{Eq. (7.65)}$$

Note that the time autocorrelation function of the signal $\tilde{x}(t)$ is equal to $\chi(\tau;0)$. Similarly, the cut along the Doppler axis is

$$|\chi(0;f_d)|^2 = \left|\frac{\sin \pi\tau_0 f_d}{\pi\tau_0 f_d}\right|^2. \qquad \text{Eq. (7.66)}$$

Figures 7.4 and 7.5, respectively, show the plots of the uncertainty function cuts defined by Eqs. (7.65) and (7.66). Since the zero Doppler cut along the time-delay axis extends between $-\tau_0$ and τ_0, close targets will be unambiguous if they are at least τ_0 seconds apart. The zero time cut along the Doppler frequency axis has a $(\sin x/x)^2$ shape. It extends from $-\infty$ to ∞. The first null occurs at $f_d = \pm 1/\tau_0$. Hence, it is possible to detect two targets that are shifted by $1/\tau_0$, without any ambiguity. Thus, a single pulse range and

Doppler resolutions are limited by the pulse width τ_0. Fine range resolution requires that a very short pulse be used. Unfortunately, using very short pulses requires very large operating bandwidths and may limit the radar average transmitted power to impractical values.

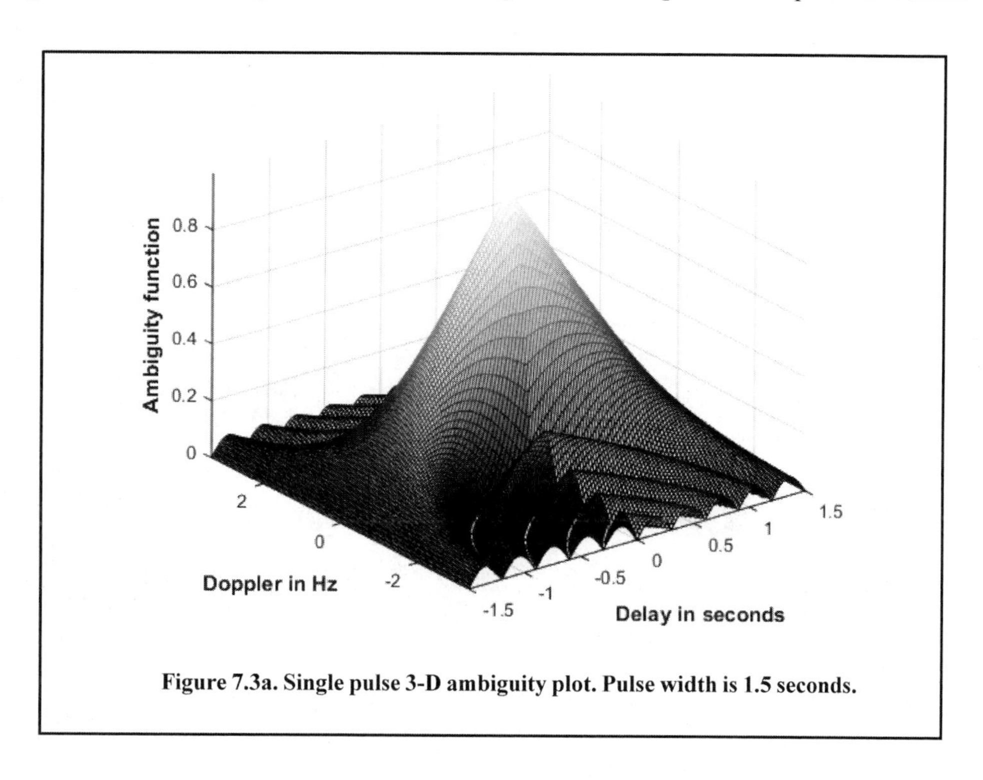

Figure 7.3a. Single pulse 3-D ambiguity plot. Pulse width is 1.5 seconds.

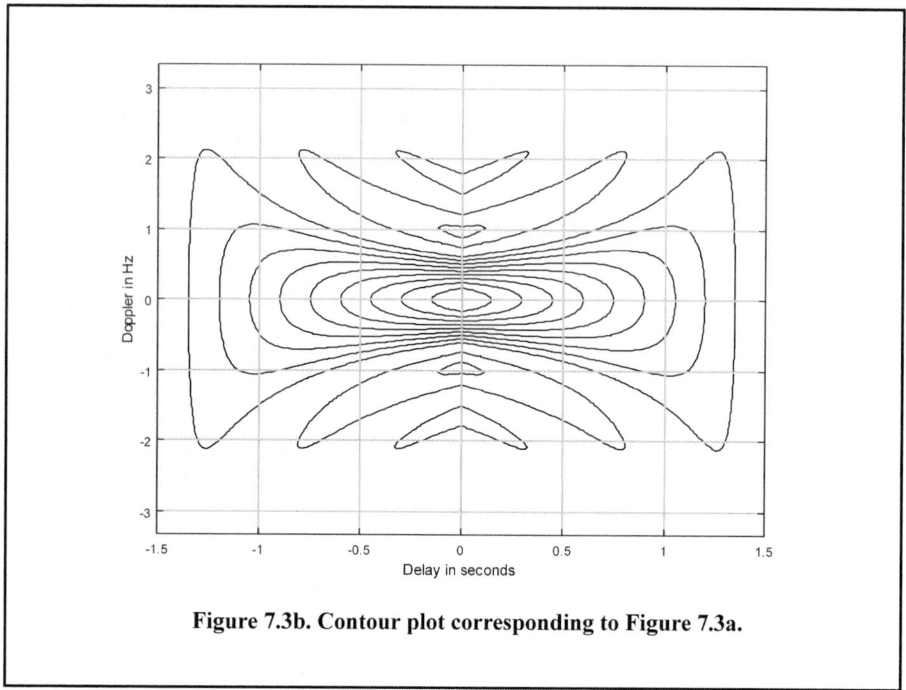

Figure 7.3b. Contour plot corresponding to Figure 7.3a.

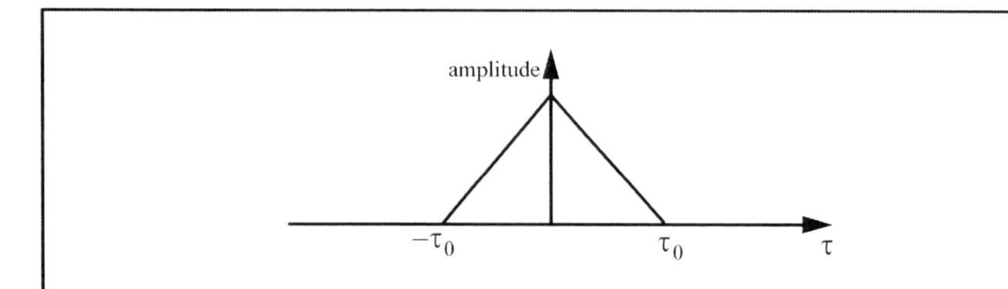

Figure 7.4. Zero Doppler ambiguity function cut along the time delay axis.

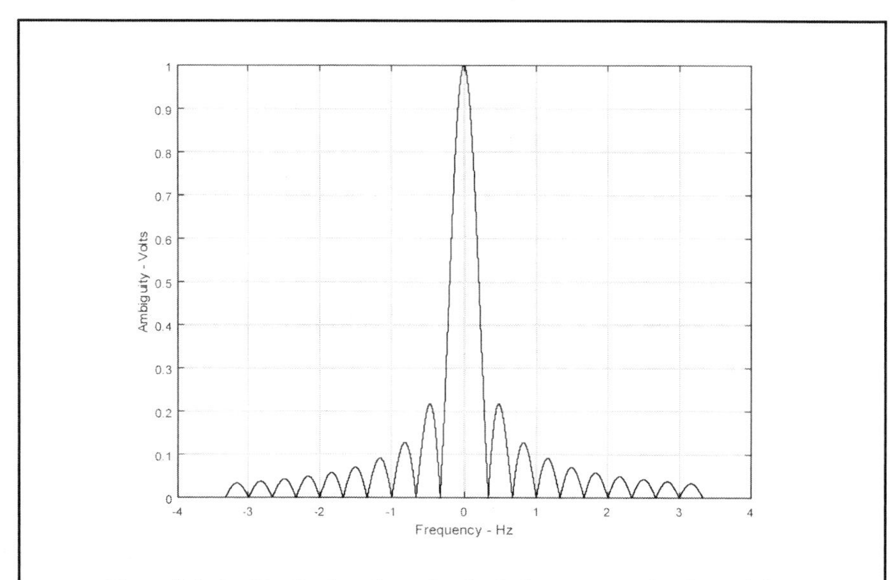

Figure 7.5. Ambiguity function of a single frequency pulse (zero delay). The pulse width is 3 seconds.

7.4.2. LFM Ambiguity Function

Consider the LFM complex envelope signal defined by

$$\tilde{x}(t) = \frac{1}{\sqrt{\tau_0}} Rect\left(\frac{t}{\tau_0}\right) e^{j\pi\mu t^2}. \qquad \text{Eq. (7.67)}$$

In order to compute the ambiguity function for the LFM complex envelope, we will first consider the case when $0 \le \tau \le \tau_0$. In this case the integration limits are from $-\tau_0/2$ to $(\tau_0/2) - \tau$. Using Eq.s (7.67) and (7.53) yields

$$\chi(\tau;f_d) = \frac{1}{\tau_0}\int_{-\infty}^{\infty} Rect\left(\frac{t}{\tau_0}\right) Rect\left(\frac{t-\tau}{\tau_0}\right) e^{j\pi\mu t^2} e^{-j\pi\mu(t-\tau)^2} e^{j2\pi f_d t} dt. \qquad \text{Eq. (7.68)}$$

It follows that

$$\chi(\tau;f_d) = \frac{e^{-j\pi\mu\tau^2}}{\tau_0} \int_{\frac{-\tau_0}{2}}^{\frac{\tau_0}{2}-\tau} e^{j2\pi(\mu\tau + f_d)t} dt. \qquad \text{Eq. (7.69)}$$

Finishing the integration process in Eq. (7.69) yields

$$\chi(\tau;f_d) = e^{j\pi\tau f_d}\left(1 - \frac{\tau}{\tau_0}\right) \frac{\sin\left(\pi\tau_0(\mu\tau + f_d)\left(1 - \frac{\tau}{\tau_0}\right)\right)}{\pi\tau_0(\mu\tau + f_d)\left(1 - \frac{\tau}{\tau_0}\right)} \qquad 0 \le \tau \le \tau_0. \qquad \text{Eq. (7.70)}$$

Similar analysis for the case when $-\tau_0 \le \tau \le 0$ can be carried out, where, in this case, the integration limits are from $(-\tau_0/2) - \tau$ to $\tau_0/2$. The same result can be obtained by using the symmetry property of the ambiguity function ($|\chi(-\tau, -f_d)| = |\chi(\tau, f_d)|$). It follows that an expression for $\chi(\tau;f_d)$ that is valid for any τ is given by

$$\chi(\tau;f_d) = e^{j\pi\tau f_d}\left(1 - \frac{|\tau|}{\tau_0}\right) \frac{\sin\left(\pi\tau_0(\mu\tau + f_d)\left(1 - \frac{|\tau|}{\tau_0}\right)\right)}{\pi\tau_0(\mu\tau + f_d)\left(1 - \frac{|\tau|}{\tau_0}\right)} \qquad |\tau| \le \tau_0 \qquad \text{Eq. (7.71)}$$

and the LFM ambiguity function is

$$|\chi(\tau;f_d)|^2 = \left|\left(1 - \frac{|\tau|}{\tau_0}\right) \frac{\sin\left(\pi\tau_0(\mu\tau + f_d)\left(1 - \frac{|\tau|}{\tau_0}\right)\right)}{\pi\tau_0(\mu\tau + f_d)\left(1 - \frac{|\tau|}{\tau_0}\right)}\right|^2 \qquad |\tau| \le \tau_0. \qquad \text{Eq. (7.72)}$$

Again the time autocorrelation function is equal to $\chi(\tau, 0)$. The reader can verify that the ambiguity function for a down-chirp LFM waveform is given by

$$|\chi(\tau;f_d)|^2 = \left|\left(1 - \frac{|\tau|}{\tau_0}\right) \frac{\sin\left(\pi\tau_0(\mu\tau - f_d)\left(1 - \frac{|\tau|}{\tau_0}\right)\right)}{\pi\tau_0(\mu\tau - f_d)\left(1 - \frac{|\tau|}{\tau_0}\right)}\right|^2 \qquad |\tau| \le \tau_0. \qquad \text{Eq. (7.73)}$$

The up-chirp ambiguity function cut along the time-delay axis τ is

$$|\chi(\tau;0)|^2 = \left|\left(1 - \frac{|\tau|}{\tau_0}\right) \frac{\sin\left(\pi\mu\tau\tau_0\left(1 - \frac{|\tau|}{\tau_0}\right)\right)}{\pi\mu\tau\tau_0\left(1 - \frac{|\tau|}{\tau_0}\right)}\right|^2 \qquad |\tau| \le \tau_0. \qquad \text{Eq. (7.74)}$$

Figure 7.6 shows the 3-D and the contour plots for the LFM uncertainty and ambiguity functions for $\tau_0 = 1$ second and $B = 5Hz$ for a down-chirp pulse. Note that the LFM ambiguity function cut along the Doppler frequency axis is similar to that of the single pulse. This should not be surprising since the pulse shape has not changed (only frequency modulation was added). However, the cut along the time-delay axis changes significantly. It is now much narrower compared to the unmodulated pulse cut. In this case, the first null occurs at

$$\tau_{n1} \approx 1/B. \qquad \text{Eq. (7.75)}$$

Figure 7.7 shows a plot for a cut in the uncertainty function corresponding to Eq. (7.74). Equation (7.75) indicates that the effective pulse width (compressed pulse width) of the matched filter output is completely determined by the radar bandwidth. It follows that the LFM ambiguity function cut along the time-delay axis is narrower than that of the unmodulated pulse by a factor

$$\xi = \frac{\tau_0}{(1/B)} = \tau_0 B \qquad \text{Eq. (7.76)}$$

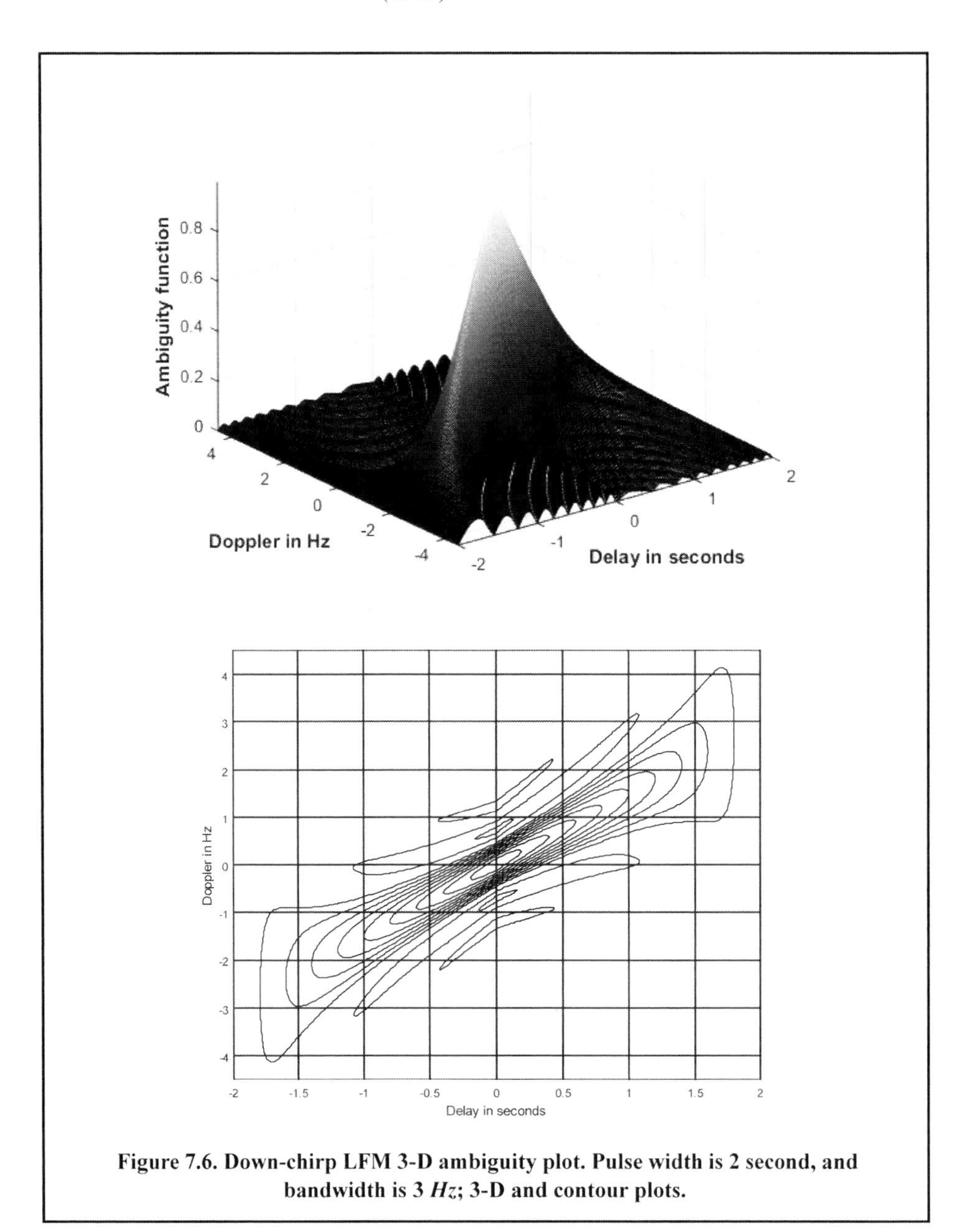

Figure 7.6. Down-chirp LFM 3-D ambiguity plot. Pulse width is 2 second, and bandwidth is 3 *Hz*; 3-D and contour plots.

where ξ is referred to as the compression ratio (also called the time-bandwidth product and compression gain). All three names can be used interchangeably to mean the same thing. As indicated by Eq. (7.76), the compression ratio also increases as the radar bandwidth is increased.

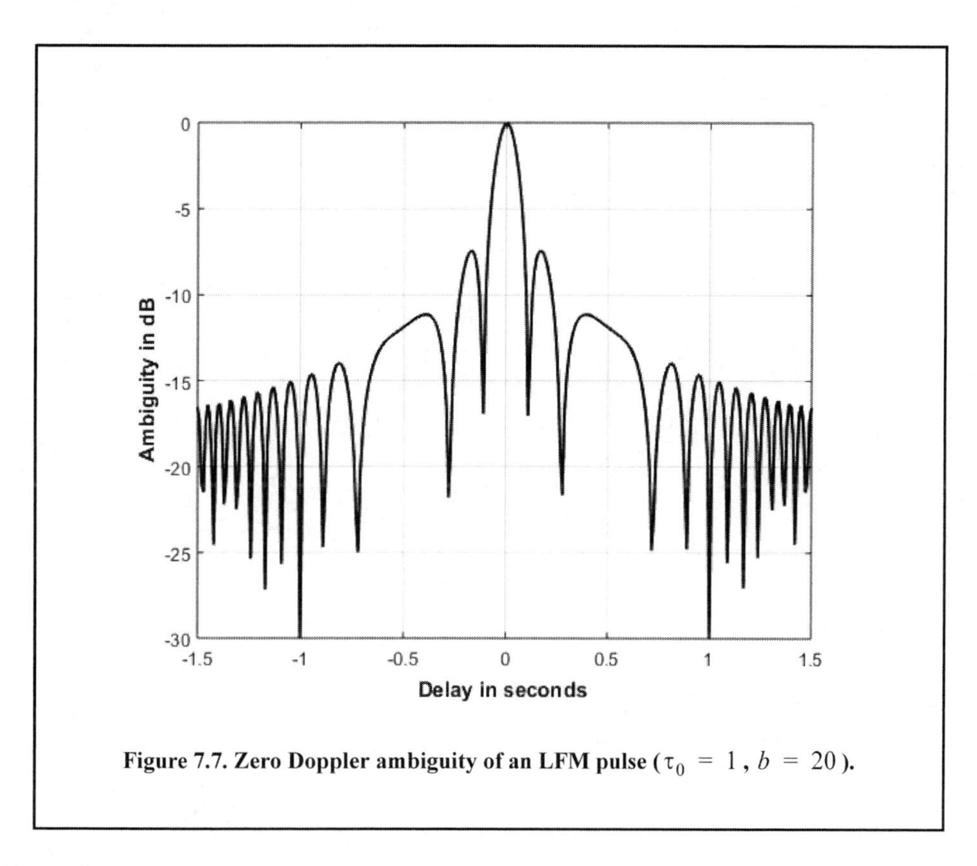

Figure 7.7. Zero Doppler ambiguity of an LFM pulse ($\tau_0 = 1$, $b = 20$).

Example:

Compute the range resolution before and after pulse compression corresponding to an LFM waveform with the following specifications: bandwidth $B = 1GHz$ and pulse width $\tau_0 = 10ms$.

Solution:

The range resolution before pulse compression is

$$\Delta R_{uncomp} = \frac{c\tau_0}{2} = \frac{3 \times 10^8 \times 10 \times 10^{-3}}{2} = 1.5 \times 10^6 \ meters.$$

Using Eq. (7.75) yields

$$\tau_{n1} = \frac{1}{1 \times 10^9} = 1 \ ns$$

$$\Delta R_{comp} = \frac{c\tau_{n1}}{2} = \frac{3 \times 10^8 \times 1 \times 10^{-9}}{2} = 15 \ cm.$$

****

7.4.3. Approximation of the Ambiguity Contours

Plots of the ambiguity function are called ambiguity diagrams. For a given waveform, the corresponding ambiguity diagram is normally used to determine the waveform properties such as the target resolution capability, measurements (time and frequency) accuracy and its response to clutter. Three-dimensional ambiguity diagrams are difficult to plot and interpret. This is the reason why contour plots of the 3-D ambiguity diagram are often used to study the characteristics of a waveform. An ambiguity contour is a 2-D plot (frequency/time) of a plane intersecting the 3-D ambiguity diagram that corresponds to some threshold value. The resultant plots are ellipses. It is customary to display the ambiguity contour plots that correspond to one-half of the peak autocorrelation value.

Figure 7.8 shows a sketch of typical ambiguity contour plots associated with a gated CW pulse. It indicates that narrow pulses provide better range accuracy than long pulses. Alternatively, the Doppler accuracy is better for a wider pulse than it is for a short one. This trade off between range and Doppler measurements comes from the uncertainty associated with the time-bandwidth product of a single sinusoidal pulse, where the product of uncertainty in time (range) and uncertainty in frequency (Doppler) cannot be much smaller than unity.

Multiple ellipses in an ambiguity contour plot indicate the presence of multiple targets. Thus, it seems that one may improve the radar resolution by increasing the ambiguity diagram threshold value. This is illustrated in Figure 7.9. However, in practice this is not possible for two reasons. First, in the presence of noise we lack knowledge of the peak correlation value; and second, targets in general will have different amplitudes.

Now consider the case of a sinusoid modulated pulse train, where the pulse width is τ', the PRI is T and the train length is $(N-1)T$ where N is the number of pulses in the train. This is illustrated in Figure 7.10 for $N = 5$. For a pulse train, range accuracy is still determined by the pulse width, the same way as in the case of a single pulse, while Doppler accuracy is determined by the train length. Thus, time and frequency measurements can be made independently of each other. However, additional peaks appear in the ambiguity diagram which may cause range and Doppler uncertainties. This is illustrated in Figure 7.11.

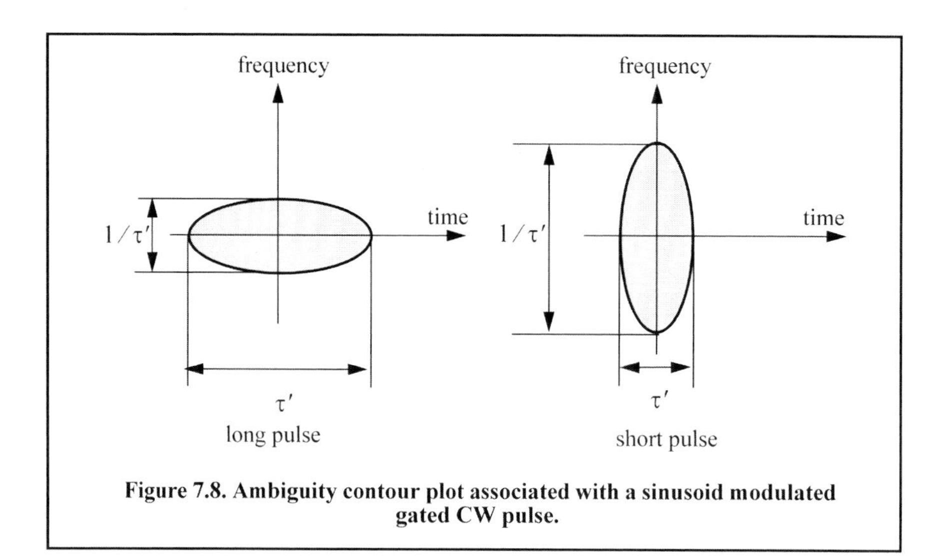

Figure 7.8. Ambiguity contour plot associated with a sinusoid modulated gated CW pulse.

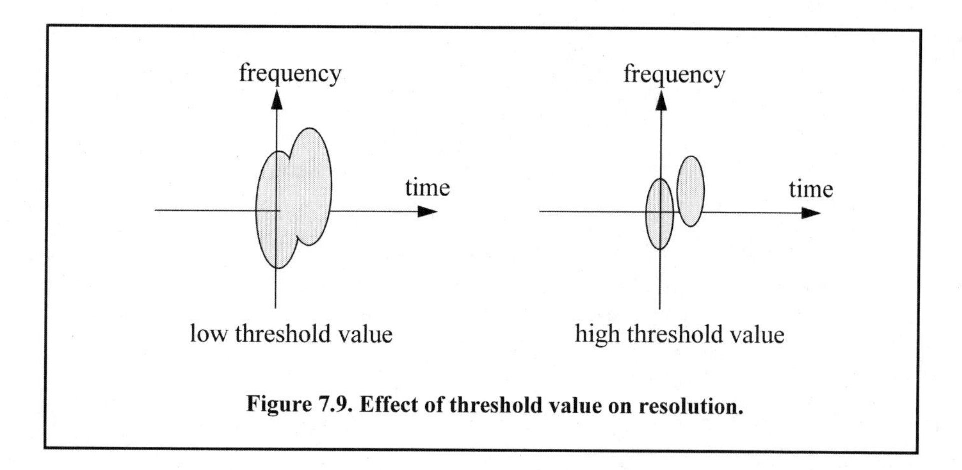

Figure 7.9. Effect of threshold value on resolution.

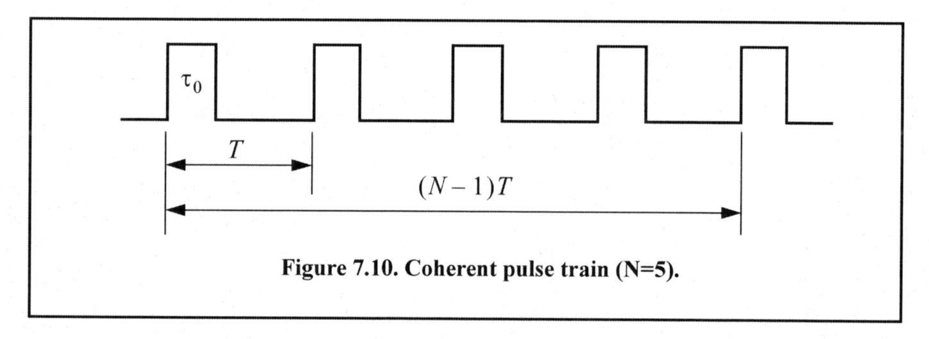

Figure 7.10. Coherent pulse train (N=5).

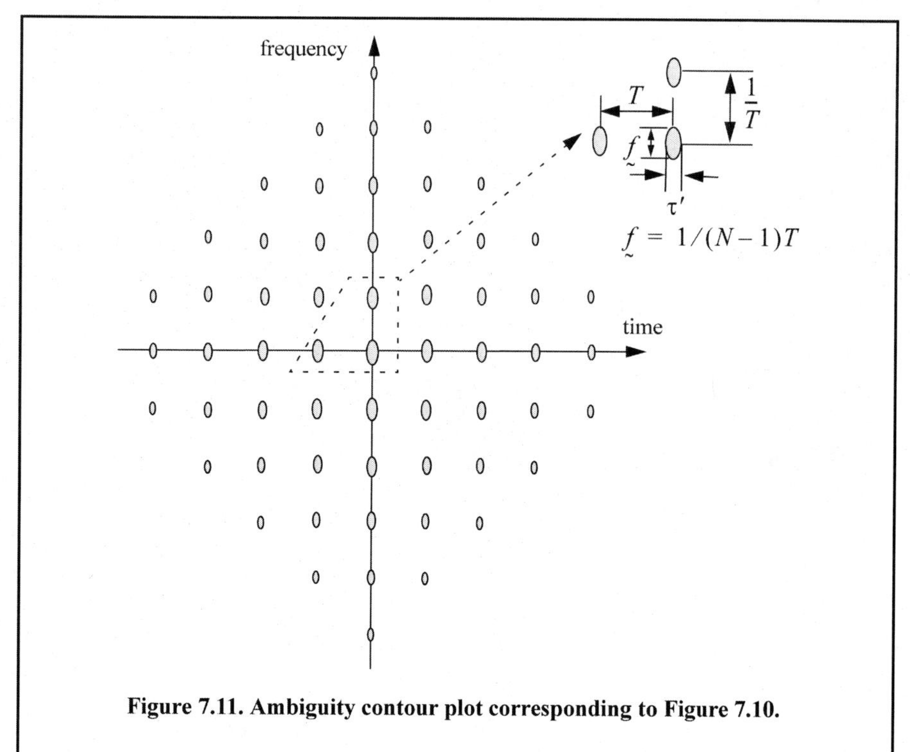

Figure 7.11. Ambiguity contour plot corresponding to Figure 7.10.

As one would expect, high PRF pulse trains (i.e., small T) lead to extreme uncertainty in range, while low PRF pulse trains have extreme ambiguity in Doppler, as shown in Figure 7.12. Medium PRF pulse trains have moderate ambiguity in both range and Doppler, which can be overcome by using multiple PRFs, as illustrated in Figure 7.13 for two medium PRFs. Note that the two diagrams (in Figure 7.13) agree only in one location (center of the plot) which corresponds to the true target location.

It is possible to avoid ambiguities caused by pulse trains and still have reasonable independent control on both range and Doppler accuracies by using a single modulated pulse with a time-bandwidth product that is much larger than unity. Figure 7.14 shows the ambiguity contour plot associated with an LFM waveform. In this case, τ' is the pulse width and B is the pulse bandwidth.

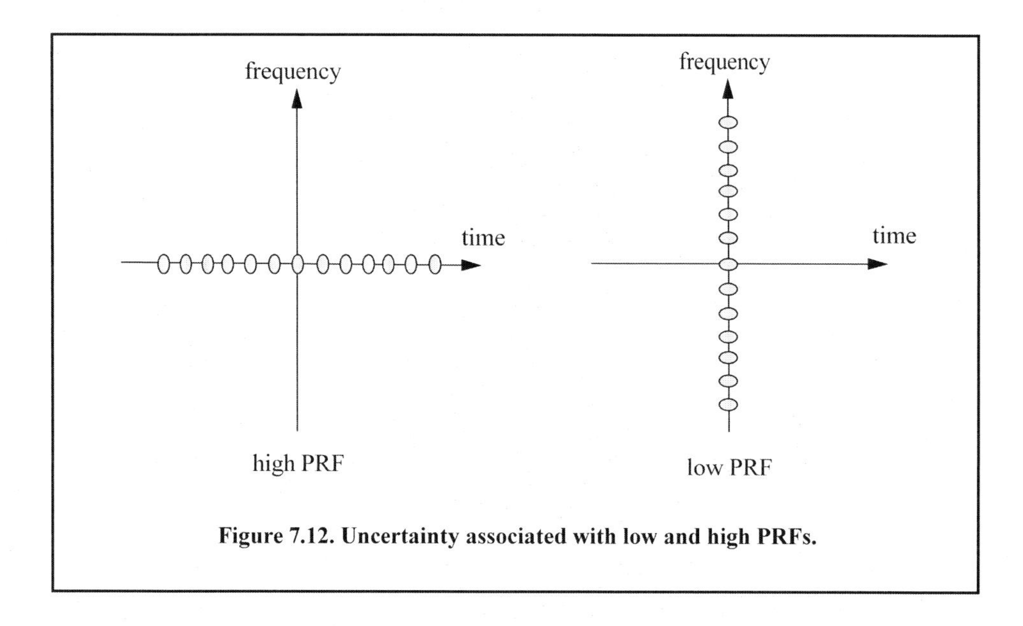

Figure 7.12. Uncertainty associated with low and high PRFs.

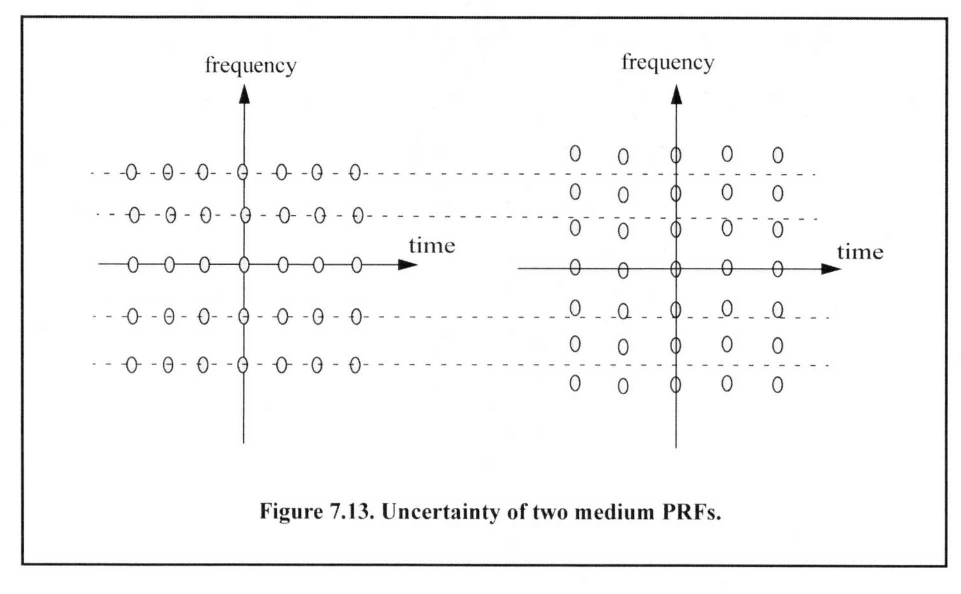

Figure 7.13. Uncertainty of two medium PRFs.

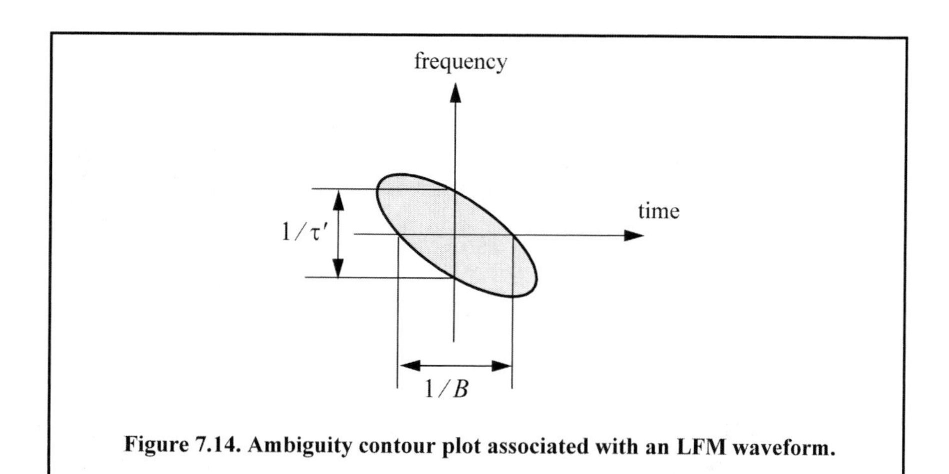

Figure 7.14. Ambiguity contour plot associated with an LFM waveform.

7.5. Phase Coded Ambiguity Function

The *goodness* of a given waveform is determined based on its range and Doppler resolutions (i.e., the zero Doppler and zero delay cuts of the ambiguity function). A few common analog radar waveforms were analyzed in the preceding section of this chapter. In this section, another type of radar waveforms based on discrete codes is analyzed. Discrete coded waveforms are more effective in improving range characteristics than Doppler (velocity) characteristics. Furthermore, in some radar applications, discrete coded waveforms are heavily favored because of their inherent anti-jamming capabilities. In this section, a quick overview of discrete coded waveforms is presented; then examples of some common phase-modulated (binary or polyphase) codes are introduced.

7.5.1. Discrete Code Signal Representation

The general form for a discrete coded signal can be written as

$$x(t) = e^{j2\pi f_0 t}\sum_{n=1}^{N} u_n(t) = e^{j2\pi f_0 t}\sum_{n=1}^{N} P_n(t)e^{j(2\pi f_n t + \theta_n)} \qquad \text{Eq. (7.77)}$$

where f_0 is the carrier frequency in Hertz, (f_n, θ_n) are constants, N is the code length (number of bits in the code), and the signal $P_n(t)$ is given by

$$P_n(t) = a_n Rect\left(\frac{t}{\tau_0}\right). \qquad \text{Eq. (7.78)}$$

The constant a_n is either (1) or (0), and

$$Rect\left(\frac{t}{\tau_0}\right) = \begin{cases} 1 & ; \quad 0 < t < \tau_0 \\ 0 & ; \quad elsewhere \end{cases}. \qquad \text{Eq. (7.79)}$$

Using this notation, the discrete code can be described through the sequence

$$U[n] = \{u_n \ , n = 1, 2, \ldots, N\} \qquad \text{Eq. (7.80)}$$

which, in general, is a complex sequence depending on the values of f_n and θ_n. The sequence $U[n]$ is called the code, and for convenience it will be denoted by U.

In general, the output of the matched filter is

$$\chi(\tau, f_d) = \int_{-\infty}^{\infty} x^*(t) x(t+\tau) e^{-j2\pi f_d t} dt . \qquad \text{Eq. (7.81)}$$

Substituting Eq. (7.77) into Eq. (7.81) yields

$$\chi(\tau, f_d) = \sum_{n=1}^{N} \sum_{k=1}^{N} \int_{-\infty}^{\infty} u_n^*(t) u_k(t+\tau) e^{-j2\pi f_d t} dt . \qquad \text{Eq. (7.82)}$$

Depending on the choice of combination for a_n, f_n, and θ_n, different classes of codes can be generated.

7.5.2. Phase Coding

The signal corresponding to this class of code is obtained from Eq. (7.77) by letting $f_n = 0$. It follows that

$$x(t) = e^{j2\pi f_0 t} \sum_{n=1}^{N} u_n(t) = e^{j2\pi_0 t} \sum_{n=1}^{N} P_n(t) e^{j\theta_n} . \qquad \text{Eq. (7.83)}$$

Two subclasses of phase codes are analyzed. They are binary phase codes and polyphase codes.

7.5.2.1. Binary Phase Codes

In this case, the phase θ_n is set equal to either (0) or (π), and hence, the term *binary* is used. For this purpose, define the coefficient D_n as

$$D_n = e^{j\theta_n} = \pm 1 . \qquad \text{Eq. (7.84)}$$

The ambiguity function for this class of code is derived by substituting Eq. (7.83) into Eq. (7.81). The resulting ambiguity function is given by

$$\chi(\tau; f_d) = \begin{cases} \chi_0(\tau', f_d) \displaystyle\sum_{n=1}^{N-k} D_n D_{n+k} e^{-j2\pi f_d (n-1)\tau_0} + \\ \chi_0(\tau_0 - \tau', f_d) \displaystyle\sum_{n=1}^{N-(k+1)} D_n D_{n+k+1} e^{-j2\pi f_d n \tau_0} \end{cases} \quad 0 < \tau < N\tau_0 \qquad \text{Eq. (7.85)}$$

where

$$\tau = k\tau_0 + \tau' \qquad \begin{cases} 0 < \tau' < \tau_0 \\ k = 0, 1, 2, \ldots, N \end{cases} \qquad \text{Eq. (7.86)}$$

$$\chi_0(\tau', f_d) = \int_0^{\tau_0 - \tau'} \exp(-j2\pi f_d t) dt \qquad 0 < \tau' < \tau_0 . \qquad \text{Eq. (7.87)}$$

The corresponding zero Doppler cut is then given by

$$\chi(\tau;0) = \tau_0\left(1 - \frac{|\tau'|}{\tau_0}\right) \sum_{n=1}^{N-|k|} D_n D_{n+k} + |\tau'| \sum_{n=1}^{N-|k+1|} D_n D_{n+k+1} , \qquad \text{Eq. (7.88)}$$

and when $\tau' = 0$ then

$$\chi(k;0) = \tau_0 \sum_{n=1}^{N-|k|} D_n D_{n+k} . \qquad \text{Eq. (7.89)}$$

Barker Code

Barker code is one of the most commonly known codes from the binary phase code class. In this case, a long pulse of width T_p is divided into N smaller pulses; each is of width $\tau_0 = T_p/N$. Then, the phase of each subpulse is chosen as either 0 or π radians relative to some code. It is customary to characterize a subpulse that has 0 phase (amplitude of +1 volt) as either "1" or "+." Alternatively, a subpulse with phase equal to π (amplitude of -1 volt) is characterized by either "0" or "-." Figure 7.15 illustrates this concept for a Barker code of length seven. A Barker code of length N is denoted as B_N. There are only seven known Barker codes that share this unique property; they are listed in Table 7.1. Note that B_2 and B_4 have complementary forms that have the same characteristics.

In general, the autocorrelation function (which is an approximation for the matched filter output) for a B_N Barker code will be $2N\tau_0$ wide. The main lobe is $2\tau_0$ wide; the peak value is equal to N. There are $(N-1)/2$ sidelobes on either side of the main lobe; this is illustrated in Figure 7.16 for a B_{13}. Notice that the main lobe is equal to 13, while all sidelobes are unity. The most sidelobe reduction offered by a Barker code is $-22.3dB$, which may not be sufficient for the desired radar application. However, Barker codes can be combined to generate much longer codes. In this case, a B_M code can be used within a B_N code (M within N) to generate a code of length MN. The compression ratio for the combined B_{MN} code is equal to MN. As an example, a combined B_{54} is given by

$$B_{54} = \{11101, 11101, 00010, 11101\} \qquad \text{Eq. (7.90)}$$

and is illustrated in Figure 7.17. Unfortunately, the sidelobes of a combined Barker code autocorrelation function are no longer equal to unity. Some sidelobes of a combined Barker code autocorrelation function can be reduced to zero if the matched filter is followed by a linear transversal filter with impulse response given by

$$h(t) = \sum_{k=-N}^{N} \beta_k \delta(t - 2k\tau_0) \qquad \text{Eq. (7.91)}$$

where N is the filter's order, the coefficients β_k ($\beta_k = \beta_{-k}$) are to be determined, $\delta(\ )$ is the delta function, and τ_0 is the Barker code subpulse width. A filter of order N produces N zero sidelobes on either side of the main lobe. The main lobe amplitude and width do not change, as illustrated in Figure 7.18.

In order to illustrate this approach, consider the case where the input to the matched filter is B_{11}, and assume $N = 4$. The autocorrelation for a B_{11} is

$$R_{11} = \{-1, 0, -1, 0, -1, 0, -1, 0, -1, 0, 11, 0, -1, 0, -1, 0, -1, 0, -1, 0, -1\}. \qquad \text{Eq. (7.92)}$$

The output of the transversal filter is the discrete convolution between its impulse response and the sequence R_{11}. At this point we need to compute the coefficients β_k that guarantee the desired filter output (i.e., unchanged main lobe and four zero sidelobe levels). Performing the discrete convolution as defined in Eq. (7.91) and collecting equal terms ($\beta_k = \beta_{-k}$) yield the following set of five linearly independent equations:

$$\begin{bmatrix} 11 & -2 & -2 & -2 & -2 \\ -1 & 10 & -2 & -2 & -1 \\ -1 & -2 & 10 & -2 & -1 \\ -1 & -2 & -1 & 11 & -1 \\ -1 & -1 & -1 & -1 & 11 \end{bmatrix} \begin{bmatrix} \beta_0 \\ \beta_1 \\ \beta_2 \\ \beta_3 \\ \beta_4 \end{bmatrix} = \begin{bmatrix} 11 \\ 0 \\ 0 \\ 0 \\ 0 \end{bmatrix}. \qquad \text{Eq. (7.93)}$$

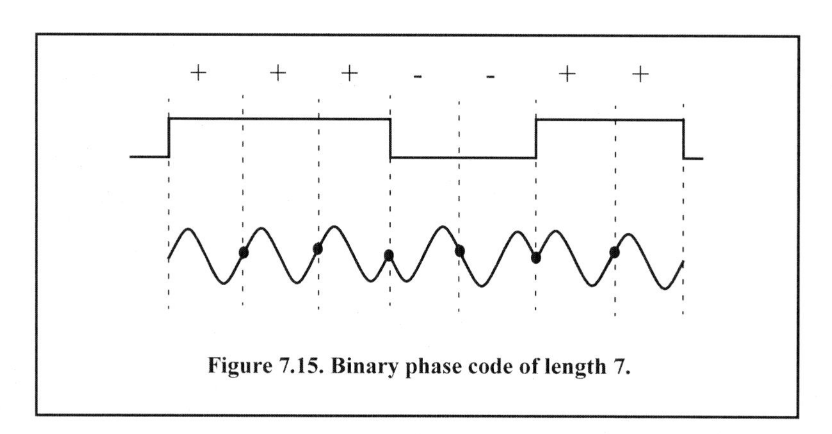

Figure 7.15. Binary phase code of length 7.

Table 7.1. Barker codes.

Code Symbol	Code Length	Code Elements	Side Lode Reduction (dB)
B_2	2	+- ; ++	6.0
B_3	3	++-	9.5
B_4	4	++-+; +++-	12.0
B_5	5	+++-+	14.0
B_7	7	+++--+-	16.9
B_{11}	11	+++---+--+-	20.8
B_{13}	13	+++++--++-+-+	22.3

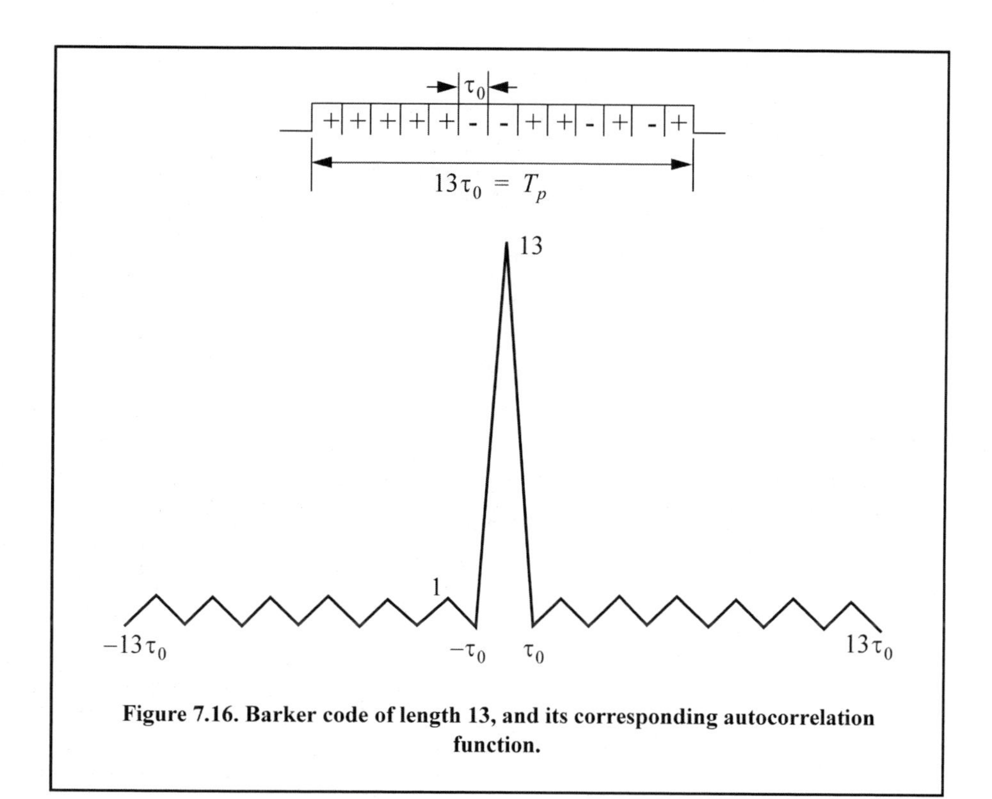

Figure 7.16. Barker code of length 13, and its corresponding autocorrelation function.

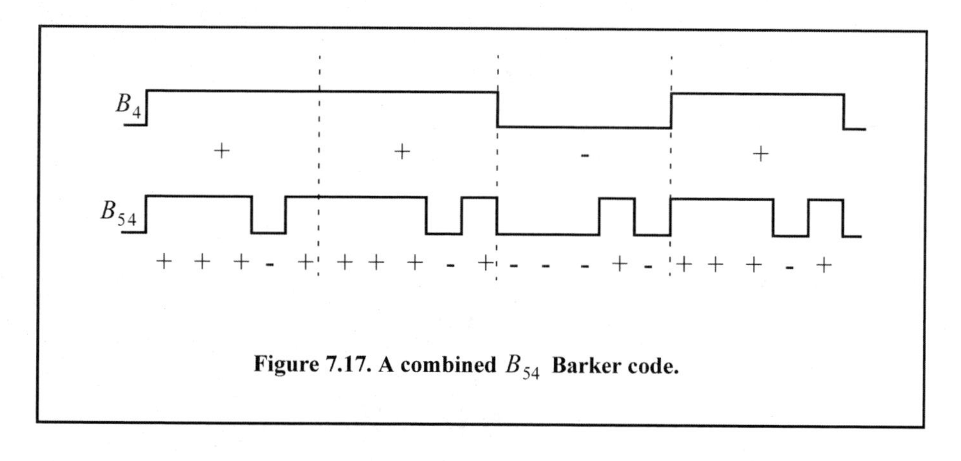

Figure 7.17. A combined B_{54} Barker code.

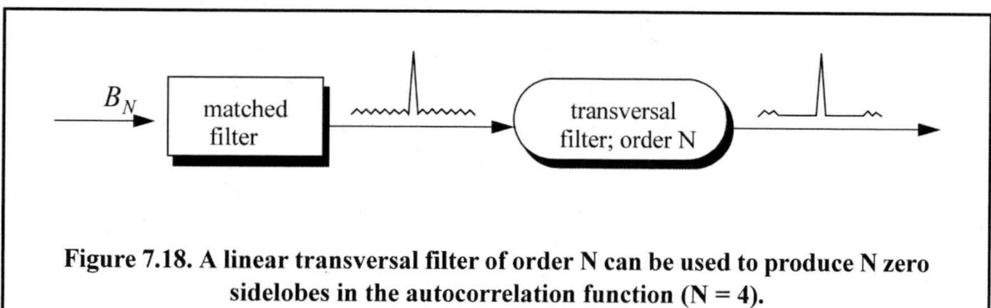

Figure 7.18. A linear transversal filter of order N can be used to produce N zero sidelobes in the autocorrelation function (N = 4).

Solving Eq. (7.93) yields

$$\begin{bmatrix} \beta_0 \\ \beta_1 \\ \beta_2 \\ \beta_3 \\ \beta_4 \end{bmatrix} = \begin{bmatrix} 1.1342 \\ 0.2046 \\ 0.2046 \\ 0.1731 \\ 0.1560 \end{bmatrix} . \qquad \text{Eq. (7.94)}$$

Note that setting the first equation equal to 11 and all other equations to 0 and then solving for β_k guarantees that the main peak remains unchanged, and that the next four sidelobes are zeros. So far we have assumed that coded pulses have rectangular shapes. Using other pulses of other shapes, such as Gaussian, may produce better sidelobe reduction and a larger compression ratio. Figure 7.19 shows the output of this function when B_{13} is used as an input.

Figure 7.20 is similar to Figure 7.19, except in this case B_7 is used as an input. Figure 7.21 shows the ambiguity function, the zero Doppler cut, and the contour plot for the combined Barker code B_{54}, defined in Figure 7.17.

7.5.2.2. Polyphase Codes

The signal corresponding to polyphase codes is that given in Eq. (7.83) and the corresponding ambiguity function was given in Eq. (7.85). The only exception is that the phase θ_n is no longer restricted to $(0, \pi)$. Hence, the coefficient D_n are no longer equal to ± 1 but can be complex depending on the value of θ_n. Polyphase Barker codes have been investigated by many scientists, and much is well documented in the literature. In this chapter the discussion will be limited to Frank codes.

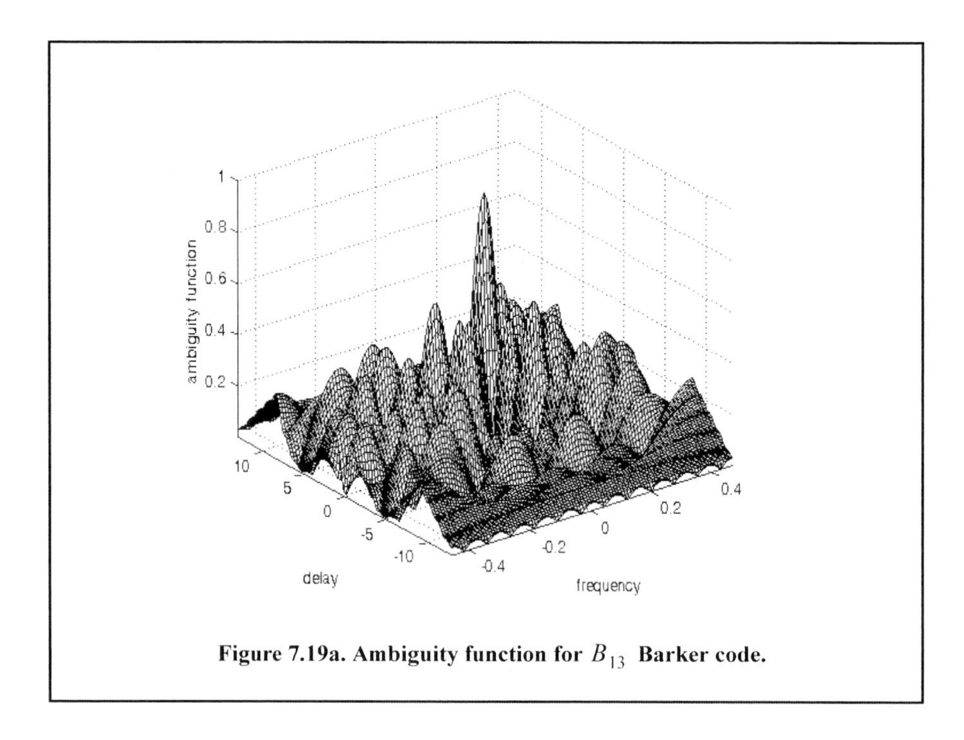

Figure 7.19a. Ambiguity function for B_{13} Barker code.

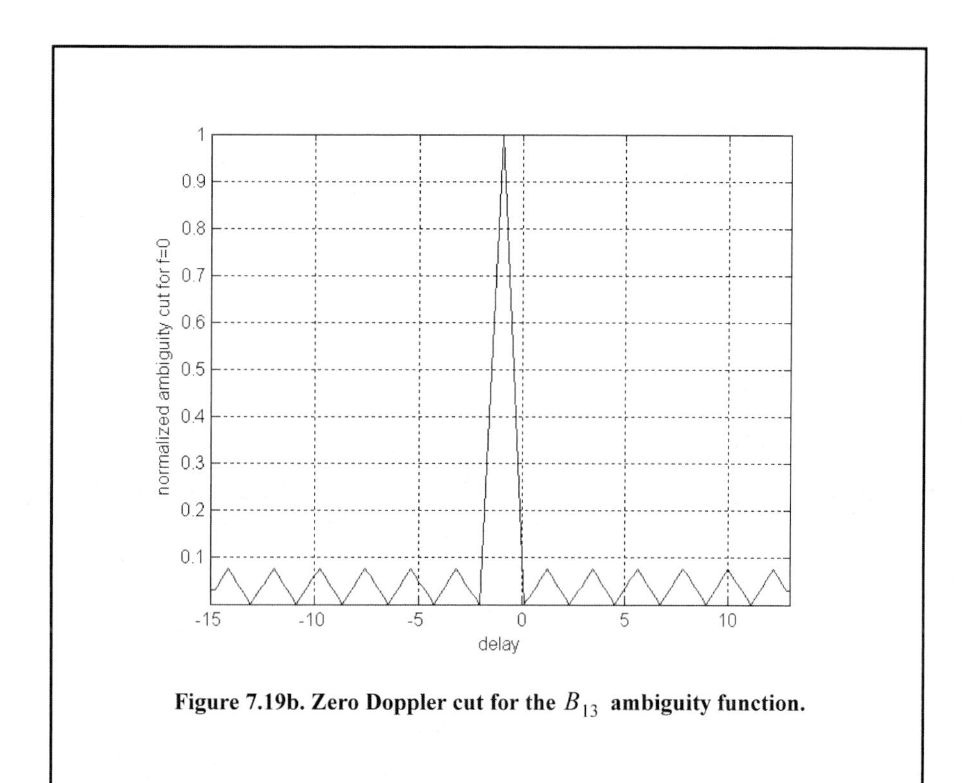

Figure 7.19b. Zero Doppler cut for the B_{13} ambiguity function.

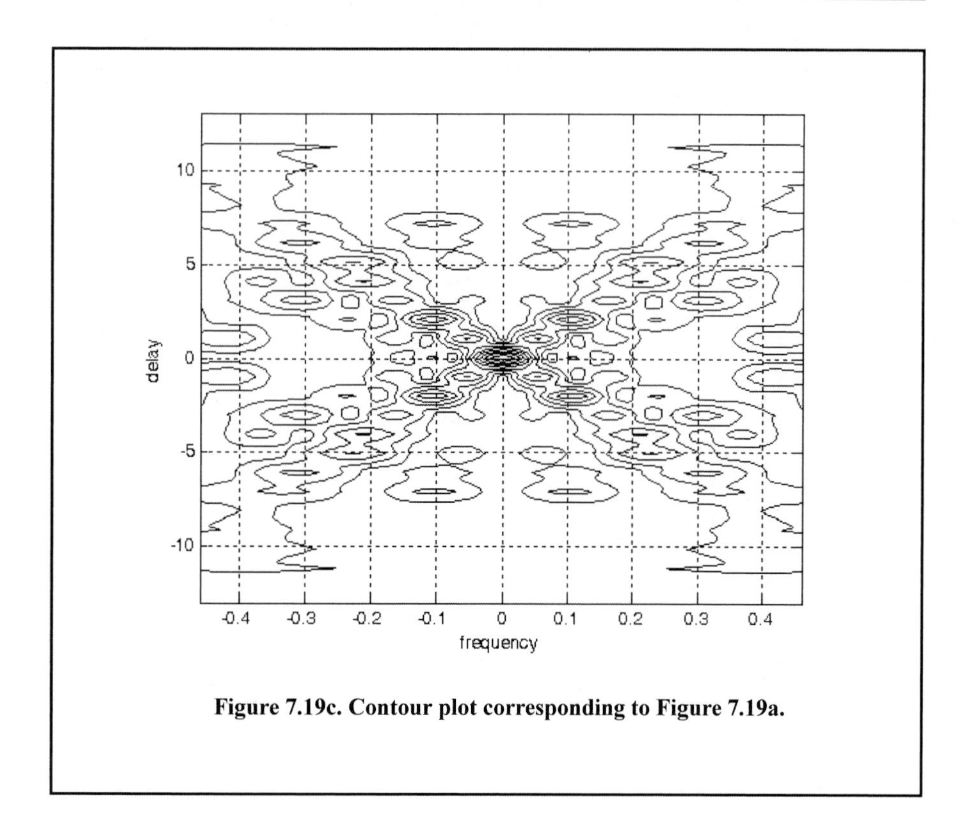

Figure 7.19c. Contour plot corresponding to Figure 7.19a.

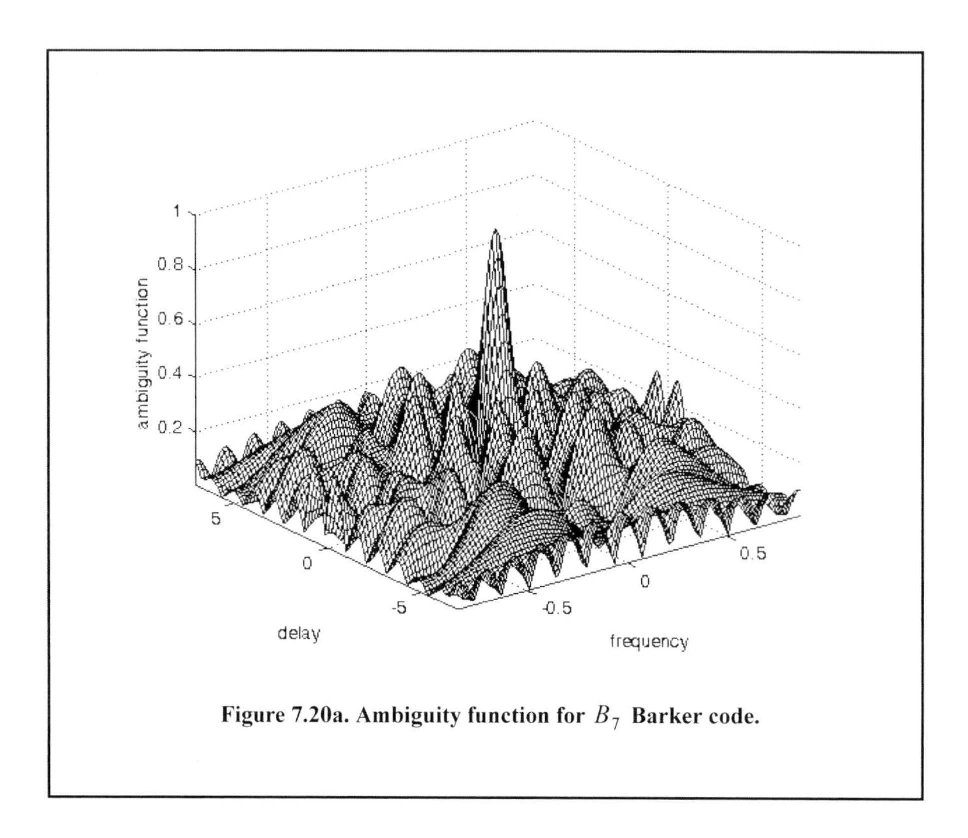

Figure 7.20a. Ambiguity function for B_7 Barker code.

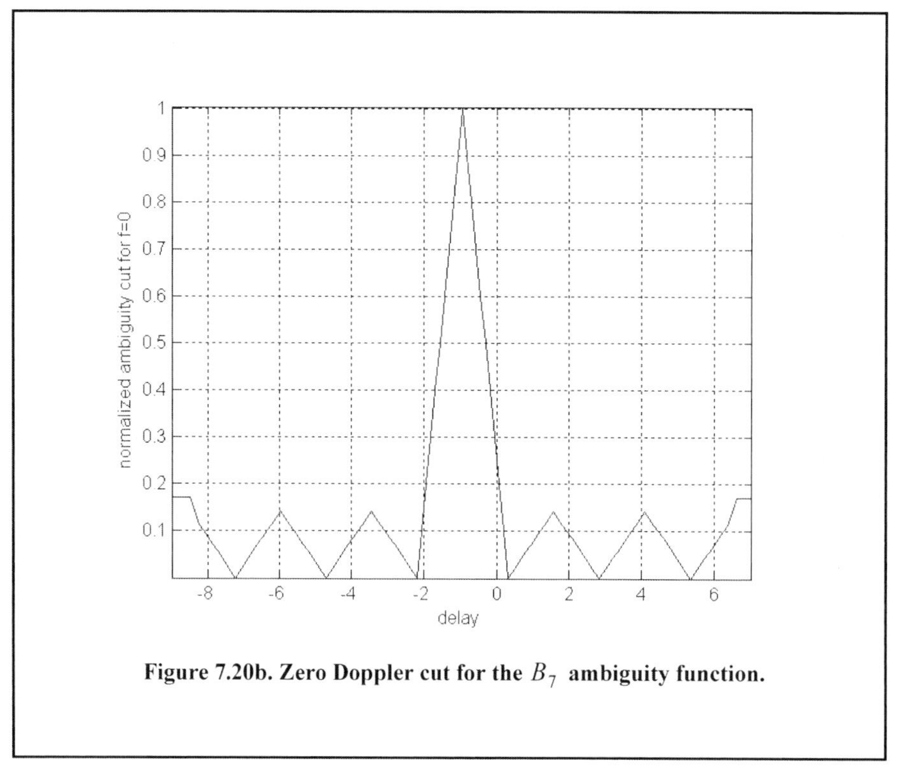

Figure 7.20b. Zero Doppler cut for the B_7 ambiguity function.

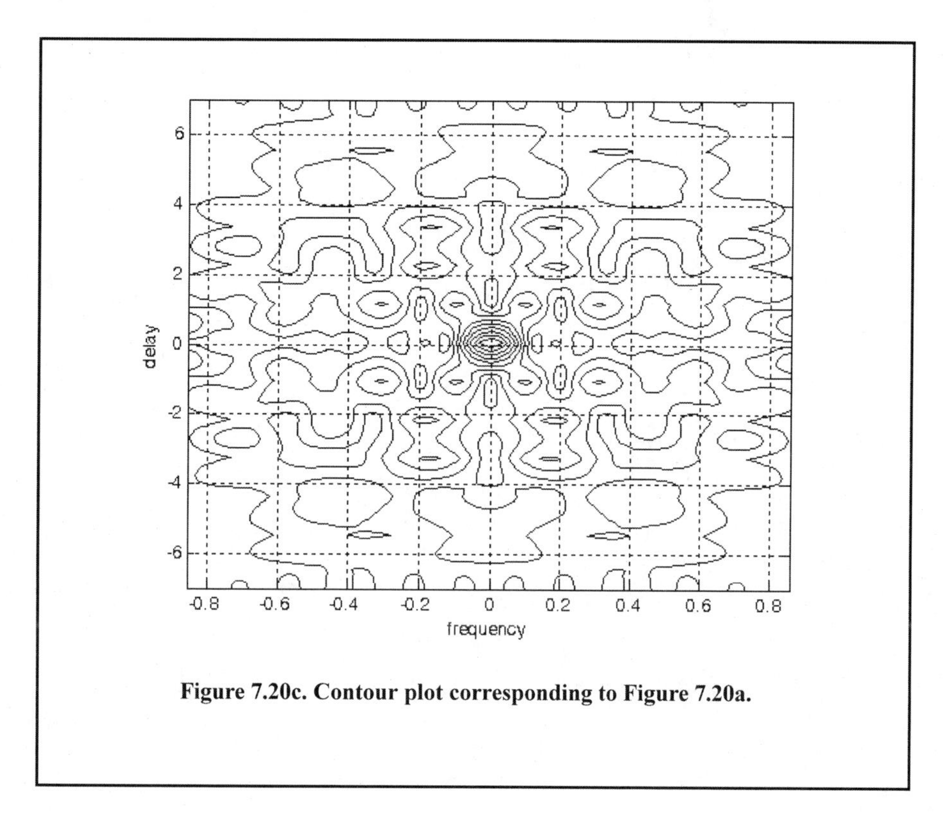

Figure 7.20c. Contour plot corresponding to Figure 7.20a.

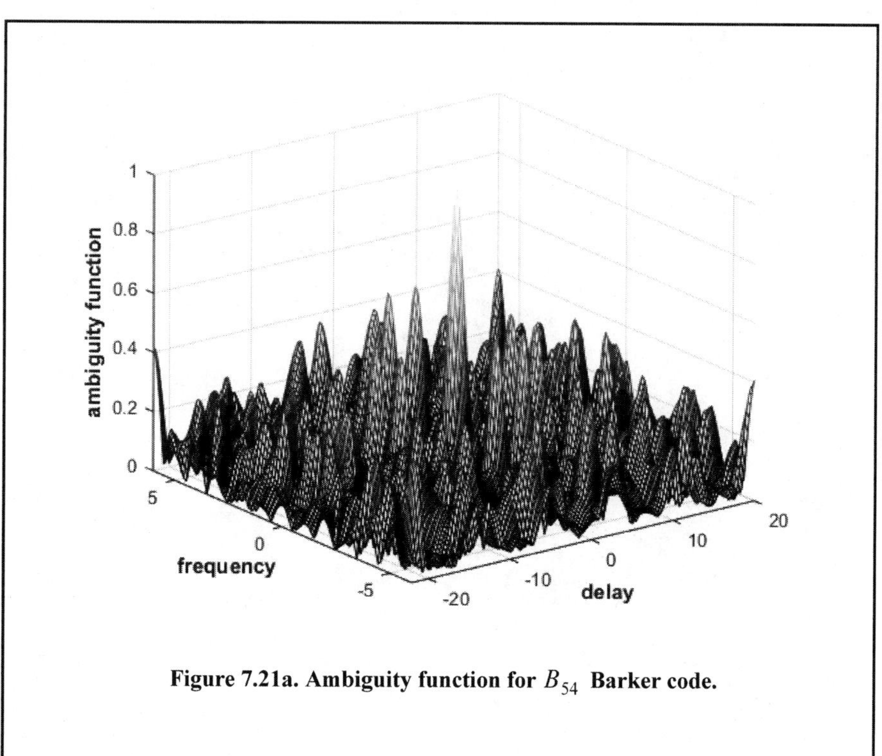

Figure 7.21a. Ambiguity function for B_{54} Barker code.

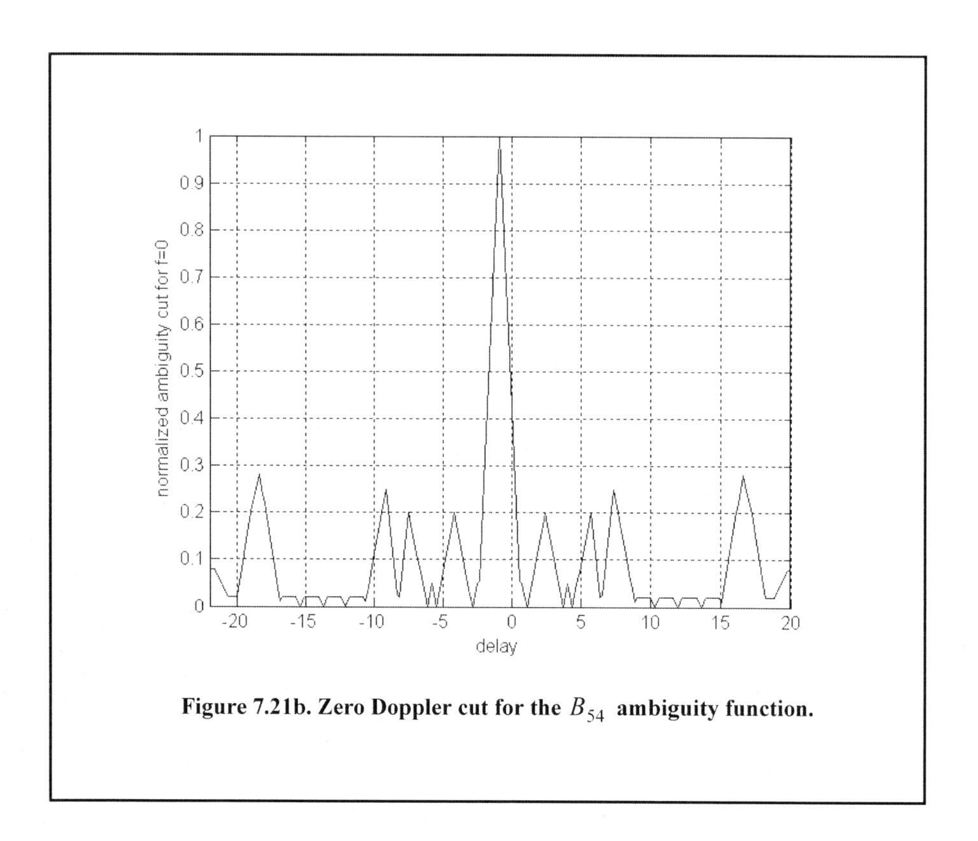

Figure 7.21b. Zero Doppler cut for the B_{54} ambiguity function.

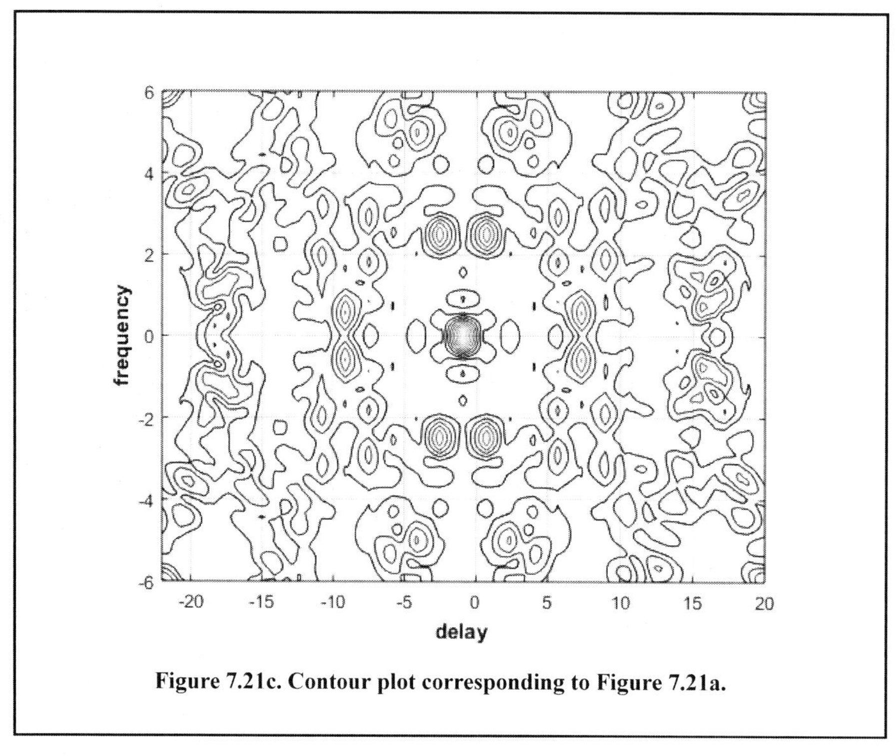

Figure 7.21c. Contour plot corresponding to Figure 7.21a.

Frank Codes

In this case, a single pulse of width T_p is divided into N equal groups; each group is subsequently divided into other N subpulses, each of width τ_0. Therefore, the total number of subpulses within each pulse is N^2, and the compression ratio is $\xi = N^2$. As previously, the phase within each subpulse is held constant with respect to some CW reference signal. A Frank code of N^2 subpulses is referred to as an N-phase Frank code. The first step in computing a Frank code is to divide $360°$ by N and define the result as the fundamental phase increment $\Delta\varphi$. More precisely,

$$\Delta\varphi = 360°/N. \qquad \text{Eq. (7.95)}$$

Note that the size of the fundamental phase increment decreases as the number of groups is increased, and because of phase stability, this may degrade the performance of very long Frank codes. For an N-phase Frank code the phase of each subpulse is computed from

$$\begin{pmatrix} 0 & 0 & 0 & 0 & \ldots & 0 \\ 0 & 1 & 2 & 3 & \ldots & N-1 \\ 0 & 2 & 4 & 6 & \ldots & 2(N-1) \\ \ldots & \ldots & \ldots & \ldots & \ldots & \ldots \\ \ldots & \ldots & \ldots & \ldots & \ldots & \ldots \\ 0 & (N-1) & 2(N-1) & 3(N-1) & \ldots & (N-1)^2 \end{pmatrix} \Delta\varphi \qquad \text{Eq. (7.96)}$$

where each row represents a group, and a column represents the subpulses for that group. For example, a 4-phase Frank code has $N = 4$, and the fundamental phase increment is $\Delta\varphi = (360°/4) = 90°$. It follows that

$$\begin{pmatrix} 0 & 0 & 0 & 0 \\ 0 & 90° & 180° & 270° \\ 0 & 180° & 0 & 180° \\ 0 & 270° & 180° & 90° \end{pmatrix} \Rightarrow \begin{pmatrix} 1 & 1 & 1 & 1 \\ 1 & j & -1 & -j \\ 1 & -1 & 1 & -1 \\ 1 & -j & -1 & j \end{pmatrix}. \qquad \text{Eq. (7.97)}$$

Therefore, a Frank code of 16 elements is given by

$$F_{16} = \{1\ 1\ 1\ 1\ 1\ j\ -1\ -j\ 1\ -1\ 1\ -1\ 1\ -j\ -1\ j\}. \qquad \text{Eq. (7.98)}$$

A plot of the ambiguity function for F_{16} is shown in Figure 7.22. Note the thumbtack shape of the ambiguity function. The phase increments within each row represent a stepwise approximation of an up-chirp LFM waveform. The phase increments for subsequent rows increase linearly versus time. Thus, the corresponding LFM chirp slopes also increase linearly for subsequent rows. This is illustrated in Figure 7.23, for F_{16}.

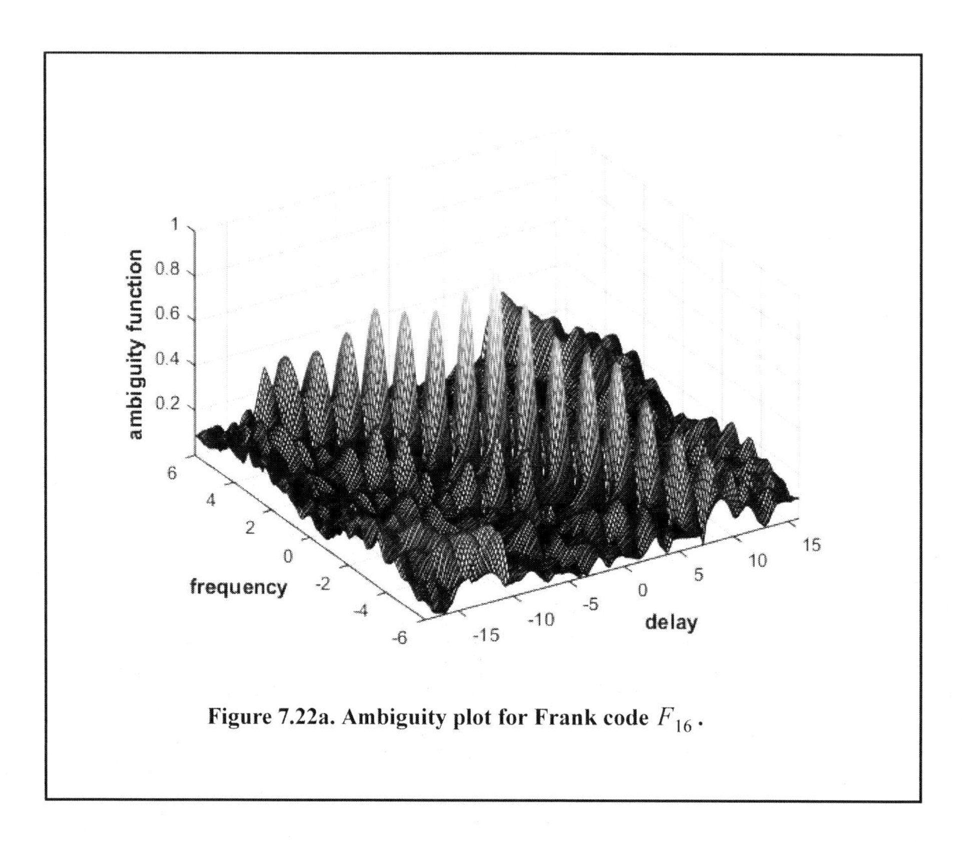

Figure 7.22a. Ambiguity plot for Frank code F_{16}.

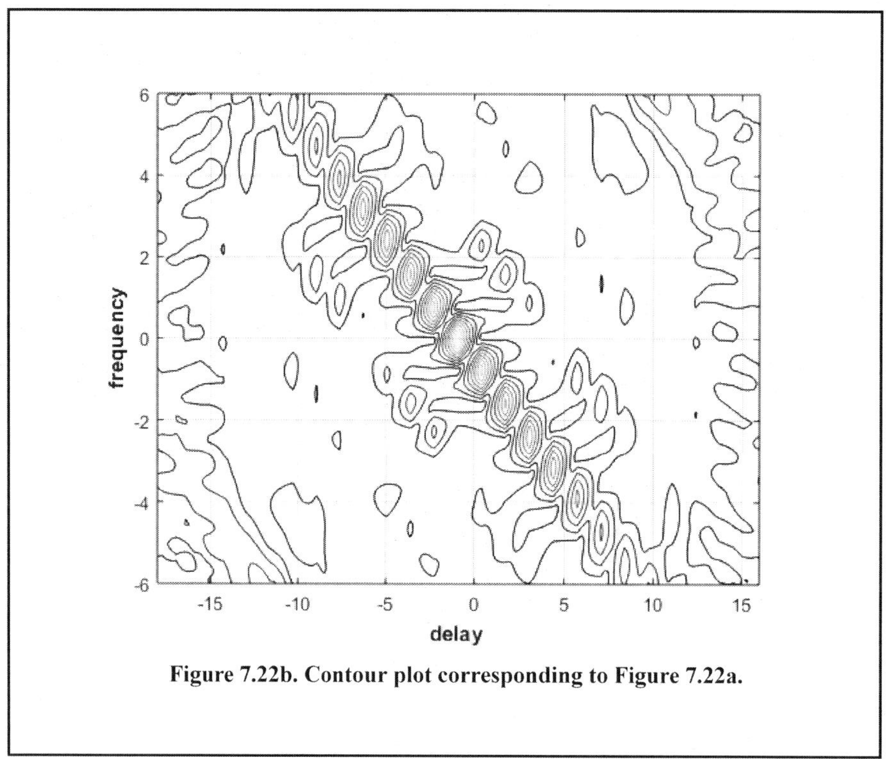

Figure 7.22b. Contour plot corresponding to Figure 7.22a.

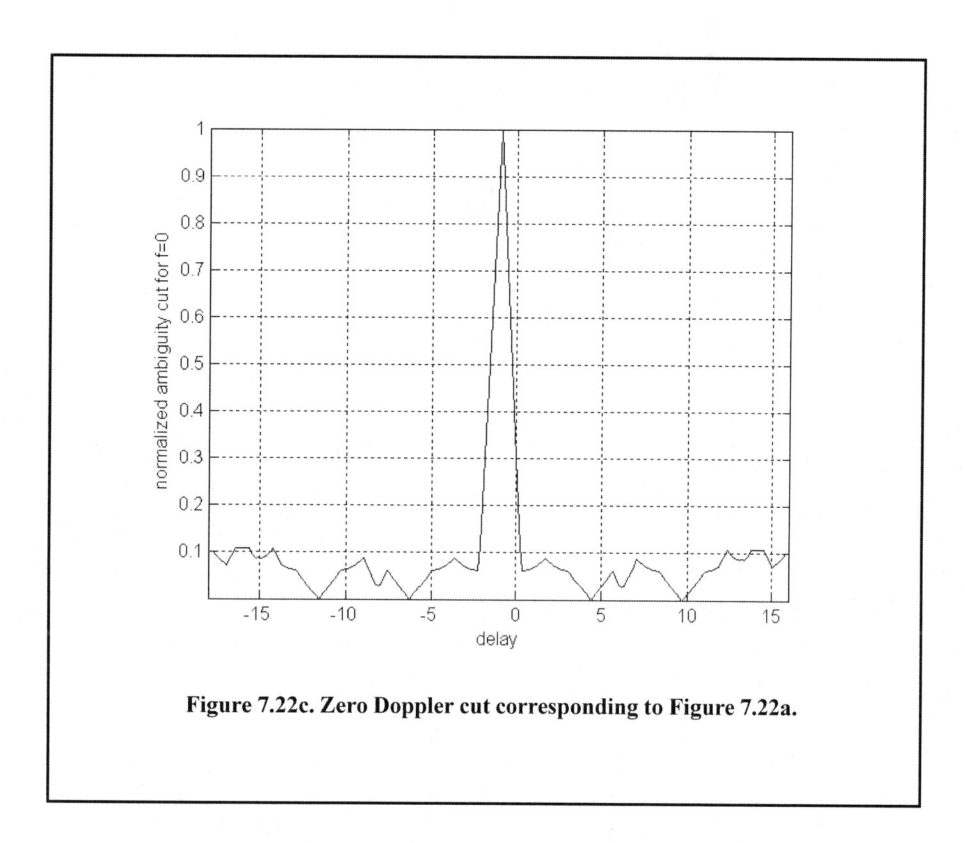

Figure 7.22c. Zero Doppler cut corresponding to Figure 7.22a.

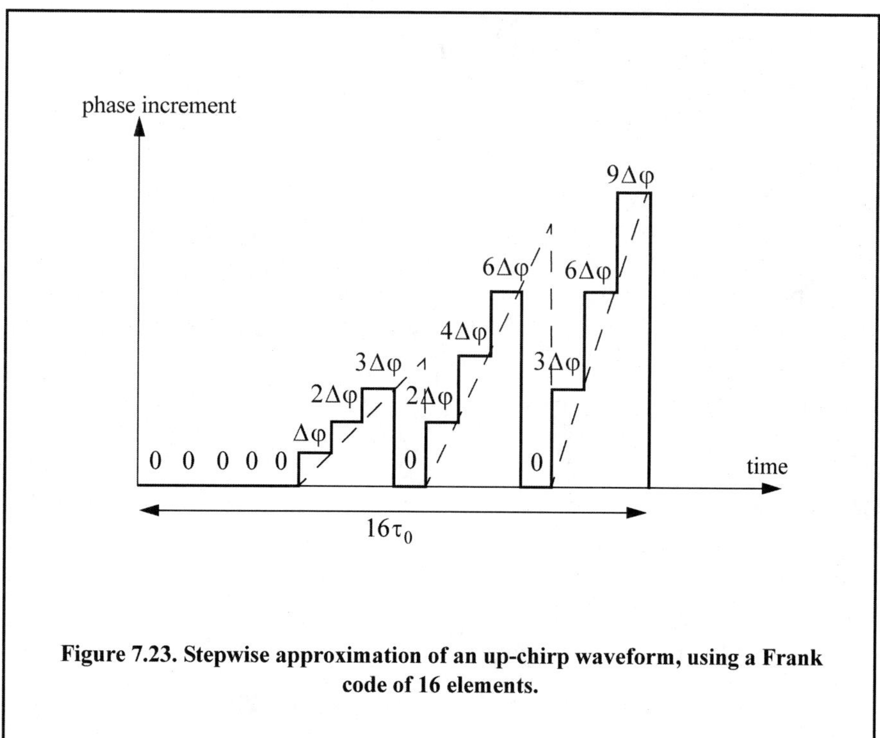

Figure 7.23. Stepwise approximation of an up-chirp waveform, using a Frank code of 16 elements.

Problems

7.1. Compute the frequency response for the filter matched to the signal

(a) $x(t) = \exp\left(\frac{-t^2}{2T}\right)$;

(b) $x(t) = u(t)\exp(-\alpha t)$ where α is a positive constant.

7.2. Repeat the example in Section 7.1 using $x(t) = u(t)\exp(-\alpha t)$.

7.3. A closed form expression for the SNR at the output of the matched filter when the input noise is white was developed in Section 7.1. Derive an equivalent formula for the non-white noise case.

7.4. A radar system uses LFM waveforms. The received signal is of the form $s_r(t) = As(t-\tau) + n(t)$, where τ is a time delay that depends on range, $s(t) = Rect(t/\tau')\cos(2\pi f_0 t - \phi(t))$, and $\phi(t) = -\pi Bt^2/\tau'$. Assume that the radar bandwidth is $B = 5MHz$, and the pulse width is $\tau' = 5\mu s$. (a) Give the quadrature components of the matched filter response that is matched to $s(t)$. (b) Write an expression for the output of the matched filter. (c) Compute the increase in SNR produced by the matched filter.

7.5. (a) Write an expression for the ambiguity function of an LFM waveform, where $\tau' = 6.4\mu s$ and the compression ratio is 32. (b) Give an expression for the matched filter impulse response.

7.6. (a) Write an expression for the ambiguity function of an LFM signal with bandwidth $B = 10MHz$, pulse width $\tau' = 1\mu s$, and wavelength $\lambda = 1cm$. (b) Plot the zero Doppler cut of the ambiguity function. (c) Assume a target moving toward the radar with radial velocity $v_r = 100m/s$. What is the Doppler shift associated with this target? (d) Plot the ambiguity function for the Doppler cut in part (c). (e) Assume that three pulses are transmitted with PRF $f_r = 2000Hz$. Repeat part (b).

7.7. (a) Give an expression for the ambiguity function for a pulse train consisting of 4 pulses, where the pulse width is $\tau' = 1\mu s$ and the pulse repetition interval is $T = 10\mu s$. Assume a wavelength of $\lambda = 1cm$. (b) Sketch the ambiguity function contour.

7.8. Consider a sonar system with range resolution $\Delta R = 4cm$. (a) A sinusoidal pulse at frequency $f_0 = 100KHz$ is transmitted. What is the pulse width, and what is the bandwidth? (b) By using an up-chirp LFM, centered at f_0, one can increase the pulse width for the same range resolution. If you want to increase the transmitted energy by a factor of 20, give an expression for the transmitted pulse. (c) Give an expression for the causal filter matched to the LFM pulse in part b.

7.9. A pulse train $y(t)$ is given by

$$y(t) = \sum_{n=0}^{2} w(n)x(t - n\tau')$$

where $x(t) = \exp(-t^2/2)$ is a single pulse of duration τ' and the weighting sequence is $\{w(n)\} = \{0.5, 1, 0.7\}$. Find and sketch the correlations R_x, R_w, and R_y.

7.10. Repeat the previous problem for $x(t) = \exp(-t^2/2)\cos 2\pi f_0 t$.

7.11. Show that

$$\int_{-\infty}^{\infty} tx^*(t)x'(t)\ dt = -\int_{-\infty}^{\infty} fX^*(f)X'(f)\ df$$

where $X(f)$, is the FT of $x(t)$ and $x'(t)$ is its derivative with respect to time. The function $X'(f)$ is the derivative of $X(f)$ with respect to frequency.

7.12. Derive an expression for the ambiguity function of a Gaussian pulse defined by

$$x(t) = \frac{1}{\sqrt{\sigma}\ ^{1/4}\sqrt{\pi}} \exp\left[\frac{-t^2}{2\sigma^2}\right] \qquad ;0 < t < T$$

where T is the pulsewidth and σ is a constant.

7.13. Derive an expression for the ambiguity function of a V-LFM waveform, illustrated in figure below. In this case, the overall complex envelope is

$$\tilde{x}(t) = \tilde{x}_1(t) + \tilde{x}_2(t) \qquad ;-T < t < T$$

where

$$\tilde{x}_1(t) = \frac{1}{\sqrt{2T}} \exp[-\mu t^2] \qquad ;-T < t < 0$$

and

$$\tilde{x}_2(t) = \frac{1}{\sqrt{2T}} \exp[\mu t^2] \qquad ;0 < t < T$$

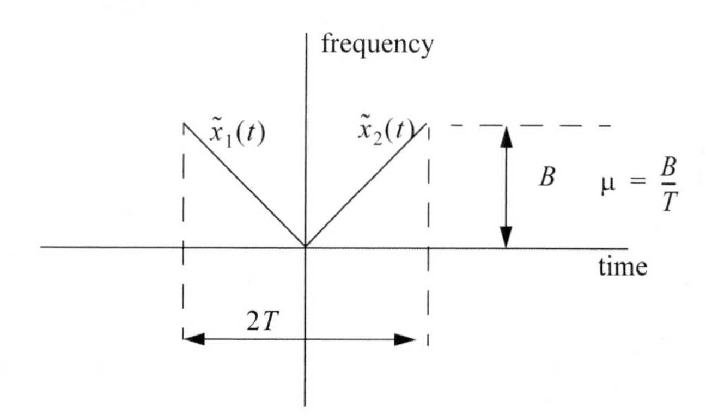

7.14. Using the stationary phase concept, find the instantaneous frequency for the waveform whose envelope and complex spectrum are, respectively, given by

$$r(t) = \frac{1}{\sqrt{T}} \exp\left[-\left(\frac{2t}{T}\right)^2\right] \qquad ;0 < t < T$$

and

$$|X(f)| = \frac{1}{\sqrt{B}} \exp\left[-\left(\frac{2f}{B}\right)^2\right].$$

7.15. Using the stationary phase concept, find the instantaneous frequency for the waveform whose envelope and complex spectrum are given respectively by

$$r(t) = \frac{1}{\sqrt{\tau_0}} Rect\left(\frac{t}{\tau_0}\right) \qquad ; \; 0 < t < \tau_0$$

and

$$|X(\omega)| = \frac{2}{\sqrt{B}} \frac{1}{\sqrt{1 + (2\omega/B)^2}}.$$

7.16. Prove that cuts in the ambiguity function are always defined by an ellipse. Hint: Approximate the ambiguity function using a Taylor series expansion about the values $(\tau, f_d) = (0, 0)$; use only the first three terms in the Taylor series expansion.

7.17. Show that the zero Doppler cut for the ambiguity function of an arbitrary phase coded pulse with a pulse width τ_p is given by $Y(f) = |\mathrm{sin}c(f\tau_p)|^2$.

7.18. Consider the 7-bit Barker code, designated by the sequence $x(n)$. (a) Compute and plot the autocorrelation of this code. (b) A radar uses binary phase-coded pulses of the form $s(t) = r(t)\cos(2\pi f_0 t)$, where $r(t) = x(0),\ for\ 0 < t < \Delta t$.

7.19. $r(t) = x(n),\ for\ n\Delta t < t < (n+1)\Delta t$, and $r(t) = 0,\ for\ t > 7\Delta t$. Assume $\Delta t = 0.5\mu s$. (a) Give an expression for the autocorrelation of the signal $s(t)$, and for the output of the matched filter when the input is $s(t - 10\Delta t)$. (b) Compute the time bandwidth product, the increase in the peak SNR, and the compression ratio.

7.20. Develop a Barker code of length 35. Consider both B_{75} and B_{57}.

7.21. Compute the discrete autocorrelation for an F_{16} Frank code.

7.22. Generate a Frank code of length 8, i.e., F_8.

Chapter 8

Pulse Compression

Range resolution for a given radar can be significantly improved by using very short pulses. Unfortunately, utilizing very short pulses decreases the average transmitted power, which can hinder the radar's normal modes of operation particularly for multi-function and surveillance radars. Since the average transmitted power is directly linked to the receiver SNR, it is often desirable to increase the pulse width (i.e., increase the average transmitted power) while simultaneously maintaining adequate range resolution (i.e., use short pulses). This can be made possible by using pulse compression techniques. Pulse compression allows us to achieve the average transmitted power of a relatively long pulse, while obtaining the range resolution corresponding to a short pulse. Two pulse compression techniques are discussed. First, we consider the correlation processing, which is predominantly used for narrowband and some medium-band radar operations. Second stretch processing (normally used for extremely wideband radar operations) is considered next.

8.1. Time-Bandwidth Product

Consider a radar system that employs a matched filter receiver. Let the matched filter receiver bandwidth be denoted as B. Then the noise power available within the matched filter bandwidth is given by

$$N_i = 2B(\eta_0/2) \qquad \text{Eq. (8.1)}$$

where the factor of two is used to account for both negative and positive frequency bands, as illustrated in Figure 8.1. The average input signal power over a pulse duration τ_0 is

$$S_i = E_x/\tau_0 . \qquad \text{Eq. (8.2)}$$

E_x is the signal energy. Consequently, the matched filter input SNR is given by

$$(SNR)_i = S_i/N_i = E/(\eta_0 B\tau_0) . \qquad \text{Eq. (8.3)}$$

The output peak instantaneous SNR to the input SNR ratio, at a specific time t_0, is

$$(SNR(t_0))/(SNR)_i = 2B\tau_0 . \qquad \text{Eq. (8.4)}$$

The quantity $B\tau_0$ is referred to as the time-bandwidth product for a given waveform or its corresponding matched filter. The factor $B\tau_0$ by which the output SNR is increased over that at the input is called the matched filter gain, or simply the compression gain.

In general, the time-bandwidth product of an unmodulated pulse approaches unity. The time-bandwidth product of a pulse can be made much greater than unity by using frequency or phase modulation. If the radar receiver transfer function is perfectly matched to that of the input waveform, then the compression gain is equal to $B\tau_0$.

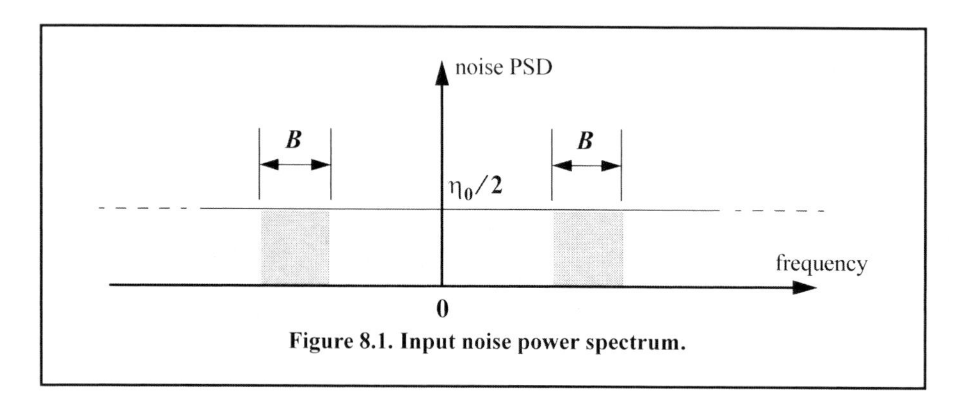

Figure 8.1. Input noise power spectrum.

8.2. Basic Principle of Pulse Compression

For this purpose, consider a long pulse with LFM modulation and assume a matched filter receiver. The output of the matched filter (along the delay axis, i.e., range) is an order of magnitude narrower than that at its input. More precisely, the matched filter output is compressed by a factor $\xi = B\tau_0$, where τ_0 is the pulse width and B is the bandwidth. Thus, by using long pulses and wideband LFM modulation, large compression ratios can be achieved. Figure 8.2 illustrates the ideal LFM pulse compression process. Part (a) shows the envelope of a pulse; part (b) shows the frequency modulation (in this case it is an upchirp LFM) with bandwidth $B = f_2 - f_1$. Part (c) shows the matched filter time-delay characteristic while part (d) shows the compressed pulse envelope. Finally, part (e) shows the matched filter input/output waveforms. Figure 8.3 illustrates the advantage of pulse compression. In this example, two received pulses (from two distinct targets) overlap before pulse compression. However, after pulse compression the two pulses are completely separated and can be resolved as two targets. In fact when using LFM, returns from neighboring targets are resolved as long as they are separated, in time, by τ_{n1} the compressed pulse width. Finally, the compressed pulse range resolution is $\Delta R = c/(2B)$.

In pulse compression, it is desirable to use modulation schemes that can accomplish a maximum pulse compression ratio, and can significantly reduce the side lobe levels of the compressed waveform. For the LFM case the first side lobe is approximately $13.4dB$ below the main peak, and for most radar applications this may not be sufficient. In practice, high side lobe levels are not preferable because noise and / or jammers located at the side lobes may interfere with target returns in the main lobe. Weighting functions (windows) can be used on the compressed pulse spectrum in order to reduce the side lobe levels. The cost associated with such an approach is a loss in the main lobe resolution, and a reduction in the peak value (i.e., loss in the SNR), as illustrated in Figure 8.4. Weighting the time domain transmitted or received signal instead of the compressed pulse spectrum will theoretically achieve the same goal. However, this approach is rarely used, since amplitude modulating the transmitted waveform introduces extra burdens on the transmitter.

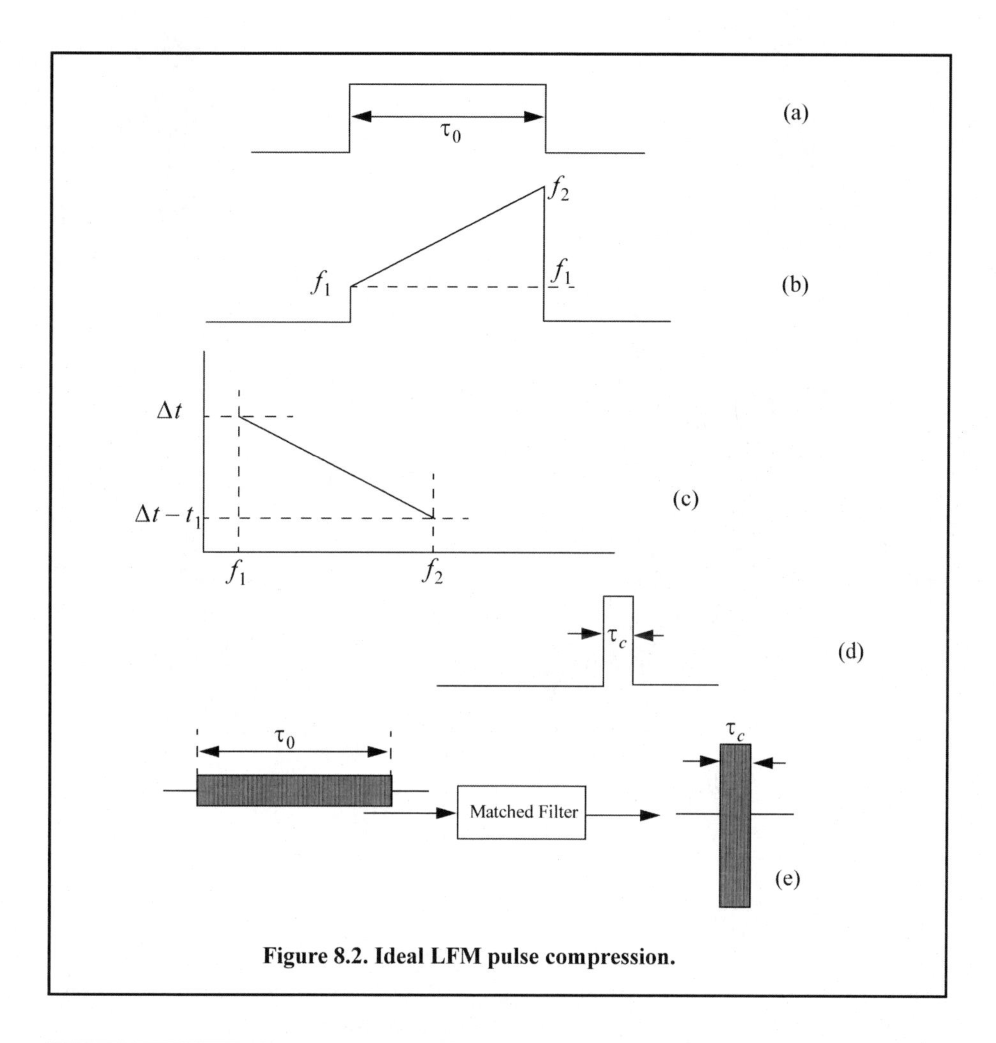

Figure 8.2. Ideal LFM pulse compression.

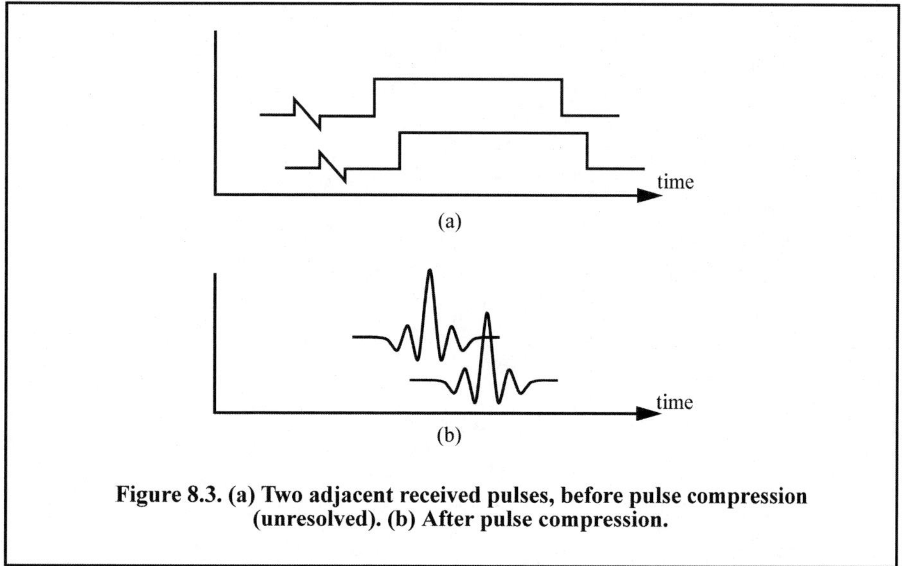

Figure 8.3. (a) Two adjacent received pulses, before pulse compression (unresolved). (b) After pulse compression.

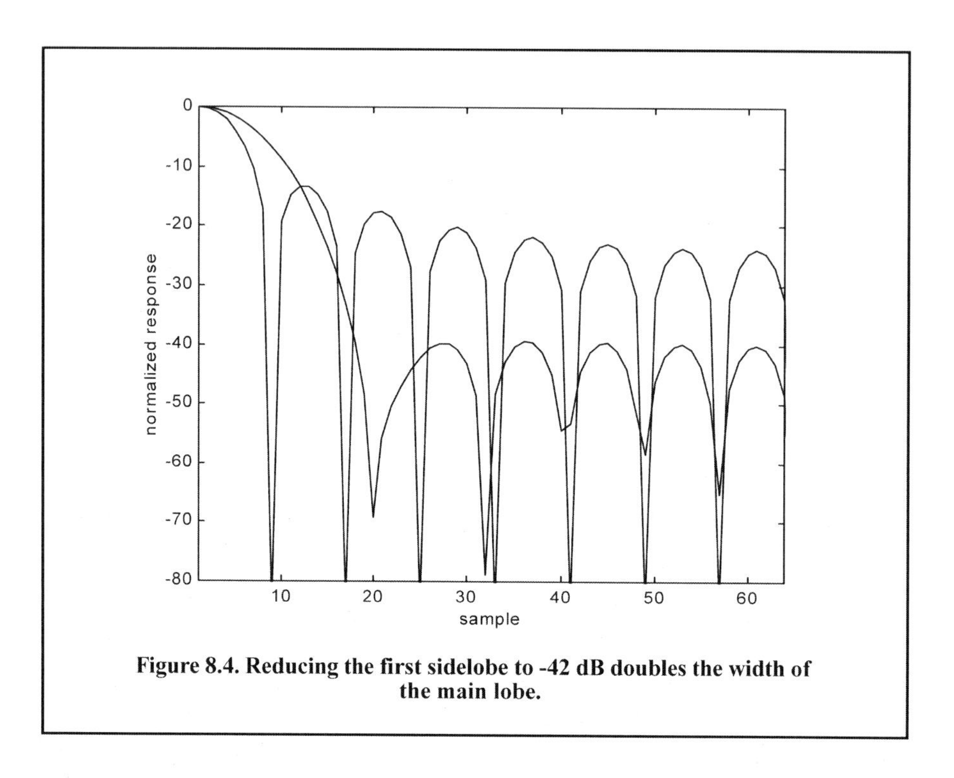

Figure 8.4. Reducing the first sidelobe to -42 dB doubles the width of the main lobe.

8.3. Correlation Processor

Radar operations (search, track, etc.) are usually carried out over a specified range window, referred to as the receive window, and defined by the difference between the radar maximum and minimum range. Returns from all targets within the receive window are collected and passed through matched filter circuitry to perform pulse compression. One implementation of such analog processors is the Surface Acoustic Wave (SAW) devices. Because of the recent advances in digital computer development, the correlation processor is often performed digitally using the FFT. This digital implementation is called Fast Convolution Processing (FCP) and can be implemented at the base band. The fast convolution process is illustrated in Figure 8.5.

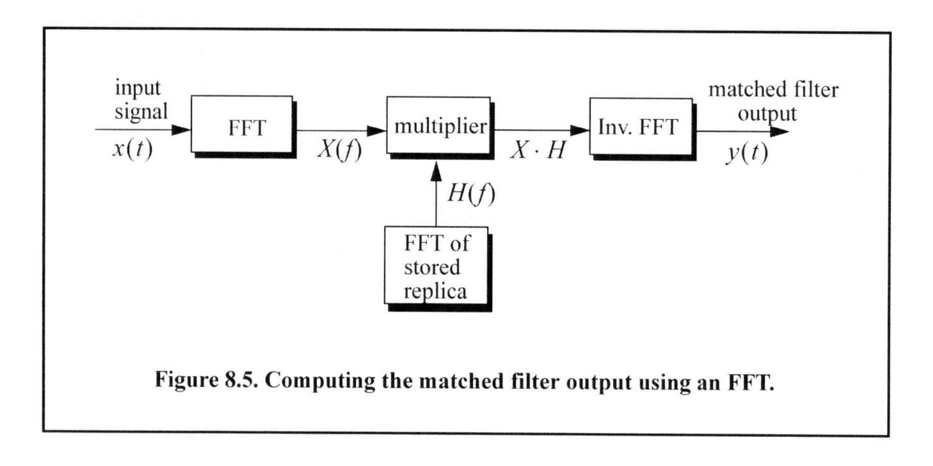

Figure 8.5. Computing the matched filter output using an FFT.

Since the matched filter is a linear time invariant system, its output can be described mathematically by the convolution between its input and its impulse response,

$$y(t) = x(t) \otimes h(t) \quad \text{Eq. (8.5)}$$

where $x(t)$ is the input signal, $h(t)$ is the matched filter impulse response (replica), and the ($\otimes$) operator symbolically represents convolution. From the Fourier transform properties,

$$FFT\{x(t) \otimes h(t)\} = X(f) \cdot H(f), \quad \text{Eq. (8.6)}$$

and when both signals are sampled properly, the compressed signal $y(t)$ can be computed from

$$y = FFT^{-1}\{X \cdot H\} \quad \text{Eq. (8.7)}$$

where FFT^{-1} is the inverse FFT. When using pulse compression, it is desirable to use modulation schemes that can accomplish a maximum pulse compression ratio and can significantly reduce the sidelobe levels of the compressed waveform. For the LFM case, the first sidelobe is approximately $13.4dB$ below the main peak, and for most radar applications this may not be sufficient. In practice, high sidelobe levels are not preferable because noise and/or jammers located at the sidelobes may interfere with target returns in the main lobe.

Weighting functions (windows) can be used on the compressed pulse spectrum in order to reduce the sidelobe levels. The cost associated with such an approach is a loss in the main lobe resolution, and a reduction in the peak value (i.e., loss in the SNR). Weighting the time domain transmitted or received signal instead of the compressed pulse spectrum will theoretically achieve the same goal. However, this approach is rarely used, since amplitude modulating the transmitted waveform introduces extra burdens on the transmitter.

Consider a radar system that utilizes a correlation processor receiver (i.e., matched filter). The receive window in meters is defined by

$$R_{rec} = R_{max} - R_{min} \quad \text{Eq. (8.8)}$$

where R_{max} and R_{min}, respectively, define the maximum and minimum range over which the radar performs detection. Typically, R_{rec} is limited to the extent of the target complex. The normalized complex transmitted signal has the form

$$x(t) = \exp\left(j2\pi\left(f_0 t + \frac{\mu}{2}t^2\right)\right) \qquad 0 \le t \le \tau_0. \quad \text{Eq. (8.9)}$$

τ_0 is the pulse width, $\mu = B/\tau_0$, and B is the bandwidth.

The radar echo signal is similar to the transmitted one with the exception of a time delay and an amplitude change that correspond to the target RCS. Consider a target at range R_1. The echo received by the radar from this target is

$$x_r(t) = a_1 \exp\left(j2\pi\left(f_0(t - t_1) + \frac{\mu}{2}(t - t_1)^2\right)\right) \quad \text{Eq. (8.10)}$$

where a_1 is proportional to target RCS, antenna gain, and range attenuation. The time delay t_1 is given by

$$t_1 = (2R_1)/c. \qquad \text{Eq. (8.11)}$$

The first step of the processing consists of removing the frequency f_0. This is accomplished by mixing $x_r(t)$ with a reference signal whose phase is $2\pi f_0 t$. The phase of the resultant signal, after lowpass filtering, is then given by

$$\phi(t) = 2\pi\left(-f_0 t_1 + \frac{\mu}{2}(t - t_1)^2\right) \qquad \text{Eq. (8.12)}$$

and the instantaneous frequency is

$$f_i(t) = \frac{1}{2\pi}\frac{d}{dt}\phi(t) = \mu(t - t_1) = \frac{B}{\tau_0}\left(t - \frac{2R_1}{c}\right). \qquad \text{Eq. (8.13)}$$

The quadrature components are

$$\begin{pmatrix} x_I(t) \\ x_Q(t) \end{pmatrix} = \begin{pmatrix} \cos\phi(t) \\ \sin\phi(t) \end{pmatrix}. \qquad \text{Eq. (8.14)}$$

Sampling the quadrature components is performed next. The number of samples, N, must be chosen so that foldover (ambiguity) in the spectrum is avoided. For this purpose, the sampling frequency, f_s (based on the Nyquist sampling rate), must be

$$f_s \geq 2B \qquad \text{Eq. (8.15)}$$

and the sampling interval is

$$\Delta t \leq 1/2B. \qquad \text{Eq. (8.16)}$$

The frequency resolution of the FFT is

$$\Delta f = 1/\tau_0. \qquad \text{Eq. (8.17)}$$

The minimum required number of samples is

$$N = \frac{1}{\Delta f \Delta t} = \frac{\tau_0}{\Delta t}. \qquad \text{Eq. (8.18)}$$

Equating Eqs. (8.16) and (8.18) yields

$$N \geq 2B\tau_0. \qquad \text{Eq. (8.19)}$$

Consequently, a total of $2B\tau_0$ real samples, or $B\tau_0$ complex samples, is sufficient to completely describe an LFM waveform of duration τ_0 and bandwidth B.

For better implementation of the FFT, N is extended to the next power of two, by zero padding. Thus, the total number of samples, for some positive integer n, is

$$N_{FFT} = 2^n \geq N. \qquad \text{Eq. (8.20)}$$

The final steps of the FCP processing include (1) taking the FFT of the sampled sequence, (2) multiplying the frequency domain sequence of the signal with the FFT of the matched filter impulse response, and (3) performing the inverse FFT of the composite frequency domain sequence in order to generate the time domain compressed pulse. Of course, weighting, antenna gain, and range attenuation compensation must also be performed.

Assume that M targets at ranges R_1, R_2, and so forth are within the receive window. From superposition, the phase of the down-converted signal is

$$\phi(t) = \sum_{m=1}^{M} 2\pi\left(-f_0 t_m + \frac{\mu}{2}(t - t_m)^2\right) . \qquad \text{Eq. (8.21)}$$

The times $\{t_m = (2R_m/c);\ m = 1, 2, \ldots, M\}$ represent the two-way time delays, where t_1 coincides with the start of the receive window.

Figure 8.6a shows the uncompressed echo of three adjacent targets; clearly the three targets are not resolved. In this example, the received window is *200* meters wide, and targets *1* and *2* are *40* meters apart, while target *2* and *3* are *50* meters apart; all *3* targets are assumed to have equal RCS. Figure 8.6b shows the compressed MF output of the same echo signal; note that after pulse compression the three targets are completely resolved. Note that the scatterer amplitude attenuation is also a function of the inverse of the scatterer's range within the receive window.

Figure 8.6c is similar to Figure 8.6b except in this case the first and second scatterers are less than 1.5 meters apart (they are at 70 and 71 meters). Since those *2* targets are separated less than the radar range resolution, they are not resolved after pulse compression.

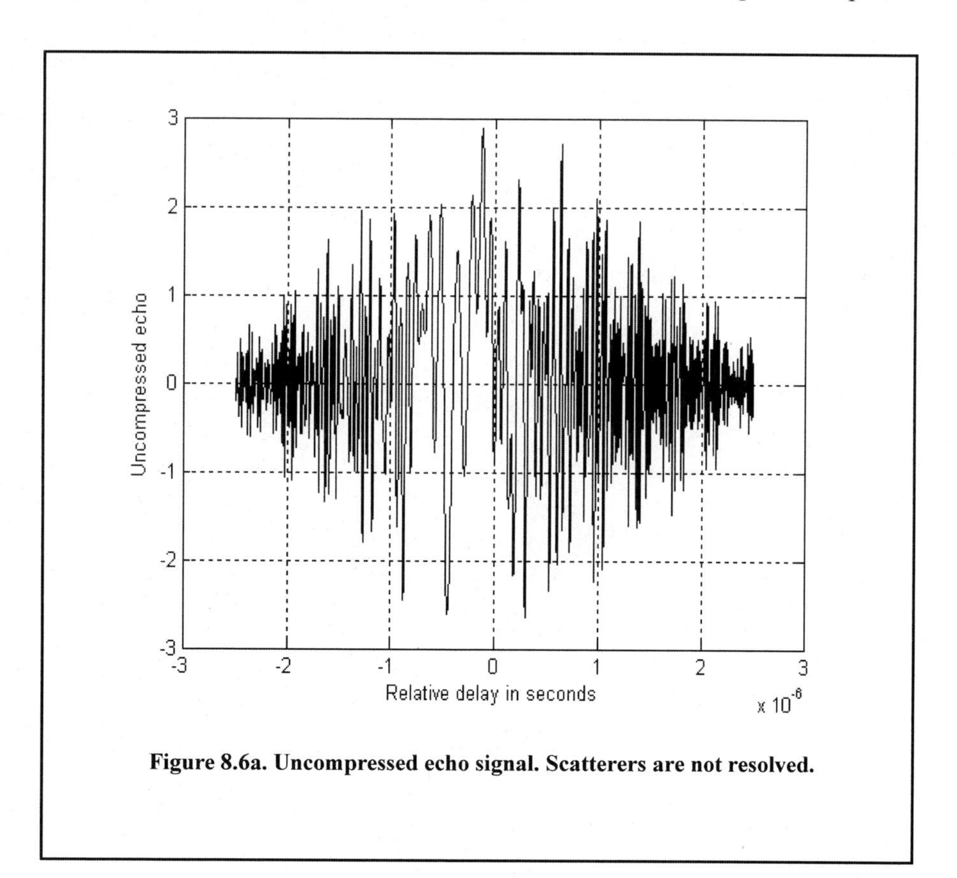

Figure 8.6a. Uncompressed echo signal. Scatterers are not resolved.

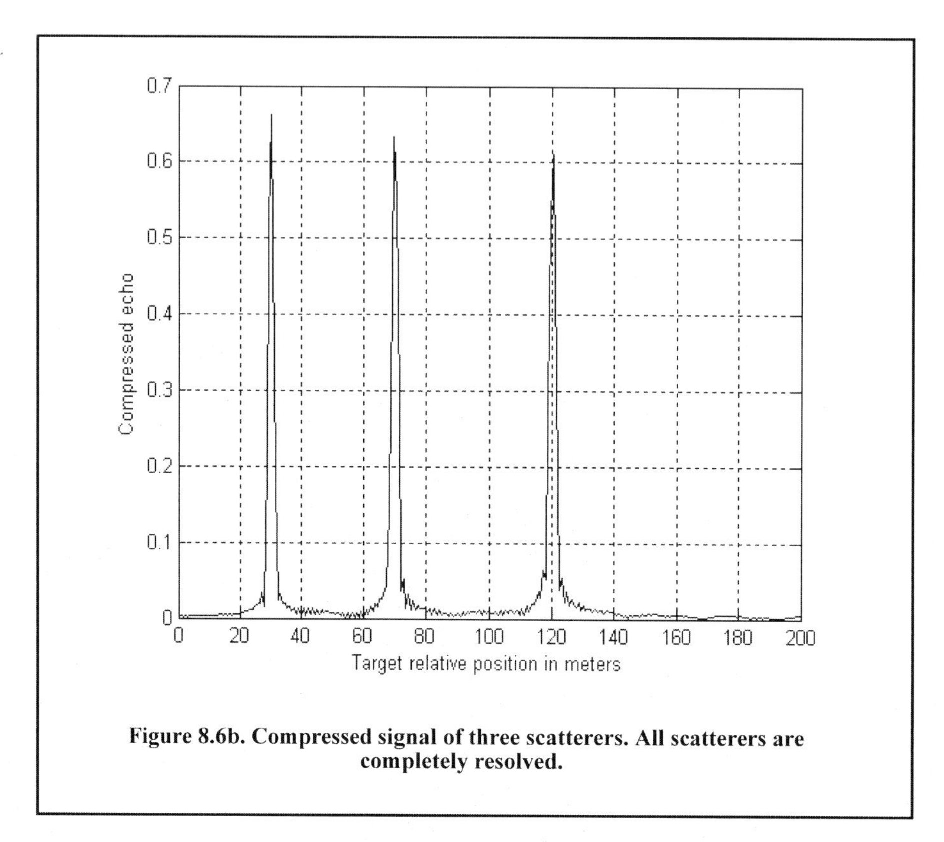

Figure 8.6b. Compressed signal of three scatterers. All scatterers are completely resolved.

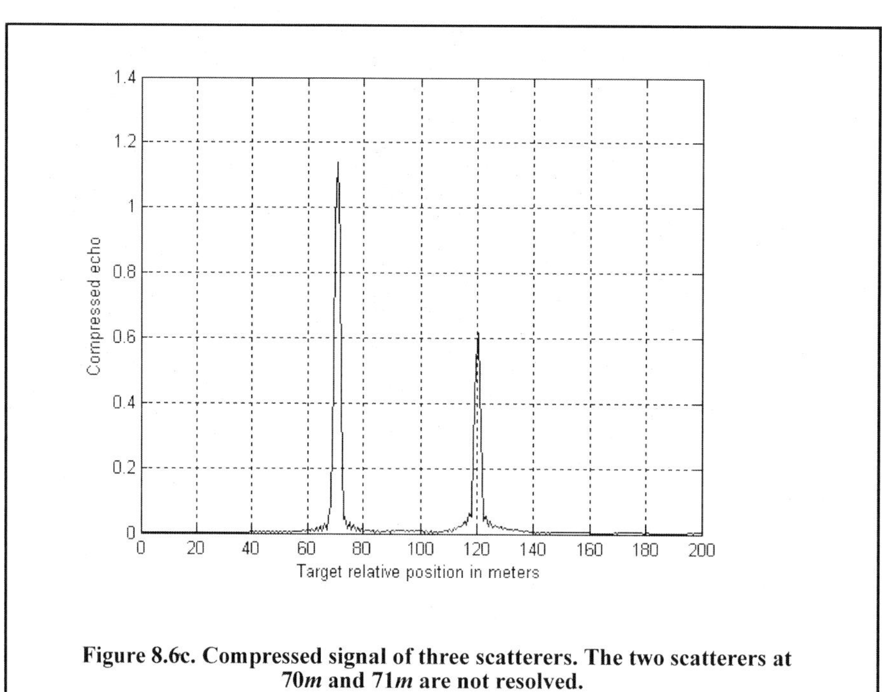

Figure 8.6c. Compressed signal of three scatterers. The two scatterers at 70*m* and 71*m* are not resolved.

8.4. Stretch Processor

Stretch processing, also known as *active correlation,* is normally used to process extremely high-bandwidth LFM waveforms. This processing technique consists of the following steps: first, the radar returns are mixed with a replica (reference signal) of the transmitted waveform. This is followed by Low Pass Filtering (LPF) and coherent detection. Next, Analog-to-Digital (A/D) conversion is performed; and finally, a bank of Narrow-Band Filters (NBFs) is used in order to extract the tones that are proportional to target range, since stretch processing effectively converts time delay into frequency. All returns from the same range bin produce the same constant frequency.

8.4.1. Single LFM Pulse

Figure 8.7 shows a block diagram for a stretch processing receiver. The reference signal is an LFM waveform that has the same LFM slope as the transmitted LFM signal. It exists over the duration of the radar "receive-window," which is computed from the difference between the radar maximum and minimum range. Denote the start frequency of the reference chirp as f_r.

Consider the case when the radar receives returns from a few close (in time or range) targets, as illustrated in Figure 8.8. Mixing with the reference signal and performing low-pass filtering are effectively equivalent to subtracting the return frequency chirp from the reference signal. Thus, the LPF output consists of constant tones corresponding to the targets' positions. The normalized transmitted signal is expressed by,

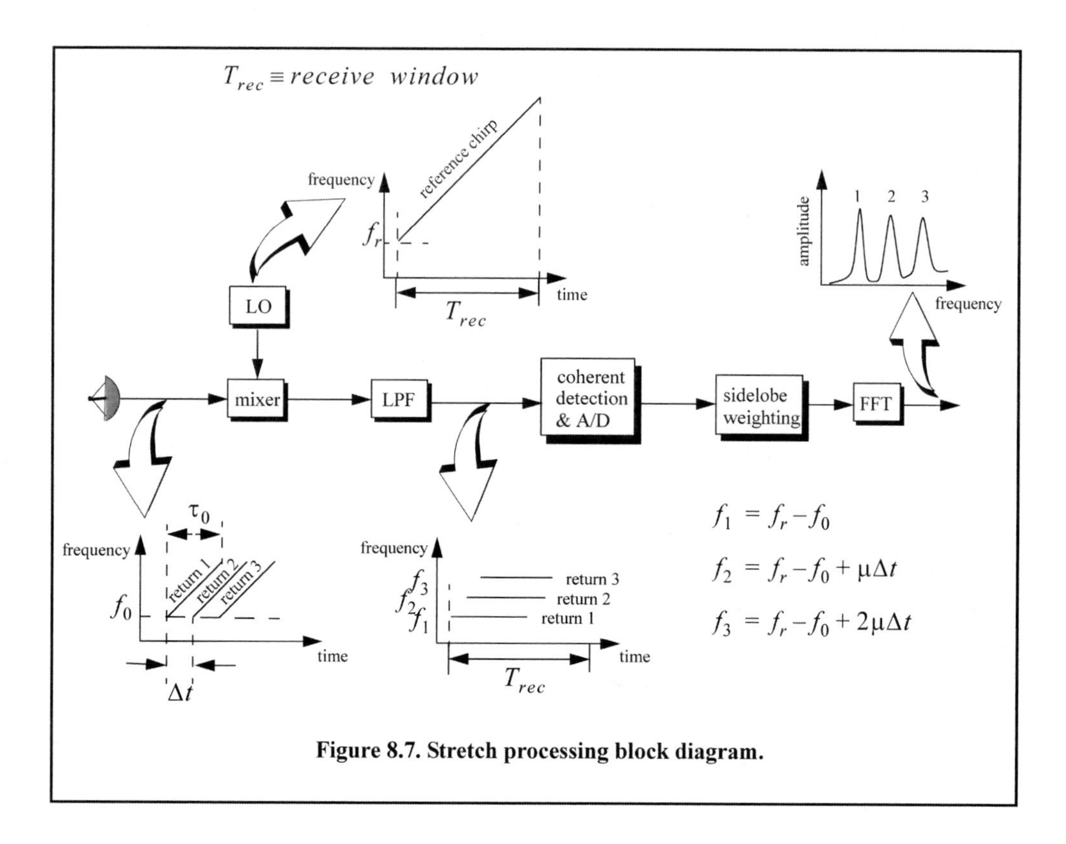

Figure 8.7. Stretch processing block diagram.

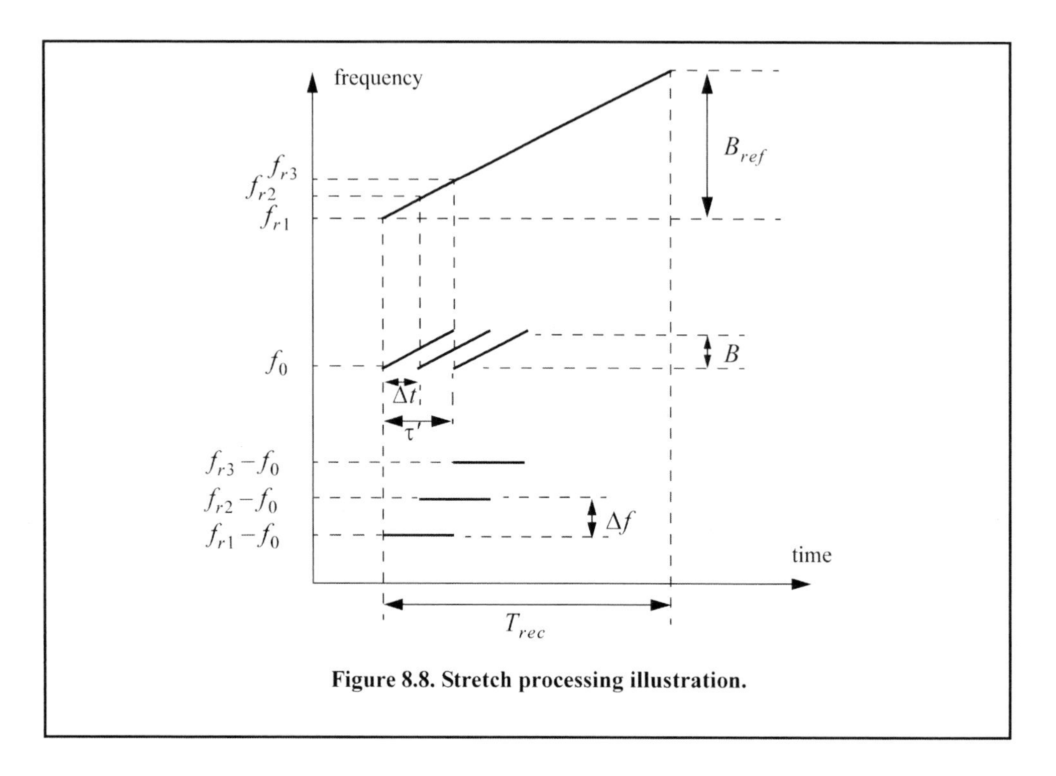

Figure 8.8. Stretch processing illustration.

$$x(t) = \cos\left(2\pi\left(f_0 t + \frac{\mu}{2}t^2\right)\right) \qquad 0 \le t \le \tau_0 \qquad \text{Eq. (8.22)}$$

where $\mu = B/\tau_0$ is the LFM coefficient and f_0 is the chirp start frequency. Assume a point scatterer at range R_1. The signal received by the radar is

$$x_r(t) = a\cos\left[2\pi\left(f_0(t-t_1) + \frac{\mu}{2}(t-t_1)^2\right)\right], \qquad \text{Eq. (8.23)}$$

where a is proportional to target RCS, antenna gain, and range attenuation. The time delay t_1 is

$$t_1 = 2R_1/c. \qquad \text{Eq. (8.24)}$$

The reference signal is

$$x_{ref}(t) = 2\cos\left(2\pi\left(f_r t + \frac{\mu}{2}t^2\right)\right) \qquad 0 \le t \le T_{rec}. \qquad \text{Eq. (8.25)}$$

The receive window in seconds is

$$T_{rec} = \frac{2(R_{max} - R_{min})}{c} = \frac{2R_{rec}}{c}. \qquad \text{Eq. (8.26)}$$

It is customary to let $f_r = f_0$. The output of the mixer is the product of the received and reference signals. After lowpass filtering, the signal is

$$x_0(t) = a\cos(2\pi f_0 t_1 + 2\pi\mu t_1 t - \pi\mu(t_1)^2). \qquad \text{Eq. (8.27)}$$

Substituting Eq. (8.24) into Eq. (8.27) and collecting terms yields

$$x_0(t) = a \cos\left[\left(\frac{4\pi BR_1}{c\tau_0}\right)t + \frac{2R_1}{c}\left(2\pi f_0 - \frac{2\pi BR_1}{c\tau_0}\right)\right], \qquad \text{Eq. (8.28)}$$

and since $\tau_0 \gg 2R_1/c$, Eq. (8.28) is approximated by

$$x_0(t) \approx a \cos\left[\left(\frac{4\pi BR_1}{c\tau_0}\right)t + \frac{4\pi R_1}{c}f_0\right]. \qquad \text{Eq. (8.29)}$$

The instantaneous frequency is

$$f_{inst} = \frac{1}{2\pi}\frac{d}{dt}\left(\left(\frac{4\pi BR_1}{c\tau_0}t + \frac{4\pi R_1}{c}f_0\right)\right) = \frac{2BR_1}{c\tau_0}, \qquad \text{Eq. (8.30)}$$

which clearly indicates that target range is proportional to the instantaneous frequency. Therefore, proper sampling of the LPF output and taking the FFT of the sampled sequence lead to the following conclusion: a peak at some frequency f_1 indicates the presence of a target at range

$$R_1 = f_1 c\tau_0/2B. \qquad \text{Eq. (8.31)}$$

Assume M close targets at ranges R_1, R_2, and so forth ($R_1 < R_2 < \ldots < R_M$). From superposition, the total signal is

$$x_r(t) = \sum_{m=1}^{M} a_m(t)\cos\left[2\pi\left(f_0(t-t_m) + \frac{\mu}{2}(t-t_m)^2\right)\right] \qquad \text{Eq. (8.32)}$$

where $\{a_m(t);\ m = 1, 2, \ldots, M\}$ are proportional to the targets' cross sections, antenna gain, and range. The times $\{t_m = (2R_m/c);\ m = 1, 2, \ldots, M\}$ represent the two-way time delays, where t_1 coincides with the start of the receive window. Using Eq. (8.28), the overall signal at the output of the LPF can then be described by

$$x_o(t) = \sum_{m=1}^{M} a_m \cos\left[\left(\frac{4\pi BR_m}{c\tau_0}\right)t + \frac{2R_m}{c}\left(2\pi f_0 - \frac{2\pi BR_m}{c\tau_0}\right)\right]. \qquad \text{Eq. (8.33)}$$

Hence, target returns appear as constant frequency tones that can be resolved using the FFT. Consequently, determining the proper sampling rate and FFT size is very critical. The rest of this section presents a methodology for computing the proper FFT parameters required for stretch processing.

Assume a radar system using a stretch processor receiver. The pulse width is τ_0 and the chirp bandwidth is B. Since stretch processing is normally used in extreme bandwidth cases (i.e., very large B), the receive window over which radar returns will be processed is typically limited to from a few meters to possibly less than 100 meters. Declare the FFT size to be N and its frequency resolution to be Δf. The frequency resolution can be computed using the following procedure: consider two adjacent point scatterers at ranges R_1 and R_2. The minimum frequency separation, Δf, between those scatterers so that they are resolved can be computed from Eq. (8.30). More precisely,

$$\Delta f = f_2 - f_1 = \frac{2B}{c\tau_0}(R_2 - R_1) = \frac{2B}{c\tau_0}\Delta R. \qquad \text{Eq. (8.34)}$$

Using $\Delta R = c/(2B)$ in Eq. (8.34) yields,

$$\Delta f = \frac{2B}{c\tau_0}\frac{c}{2B} = \frac{1}{\tau_0}. \qquad \text{Eq. (8.35)}$$

The maximum frequency resolvable by the FFT is limited to the region $\pm N\Delta f/2$. Thus, the maximum resolvable frequency is

$$\frac{N\Delta f}{2} > \frac{2B(R_{max} - R_{min})}{c\tau_0} = \frac{2BR_{rec}}{c\tau_0}. \qquad \text{Eq. (8.36)}$$

Using Eqs. (8.26) and (8.35) into Eq. (8.36) and collecting terms yield

$$N > 2BT_{rec}. \qquad \text{Eq. (8.37)}$$

For better implementation of the FFT, choose an FFT of size

$$N_{FFT} \geq N = 2^n \qquad \text{Eq. (8.38)}$$

where n is a nonzero positive integer. The sampling interval is then given by

$$\Delta f = \frac{1}{T_s N_{FFT}} \Rightarrow T_s = \frac{1}{\Delta f N_{FFT}}. \qquad \text{Eq. (8.39)}$$

Example:

A certain radar uses stretch processing. The desired time displacement is $\Delta t = 2ns$, the reference chirp starts at frequency $f_r = 6630MHz$, and the receive window is $T_{rec} = 10\mu s$. The transmitted chirp starts at frequency $f_0 = 5410MHz$, and has bandwidth $B_{ref} = B = 500MHz$. Assume three target returns, as in Figure 8.8. Compute N_{FFT}, Δf, f_1, f_2, and f_3.

Solution:

First, we compute

$$N_{FFT} \geq 2BT_{rec} = 10,000 \textit{ then } \Delta f = \frac{B}{T_{ref}}\Delta t = 100KHz.$$

Next, we compute

$$f_1 = (f_r - f_0) + (i-1)|_{i=1} \times \Delta f = 1.2200GHz$$

$$f_2 = (f_r - f_0) + (i-1)|_{i=2} \times \Delta f = 1.2201GHz$$

$$f_3 = (f_r - f_0) + (i-1)|_{i=3} \times \Delta f = 1.2202GHz.$$

****

To illustrate stretch processing further, consider the following example, where

# Targets	*3*
Pulse Width	*10ms*
Center Frequency	*5.6GHz*
Bandwidth	*1GHz*
Receive Window	*30m*
Relative Target's Range	*[2 5 10]m*
Target's RCS	*[1, 1, 2]m^2*
Window	*2 (Kaiser)*

The compressed pulse range resolution, without using a window, is $\Delta R = 0.15m$. Figure 8.9a and Figure 8.9b, respectively, show the uncompressed and compressed echo signals corresponding to this example. Figure 8.10 is similar to Figure 8.9b except in this case two of the scatterers are less than 15 cm apart (i.e., unresolved targets at $R_{relative} = [3, 3.1]m$).

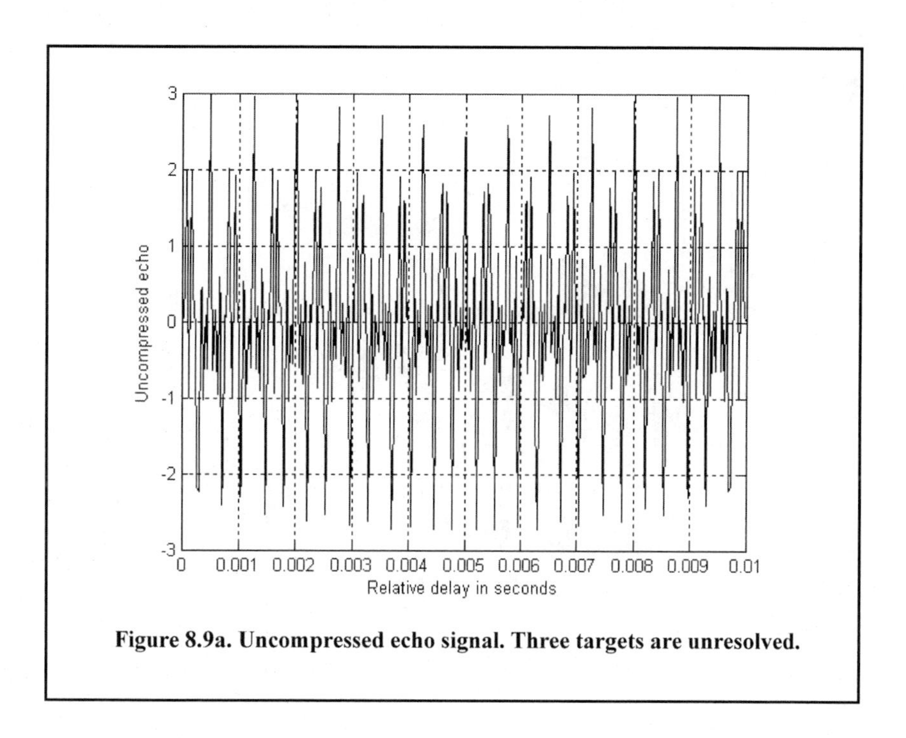

Figure 8.9a. Uncompressed echo signal. Three targets are unresolved.

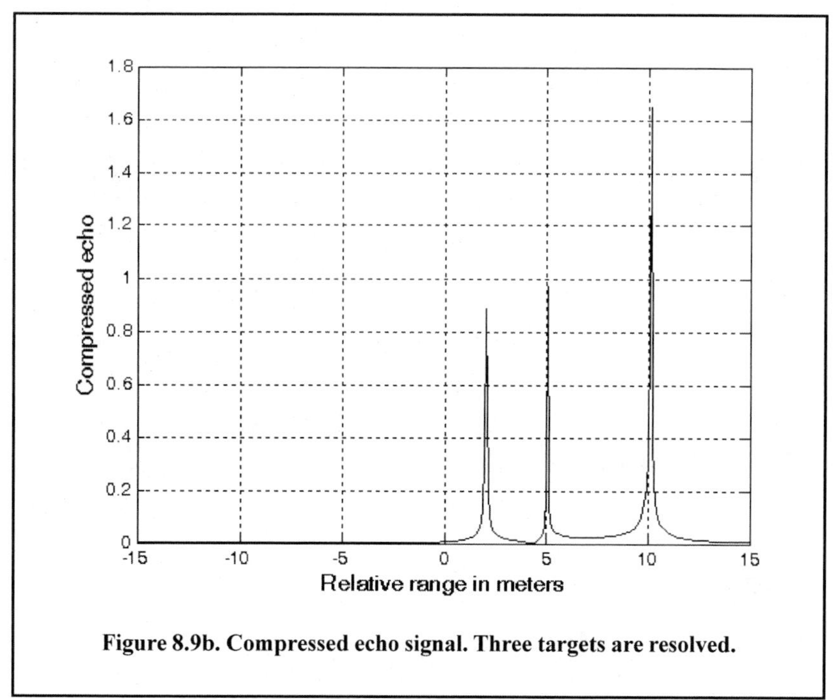

Figure 8.9b. Compressed echo signal. Three targets are resolved.

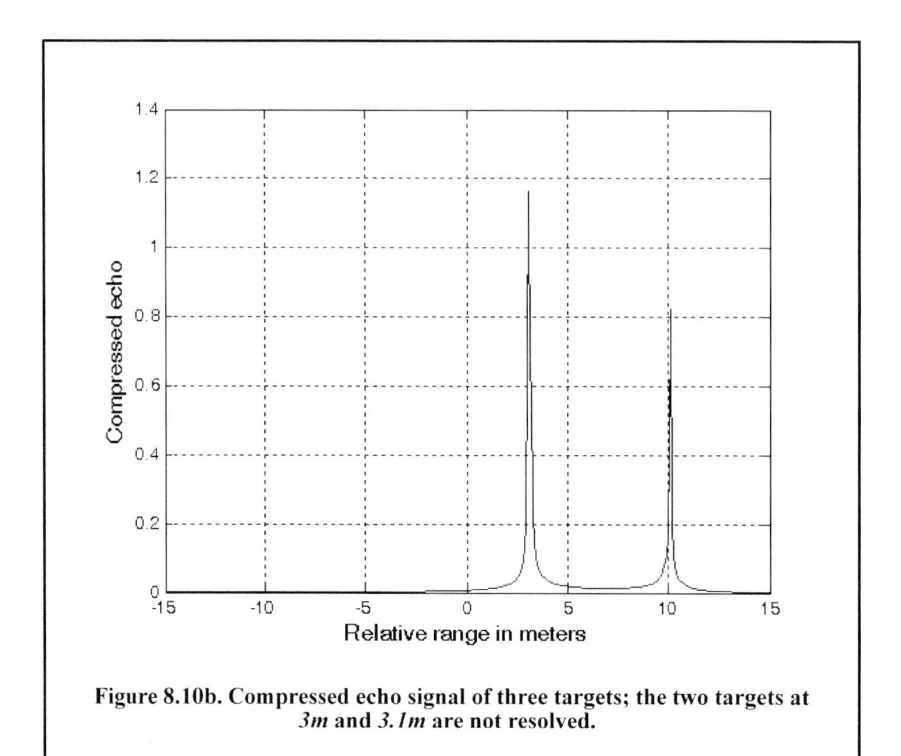

Figure 8.10b. Compressed echo signal of three targets; the two targets at *3m* and *3.1m* are not resolved.

Problems

8.1. Starting with Eq. (8.13) derive Eq. (8.17).

8.2. Using MATLAB®[1], generate a baseband (complex-valued) LFM waveform having a time duration of $10\mu s$ and bandwidth of $200MHz$ using a sampling step of $1ns$. Plot the real part, the imaginary part, and the modulus of the FFT of this waveform.

8.3. The Synthetic Aperture Radar (SAR) ambiguity function can be approximated by

$$\chi = \frac{\sin kx}{x} \frac{\sin Nry}{\sin ry} \qquad \text{Eq. (8.40)}$$

where x is the variable for the range-compressed axis, y is the azimuth-compressed axis, and k and r are related to the SAR range and azimuth resolutions. (a) Generate the x-axis from $-40m$ to $40m$ using a sampling interval of $0.1m$. Assume $k = 1$. Plot the magnitude of this range profile. (b) Generate the y-axis from $-40m$ to $40m$ using a sampling interval of $0.1m$. Assume $r = 0.00015$ and $N = 1000$. Plot the magnitude of this azimuth profile. (c) Use the findings in (a) and (b) to generate the two-dimensional ambiguity surface plot.

1. MATLAB® is a registered trademark of the The MathWorks, Inc. For product information, please contact: The MathWorks, Inc., 3 Apple Hill Drive, Natick, MA 01760-2098 USA. Tel: 508-647-7000, E-mail: information@mathworks.com, Web: *www.mathworks.com*.

Part III

SPECIAL TOPICS IN RADAR SYSTEMS

Chapter 9

Radar Clutter

9.1. Clutter Definition

Clutter is a term used to describe any object that may generate unwanted radar returns that may interfere with normal radar operations. Parasitic returns that enter the radar through the antenna's mainlobe are called mainlobe clutter; otherwise they are called side-lobe clutter. Clutter can be classified into two main categories: surface clutter and airborne or volume clutter. Surface clutter includes trees, vegetation, ground terrain, man-made structures, and sea surface (sea clutter). Volume clutter normally has a large extent (size) and includes chaff, rain, birds, and insects. Surface clutter changes from one area to another, while volume clutter may be more predictable.

Clutter echoes are random and have thermal noise-like characteristics because the individual clutter components (scatterers) have random phases and amplitudes. In many cases, the clutter signal level is much higher than the receiver noise level. Thus, the radar's ability to detect targets embedded in high clutter background depends on the Signal-to-Clutter Ratio (SCR) rather than the SNR. White noise normally introduces the same amount of noise power across all radar range bins, while clutter power may vary within a single range bin. Since clutter returns are target-like echoes, the only way a radar can distinguish target returns from clutter echoes is based on the target RCS σ_t, and the anticipated clutter RCS σ_c (via clutter map). Clutter RCS can be defined as the equivalent radar cross section attributed to reflections from a clutter area, A_c.

The average clutter RCS is given by

$$\sigma_c = \sigma^0 A_c \qquad \text{Eq. (9.1)}$$

where $\sigma^0(m^2/m^2)$ is the clutter scattering coefficient, a dimensionless quantity that is often expressed in dB. Some radar engineers express σ^0 in terms of squared centimeters per squared meter. In these cases, σ^0 is $40dB$ higher than normal.

9.2. Surface Clutter

Surface clutter includes both land and sea clutter, and is often called area clutter. Area clutter manifests itself in airborne radars in the look-down mode. It is also a major concern for ground-based radars when searching for targets at low grazing angles. The grazing angle ψ_g is the angle from the surface of the earth to the main axis of the illuminating beam, as illustrated in Figure 9.1.

Three factors affect the amount of clutter in the radar beam. They are the grazing angle, the surface roughness, and the radar wavelength. Typically, the clutter scattering coefficient σ^0 is larger for smaller wavelengths. Figure 9.2 shows a sketch describing the dependency of σ^0 on the grazing angle. Three regions are identified; they are the low grazing angle region, flat or plateau region, and the high grazing angle region.

The low grazing angle region extends from zero to about the critical angle. The critical angle is defined by Rayleigh as the angle below which a surface is considered to be smooth, and above which a surface is considered to be rough. Denote the root mean square (rms) of a surface height irregularity as h_{rms}, then according to the Rayleigh criteria, the surface is considered to be smooth if

$$\frac{\{(4\pi h_{rms})\sin\psi_g\}}{\lambda} < \frac{\pi}{2}. \qquad \text{Eq. (9.2)}$$

Consider a wave incident on a rough surface, as shown in Figure 9.3. Due to surface height irregularity (surface roughness), the "rough path" is longer than the "smooth path" by a distance $2h_{rms}\sin\psi_g$. This path difference translates into a phase differential $\Delta\psi$:

$$\Delta\psi = \{(2\pi)\ 2h_{rms}\sin\psi_g\}/\lambda. \qquad \text{Eq. (9.3)}$$

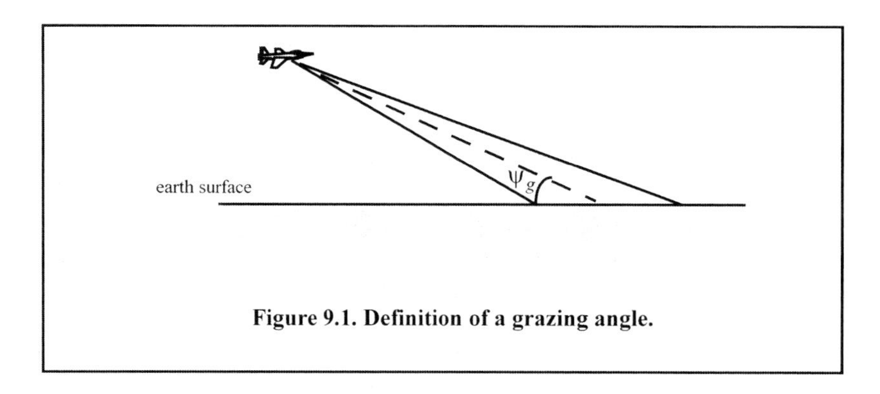

Figure 9.1. Definition of a grazing angle.

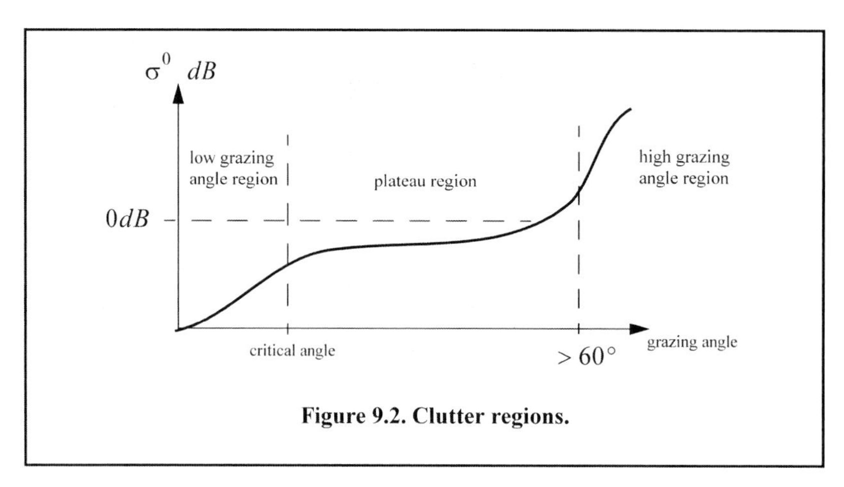

Figure 9.2. Clutter regions.

The critical angle ψ_{gc} is then computed when $\Delta\psi = \pi$ (first null), thus

$$\frac{(4\pi h_{rms})\ \sin\psi_{gc}}{\lambda} = \pi\,, \qquad \text{Eq. (9.4)}$$

or equivalently,

$$\psi_{gc} = \operatorname{asin}\frac{\lambda}{4h_{rms}}\,. \qquad \text{Eq. (9.5)}$$

In the case of sea clutter, for example, the rms surface height irregularity is

$$h_{rms} \approx 0.025 + 0.046\ S_{state}^{1.72} \qquad \text{Eq. (9.6)}$$

where S_{state} is the sea state, which is tabulated in several cited references. The sea state is characterized by the wave height, period, length, particle velocity, and wind velocity.

For example, $S_{state} = 3$ refers to a moderate sea state, where in this case the wave height is approximately between $0.9144\ to\ 1.2192\ m$, the wave period 6.5 *to* 4.5 seconds, wave length $1.9812\ to\ 33.528\ m$, wave velocity $20.372\ to\ 25.928\ Km/hr$, and wind velocity $22.224\ to\ 29.632\ Km/hr$.

Clutter at low grazing angles is often referred to as diffuse clutter, where there are a large number of clutter returns in the radar beam (noncoherent reflections). In the flat region the dependency of σ^0 on the grazing angle is minimal. Clutter in the high grazing angle region is more specular (coherent reflections) and the diffuse clutter components disappear. In this region the smooth surfaces have larger σ^0 than rough surfaces, the opposite of the low grazing angle region.

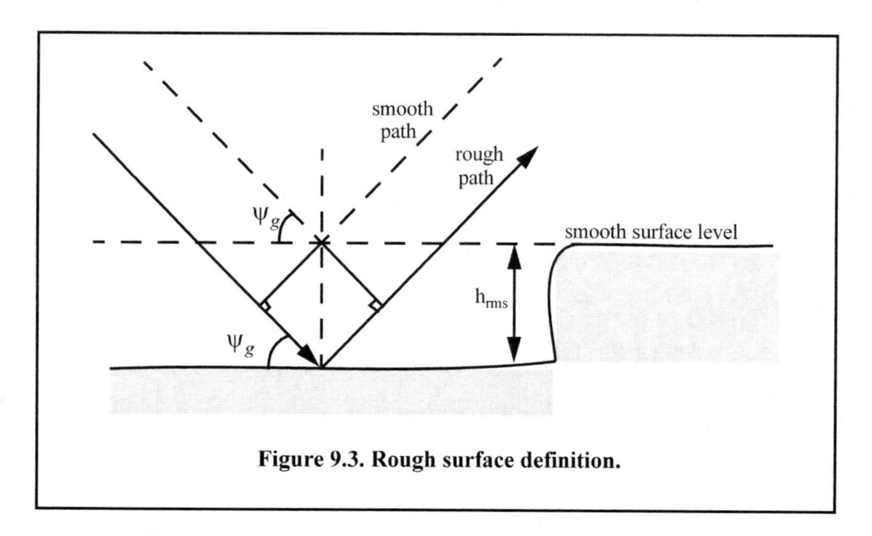

Figure 9.3. Rough surface definition.

9.2.1. Radar Equation for Area Clutter - Airborne Radar

Consider an airborne radar in the look-down mode shown in Figure 9.4. The intersection of the antenna beam with the ground defines an elliptically shaped footprint. The size of the footprint is a function of the grazing angle and the antenna $3dB$ beamwidth θ_{3dB}, as illustrated in Figure 9.5. The footprint is divided into many ground range bins each of size $(c\tau/2)\sec\psi_g$, where τ is the pulse width.

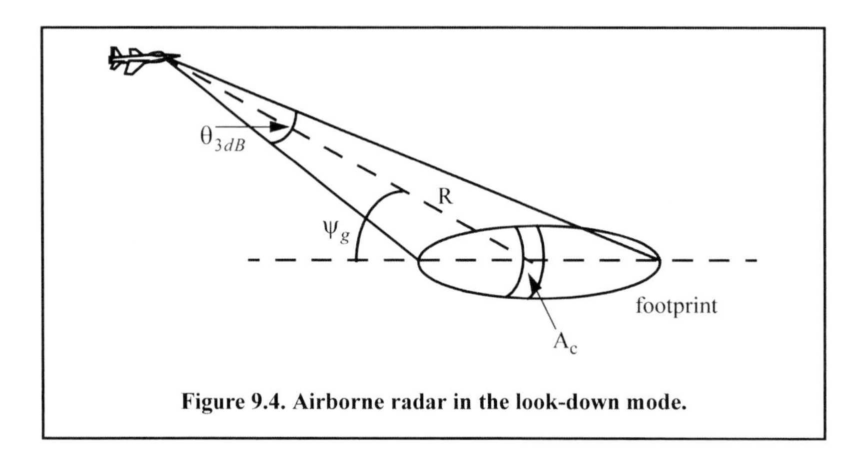

Figure 9.4. Airborne radar in the look-down mode.

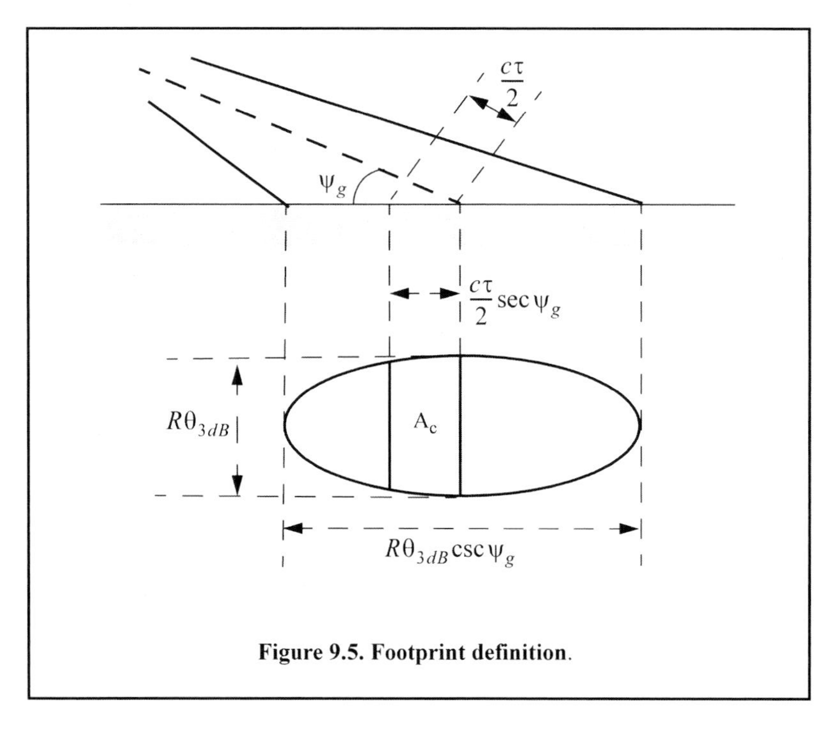

Figure 9.5. Footprint definition.

From Figure 9.5, the clutter area A_c is

$$A_c \approx R\theta_{3dB}\frac{c\tau}{2}\sec\psi_g . \qquad \text{Eq. (9.7)}$$

The power received by the radar from a scatterer within A_c is given by the radar equation as

$$S_t = \frac{P_t G^2 \lambda^2 \sigma_t}{(4\pi)^3 R^4} \qquad \text{Eq. (9.8)}$$

where, as usual, P_t is the peak transmitted power, G is the antenna gain, λ is the wavelength, and σ_t is the target RCS. Similarly, the received power from clutter is

$$S_C = \frac{P_t G^2 \lambda^2 \sigma_c}{(4\pi)^3 R^4}$$ Eq. (9.9)

where the subscript C is used for area clutter. Substituting Eq. (9.1) for σ_c and Eq. (9.7) for A_c into Eq. (9.9), one can then obtain the SCR for area clutter by dividing Eq. (9.8) by Eq. (9.9). More precisely,

$$(SCR)_C = \frac{2\sigma_t \cos\psi_g}{\sigma^0 \theta_{3dB} R c \tau}\,.$$ Eq. (9.10)

Example:

Consider an airborne radar shown in Figure 9.4. Let the antenna 3dB beamwidth be $\theta_{3dB} = 0.02 rad$, the pulse width $\tau = 2\mu s$, range $R = 20Km$, and grazing angle $\psi_g = 20°$. The target RCS is $\sigma_t = 1m^2$. Assume that the clutter reflection coefficient is $\sigma^0 = 0.0136$. Compute the SCR.

Solution:

The SCR is given by Eq. (9.10) as

$$(SCR)_C = \frac{2\sigma_t \cos\psi_g}{\sigma^0 \theta_{3dB} R c \tau} \Rightarrow$$

$$(SCR)_C = \frac{(2)(1)(\cos 20°)}{(0.0136)(0.02)(20000)(3\times 10^8)(2\times 10^{-6})} = 5.76\times 10^{-4}\,.$$

It follows that

$$(SCR)_C = -32.4dB\,.$$

Thus, for reliable detection, the radar must somehow increase its SCR by at least $(32 + X)dB$, where X is on the order of 13 to 15dB or better.

****

9.3. Volume Clutter

Volume clutter has large extents and includes rain (weather), chaff, birds, and insects. The volume clutter coefficient is normally expressed in square meters (RCS per resolution volume). Birds, insects, and other flying particles are often referred to as angle clutter or biological clutter. Weather or rain clutter can be suppressed by treating the rain droplets as perfect small spheres. We can use the Rayleigh approximation of a perfect sphere to estimate the rain droplets' RCS. The Rayleigh approximation, without regard to the propagation medium index of refraction is

$$\sigma = 9\pi r^2 (kr)^4 \qquad r \ll \lambda$$ Eq. (9.11)

where $k = 2\pi/\lambda$, and r is radius of a rain droplet.

Electromagnetic waves, when reflected from a perfect sphere, become strongly co-polarized (have the same polarization as the incident waves). Consequently, if the radar transmits, for example, a right-hand-circular (RHC) polarized wave, then the received waves are left-hand-circular (LHC) polarized because they are propagating in the opposite direction. Therefore, the back-scattered energy from rain droplets retains the same wave rotation (polarization) as the incident wave, but has a reversed direction of propagation. It follows that radars can suppress rain clutter by co-polarizing the radar transmit and receive antennas.

Denote σ_w as RCS per unit resolution volume V_w. It is computed as the sum of all individual scatterers RCS within the volume

$$\sigma_w = \sum_{i=1}^{N} \sigma_i \qquad \text{Eq. (9.12)}$$

where N is the total number of scatterers within the resolution volume. Thus, the total RCS of a single resolution volume is

$$\sigma_w = \sum_{i=1}^{N} \sigma_i V_w . \qquad \text{Eq. (9.13)}$$

A resolution volume is shown in Figure 9.6 and is approximated by

$$V_w \approx \frac{\pi}{8}\theta_a\theta_e R^2 c\tau \qquad \text{Eq. (9.14)}$$

where θ_a and θ_e are, respectively, the antenna azimuth and elevation beamwidths in radians, τ is the pulse width in seconds, c is the speed of light, and R is range.

Consider a propagation medium with an index of refraction m. The ith rain droplet RCS approximation in this medium is

$$\sigma_i \approx \frac{\pi^5}{\lambda^4} K^2 D_i^6 \qquad \text{Eq. (9.15)}$$

where

$$K^2 = \left|\frac{m^2 - 1}{m^2 + 2}\right|^2 \qquad \text{Eq. (9.16)}$$

D_i is the ith droplet diameter. For example, temperatures between $32°F$ and $68°F$ yield

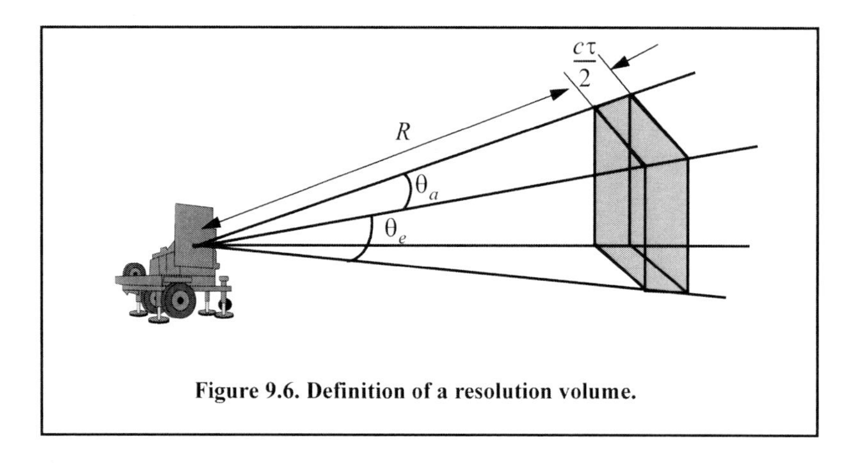

Figure 9.6. Definition of a resolution volume.

$$\sigma_i \approx 0.93\frac{\pi^5}{\lambda^4}D_i^6. \qquad \text{Eq. (9.17)}$$

and for ice, Eq. (9.17) can be approximated by

$$\sigma_i \approx 0.2\frac{\pi^5}{\lambda^4}D_i^6. \qquad \text{Eq. (9.18)}$$

Substituting Eq. (9.18) into Eq. (9.13) yields

$$\sigma_w = \frac{\pi^5}{\lambda^4}K^2 Z \qquad \text{Eq. (9.19)}$$

where the weather clutter backscatter coefficient Z is defined as

$$Z = \sum_{i=1}^{N} D_i^6. \qquad \text{Eq. (9.20)}$$

In general, a rain droplet diameter is given in millimeters and the radar resolution volume is expressed in cubic meters; thus the units of Z are often expressed in $millimeter^6/m^3$.

9.3.1. Radar Equation for Volume Clutter

The radar equation gives the total power received by the radar from a σ_t target at range R as

$$S_t = \frac{P_t G^2 \lambda^2 \sigma_t}{(4\pi)^3 R^4} \qquad \text{Eq. (9.21)}$$

where all parameters in Eq. (9.21) have been defined earlier. The weather clutter power received by the radar is

$$S_w = \frac{P_t G^2 \lambda^2 \sigma_w}{(4\pi)^3 R^4}. \qquad \text{Eq. (9.22)}$$

It follows that

$$S_w = \frac{P_t G^2 \lambda^2}{(4\pi)^3 R^4} \frac{\pi}{8} R^2 \theta_a \theta_e c\tau \sum_{i=1}^{N} \sigma_i. \qquad \text{Eq. (9.23)}$$

The SCR for weather clutter is then computed by dividing Eq. (9.21) by Eq. (9.23). More precisely,

$$(SCR)_V = \frac{S_t}{S_w} = \frac{(8\sigma_t)}{\left(\pi\theta_a\theta_e c\tau R^2 \sum_{i=1}^{N} \sigma_i\right)} \qquad \text{Eq. (9.24)}$$

where the subscript V is used to denote volume clutter.

Example:

A certain radar has target RCS $\sigma_t = 0.1m^2$, *pulse width* $\tau = 0.2\mu s$, *antenna beamwidth* $\theta_a = \theta_e = 0.02radians$. *Assume the detection range to be* $R = 50Km$, *and compute the SCR if* $\sum \sigma_i = 1.6 \times 10^{-8}(m^2/m^3)$.

Solution:

From Eq. (9.24) we have

$$(SCR)_V = \frac{8\sigma_t}{\pi\theta_a\theta_e c\tau R^2 \sum\limits^{N} \sigma_i}.$$

Substituting the proper values we get

$$(SCR)_V = \frac{(8)(0.1)}{\pi(0.02)^2(3 \times 10^8)(0.2 \times 10^{-6})(50 \times 10^3)^2(1.6 \times 10^{-8})} = 0.265$$

$$(SCR)_V = -5.76dB.$$

****

9.4. Surface Clutter RCS

The received power from clutter is calculated using Eq. (9.9). However, the clutter RCS σ_c is now computed differently. It is

$$\sigma_c = \sigma_{MBc} + \sigma_{SLc} \quad \text{Eq. (9.25)}$$

where σ_{MBc} is the main-beam clutter RCS and σ_{SLc} is the sidelobe clutter RCS, as illustrated in Figure 9.7.

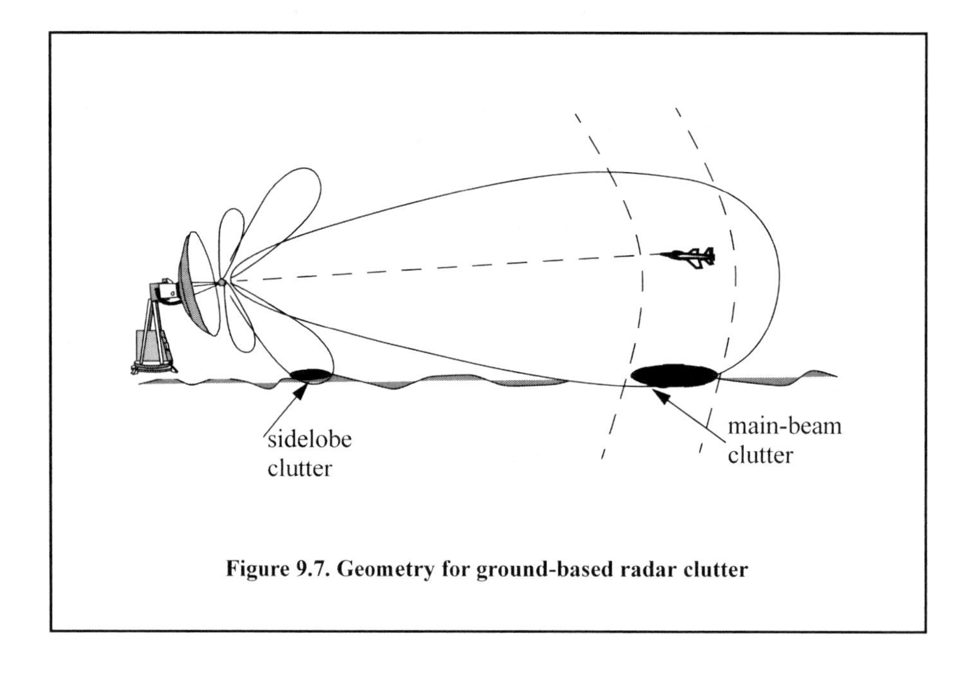

Figure 9.7. Geometry for ground-based radar clutter

In order to calculate the total clutter RCS given in Eq. (9.25), one must first compute the corresponding clutter areas for both the main beam and the sidelobes. For this purpose, consider the geometry shown in Figure 9.8. The angles θ_A *and* θ_E represent the antenna $3dB$ azimuth and elevation beamwidths, respectively.

The radar height (from the ground to the phase center of the antenna) is denoted by h_r, while the target height is denoted by h_t. The radar slant range is R, and its ground projection is R_g. The range resolution is ΔR and its ground projection is ΔR_g. The main beam clutter area is denoted by A_{MBc} and the sidelobe clutter area is denoted by A_{SLc}. From Figure 9.8, the following relations can be derived

$$\theta_r = \operatorname{asin}(h_r/R) \qquad \text{Eq. (9.26)}$$

$$\theta_e = \operatorname{asin}((h_t - h_r)/R) \qquad \text{Eq. (9.27)}$$

$$\Delta R_g = \Delta R \cos\theta_r \qquad \text{Eq. (9.28)}$$

where ΔR is the radar range resolution. The slant range ground projection is

$$R_g = R\cos\theta_r . \qquad \text{Eq. (9.29)}$$

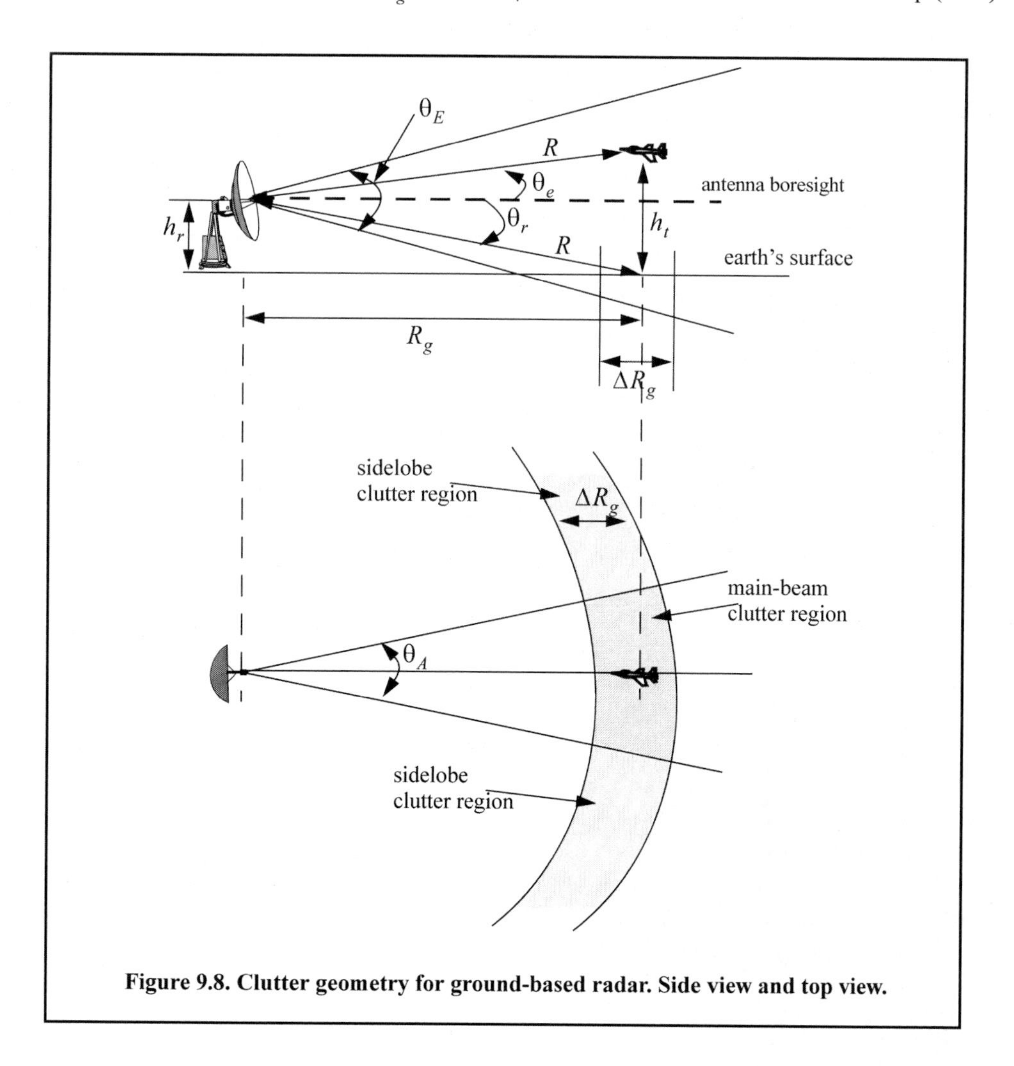

Figure 9.8. Clutter geometry for ground-based radar. Side view and top view.

It follows that the main beam and the sidelobe clutter areas are

$$A_{MBc} = \Delta R_g \; R_g \; \theta_A \qquad \text{Eq. (9.30)}$$

$$A_{SLc} = \Delta R_g \; \pi R_g . \qquad \text{Eq. (9.31)}$$

Denote the radar antenna beam by $G(\theta)$; then the main-beam clutter RCS is

$$\sigma_{MBc} = \sigma^0 A_{MBc} G^2(\theta_e + \theta_r) = \sigma^0 \Delta R_g \; R_g \; \theta_A G^2(\theta_e + \theta_r) \qquad \text{Eq. (9.32)}$$

and the sidelobe clutter RCS is

$$\sigma_{SLc} = \sigma^0 A_{SLc} (SL_{rms})^2 = \sigma^0 \Delta R_g \; \pi R_g (SL_{rms})^2 \qquad \text{Eq. (9.33)}$$

where the quantity SL_{rms} is the rms for the antenna sielobe level.

The radar SNR due to a target at range R is

$$SNR = \frac{P_t G^2 \lambda^2 \sigma_t}{(4\pi)^3 R^4 k T_o B F L} \qquad \text{Eq. (9.34)}$$

where, as usual, P_t is the peak transmitted power, G is the antenna gain, λ is the wavelength, σ_t is the target RCS, k is Boltzmann's constant, T_0 is the effective noise temperature, B is the radar operating bandwidth, F is the receiver noise figure, and L is the total radar losses. Similarly, the Clutter-to-Noise Ratio (CNR) at the radar is

$$CNR = \frac{P_t G^2 \lambda^2 \sigma_c}{(4\pi)^3 R^4 k T_o B F L} \qquad \text{Eq. (9.35)}$$

where the σ_c is the clutter RCS from Eq. (9.25).

As an example consider a case with the following parameters

clutter back scatterer coefficient	*-20 dB*
antenna 3dB elevation beamwidth	*1.5 degrees*
antenna 3dB azimuth beamwidth	*2 degrees*
antenna sidelobe level	*-25 dB*
radar height	*3 meters*
target height	*150 meters*
pulse width	*1 micro sec*
range	*2 - 45Km*
target RCS	*-10 dBsm*
radar center frequency	*5 GHz*

Figure 9.9 shows the corresponding clutter RCS versus range for two antenna pattern types, *sin(x)/x* and Gaussian shape. Figure 9.10 shows the corresponding plots for the SNR, CNR, and SCR for the *sin(x)/x* case, while Figure 9.11 shows the same results using a Gaussian antenna pattern.

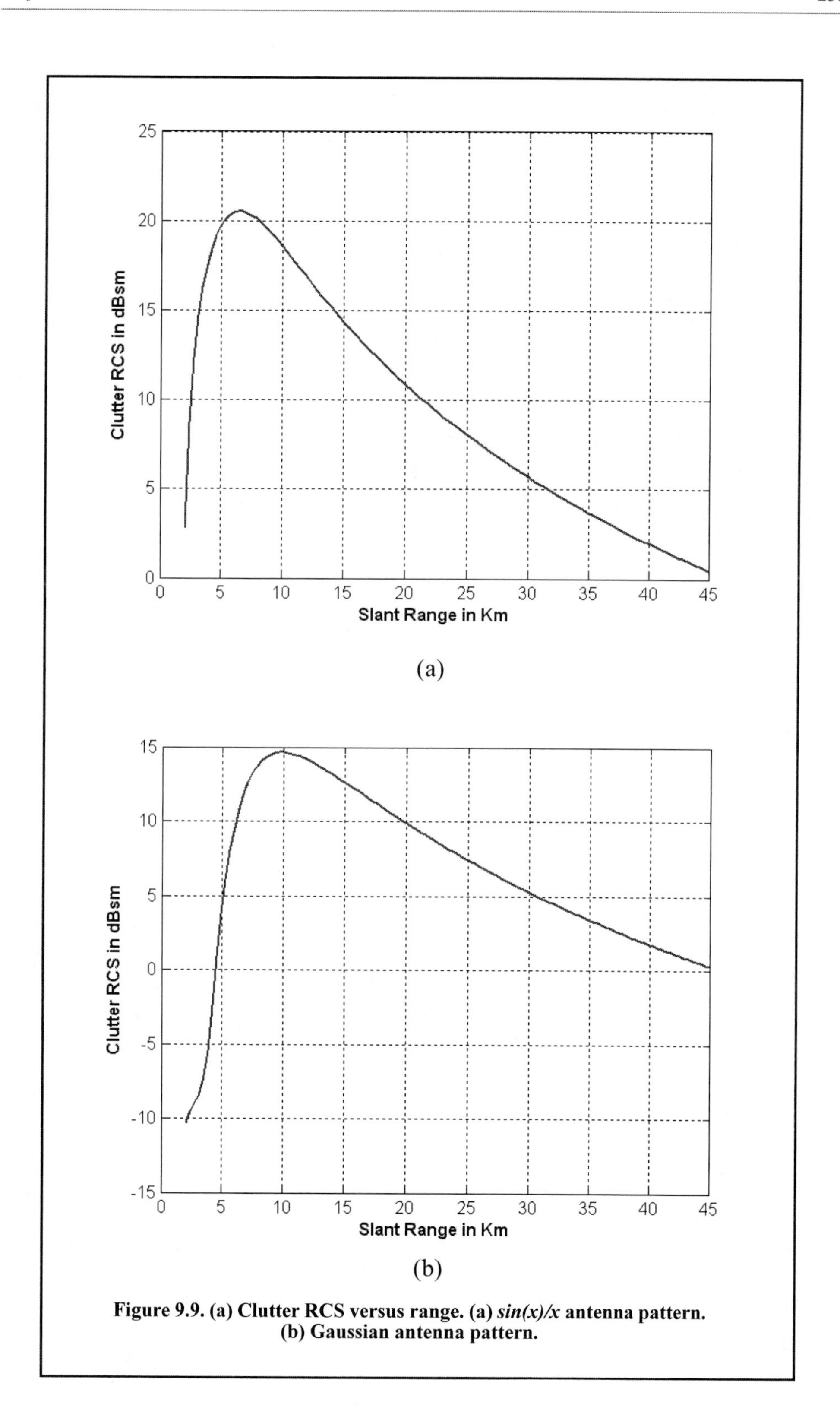

Figure 9.9. (a) Clutter RCS versus range. (a) *sin(x)/x* antenna pattern. (b) Gaussian antenna pattern.

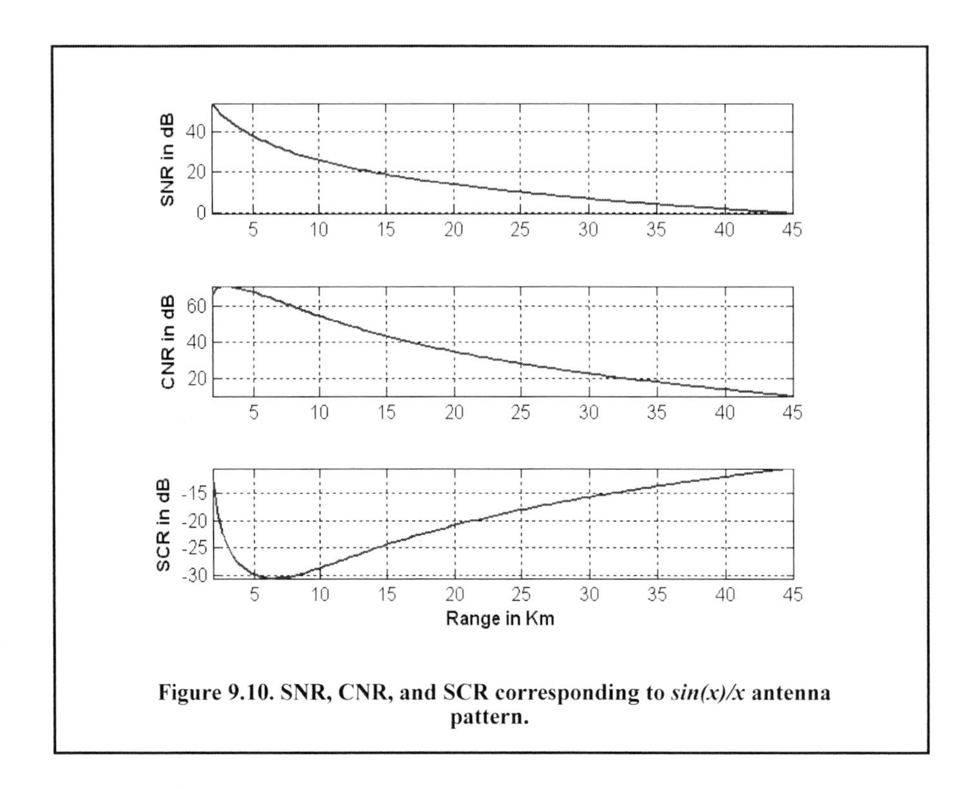

Figure 9.10. SNR, CNR, and SCR corresponding to *sin(x)/x* antenna pattern.

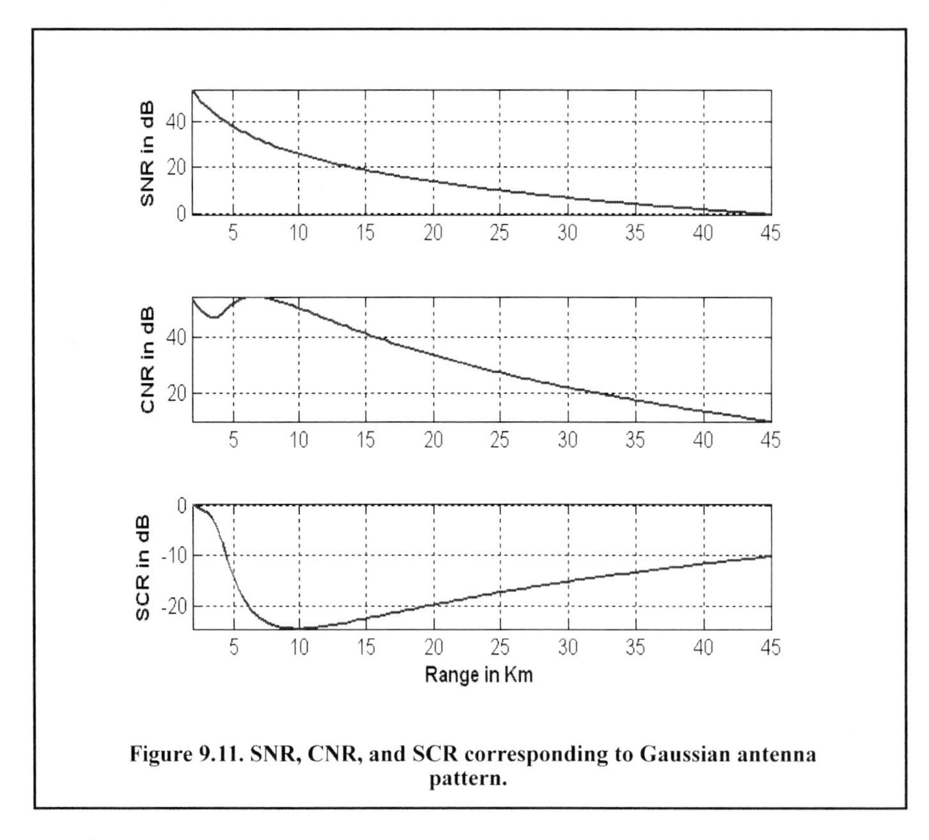

Figure 9.11. SNR, CNR, and SCR corresponding to Gaussian antenna pattern.

9.5. Clutter Components

It was established earlier that the complex envelope of the signal received by the radar comprises the target returns and additive band-limited white noise. In the presence of clutter, the complex envelope is now composed of target, noise, and clutter returns. That is,

$$\tilde{x}(t) = \tilde{s}(t) + \tilde{n}(t) + \tilde{w}(t) \qquad \text{Eq. (9.36)}$$

where $\tilde{s}(t)$, $\tilde{n}(t)$, and $\tilde{w}(t)$ are, respectively, the target, noise, and clutter complex envelope echoes. Noise is typically modeled (as discussed in earlier chapters) as a bandlimited white Gaussian random process. Furthermore, noise samples are considered statistically independent of each other and of clutter measurements.

Clutter arises from reflections of unwanted objects within the radar beam. Since many objects comprise the clutter returns, clutter may also be molded as a Gaussian random process. In other words, clutter samples from one radar measurement to another constitute a joint set of Gaussian random variables. However, because of the clutter fluctuation and due to antenna mechanical scanning, wind speed, and radar platform motion (if applicable), these random variables are not statistically independent.

More precisely, because of the antenna mechanical scanning, clutter returns in the radar main-beam do not have the same amplitude from pulse to pulse. This will effectively add amplitude modulation to the clutter returns. This additional modulation is governed by the shape of the antenna pattern, the rate of mechanical scanning, and the radar PRF. Denote the antenna two-way azimuth $3dB$ beamwidth as θ_a and the antenna scan rate as $\dot{\theta}_{scan}$. It follows that the contribution of antenna scanning to the standard deviation of the clutter fluctuation is

$$\sigma_s = 0.399\frac{\dot{\theta}_{scan}}{\theta_a}. \qquad \text{Eq. (9.37)}$$

Another contributor to the clutter spectral spreading is caused by motion of the clutter itself, due to wind. Trees, vegetation, and sea waves are the main contributors to this effect. This relative motion, although relatively small, introduces additional Doppler shift in the clutter returns. The Doppler frequency due to a relative velocity v is given by

$$f_d = (2v)/\lambda \qquad \text{Eq. (9.38)}$$

where λ is the radar operating wavelength. It follows that if the apparent rms velocity due to wind is v_{rms}, then the standard deviation is

$$\sigma_w = 2v_{rms}/\lambda. \qquad \text{Eq. (9.39)}$$

Finally, if the radar platform is in motion, then the relative motion between the platform and the stationary clutter will cause a Doppler shift given by

$$f_c = (2v_{radar}\cos\theta)/\lambda \qquad \text{Eq. (9.40)}$$

where $v_{radar}\cos\theta$ is the radial velocity component of the platform in the direction of clutter. Since the radar beam has a finite width, not all clutter components have the same radial velocity at all times. More specifically, if the angles θ_1 and θ_2 represent the edges of the radar beam, then Eq. (9.40) can be written as

$$f_c = \frac{2v_{radar}}{\lambda}(\cos\theta_2 - \cos\theta_1) \approx \frac{2v_{radar}}{\lambda}\theta_a \sin\theta \qquad \text{Eq. (9.41)}$$

and the standard deviation due to platform motion is given by

$$\sigma_v = \frac{v_{radar}}{\lambda}\sin\theta\,. \qquad \text{Eq. (9.42)}$$

Finally, the overall clutter spreading is denoted by σ_f, where

$$\sigma_f^2 = \sigma_v^2 + \sigma_s^2 + \sigma_w^2\,. \qquad \text{Eq. (9.43)}$$

9.6. Clutter Backscatter Coefficient Statistical Models

Assessing radar performance in the presence of clutter depends heavily on one's ability to accurately estimate or measure the backscatter coefficient $\sigma°$. Since clutter within a resolution or volume cell is composed of a large number of scatterers with random phases and amplitudes, the backscatter coefficient is typically described statistically by a probability distribution function. The type of distribution depends on the nature of clutter itself (sea, land, volume), the radar operating frequency, and the grazing angle.

9.6.1. Surface Clutter Case

The most common statistical model used to describe $\sigma°$ for surface clutter is the log-normal and exponential (i.e., Rayleigh amplitude) probability density functions. Although the log-normal distribution will provide an accurate measure of $\sigma°$ at large grazing angles, it is not as accurate at low grazing angles less than *5* to *7* degrees. In this case, the Rayleigh distribution (which is a special case of the Weibull distribution) provides more accurate statistical estimates of $\sigma°$. Another probability density function widely used to estimate $\sigma°$ is the Weibull distribution.

The Weibull probability density function can be written as

$$f(\sigma°) = \frac{b(\sigma°)^{b-1}}{\alpha}\exp\left(-\frac{(\sigma°)^b}{\alpha}\right); \quad \sigma° \geq 0 \qquad \text{Eq. (9.44)}$$

where b, α are the Weibull distribution parameters. Define the Weibull distribution slope a as $1/b$, and the parameter α as

$$\alpha = \frac{(\sigma_m^o)^b}{\ln 2} \qquad \text{Eq. (9.45)}$$

where σ_m^o is the median value for σ^o. The proof of Eq. (9.55) is left as an exercise (see Problem 9.6). It follows that Eq. (9.52) can be written as

$$f(\sigma°) = \frac{\ln 2\ (\sigma°)^{\frac{1}{a}-1}}{a(\sigma_m^o)^{1/a}}\exp\left(-\ln 2\left(\frac{\sigma°}{\sigma_m^o}\right)^{1/a}\right); \quad \sigma° \geq 0\,. \qquad \text{Eq. (9.46)}$$

Note that when $b=1$, then Eq. (9.44) becomes the exponential (or Rayleigh amplitude) probability density function,

$$f(\sigma^\circ) = \frac{1}{\overline{\sigma^\circ}}\exp\left(-\frac{\sigma^\circ}{\overline{\sigma^\circ}}\right);\ \sigma^\circ \geq 0 \qquad \text{Eq. (9.47)}$$

where $\alpha = \overline{\sigma^\circ}$ is the average value for σ^o.

The mean value for σ^o can be determined from the integral

$$\overline{\sigma^\circ} = \int_0^\infty \sigma^\circ f(\sigma^\circ)d\sigma^\circ \qquad \text{Eq. (9.48)}$$

by making the change of variable $q = \sigma^{\circ b}/\alpha$, and by using $a = 1/b$ yields

$$\overline{\sigma^\circ} = \alpha^a \int_0^\infty q^a e^{-q} dq, \qquad \text{Eq. (9.49)}$$

which is the incomplete Gamma integral. More precisely,

$$\overline{\sigma^\circ} = \alpha^a \Gamma(1 + a). \qquad \text{Eq. (9.50)}$$

The probability that an actual clutter radar cross section per unit area will not exceed the value σ^o is

$$Pr(\sigma_c^o \leq \sigma^o) = \int_0^{\sigma^o} f(\sigma^\circ)d\sigma^\circ. \qquad \text{Eq. (9.51)}$$

Substituting Eq. (9.47) into Eq. (9.51) and performing the integration yields

$$Pr(\sigma_c^o \leq \sigma^o) = 1 - \exp\left(\frac{\sigma^{\circ b}}{\alpha}\right). \qquad \text{Eq. (9.52)}$$

Eq. (9.52) can now be used to solve for σ^o, that is

$$\sigma^o = \left[\alpha \ln\left(\frac{1}{1 - Pr(\sigma_c^o \leq \sigma^o)}\right)\right]^{1/b} = \left[\alpha \ln\left(\frac{1}{1 - Pr(\sigma_c^o \leq \sigma^o)}\right)\right]^a. \qquad \text{Eq. (9.53)}$$

The median value for σ^o is computed by setting $Pr(\sigma_c^o \leq \sigma^o) = 0.5$ in Eq. (9.53). In this case,

$$\sigma_m^o = [\alpha \ln 2]^a. \qquad \text{Eq. (9.54)}$$

Using Eqs. (9.45) and (9.54) into Eq. (9.52) yields

$$Pr(\sigma_c^o \leq \sigma^o) = 1 - \exp\left(\ln 2\left(\frac{\sigma^o}{\sigma_m^o}\right)^b\right). \qquad \text{Eq. (9.55)}$$

To obtain a simpler formula for σ^o in decibels, substitute Eq. (9.45) into Eq. (9.55) to get

$$\sigma^o\big|_{dB} = 10\log\sigma_m^o - 10a\log(\ln 2) + 10a\log\left(\ln\left(\frac{1}{1 - Pr(\sigma_c^o \leq \sigma^o)}\right)\right), \qquad \text{Eq. (9.56)}$$

which can be rewritten as

$$\sigma^o\big|_{dB} = \sigma^o_m\big|_{dB} - 10\log(\Gamma(1+a)) + 10a\log\{-\ln[1 - Pr(\sigma^o_c \le \sigma^o)]\}\,. \qquad \text{Eq. (9.57)}$$

Figure 9.12 shows some typical plots for σ^o against the probability defined in Eq. (9.52). Note that only values where $Pr > 0.2$ and $Pr < 0.9$ are used because values of σ^o corresponding to very low probabilities are typically below the radar's noise level. Alternatively, values for σ^o corresponding to high probabilities are typically too high for an MTI radar to suppress (the next chapter addresses MTI radars in details).

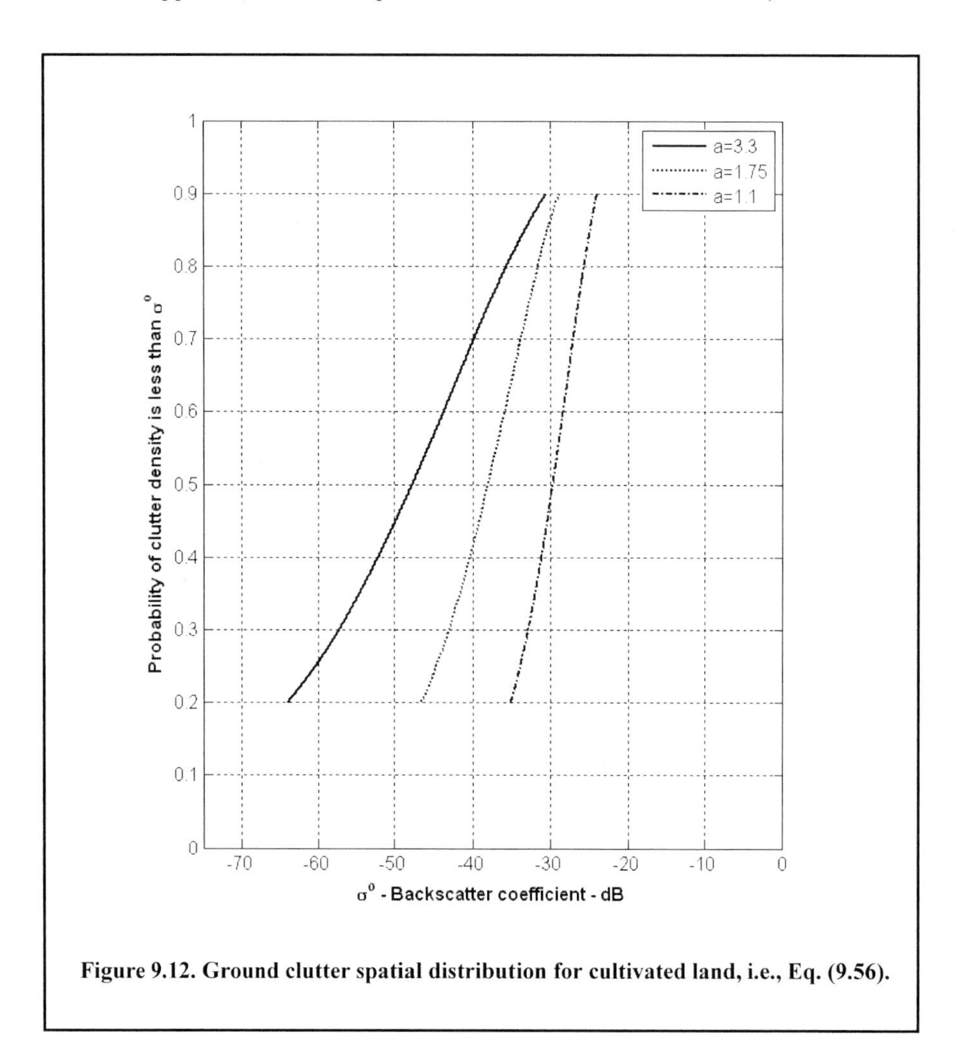

Figure 9.12. Ground clutter spatial distribution for cultivated land, i.e., Eq. (9.56).

9.6.2. Volume Clutter Case

The backscatter coefficient, Z, defined in Eq. (9.20), is often used by meteorologists and less often by radar engineers. In radar applications, it is more meaningful to use a precipitation backscatter coefficient that is measured in squared meters per cubic meter instead of $millimeter^6/m^3$. For this purpose, define a new precipitation backscatter coefficient η as

$$\eta = \frac{5.63 \times 10^{-14} r^{1.6}}{\lambda} \quad m^2/m^3 \qquad \text{Eq. (9.58)}$$

where r is the rate of precipitation in *millimeter/hr* and λ is the radar operating wavelength in *meters*.

The value of the exponent in Eq. (9.58) varies from 0.95 at tropical latitudes and frequencies above $10GHz$ to about 1.6, which is more applicable to temperate latitudes. Additionally, radar waves using circular polarization and wavelengths comparable to the rain droplets' average diameter will result in less backscattering than is the case for linearly polarized waves. To explain this observation further, consider a right circular polarized radar whose wavelength is comparable to the average rain droplet diameter. The reflected waves from the rain droplets will also be right circularly polarized waves but traveling in the opposite direction (i.e., from the point view of the radar they will be left circularly polarized). Therefore, most of the reflected energy will be denied entry into the radar receiver by its antenna, resulting in less backscatter energy in the radar signal and data processors. The average ratio of a circularly polarized to a linearly polarized backscatter coefficient is

$$\left.\frac{\eta_{cp}}{\eta_{lp}}\right|_{dB} \approx \begin{cases} -15 & \text{for rain} \\ -10 & \text{for bright land} \end{cases} \qquad \text{Eq. (9.59)}$$

where bright land is defined as the transitional region between ice or snow and water resulting from melting.

Problems

9.1. Compute the signal-to-clutter ratio (SCR) for the radar described in Section 9.2.1. In this case, assume antenna $3dB$ beam width $\theta_{3dB} = 0.03rad$, pulse width $\tau = 10\mu s$, range $R = 50Km$, grazing angle $\psi_g = 15°$, target RCS $\sigma_t = 0.1m^2$, and clutter reflection coefficient $\sigma^0 = 0.02(m^2/m^2)$.

9.2. Repeat the example of Section 9.3 for target RCS $\sigma_t = 0.15m^2$, pulse width $\tau = 0.1\mu s$, antenna beam width $\theta_a = \theta_e = 0.03radians$; the detection range is $R = 100Km$, and $\sum \sigma_i = 1.6 \times 10^{-9}(m^2/m^3)$.

9.3. The quadrature components of the clutter power spectrum are, respectively, given by

$$\bar{S}_I(f) = \delta(f) + \frac{C}{\sqrt{2\pi}\sigma_c} \exp(-f^2/2\sigma_c^2)$$

$$\overline{S_Q}(f) = \frac{C}{\sqrt{2\pi}\sigma_c} \exp(-f^2/2\sigma_c^2).$$

Compute the D.C. and A.C. power of the clutter. Let $\sigma_c = 10Hz$.

9.4. A certain radar has the following specifications: pulse width $\tau' = 1\mu s$, antenna beam width $\Omega = 1.5°$, and wavelength $\lambda = 3cm$. The radar antenna is $7.5m$ high. A certain target is simulated by two point targets (scatterers). The first scatterer is $4m$ high and has RCS $\sigma_1 = 20m^2$. The second scatterer is $12m$ high and has RCS $\sigma_2 = 1m^2$. If the target is detected at $10Km$, compute (a) the SCR when both scatterers are observed by the radar; (b) the SCR when only the first scatterer is observed by the radar. Assume a reflection coefficient of -1, and $\sigma^0 = -30dB$.

9.5. A certain radar has range resolution of $300m$ and is observing a target somewhere in a line of high towers each having RCS $\sigma_{tower} = 10^6 m^2$. If the target has RCS $\sigma_t = 1m^2$, (a) how much signal-to-clutter ratio should the radar have? (b) Repeat part (a) for range resolution of $30m$.

9.6. Prove that the Weibull distribution α is given by $\alpha = \frac{(\sigma_m^o)^b}{\ln 2}$ where σ_m^o is the median value for σ^o.

Chapter 10

Moving Target Indicator

10.1. Clutter Power Spectrum Density

Clutter primarily comprises unwanted stationary ground reflections with limited relative motion with respect to the radar. Therefore, its power spectrum density will be concentrated around $f = 0$. However, because the overall clutter spreading σ_f is not always zero, clutter actually exhibits some Doppler frequency spread. The overall clutter spreading is denoted by σ_f, and is given by

$$\sigma_f^2 = \sigma_v^2 + \sigma_s^2 + \sigma_w^2 . \qquad \text{Eq. (10.1)}$$

σ_v accounts for clutter spread due to platform motion, σ_s accounts for the antenna scan rate, and σ_w accounts for the clutter spread due to wind.

The clutter power spectrum can be written as the sum of fixed (stationary) and random (due to frequency spreading) components, as

$$S_c(f) = \frac{P_c}{T\sigma_f\sqrt{2\pi}} \sum_{k=-\infty}^{\infty} \exp\left(-\frac{(f-k/T)^2}{2\sigma_f^2}\right) \qquad \text{Eq. (10.2)}$$

where T is the PRI (i.e., $1/f_r$, f_r is the PRF), P_c is the clutter power or clutter mean square value, and σ_f is the clutter spectral spreading parameter as defined in Eq. (10.1). As clearly indicated by Eq. (10.2), the clutter PSD is periodic with period equal to f_r. Furthermore, the clutter PSD extends about each multiple integer of the PRF. It must be noted that this spread is relatively small and thus the relation $\sigma_f \ll f_r$ is always true. This is illustrated in Figure 10.1. The mean square value can be calculated from

$$P_c = T\int_{-f_r/2}^{f_r/2} S_c(f)df . \qquad \text{Eq. (10.3)}$$

Let $S_{c0}(f)$ denote the central portion of Eq. (10.2) (i.e., $k = 0$); then P_c is expressed as

$$P_c = T\int_{-\infty}^{\infty} S_{c0}(f)df \qquad \text{Eq. (10.4)}$$

where $S_{c0}(f)$ is a Gaussian shape function given by

$$S_{c0}(f) = \frac{k}{\sigma_f\sqrt{2\pi}} \exp\left(-\frac{f^2}{2\sigma_f^2}\right) \qquad \text{Eq. (10.5)}$$

and $k = P_c/T$.

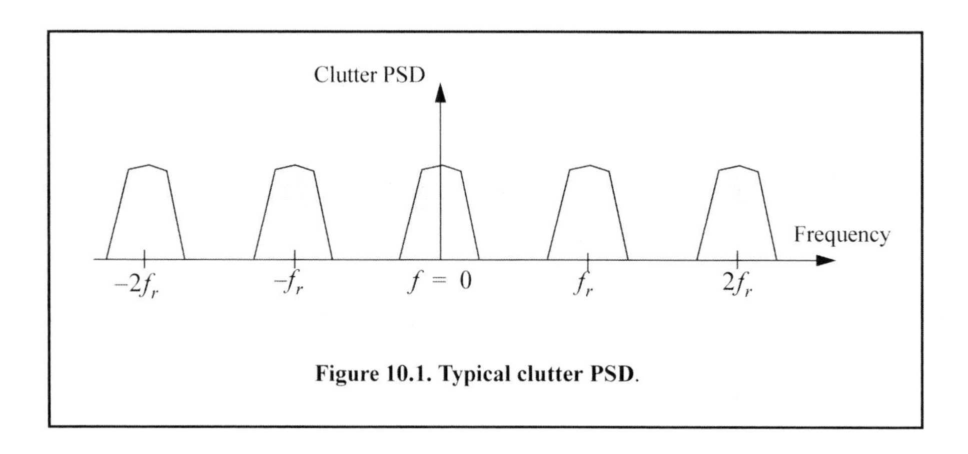

Figure 10.1. Typical clutter PSD.

10.2. Concept of a Moving Target Indicator (MTI)

The clutter spectrum is concentrated around DC ($f = 0$) and multiple integers of the radar PRF f_r, as was illustrated in Figure 10.1. In CW radars, clutter is avoided or suppressed by ignoring the receiver output around DC, since most of the clutter power is concentrated about the zero frequency band. Pulsed radar systems may utilize special filters that can distinguish between slow-moving or stationary targets and fast-moving ones. This class of filter is known as the Moving Target Indicator (MTI). In simple words, the purpose of an MTI filter is to suppress target-like returns produced by clutter and allow returns from moving targets to pass through with little or no degradation. In order to effectively suppress clutter returns, an MTI filter needs to have a deep stopband at DC and at integer multiples of the PRF. Figure 10.2b shows a typical sketch of an MTI filter response, while Figure 10.2c shows its output when the PSD shown in Figure 10.2a is the input. MTI filters can be implemented using delay line cancelers. As we will show later in this chapter, the frequency response of this class of MTI filter is periodic, with nulls at integer multiples of the PRF. Thus, targets with Doppler frequencies equal to nf_r are severely attenuated. Since Doppler is proportional to target velocity ($f_d = 2v/\lambda$), target speeds that produce Doppler frequencies equal to integer multiples of f_r are known as blind speeds. More precisely,

$$v_{blind} = (n\lambda f_r)/2;\ n \geq 0\,. \qquad \text{Eq. (10.6)}$$

Radar systems can minimize the occurrence of blind speeds either by employing multiple PRF schemes (PRF staggering) or by using high PRFs in which the radar may become range ambiguous. The main difference between PRF staggering and PRF agility is that the pulse repetition interval (within an integration interval) can be changed between consecutive pulses for the case of PRF staggering.

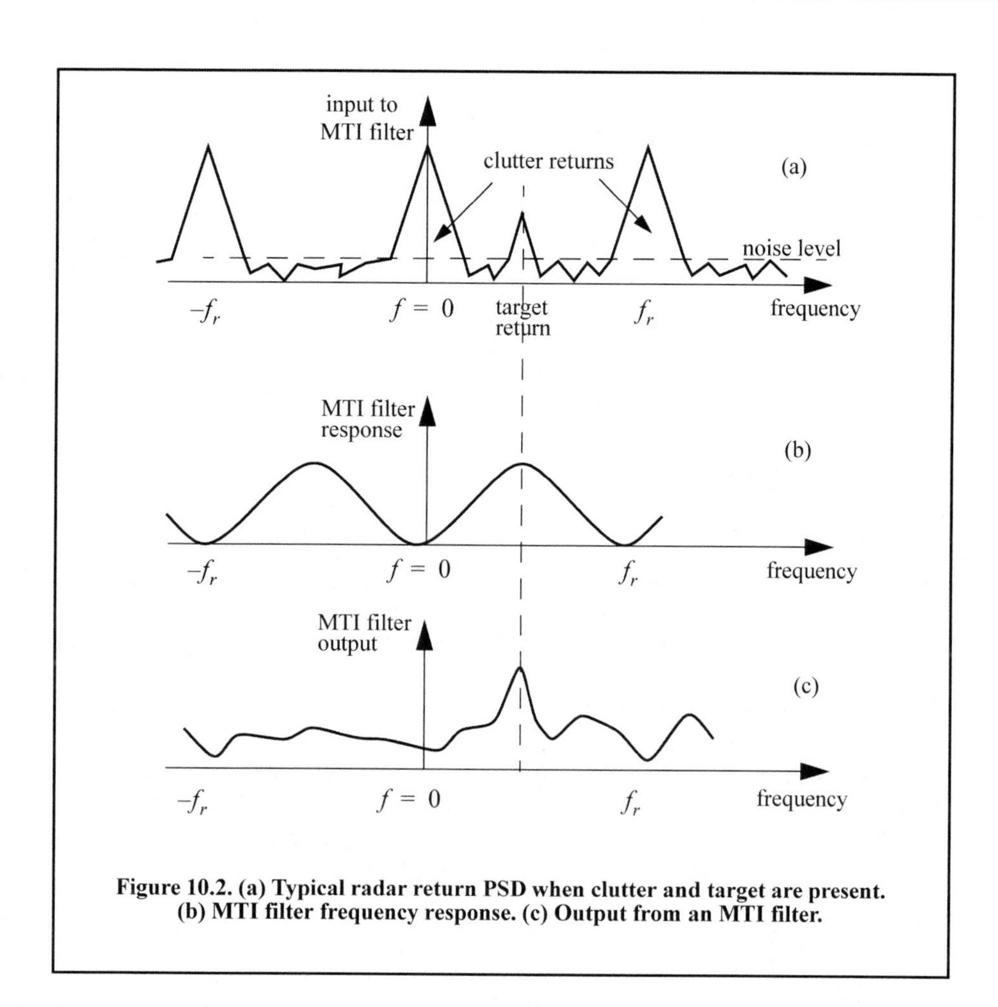

Figure 10.2. (a) Typical radar return PSD when clutter and target are present. (b) MTI filter frequency response. (c) Output from an MTI filter.

10.2.1. Single Delay Line Canceler

A single delay line canceler can be implemented as shown in Figure 10.3. The canceler's impulse response is denoted as $h(t)$. The output $y(t)$ is equal to the convolution between the impulse response $h(t)$ and the input $x(t)$. The single delay canceler is often called a two-pulse canceler since it requires two distinct input pulses before an output can be read.

The delay T is equal to the radar PRI ($1/f_r$). The output signal $y(t)$ is

$$y(t) = x(t) - x(t-T) . \qquad \text{Eq. (10.7)}$$

The impulse response of the canceler is given by

$$h(t) = \delta(t) - \delta(t-T) \qquad \text{Eq. (10.8)}$$

where $\delta(\ )$ is the delta function. It follows that the Fourier transform (FT) of $h(t)$ is

$$H(\omega) = 1 - e^{-j\omega T} \qquad \text{Eq. (10.9)}$$

where $\omega = 2\pi f$. In the z-domain, the single delay line canceler response is

$$H(z) = 1 - z^{-1} . \qquad \text{Eq. (10.10)}$$

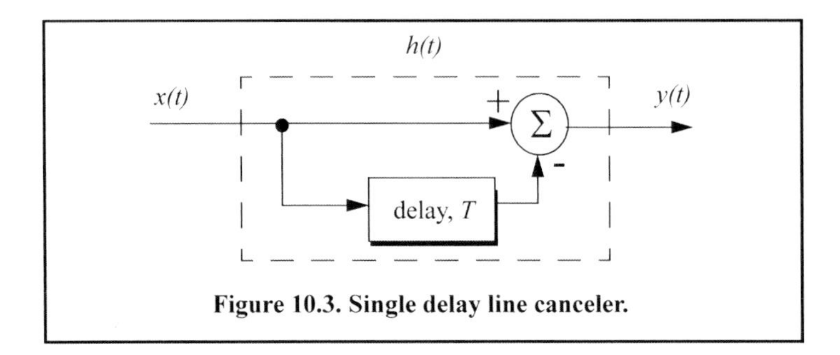

Figure 10.3. Single delay line canceler.

The power gain for the single delay line canceler is given by

$$|H(\omega)|^2 = H(\omega)H^*(\omega) = (1 - e^{-j\omega T})(1 - e^{j\omega T}). \qquad \text{Eq. (10.11)}$$

It follows that

$$|H(\omega)|^2 = 1 + 1 - (e^{j\omega T} + e^{-j\omega T}) = 2(1 - \cos\omega T) \qquad \text{Eq. (10.12)}$$

and using the trigonometric identity $(2 - 2\cos 2\vartheta) = 4(\sin\vartheta)^2$ yields

$$|H(\omega)|^2 = 4(\sin(\omega T/2))^2. \qquad \text{Eq. (10.13)}$$

The amplitude frequency response for a single delay line canceller is shown in Figure 10.4. Clearly, the frequency response of a single canceler is periodic with a period equal to f_r. The peaks occur at $f = (2n + 1)/(2f_r)$, and the nulls are at $f = nf_r$, where $n \geq 0$. In most radar applications the response of a single canceler is not acceptable since it does not have a wide notch in the stopband. A double delay line canceler has better response in both the stop- and passbands, and thus it is more frequently used than a single canceler. The names *single delay line canceler* and *single canceler* will be used interchangeably.

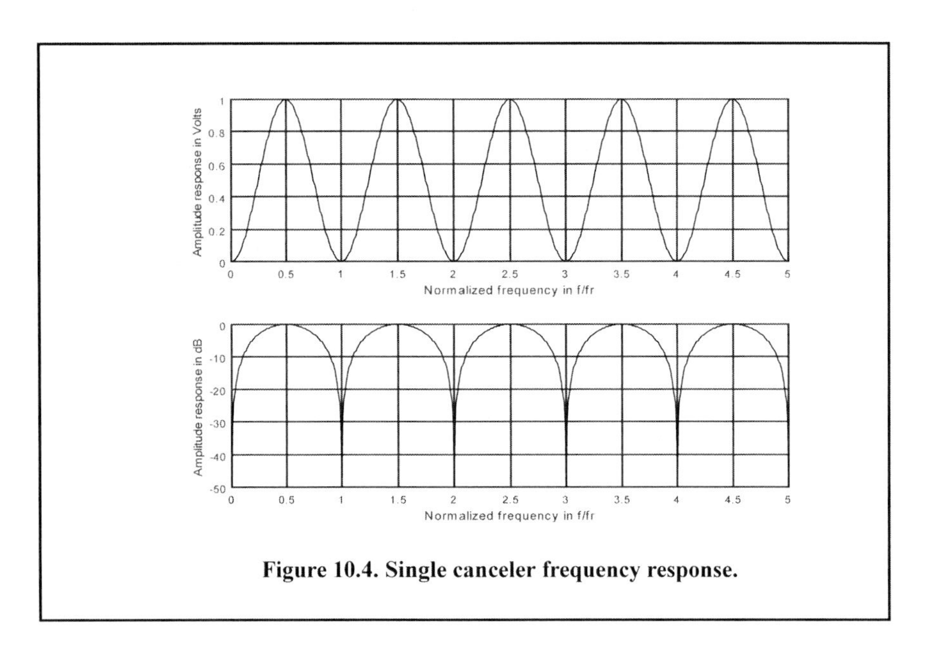

Figure 10.4. Single canceler frequency response.

10.2.2. Double Delay Line Canceler

Two basic configurations of a double delay line canceler are shown in Figure 10.5. Double cancelers are often called three-pulse cancelers since they require three distinct input pulses before an output can be read. The double line canceler impulse response is given by

$$h(t) = \delta(t) - 2\delta(t-T) + \delta(t-2T). \quad \text{Eq. (10.14)}$$

Again, the names *double delay line canceler* and *double canceler* will be used interchangeably. The power gain for the double delay line canceler is

$$|H(\omega)|^2 = |H_1(\omega)|^2 |H_1(\omega)|^2. \quad \text{Eq. (10.15)}$$

$|H_1(\omega)|^2$ is the single line canceler power gain given in Eq. (10.13). It follows that

$$|H(\omega)|^2 = 16\left(\sin\left(\omega\frac{T}{2}\right)\right)^4. \quad \text{Eq. (10.16)}$$

And in the z-domain, we have

$$H(z) = (1-z^{-1})^2 = 1 - 2z^{-1} + z^{-2}. \quad \text{Eq. (10.17)}$$

Figure 10.6 shows typical output from this function. Note that the double canceler has a better response than the single canceler (deeper notch and flatter passband response).

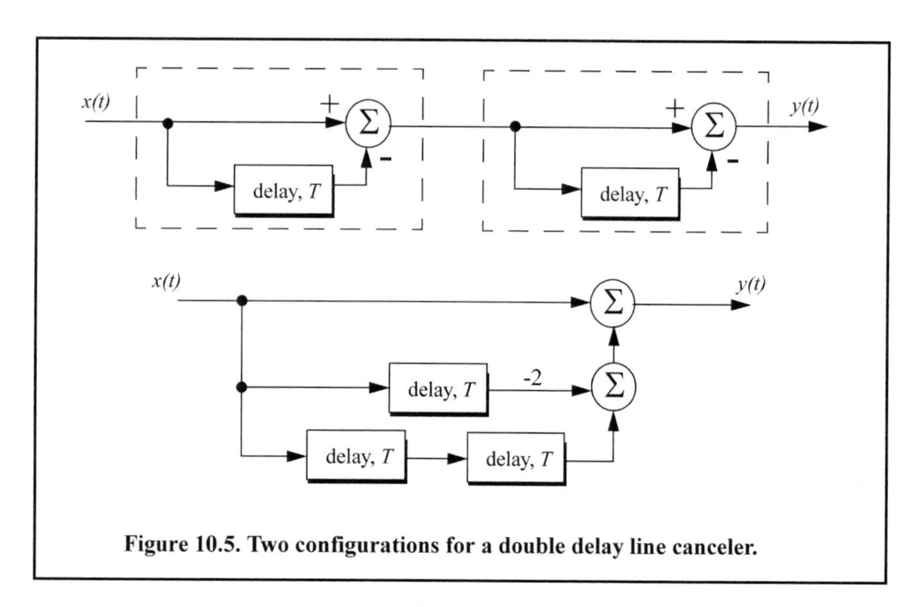

Figure 10.5. Two configurations for a double delay line canceler.

10.2.3. Delay Lines with Feedback (Recursive Filters)

Delay line cancelers with feedback loops are known as recursive filters. The advantage of a recursive filter is that through a feedback loop, we will be able to shape the frequency response of the filter. As an example, consider the single canceler shown in Figure 10.7. From the figure we can write

$$y(t) = x(t) - (1-K)w(t) \quad \text{Eq. (10.18)}$$

$$v(t) = y(t) + w(t) \quad \text{Eq. (10.19)}$$

$$w(t) = v(t-T). \quad \text{Eq. (10.20)}$$

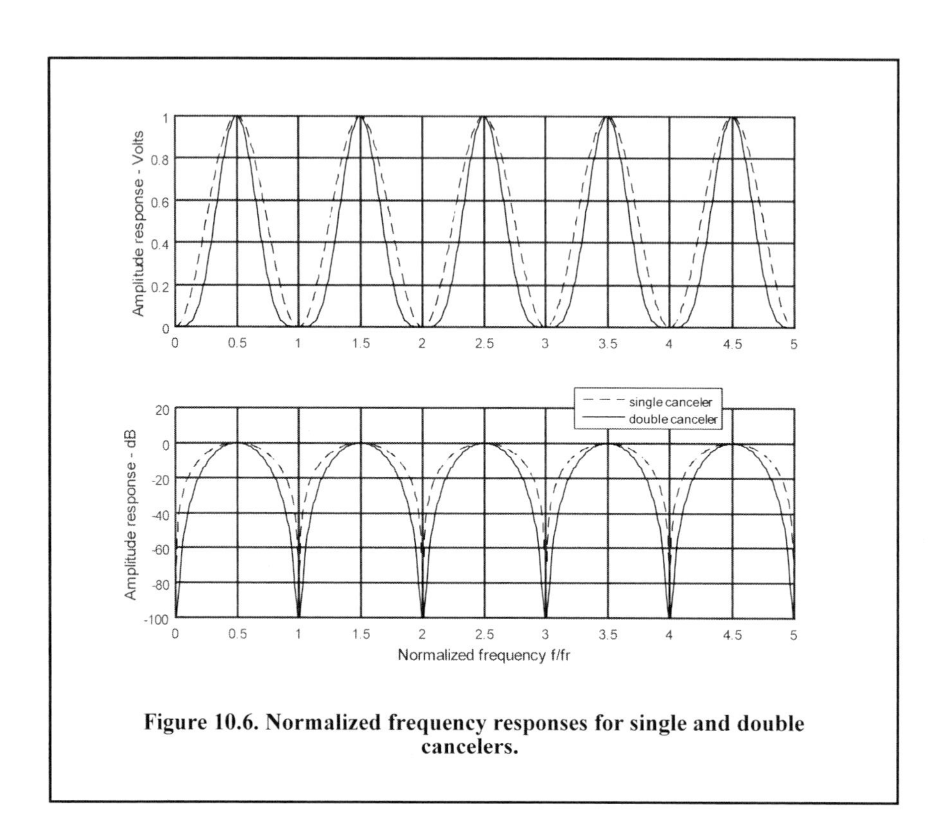

Figure 10.6. Normalized frequency responses for single and double cancelers.

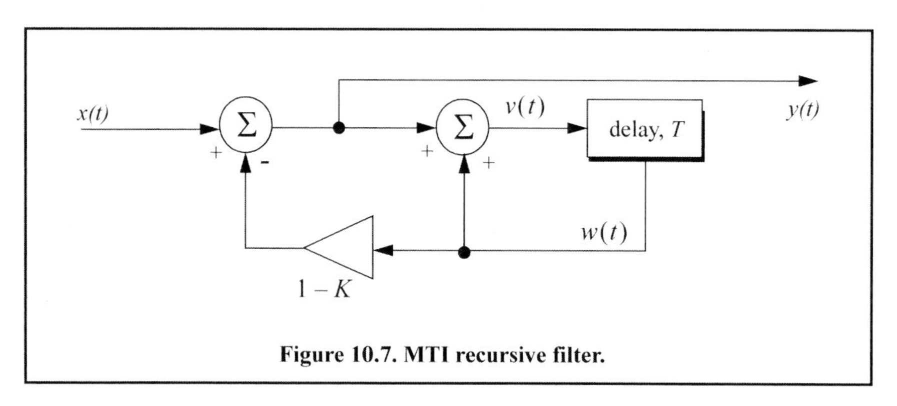

Figure 10.7. MTI recursive filter.

Applying the z-transform to the above three equations yields

$$Y(z) = X(z) - (1 - K)W(z) \qquad \text{Eq. (10.21)}$$

$$V(z) = Y(z) + W(z) \qquad \text{Eq. (10.22)}$$

$$W(z) = z^{-1}V(z). \qquad \text{Eq. (10.23)}$$

Solving for the transfer function $H(z) = Y(z)/X(z)$ yields

$$H(z) = \frac{1 - z^{-1}}{1 - Kz^{-1}}. \qquad \text{Eq. (10.24)}$$

The modulus square of $H(z)$ is then equal to

$$|H(z)|^2 = \frac{(1-z^{-1})(1-z)}{(1-Kz^{-1})(1-Kz)} = \frac{2-(z+z^{-1})}{(1+K^2)-K(z+z^{-1})}. \qquad \text{Eq. (10.25)}$$

Using the transformation $z = e^{j\omega T}$ yields

$$z + z^{-1} = 2\cos\omega T. \qquad \text{Eq. (10.26)}$$

Thus, Eq. (10.24) can now be rewritten as

$$|H(e^{j\omega T})|^2 = \frac{2(1-\cos\omega T)}{(1+K^2)-2K\cos(\omega T)}. \qquad \text{Eq. (10.27)}$$

Note that when $K = 0$, Eq. (10.27) collapses to Eq. (10.11) (single line canceler). Figure 10.8 shows a plot of Eq. (10.27) for $K = 0.15, 0.25, 0.5$. Clearly, by changing the gain factor K, one can control the filter response.

In order to avoid oscillation due to the positive feedback, the value of K should be less than unity. The value $(1-K)^{-1}$ is normally equal to the number of pulses received from the target. For example, $K = 0.9$ corresponds to ten pulses, while $K = 0.98$ corresponds to about fifty pulses.

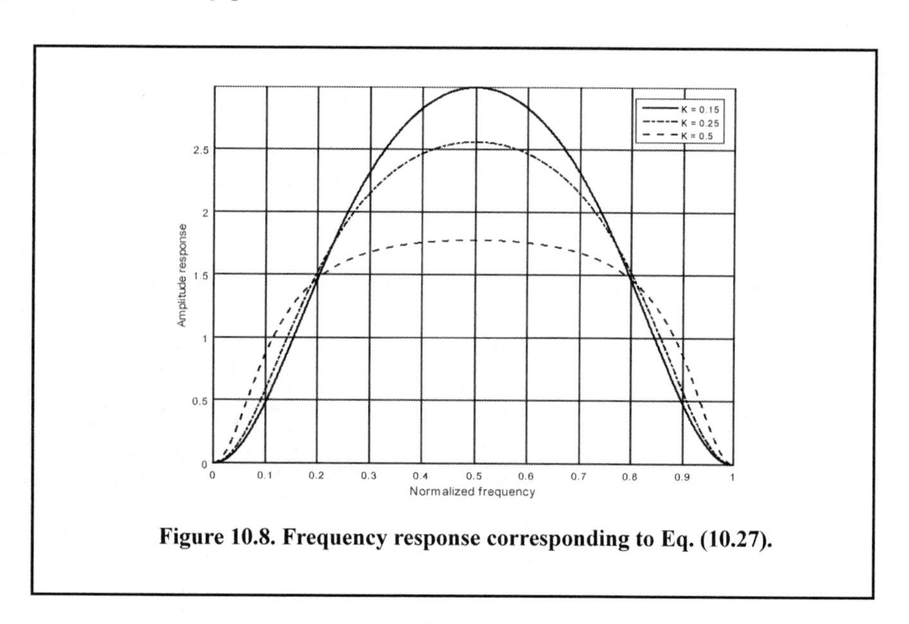

Figure 10.8. Frequency response corresponding to Eq. (10.27).

10.3. PRF Staggering

Target velocities that correspond to multiple integers of the PRF are referred to as blind speeds. This terminology is used since an MTI filter response is equal to zero at these values. Blind speeds can pose serious limitations on the performance of MTI radars and their ability to perform adequate target detection. Using PRF agility by changing the pulse repetition interval between consecutive pulses can extend the first blind speed to more tolerable values. In order to show how PRF staggering can alleviate the problem of blind speeds, let us first assume that two radars with distinct PRFs are utilized for detection.

Since blind speeds are proportional to the PRF, the blind speeds of the two radars would be different. However, using two radars to alleviate the problem of blind speeds is a very costly option. A more practical solution is to use a single radar with two or more different PRFs.

For example, consider a radar system with two interpulse periods T_1 and T_2, such that

$$\frac{T_1}{T_2} = \frac{n_1}{n_2} \qquad \text{Eq. (10.28)}$$

where n_1 and n_2 are integers. The first true blind speed occurs when

$$\frac{n_1}{T_1} = \frac{n_2}{T_2}. \qquad \text{Eq. (10.29)}$$

This is illustrated in Figure 10.9 for $n_1 = 5$ and $n_2 = 7$.

The ratio

$$k_s = n_1/n_2 \qquad \text{Eq. (10.30)}$$

is known as the stagger ratio. Using staggering ratios closer to unity pushes the first true blind speed farther out. However, the dip in the vicinity of $1/T_1$ becomes deeper. In general, if there are N PRFs related by

$$\frac{n_1}{T_1} = \frac{n_2}{T_2} = \ldots = \frac{n_N}{T_N}, \qquad \text{Eq. (10.31)}$$

and if the first blind speed to occur for any of the individual PRFs is v_{blind1}, then the first true blind speed for the staggered waveform is

$$v_{blind} = \frac{n_1 + n_2 + \ldots + n_N}{N} v_{blind1}. \qquad \text{Eq. (10.32)}$$

To better determine the frequency response of an MTI filter with staggered PRFs, consider a three-pulse canceler with two PRFs, or equivalently two PRIs, T_1 and T_2. In this case, the impulse response will be given by

$$h(t) = [\delta(t) - \delta(t - T_1)] - [\delta(t - T_1) - \delta(t - T_1 - T_2)] \qquad \text{Eq. (10.33)}$$

which can be written as

$$h(t) = \delta(t) - 2\delta(t - T_1) + \delta(t - T_1 - T_2). \qquad \text{Eq. (10.34)}$$

Note that PRF staggering requires a minimum of two PRFs.

Make the change of variables $u = t - T_1$ in Eq. (10.34), and it follows that

$$h(u + T_1) = \delta(u + T_1) - 2\delta(u) + \delta(u - T_2). \qquad \text{Eq. (10.35)}$$

The Z-transform of the impulse response in Eq. (10.35) is then given by

$$H(z)z^{-T_1} = z^{T_1} - 2 + z^{-T_2} \qquad \text{Eq. (10.36)}$$

and the amplitude frequency response for the staggered double delay line canceller is then given by

$$|H(z)|^2\Big|_{z = e^{j\omega T}} = (z^{T_1} - 2 + z^{-T_2})(z^{-T_1} - 2 + z^{T_2}). \qquad \text{Eq. (10.37)}$$

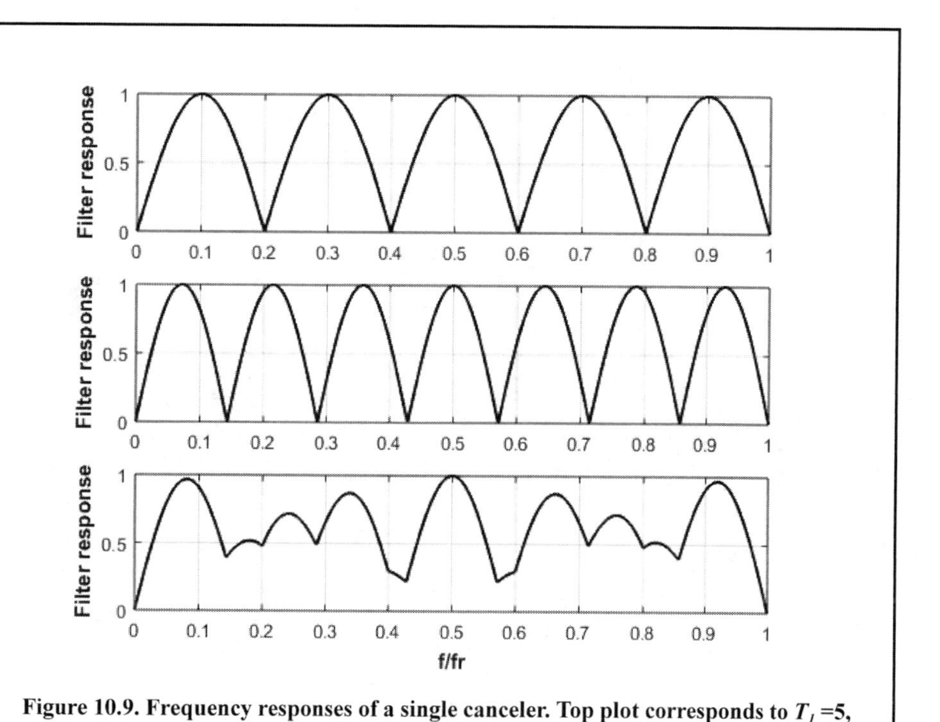

Figure 10.9. Frequency responses of a single canceler. Top plot corresponds to $T_1 = 5$, middle plot corresponds to $T_2 = 7$, bottom plot corresponds to stagger ratio $T_2/T_1 = 7/5$.

Performing the algebraic manipulation in Eq. (10.37) and using the trigonometric identity $(e^{j\omega T} + e^{-j\omega T}) = 2\cos\omega T$ yields

$$|H(\omega)|^2 = 6 - 4\cos(2\pi f T_1) - 4\cos(2\pi f T_2) + 2\cos(2\pi f(T_1 + T_2)). \qquad \text{Eq. (10.38)}$$

It is customary to normalize the amplitude frequency response, thus

$$|H(\omega)|^2 = 1 - \frac{2}{3}\cos(2\pi f T_1) - \frac{2}{3}\cos(2\pi f T_2) + \frac{1}{3}\cos(2\pi f(T_1 + T_2)). \qquad \text{Eq. (10.39)}$$

To determine the characteristics of higher stagger ratio MTI filters, adopt the notion of having several MTI filters, one for each combination of two staggered PRFs. Then the overall filter response is computed as the average of all individual filters. For example, consider the case where a PRF stagger is required with PRIs T_1, T_2, T_3, and T_4. First, compute the filter response using T_1 T_2 and denote by H_1. Then compute H_2 using T_2 and T_3, the filter H_3 is computed using T_3 T_4, and the filter H_4 is computed using T_4 and T_1. Finally compute the overall response as

$$H(f) = \frac{1}{4}[H_1(f) + H_2(f) + H_3(f) + H_4(f)]. \qquad \text{Eq. (10.40)}$$

Figure 10.10 shows the MTI filter response for a 4-stagger-ratio defined. The overall response is computed as the average of 4 individual filters, each corresponding to one combination of the stagger ratio. In the top portion of the figure the individual filters used were 2-pulse MTIs, while the bottom portion used 4-pulse individual MTI filters.

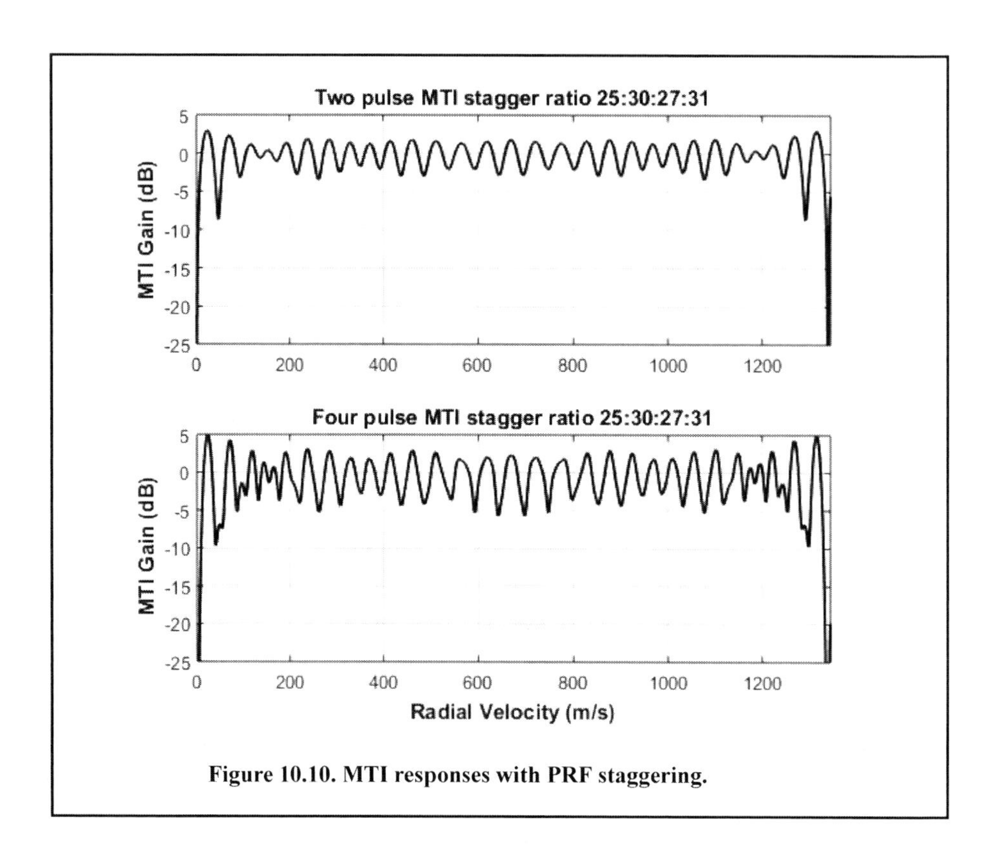

Figure 10.10. MTI responses with PRF staggering.

10.4. MTI Improvement Factor

In this section, two quantities that are normally used to define the performance of MTI systems are introduced. They are Clutter Attenuation (CA) and the Improvement Factor. The MTI CA is defined as the ratio between the MTI filter input clutter power C_i to the output clutter power C_o,

$$CA = C_i / C_o . \qquad \text{Eq. (10.41)}$$

The MTI improvement factor is defined as the ratio of the SCR at the output to the SCR at the input,

$$I = \left(\frac{S_o}{C_o}\right) / \left(\frac{S_i}{C_i}\right) , \qquad \text{Eq. (10.42)}$$

which can be rewritten as

$$I = \frac{S_o}{S_i} CA . \qquad \text{Eq. (10.43)}$$

The ratio S_o / S_i is the average power gain of the MTI filter, and it is equal to $|H(\omega)|^2$. In this section, a closed form expression for the improvement factor using a Gaussian-shaped power spectrum is developed. A Gaussian-shaped clutter power spectrum is given by

$$S(f) = \frac{P_c}{\sqrt{2\pi}\ \sigma_f} \exp(-f^2 / 2\sigma_f^2) \qquad \text{Eq. (10.44)}$$

where P_c is the clutter power (constant), and σ_f is the clutter rms frequency (which describes the clutter spectrum spread in the frequency domain, see Eq. (10.1)).

The clutter power at the input of an MTI filter is

$$C_i = \int_{-\infty}^{\infty} \frac{P_c}{\sqrt{2\pi}\ \sigma_f} \exp\left(-\frac{f^2}{2\sigma_f^2}\right) df. \qquad \text{Eq. (10.45)}$$

Factoring out the constant P_c yields

$$C_i = P_c \int_{-\infty}^{\infty} \frac{1}{\sqrt{2\pi}\sigma_f} \exp\left(-\frac{f^2}{2\sigma_f^2}\right) df. \qquad \text{Eq. (10.46)}$$

It follows that

$$C_i = P_c. \qquad \text{Eq. (10.47)}$$

The clutter power at the output of an MTI is

$$C_o = \int_{-\infty}^{\infty} S(f)|H(f)|^2\ df. \qquad \text{Eq. (10.48)}$$

10.4.1. Two-Pulse MTI Case

In this section we will continue the analysis using a single delay line canceler. The frequency response for a single delay line canceler is

$$|H(f)|^2 = 4\left(\sin\left(\frac{\pi f}{f_r}\right)\right)^2. \qquad \text{Eq. (10.49)}$$

It follows that

$$C_o = \int_{-\infty}^{\infty} \frac{P_c}{\sqrt{2\pi}\ \sigma_f} \exp\left(-\frac{f^2}{2\sigma_f^2}\right) 4\left(\sin\left(\frac{\pi f}{f_r}\right)\right)^2\ df. \qquad \text{Eq. (10.50)}$$

Since clutter power will only be significant for small f, the ratio f/f_r is very small. Consequently, by using the small angle approximation, Eq. (10.50) is approximated by

$$C_o \approx \int_{-\infty}^{\infty} \frac{P_c}{\sqrt{2\pi}\ \sigma_f} \exp\left(-\frac{f^2}{2\sigma_f^2}\right) 4\left(\frac{\pi f}{f_r}\right)^2\ df, \qquad \text{Eq. (10.51)}$$

which can be rewritten as

$$C_o = \frac{4P_c\pi^2}{f_r^2} \int_{-\infty}^{\infty} \frac{1}{\sqrt{2\pi\sigma_f^2}} \exp\left(-\frac{f^2}{2\sigma_f^2}\right) f^2\ df. \qquad \text{Eq. (10.52)}$$

The integral part in Eq. (10.52) is the second moment of a zero-mean Gaussian distribution with variance σ_f^2. Replacing the integral in Eq. (10.52) by σ_f^2 yields

$$C_o = \frac{4P_c\pi^2}{f_r^2}\sigma_f^2. \qquad \text{Eq. (10.53)}$$

Substituting Eq. (10.53) and Eq. (10.47) into Eq. (10.41) produces

$$CA = \frac{C_i}{C_o} = \left(\frac{f_r}{2\pi\sigma_f}\right)^2. \qquad \text{Eq. (10.54)}$$

It follows that the improvement factor for a single canceler is

$$I = \left(\frac{f_r}{2\pi\sigma_f}\right)^2 \frac{S_o}{S_i}. \qquad \text{Eq. (10.55)}$$

The power gain ratio for a single canceler is (remember that $|H(f)|$ is periodic with period f_r)

$$\frac{S_o}{S_i} = |H(f)|^2 = \frac{1}{f_r}\int_{-f_r/2}^{f_r/2} 4\left(\sin\frac{\pi f}{f_r}\right)^2 df. \qquad \text{Eq. (10.56)}$$

Using the trigonometric identity $(2 - 2\cos 2\vartheta) = 4(\sin\vartheta)^2$ yields

$$|H(f)|^2 = \frac{1}{f_r}\int_{-f_r/2}^{f_r/2}\left(2 - 2\cos\frac{2\pi f}{f_r}\right)df = 2. \qquad \text{Eq. (10.57)}$$

It follows that

$$I = 2(f_r/2\pi\sigma_f)^2. \qquad \text{Eq. (10.58)}$$

The expression given in Eq. (10.58) is an approximation valid only for $\sigma_f \ll f_r$. When the condition $\sigma_f \ll f_r$ is not true, then the autocorrelation function needs to be used in order to develop an exact expression for the improvement factor. Furthermore, when taking into account Eq. (10.1) (i.e., account for antenna scan rate, wind, and platform motion) the improvement factor is reduced since σ_f becomes larger.

Example:

A certain radar has $f_r = 800Hz$. If the clutter rms is $\sigma_f = 6.4Hz$, find the improvement factor when a single delay line canceler is used.

Solution:

The clutter attenuation CA is

$$CA = \left(\frac{f_r}{2\pi\sigma_f}\right)^2 = \left(\frac{800}{(2\pi)(6.4)}\right)^2 = 395.771 = 25.974dB$$

and since $S_o/S_i = 2 = 3dB$ one gets

$$I_{dB} = (CA + S_o/S_i)_{dB} = 3 + 25.97 = 28.974dB.$$

****

10.5. Subclutter Visibility (SCV)

Subclutter Visibility (SCV) describes the radar's ability to detect nonstationary targets embedded in a strong clutter background, for some probabilities of detection and false alarm. It is often used as a measure of MTI performance. For example, a radar with $10dB$ SCV will be able to detect moving targets whose returns are ten times smaller than those of clutter. A sketch illustrating the concept of SCV is shown in Figure 10.11. If a radar system can resolve the areas of strong and weak clutter within its field of view, then Interclutter Visibility (ICV) describes the radar's ability to detect non-stationary targets between strong clutter points. The subclutter visibility is expressed as the ratio of the improvement factor to the minimum MTI output SCR required for proper detection for a given probability of detection. More precisely,

$$SCV = I/(SCR)_o . \qquad \text{Eq. (10.59)}$$

When comparing the performance of different radar systems on the basis of SCV, one should use caution since the amount of clutter power is dependent on the radar resolution cell (or volume), which may be different from one radar to another. Thus, only if the different radars have the same beamwidths and the same pulse widths can SCV be used as a basis of performance comparison.

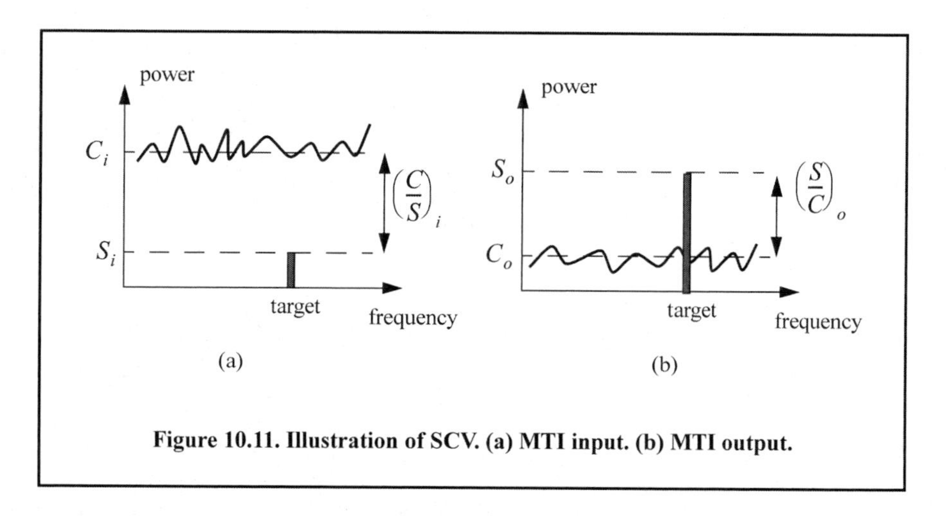

Figure 10.11. Illustration of SCV. (a) MTI input. (b) MTI output.

10.6. Delay Line Cancelers with Optimal Weights

The delay line cancelers discussed in this chapter belong to a family of transversal Finite Impulse Response (FIR) filters widely known as the "tapped delay line" filters. Figure 10.12 shows an N-stage tapped delay line implementation. When the weights are chosen such that they are the binomial coefficients (i.e., the coefficients of the expansion $(1-x)^N$) with alternating signs, then the resultant MTI filter is equivalent to N-stage cascaded single line cancelers. This is illustrated in Figure 10.13 for $N = 4$. In general, the binomial coefficients are given by

$$w_i = (-1)^{i-1} \frac{N!}{(N-i+1)!(i-1)!} \quad ; \quad i = 1, \dots, N+1 . \qquad \text{Eq. (10.60)}$$

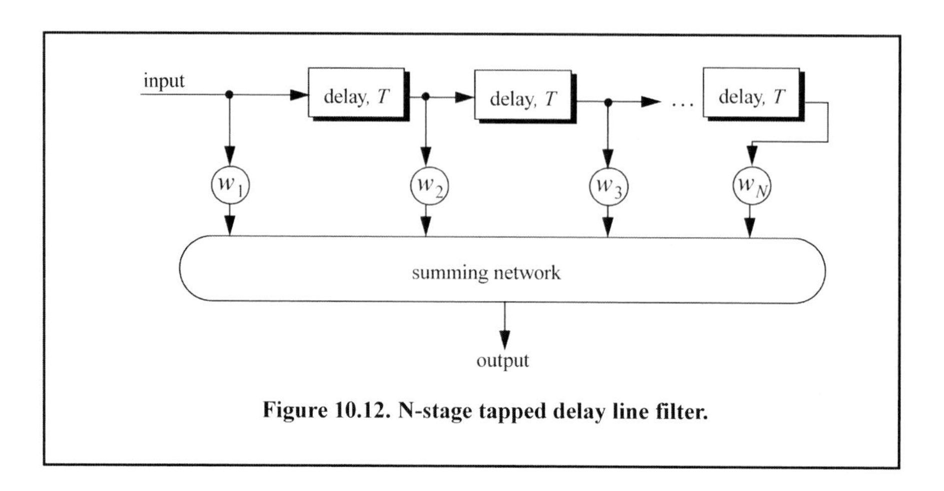

Figure 10.12. N-stage tapped delay line filter.

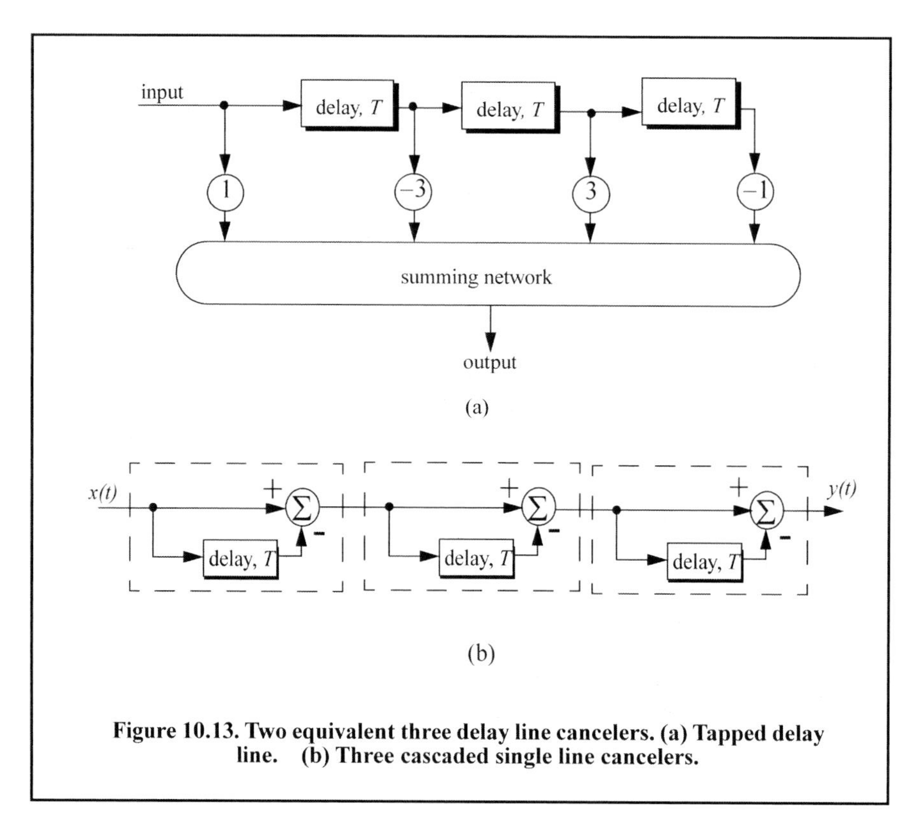

Figure 10.13. Two equivalent three delay line cancelers. (a) Tapped delay line. (b) Three cascaded single line cancelers.

Using the binomial coefficients with alternating signs produces an MTI filter that closely approximates the optimal filter in the sense that it maximizes the improvement factor, as well as the probability of detection. In fact, the difference between an optimal filter and one with binomial coefficients is so small that the latter one is considered to be optimal by most radar designers. However, being optimal in the sense of the improvement factor does not guarantee a deep notch or a flat passband in the MTI filter response. Consequently, many researchers have been investigating other weights that can produce a deeper notch around DC, as well as a better passband response.

In general, the average power gain for an N-stage delay line canceler is

$$\frac{S_o}{S_i} = \prod_{i=1}^{N} |H_1(f)|^2 = \prod_{i=1}^{N} 4\left(\sin\left(\frac{\pi f}{f_r}\right)\right)^2. \qquad \text{Eq. (10.61)}$$

For example, $N = 2$ (double delay line canceler) gives

$$\frac{S_o}{S_i} = 16\left(\sin\left(\frac{\pi f}{f_r}\right)\right)^4. \qquad \text{Eq. (10.62)}$$

Equation (10.61) can be rewritten as

$$\frac{S_o}{S_i} = |H_1(f)|^{2N} = 2^{2N}\left(\sin\left(\frac{\pi f}{f_r}\right)\right)^{2N}. \qquad \text{Eq. (10.63)}$$

As indicated by Eq. (10.63), blind speeds for an N-stage delay canceler are identical to those of a single canceler. It follows that blind speeds are independent from the number of cancelers used. It is possible to show that Eq. (10.63) can be written as

$$\frac{S_o}{S_i} = 1 + N^2 + \left(\frac{N(N-1)}{2!}\right)^2 + \left(\frac{N(N-1)(N-2)}{3!}\right)^2 + \ldots. \qquad \text{Eq. (10.64)}$$

A general expression for the improvement factor of an N-stage tapped delay line canceler is

$$I = \frac{(S_o/S_i)}{\sum_{k=1}^{N} \sum_{j=1}^{N} w_k w_j{}^* \rho\left(\frac{(k-j)}{f_r}\right)} \qquad \text{Eq. (10.65)}$$

where the weights w_k and w_j are those of a tapped delay line canceler, and $\rho((k-j)/f_r)$ is the correlation coefficient between the kth and jth samples. For example, $N = 2$ produces

$$I = \frac{1}{1 - \frac{4}{3}\rho T + \frac{1}{3}\rho 2T}. \qquad \text{Eq. (10.66)}$$

10.7. Phase Noise

It was determined in earlier chapters that the radar performance is improved as the SNR becomes larger. It was also established that in the presence of clutter, the radar performance will degrade beyond the impact of thermal noise alone, and in this case, the SCR becomes more critical than the SNR. Another source of noise that greatly limits MTI and pulsed Doppler radar performance is known as phase noise. Phase noise, sometimes called flicker noise, is random in nature and is caused by instabilities within the radar's local oscillator.

Phase noise may limit, depending on its actual value, pulsed Doppler radars' ability to detect very slow moving targets whose RCS is relatively small. This is illustrated in Figure 10.14. In this case, when the slow-moving target return signal is close to zero (or multiple integers of the PRF) it will likely be masked by the phase noise power caused by a noisy radar local oscillator. In addition to the masking problem illustrated in Figure 10.14, the MTI improvement factor will also be reduced by some appreciable values corresponding to the amount of phase noise present in the radar receiver, as will be explained later in this section.

Simply put, phase noise is a term used to describe the random frequency perturbation (relatively small in nature) occurring around the signal carrier frequency, thus causing a new instantaneous signal frequency that is slightly different from the original value. For example, consider the signal defined by the simple sinusoid

$$x(t) = r(t)\sin(2\pi f_c t) \qquad \text{Eq. (10.67)}$$

where $r(t)$ is the amplitude modulation and f_c is the carrier frequency. The perturbed signal due to amplitude and phase instabilities of the local oscillator will take on the form

$$x(t) = r(t + a(t))\sin(2\pi f_c t + \phi(t)) \qquad \text{Eq. (10.68)}$$

where $a(t)$ is the amplitude perturbation and $\phi(t)$ is the phase perturbation or fluctuation. Recalling that the instantaneous frequency is the derivative of phase with respect to time divided by 2π, then the phase perturbation will change the center frequency from f_c to $f_c + \delta f$ where δf is a fractional frequency deviation away from the carrier. More specifically,

$$\delta f = \frac{1}{2\pi}\frac{d}{dt}\phi(t). \qquad \text{Eq. (10.69)}$$

The notation commonly used in the literature to describe the phase noise power spectrum density is

$$S_\phi(f) = 2L(f). \qquad \text{Eq. (10.70)}$$

The factor of 2 accounts for both sidebands of the spectrum (lower and upper sidebands around f_c) and $L(f)$ is the ratio of noise power in a $1Hz$ bandwidth at an offset from the carrier signal power measured in dBc/Hz. In this notation, $L(f)$ is often used to denote phase noise. For example, consider a frequency-modulated signal as described in Eq. (3.47) in Chapter 3, which is repeated here for convenience as Eq. (10.71),

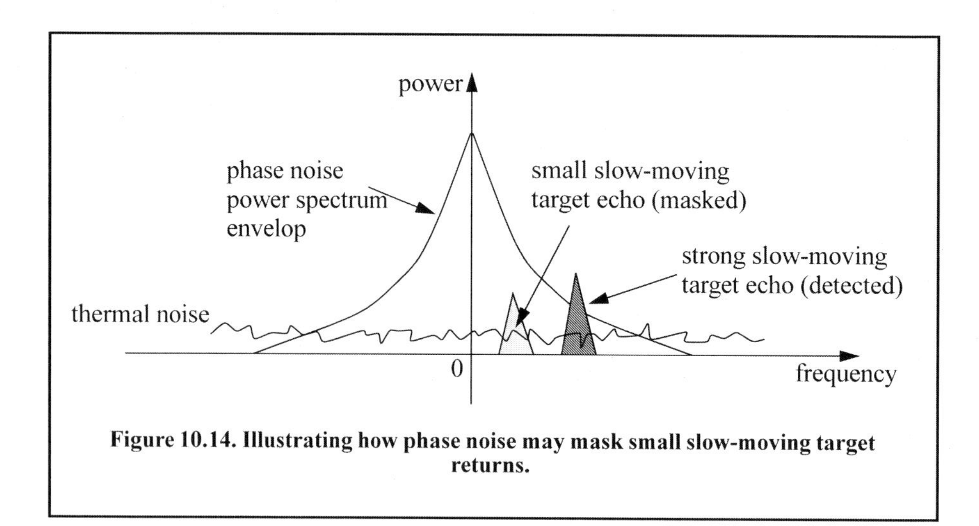

Figure 10.14. Illustrating how phase noise may mask small slow-moving target returns.

$$x(t) = A_c\left\{J_0(\beta)\cos 2\pi f_c t + \sum_{n = even}^{\infty} J_n(\beta)[\cos(2\pi f_c + 2n\pi f_m)t + \right. \quad \text{Eq. (10.71)}$$

$$\left.\cos(2\pi f_c - 2n\pi f_m)t] + \sum_{q = odd}^{\infty} J_q(\beta)[\cos(2\pi f_c + 2q\pi f_m)t - \cos(2\pi f_c - 2q\pi f_m)t]\right\}$$

where β is the modulation index and A is the amplitude. In this case, phase noise is defined as the ratio of the sideband power to the carrier power at a certain modulation frequency offset from the carrier. More specifically,

$$L(f)\big|_{dBc/Hz} = 10 \times \log\left\{\frac{|J_1(\beta)|^2}{|J_0(\beta)|^2}\right\}. \quad \text{Eq. (10.72)}$$

In general, the phase noise $L(f)$ decreases with frequency as a function of $(1/f^3)$, $(1/f^2)$, and $(1/f)$. Figure 10.15 shows an illustration plot for $L(f)$ versus the *log* of the frequency. Typically, the manufacturer of a given oscillator will measure and publish the phase noise characteristics as part of their product documentation. Observation of Figure 10.15 shows that this plot is a piece-wise linear function. It follows that the formula for a given segment of this plot is given by

$$L(f;f_{i+1} - f_i)\big|_{dBc/Hz} = m_i\log(f) - m_i\log(f_i) + L(f_i), \quad \text{Eq. (10.73)}$$

where m_i is the slope of the ith segment defined by

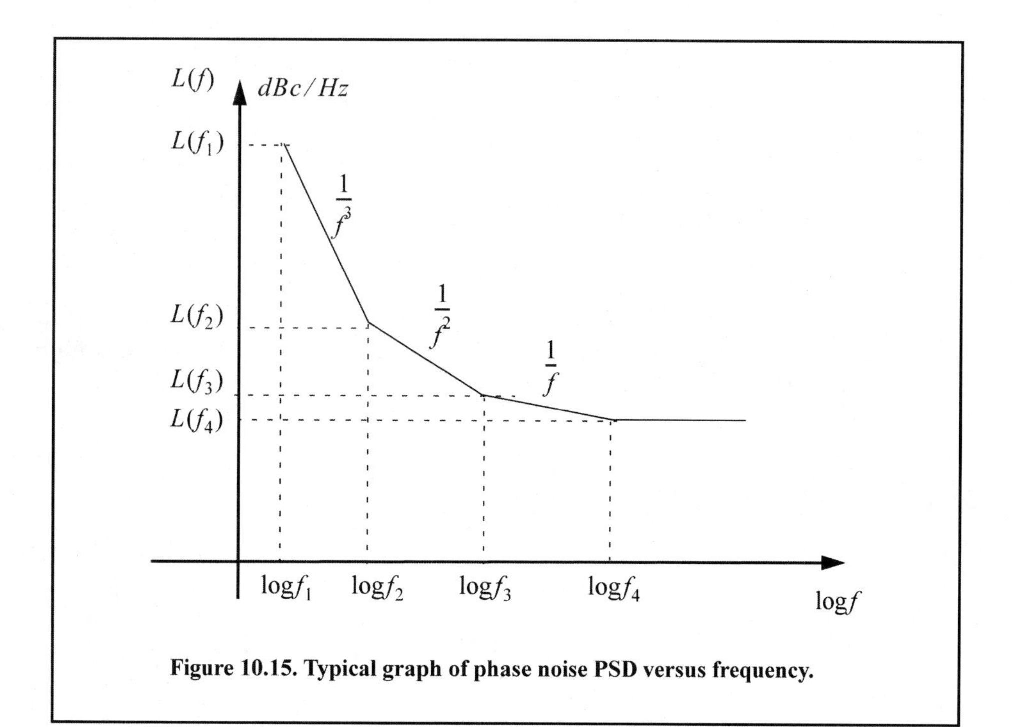

Figure 10.15. Typical graph of phase noise PSD versus frequency.

$$m_i = \frac{L(f_{i+1}) - L(f_i)}{\log(f_{i+1}) - \log(f_i)}\ .\qquad \text{Eq. (10.74)}$$

Eq. (10.73) can be written in a more compact form as

$$L(f) = L(f_i) \times 10^{\left[\frac{m_i}{10}\log\left(\frac{f}{f_i}\right)\right]}\ .\qquad \text{Eq. (10.75)}$$

Figure 10.16 shows an actual plot for $L(f)$ using the following values:

$$\{L(f_1), L(f_2), L(f_3), L(f_4), L(f_5)\} = \{-55, -85, -105, -115, -115\}\ .$$

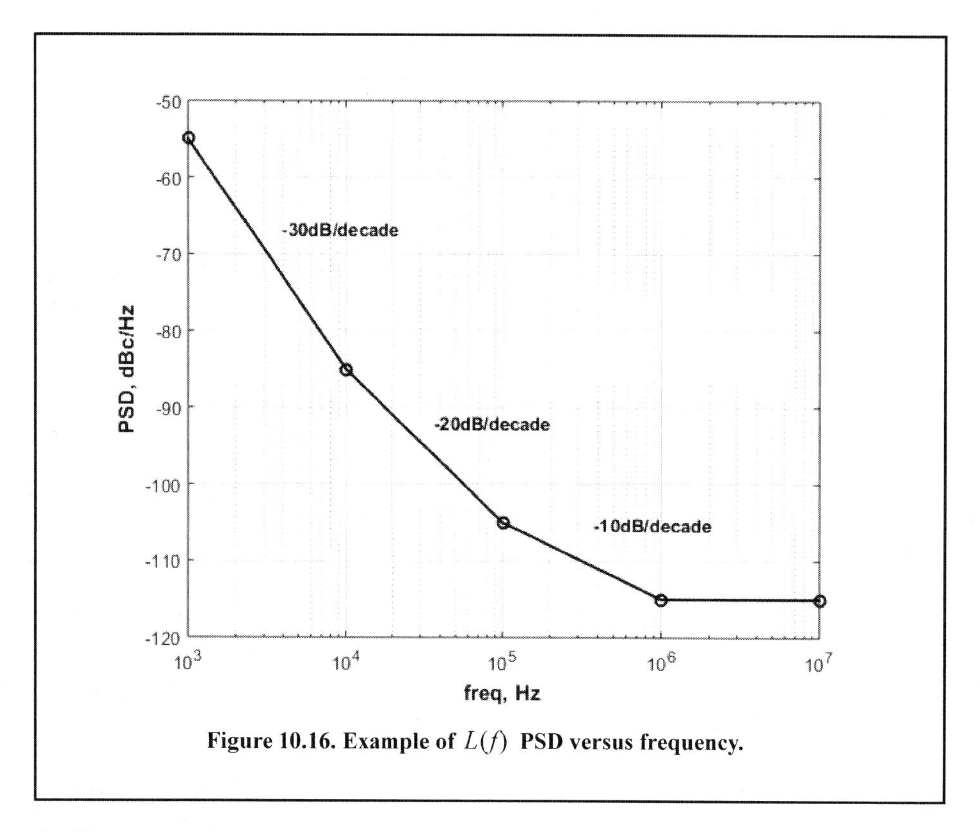

Figure 10.16. Example of $L(f)$ PSD versus frequency.

The literature is flooded with sources on phase noise. Different users use slightly different formulas to express phase noise in their particular application. In this book, the following formula for phase noise is adopted,

$$L(f) = \frac{1}{\pi} \frac{c_p \pi f_c^2}{(f - c_p \pi f_c)^2 + (c_p \pi f_c^2)^2}\qquad \text{Eq. (10.76)}$$

where f_c is the carrier frequency and c_p is a constant that describes phase noise of the oscillator. Figure 10.17 shows some typical plots of the ratio of noise power to the carrier power (as defined in Eq. (10.76)). Note that when the constant c_p becomes larger, the noise ratio spectrum becomes wider with lower amplitude of the main beam, and hence less phase noise in the system.

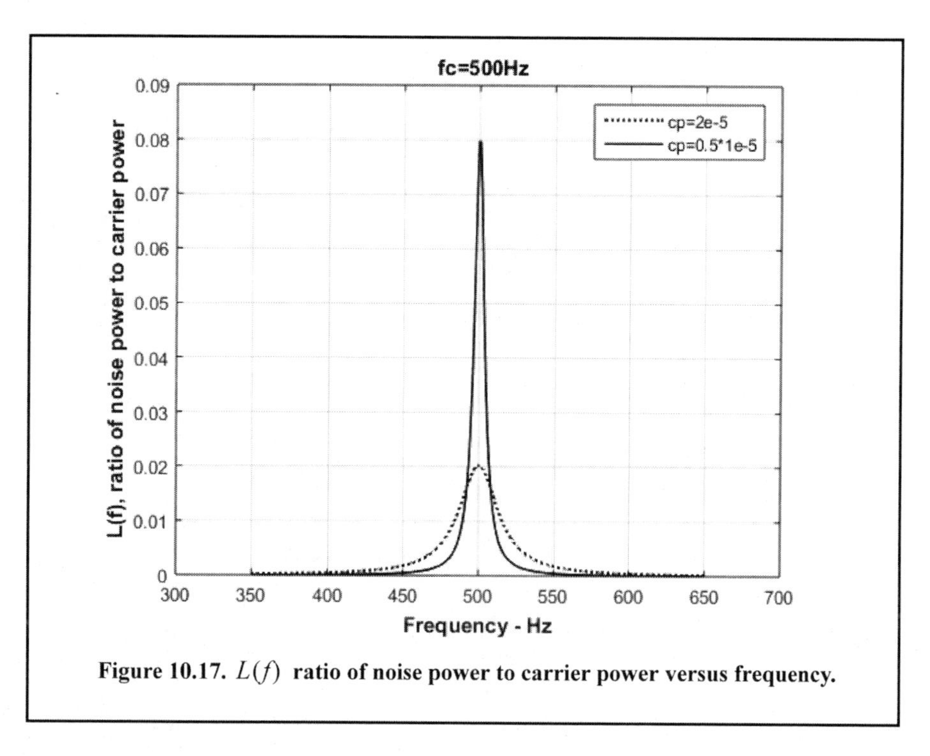

Figure 10.17. $L(f)$ **ratio of noise power to carrier power versus frequency.**

The normalized phase noise power spectrum density can be computed using Eqs. (10.70) and (10.75), and can be approximated as

$$L(f) = \begin{cases} \dfrac{1}{\pi f_b} & ;f \text{ approaches } f_c \\ \dfrac{1}{2\pi f_b} & ;f = f_c + f_b \\ \dfrac{f_b}{\pi f^2} & ;f \gg f_c + f_b \end{cases} \qquad \text{Eq. (10.77)}$$

where $f_b = c_p \pi f_c^2$. Figure 10.18 shows the corresponding normalized phase noise power spectrum density versus frequency. As indicated by Figure 10.18, quiet oscillators will almost always have phase noise of less than $0 dBc/Hz$ at frequency offsets of more than $1 Hz$ away from the carrier. However, at some small frequency bandwidth of less than $1 Hz$, phase noise may be greater than $0 dBc/Hz$.

The power spectral density function of phase noise at the output of the radar's matched filter can be expressed as

$$S_\phi(f) = L_0 P_c \left(\frac{\sin(\pi f \tau)}{\pi f \tau} \right)^2 \qquad \text{Eq. (10.78)}$$

where L_0 is the phase noise ratio relative to the carrier, P_c is the clutter power, and τ is the radar pulsewidth. L_0 can be computed from the analysis presented earlier; however, an acceptable range for L_0 varies between 10^{-9} to 10^{-15} dBc/Hz.

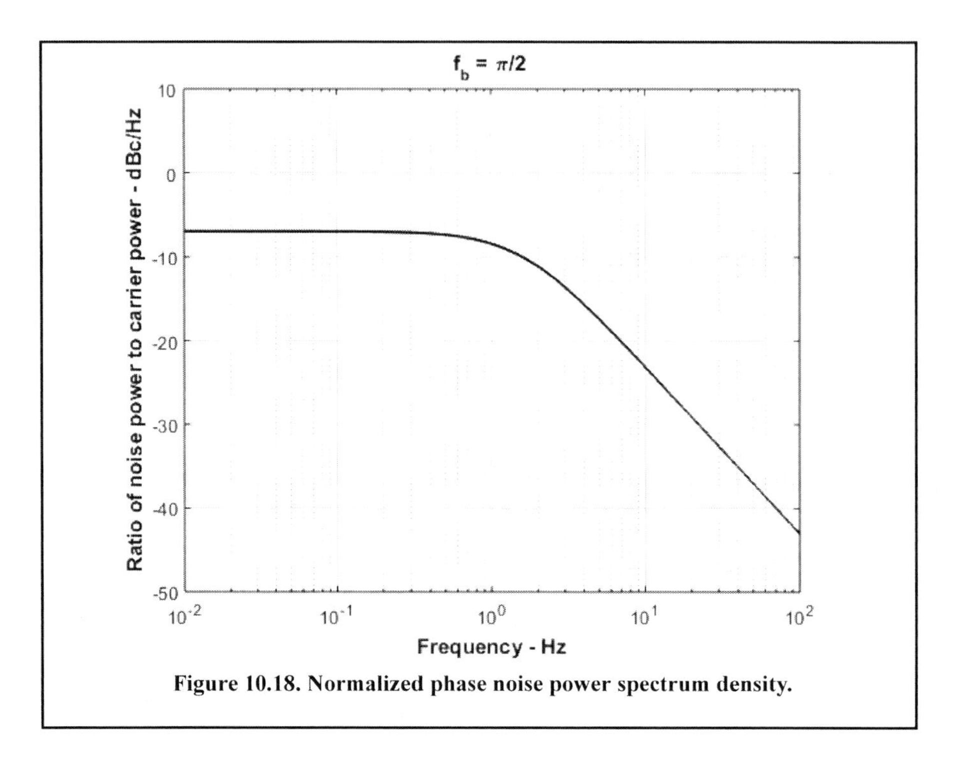

Figure 10.18. Normalized phase noise power spectrum density.

Recall the clutter attenuation which was defined as

$$CA = C_i/C_o \qquad \text{Eq. (10.79)}$$

where C_i is equal to P_c. Ignoring the phase noise, the clutter power spectrum was given in Eq. (10.44), but when phase noise is taken into consideration, Eq. (10.44) is modified to

$$S_t(f) = S(f) + S_\phi(f) = \frac{P_c}{\sqrt{2\pi}\ \sigma_f}\exp(-f^2/2\sigma_f^2) + L_0 P_c\left(\frac{\sin(\pi f\tau)}{\pi f\tau}\right)^2 \qquad \text{Eq. (10.80)}$$

In this case $S(f)$ is replaced by $S_t(f)$ in Eq. (10.102). Performing the integration and collecting terms (assuming a 2-pulse MTI filter), yields

$$C_o = P_c\frac{1}{2}\left(\frac{2\pi\sigma_f}{f_r}\right)^2 + P_c\frac{L_0}{\tau} \qquad \text{Eq. (10.81)}$$

where f_r is the PRF. It follows that the clutter attenuation in the presence of phase noise is given by

$$CA = \frac{C_i}{C_o} = \frac{1}{\frac{1}{2}\left(\frac{2\pi\sigma_f}{f_r}\right)^2 + \frac{L_0}{\tau}}. \qquad \text{Eq. (10.82)}$$

Figure 10.19 shows the clutter attenuation for two values of σ_f, with $f_r = 2.5KHz$ and $\tau = 1\mu s$ using Eq. (10.82). Observation of Figure 10.19 leads to the following conclusions: Larger values of σ_f will result in less clutter attenuation, so if more clutter attenuation is desired then a 3-pulse or higher order MTI filter ought to be used. Next, phase noise does not start to affect the performance of the MTI filter until it becomes higher than -100*dBc/Hz*. Clearly, using higher-order MTI filters will increase the amount of clutter attenuation; however, the question that remain is how phase noise affects the MTI performance when higher-order filters are used. This analysis is left as an exercise (see Problem 10.15).

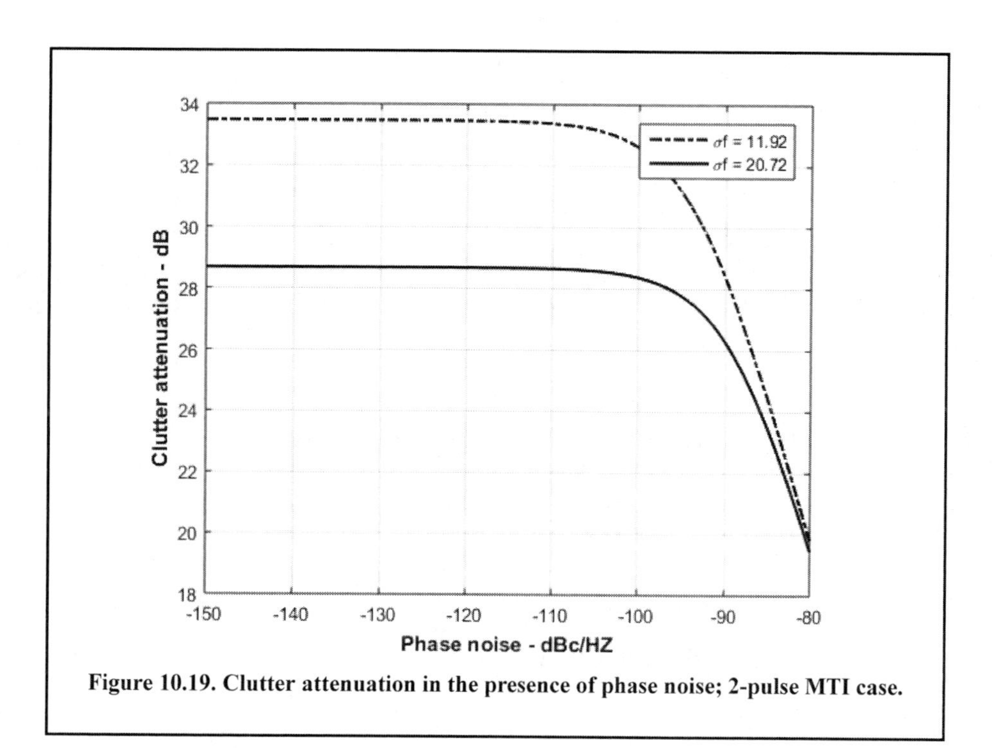

Figure 10.19. Clutter attenuation in the presence of phase noise; 2-pulse MTI case.

Problems

10.1. (a) Derive an expression for the impulse response of a single delay line canceler. (b) Repeat for a double delay line canceler.

10.2. Plot the frequency response for the filter described in the previous problem for $K = -0.5, 0$ and 0.5.

10.3. Consider a single delay line canceler. Calculate the clutter attenuation and the improvement factor. Assume that $\sigma_c = 4Hz$ and PRF $f_r = 450Hz$.

10.4. One implementation of a single delay line canceler with feedback is shown below.

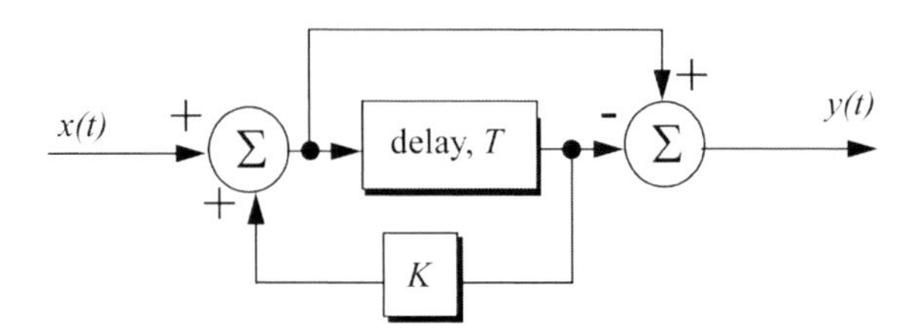

(a) What is the transfer function, $H(z)$? (b) If the clutter power spectrum is $W(f) = w_0 \exp(-f^2/2\sigma_c^2)$, find an exact expression for the filter power gain. (c) Repeat part (b) for small values of frequency, f. (d) Compute the clutter attenuation and the improvement factor in terms of K and σ_c.

10.5. An implementation of a double delay line canceler with feedback is shown below.

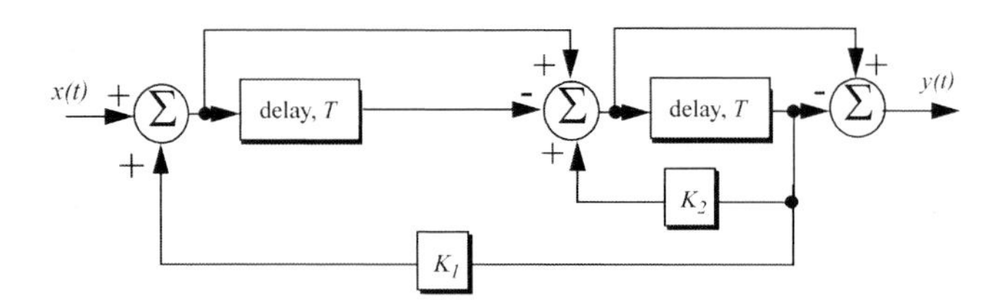

(a) What is the transfer function, $H(z)$? (b) Plot the frequency response for $K_1 = 0 = K_2$, and $K_1 = 0.2, K_2 = 0.5$.

10.6. Develop an expression for the improvement factor of a double delay line canceler.

10.7. Repeat Problem 10.6 for a double delay line canceler.

10.8. An experimental expression for the clutter power spectrum density is

$$W(f) = w_0 \exp(-f^2/2\sigma_c^2)$$

where w_0 is a constant. Show that using this expression leads to the same result obtained for the improvement factor as developed in this chapter.

10.9. A certain radar uses two PRFs with a stagger ratio 63/64. If the first PRF is $f_{r1} = 500Hz$. Compute the blind speeds for both PRFs and for the resultant composite PRF. Assume $\lambda = 3cm$.

10.10. Using PRI ratios 25:30:27:31, generate the MTI response for a 3-pulse MTI.

10.11. A certain filter used for clutter rejection has an impulse response $h(n) = \delta(n) - 3\delta(n-1) + 3\delta(n-2) - \delta(n-3)$. (a) Show an implementation of this filter using delay lines and adders. (b) What is the transfer function? (c) Plot the frequency response of this filter. (d) Calculate the output when the input is the unit step sequence.

10.12. The quadrature components of the clutter power spectrum are given in Problem 9.3. Let $\sigma_c = 10Hz$ and $f_r = 500Hz$. Compute the improvement of the signal-to-clutter ratio when a double delay line canceler is utilized.

10.13. The quadrature components of the clutter power spectrum are

$$\bar{S}_I(f) = \delta(f) + \frac{C}{\sqrt{2\pi}\sigma_c}\exp(-f^2/2\sigma_c^2)$$

$$\overline{S_Q}(f) = \frac{C}{\sqrt{2\pi}\sigma_c}\exp(-f^2/2\sigma_c^2).$$

Let $\sigma_c = 10Hz$ and $f_r = 500Hz$. Compute the improvement of the signal-to-clutter ratio when a double delay line canceler is utilized.

10.14. Develop an expression for the clutter improvement factor for single and double line cancelers using the clutter autocorrelation function.

10.15. Starting with Eq. (10.72), derive a closed form expression for the phase noise of an FM modulated waveform.

Chapter 11

Radar Detection

11.1. Single Pulse with Known Parameters

In its simplest form, a radar signal can be represented by a single pulse comprising a sinusoid of known amplitude and phase. Consequently, a retuned signal will also comprise a sinusoid. Under the assumption of completely known signal parameters, a returned pulse from a target has known amplitude and known phase with no random components; and the radar signal processor will attempt to maximize the probability of detection for a given probability of false alarm. In this case, detection is referred to as coherent detection or coherent demodulation. A radar system will declare detection with a certain probability of detection if the received voltage signal envelope exceeds a pre-set threshold value. For this purpose, the radar receiver is said to employ an envelope detector.

Figure 11.1 shows a simplified block diagram of a radar matched filter receiver followed by a threshold decision logic. The signal at the input of the matched filter $s(t)$ is composed of the target echo signal $x(t)$ and additive zero mean Gaussian noise (white noise is assumed in the analysis presented in this chapter) random process $n(t)$, with variance σ^2. The input noise is assumed to be spatially incoherent and uncorrelated with the signal. The matched filter impulse response is $h(t)$, and its output is denoted by the signal $v(t)$; it is given by

$$v(t) = \int_{-\infty}^{\infty} s(t)h(t-u)\ du . \qquad \text{Eq. (11.1)}$$

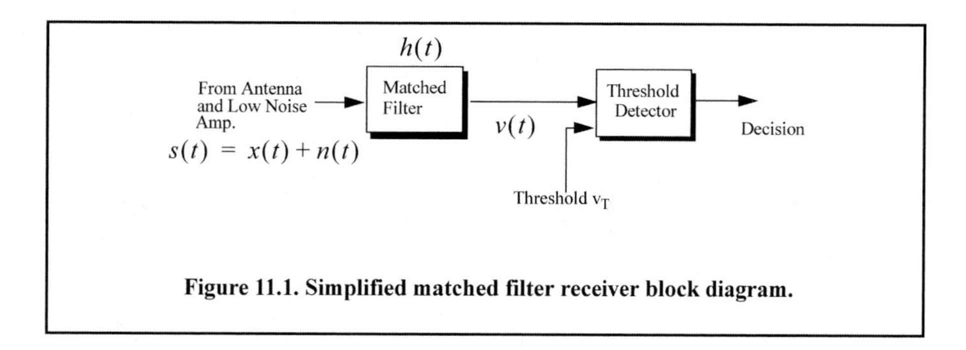

Figure 11.1. Simplified matched filter receiver block diagram.

Since the noise $n(t)$ is a Gaussian random process, then so is $s(t)$ since $x(t)$ is a deterministic signal; its only effect is a shift of the mean of the random process. It follows that the signal $v(t)$ is also a Gaussian random process, and over a coherent processing interval $\{0, T\}$, two hypotheses are considered; they are H_0 when the signal $s(t)$ is made of noise only, and H_1 when the signal $s(t)$ is made of signal plus noise.

More specifically,

$$H_0 \Leftrightarrow s = n \qquad ;0 < t < T \qquad \text{Eq. (11.2)}$$

$$H_1 \Leftrightarrow s = n + x \qquad ;0 < t < T. \qquad \text{Eq. (11.3)}$$

The statistics associated with the random process $v(t)$ over the interval $\{0, T\}$ is Gaussian. In general, a Gaussian *pdf* function is given by

$$f_V(v) = \frac{1}{\sigma\sqrt{2\pi}} \exp\left(-\frac{(v - \bar{V})^2}{2\sigma^2}\right) \qquad \text{Eq. (11.4)}$$

where σ^2 is the variance and $\bar{V}$ is the mean value. It follows that

$$E[V/H_0] = 0 \qquad \text{Eq. (11.5)}$$

$$Var[V/H_0] = \frac{E_x\eta_o}{2} = \sigma^2 \qquad \text{Eq. (11.6)}$$

$$E[V/H_1] = \int_0^T x^*(t)x(t)\ dt = E_x \qquad \text{Eq. (11.7)}$$

$$Var[V/H_1] = \frac{E_x\eta_o}{2} = \sigma^2 \qquad \text{Eq. (11.8)}$$

where E_x is the signal's energy.

Assuming the H_0 hypothesis, then the probability of a false P_{fa} alarm is computed from Eq. (11.4) when the signal $v(t)$ exceeds a set threshold value V_T. More specifically,

$$P_{fa} = Pr\{v(t) > V_T/H_0\} = \int_{V_T}^{\infty} \frac{1}{\sigma\sqrt{2\pi}} \exp\left(-\frac{v^2}{2\sigma^2}\right) dv. \qquad \text{Eq. (11.9)}$$

Substituting the variance as computed in Eq. (11.6) into Eq. (11.9) yields,

$$P_{fa} = \int_{V_T}^{\infty} \frac{1}{\sqrt{\pi E_x\eta_o}} \exp\left(-\frac{v^2}{E_x\eta_o}\right) dv. \qquad \text{Eq. (11.10)}$$

Making the change of variable $\zeta = v/(\sqrt{E_x\eta_o})$ yields

$$P_{fa} = \int_{\frac{V_T}{\sqrt{E_x\eta_o}}}^{\infty} \frac{1}{\sqrt{\pi}} e^{-\zeta^2} d\zeta. \qquad \text{Eq. (11.11)}$$

Multiplying and dividing Eq. (11.11) by *2* yields

$$P_{fa} = \frac{2}{2}\frac{1}{\sqrt{\pi}}\int_{\frac{V_T}{\sqrt{E_x\eta_o}}}^{\infty} e^{-\zeta^2} d\zeta = \frac{1}{2}erfc\left(\frac{V_T}{\sqrt{E_x\eta_o}}\right) \qquad \text{Eq. (11.12)}$$

where $erfc$ is the complementary error function defined by

$$erfc(z) = \frac{2}{\sqrt{\pi}}\int_{z}^{\infty} e^{-\zeta^2} d\zeta . \qquad \text{Eq. (11.13)}$$

The error function erf is related to the complementary error function using the relation

$$erfc(z) = 1 - erf(z) = 1 - \frac{2}{\sqrt{\pi}}\int_{0}^{z} e^{-\zeta^2} d\zeta . \qquad \text{Eq. (11.14)}$$

Using similar analysis, one can derive the probability of detection as

$$P_D = Pr\{v(t) > V_T/H_1\} = \frac{1}{2}erfc\left(\frac{V_T - Ex}{\sqrt{E_x\eta_o}}\right) . \qquad \text{Eq. (11.15)}$$

Figure 11.2 shows a sketch of the P_D versus the single pulse SNR, with P_{fa} as a parameter. Table 11.1 gives samples of the single pulse SNR corresponding to few values of P_D and P_{fa}, using Eq. (11.15). For example, if $P_D = 0.99$ and $P_{fa} = 10^{-10}$, then the minimum single pulse SNR required to accomplish this combination of P_D and P_{fa} is $SNR = 16.12dB$.

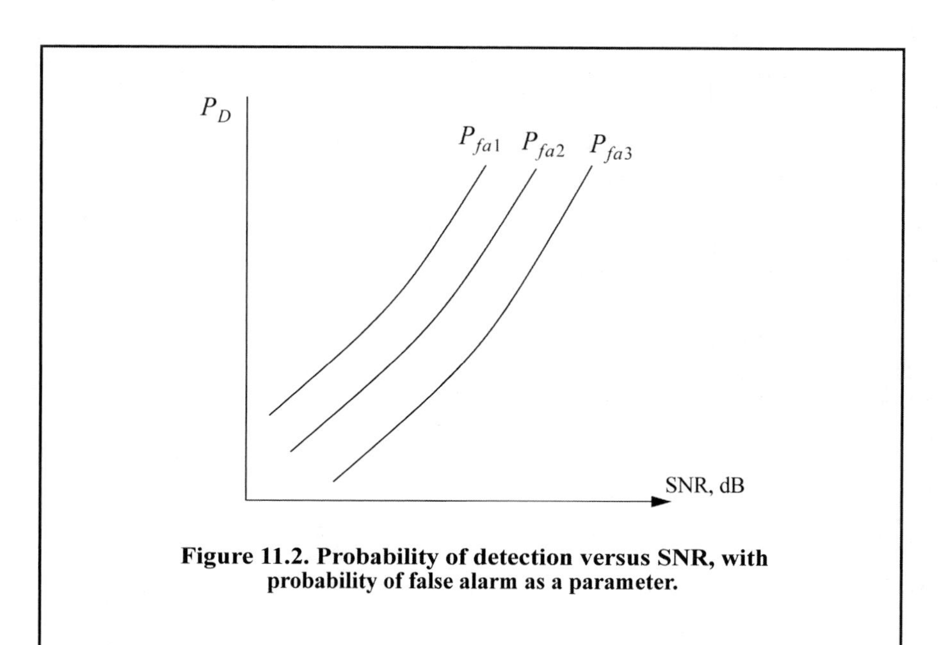

Figure 11.2. Probability of detection versus SNR, with probability of false alarm as a parameter.

Table 11.1. Single Pulse SNR (dB).

P_D	P_{fa} 10^{-3}	10^{-4}	10^{-5}	10^{-6}	10^{-7}	10^{-8}	10^{-9}	10^{-10}	10^{-11}	10^{-12}
.1	4.00	6.19	7.85	8.95	9.94	10.44	11.12	11.62	12.16	12.65
.2	5.57	7.35	8.75	9.81	10.50	11.19	11.87	12.31	12.85	13.25
.3	6.75	8.25	9.50	10.44	11.10	11.75	12.37	12.81	13.25	13.65
.4	7.87	8.85	10.18	10.87	11.56	12.18	12.75	13.25	13.65	14.00
.5	8.44	9.45	10.62	11.25	11.95	12.60	13.11	13.52	14.00	14.35
.6	8.75	9.95	11.00	11.75	12.37	12.88	13.50	13.87	14.25	14.62
.7	9.56	10.50	11.50	12.31	12.75	13.31	13.87	14.20	14.59	14.95
.8	10.18	11.12	12.05	12.62	13.25	13.75	14.25	14.55	14.87	15.25
.9	10.95	11.85	12.65	13.31	13.85	14.25	14.62	15.00	15.45	15.75
.95	11.50	12.40	13.12	13.65	14.25	14.64	15.10	15.45	15.75	16.12
.98	12.18	13.00	13.62	14.25	14.62	15.12	15.47	15.85	16.25	16.50
.99	12.62	13.37	14.05	14.50	15.00	15.38	15.75	16.12	16.47	16.75
.995	12.85	13.65	14.31	14.75	15.25	15.71	16.06	16.37	16.65	17.00
.998	13.31	14.05	14.62	15.06	15.53	16.05	16.37	16.7	16.89	17.25
.999	13.62	14.25	14.88	15.25	15.85	16.13	16.50	16.85	17.12	17.44
.9995	13.84	14.50	15.06	15.55	15.99	16.35	16.70	16.98	17.35	17.55
.9999	14.38	14.94	15.44	16.12	16.50	16.87	17.12	17.35	17.62	17.87

Example:

A pulsed radar has the following specification: Time of false alarm $T_{fa} = 16.67$ *minutes; probability of detection* $P_D = 0.9$ *and bandwidth* $B = 1\ GHz$. *Find the radar integration time* t_{int}, *the probability of false alarm* P_{fa}, *and the SNR of a single pulse.*

Solution:

$$t_{int} = \frac{1}{B} = \frac{1}{10^9} = 1 n\sec$$

$$P_{fa} = \frac{1}{T_{fa}B} = \frac{1}{10^9 \times 16.67 \times 60} \approx 10^{-12}$$

and from Table 11.1, we read

$$(SNR)_1 \approx 15.75 dB.$$

****

11.2. Single Pulse with Known Amplitude and Unknown Phase

In this case, the retuned radar signal comprises a sinusoid of a deterministic amplitude and random phase whose *pdf* is uniform over the interval $\{0, 2\pi\}$. The output of the matched filter receiver that employs an envelope detector is denoted by $v(t)$ (see Figure 11.3), and it can be written as a bandpass random process as

$$v(t) = v_I(t)\cos\omega_0 t + v_Q(t)\sin\omega_0 t = r(t)\cos(\omega_0 t - \Phi(t))$$

$$v_I(t) = r(t)\cos\Phi(t) \qquad \text{Eq. (11.16)}$$

$$v_Q(t) = r(t)\sin\Phi(t)$$

$$r(t) = \sqrt{[v_I(t)]^2 + [v_Q(t)]^2}$$

$$\Phi(t) = \left[\tan\left(\frac{v_Q(t)}{v_I(t)}\right)\right]^{-1} \qquad \text{Eq. (11.17)}$$

where $\omega_0 = 2\pi f_0$ is the radar operating frequency, $r(t)$ is the envelope of $v(t)$, the phase is $\Phi(t) = \operatorname{atan}(v_Q/v_I)$, and the subscripts I, and Q, respectively, refer to the in-phase and quadrature components. A target is detected when $r(t)$ exceeds the threshold value v_T, where the decision hypotheses are

$$\begin{array}{lll} H_0 \Leftrightarrow s(t) = n(t) & and & r(t) > v_T \Rightarrow False \ \ alarm \\ H_1 \Leftrightarrow s(t) = x(t) + n(t) & and & r(t) > v_T \Rightarrow Detection \end{array} . \qquad \text{Eq. (11.18)}$$

The case when the noise subtracts from the signal (while a target is present) to make $r(t)$ smaller than the threshold is called a miss. The matched filter output is a complex random variable that comprises either noise alone or noise plus target returns (i.e., sine wave of amplitude A and random phase). The quadrature components corresponding to the case of noise alone are

$$v_I(t) = n_I(t) \qquad \text{Eq. (11.19)}$$
$$v_Q(t) = n_Q(t)$$

where the noise quadrature components $n_I(t)$ and $n_Q(t)$ are uncorrelated zero mean low-pass Gaussian noise with equal variances, σ^2. In the second case the quadrature components are

$$v_I(t) = A + n_I(t) = r(t)\cos\Phi(t) \Rightarrow n_I(t) = r(t)\cos\Phi(t) - A$$
$$v_Q(t) = n_Q(t) = r(t)\sin\Phi(t) \qquad . \qquad \text{Eq. (11.20)}$$

The joint probability density function (*pdf*) of the two random variables $n_I;n_Q$ is

$$f_{n_I n_Q}(n_I, n_Q) = \frac{1}{2\pi\sigma^2}\exp\left(-\frac{n_I^2 + n_Q^2}{2\sigma^2}\right) = \frac{1}{2\pi\sigma^2}\exp\left(-\frac{(r\cos\varphi - A)^2 + (r\sin\varphi)^2}{2\sigma^2}\right), \qquad \text{Eq. (11.21)}$$

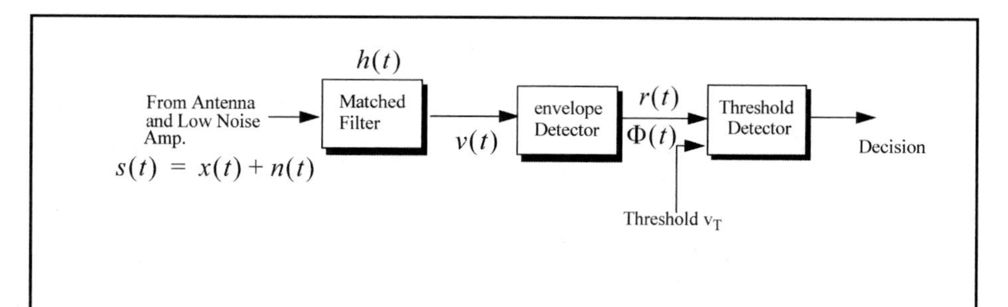

Figure 11.3. Simplified matched filter receiver employing an envelope detector.

which can be written as

$$f_{n_I n_Q}(n_I, n_Q) = \frac{1}{2\pi\sigma^2}\exp\left(-\frac{(r\cos\varphi - A)^2 + (r\sin\varphi)^2}{2\sigma^2}\right). \qquad \text{Eq. (11.22)}$$

The *pdfs* of the random variables $r(t)$ and $\Phi(t)$, respectively, represent the modulus and phase of $v(t)$. The joint *pdf* for the two random variables $r(t);\Phi(t)$ is

$$f_{R\Phi}(r, \varphi) = f_{n_I n_Q}(n_I, n_Q)|\mathbf{J}| \qquad \text{Eq. (11.23)}$$

where $|\mathbf{J}|$ is determinant of the matrix of derivatives $\mathbf{J}$ and referred to as the Jacobian. The matrix of derivatives is given by

$$\mathbf{J} = \begin{bmatrix} \dfrac{\partial n_I}{\partial r} & \dfrac{\partial n_I}{\partial \varphi} \\ \dfrac{\partial n_Q}{\partial r} & \dfrac{\partial n_Q}{\partial \varphi} \end{bmatrix} = \begin{bmatrix} \cos\varphi & -r\sin\varphi \\ \sin\varphi & r\cos\varphi \end{bmatrix}. \qquad \text{Eq. (11.24)}$$

It follows that the Jacobian is

$$|\mathbf{J}| = r(t). \qquad \text{Eq. (11.25)}$$

Substituting Eq. (11.22) and Eq. (11.25) into Eq. (11.23) and collecting terms yield

$$f_{R\Phi}(r, \varphi) = \frac{r}{2\pi\sigma^2}\exp\left(-\frac{r^2 + A^2}{2\sigma^2}\right)\exp\left(\frac{rA\cos\varphi}{\sigma^2}\right). \qquad \text{Eq. (11.26)}$$

The *pdf* for $r(t)$ alone is obtained by integrating Eq. (11.26) over φ. That is,

$$f_R(r) = \int_0^{2\pi} f_{R\Phi}(r, \varphi)d\varphi = \frac{r}{\sigma^2}\exp\left(-\frac{r^2 + A^2}{2\sigma^2}\right)\frac{1}{2\pi}\int_0^{2\pi}\exp\left(\frac{rA\cos\varphi}{\sigma^2}\right)d\varphi \qquad \text{Eq. (11.27)}$$

where the integral inside Eq. (11.27) is known as the modified Bessel function of zero order,

$$I_0(\beta) = \frac{1}{2\pi}\int_0^{2\pi} e^{\beta\cos\theta}\, d\theta. \qquad \text{Eq. (11.28)}$$

Thus,

$$f_R(r) = \frac{r}{\sigma^2} I_0\left(\frac{rA}{\sigma^2}\right)\exp\left(-\frac{r^2 + A^2}{2\sigma^2}\right), \qquad \text{Eq. (11.29)}$$

which is the Rician probability density function. The case when $A/\sigma^2 = 0$ (noise alone) and the resulting *pdf* is a Rayleigh probability density function

$$f_R(r) = \frac{r}{\sigma^2}\exp\left(-\frac{r^2}{2\sigma^2}\right). \qquad \text{Eq. (11.30)}$$

When (A/σ^2) is very large, Eq. (11.29) becomes a Gaussian probability density function of mean A and variance σ^2:

$$f_R(r) \approx \frac{1}{\sqrt{2\pi\sigma^2}} \exp\left(-\frac{(r-A)^2}{2\sigma^2}\right). \qquad \text{Eq. (11.31)}$$

Figure 11.4 shows plots for the Rayleigh and Gaussian densities. The density function for the random variable Φ is obtained from

$$f_\Phi(\varphi) = \int_0^r f_{R\Phi}(r, \varphi)\ dr. \qquad \text{Eq. (11.32)}$$

While the detailed derivation is left as an exercise, the result is

$$f_\Phi(\varphi) = \frac{1}{2\pi} \exp\left(\frac{-A^2}{2\sigma^2}\right) + \frac{A\cos\varphi}{\sqrt{2\pi\sigma^2}} \exp\left(\frac{-(A\sin\varphi)^2}{2\sigma^2}\right) F\left(\frac{A\cos\varphi}{\sigma}\right) \qquad \text{Eq. (11.33)}$$

where

$$F(x) = \int_{-\infty}^{x} \frac{1}{\sqrt{2\pi}}\ e^{-\zeta^2/2}\ d\xi. \qquad \text{Eq. (11.34)}$$

The function $F(x)$ can be found tabulated in most mathematical formula reference books. Note that for the case of noise alone ($A = 0$), Eq. (11.33) collapses to a uniform *pdf* over the interval $\{0, 2\pi\}$. One excellent approximation for the function $F(x)$ is

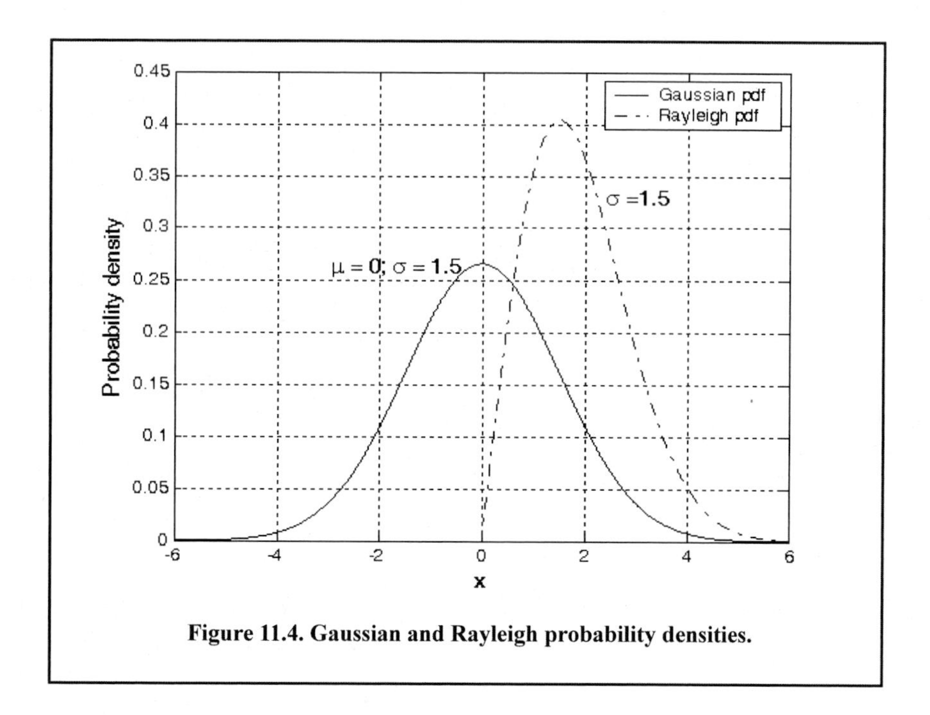

Figure 11.4. Gaussian and Rayleigh probability densities.

$$F(x) = 1 - \left(\frac{1}{0.661x + 0.339\sqrt{x^2 + 5.51}}\right)\frac{1}{\sqrt{2\pi}}e^{-x^2/2} \qquad x \ge 0, \qquad \text{Eq. (11.35)}$$

and for negative values of x

$$F(-x) = 1 - F(x). \qquad \text{Eq. (11.36)}$$

11.2.1. Probability of False Alarm

The probability of false alarm P_{fa} is defined as the probability that a sample r of the signal $r(t)$ will exceed the threshold voltage v_T when noise alone is present in the radar:

$$P_{fa} = \int_{v_T}^{\infty} \frac{r}{\sigma^2}\exp\left(-\frac{r^2}{2\sigma^2}\right) dr = \exp\left(\frac{-v_T^2}{2\sigma^2}\right) \qquad \text{Eq. (11.37)}$$

$$v_T = \sqrt{2\sigma^2 \ln\left(\frac{1}{P_{fa}}\right)}. \qquad \text{Eq. (11.38)}$$

Figure 11.5 shows a plot of the normalized threshold versus the probability of false alarm. It is evident from this figure that P_{fa} is very sensitive to small changes in the threshold value. The false alarm time T_{fa} is related to the probability of false alarm by

$$T_{fa} = \frac{t_{int}}{P_{fa}} \qquad \text{Eq. (11.39)}$$

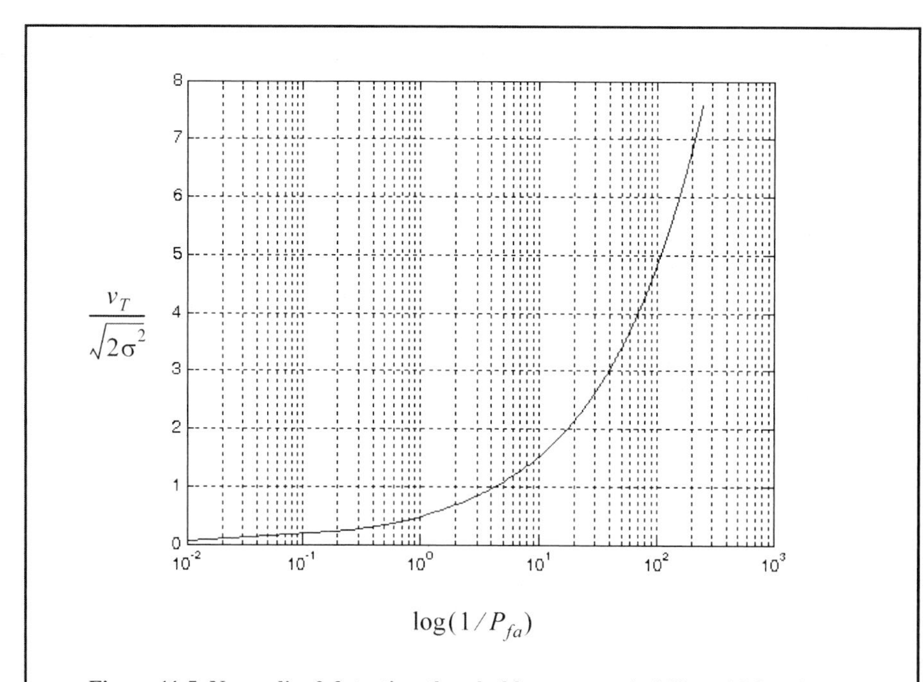

Figure 11.5. Normalized detection threshold versus probability of false alarm.

where t_{int} represents the radar integration time, or the average time that the output of the envelope detector will pass the threshold voltage. Since the radar operating bandwidth B is the inverse of t_{int}, by using the right-hand side of Eq. (11.37) and Eq. (11.38), one can rewrite T_{fa} as

$$T_{fa} = \frac{1}{B}\exp\left(\frac{v_T^2}{2\sigma^2}\right). \qquad \text{Eq. (11.40)}$$

Minimizing T_{fa} means increasing the threshold value, and as a result, the radar maximum detection range is decreased. The choice of an acceptable value for T_{fa} becomes a compromise depending on the radar mode of operation. The false alarm number is

$$n_{fa} = \frac{-\ln(2)}{\ln(1-P_{fa})} \approx \frac{\ln(2)}{P_{fa}}. \qquad \text{Eq. (11.41)}$$

Other slightly different definitions for the false alarm number exist in the literature, causing a source of confusion for many non-expert readers. Other than the definition in Eq. (11.41), the most commonly used definition for the false alarm number is the one introduced by Marcum (1960). Marcum defines the false alarm number as the reciprocal of P_{fa}. In this text, the definition given in Eq. (11.41) is always assumed.

11.2.2. Probability of Detection

The probability of detection P_D is the probability that a sample r of $r(t)$ will exceed the threshold voltage in the case of noise plus signal,

$$P_D = \int_{v_T}^{\infty} \frac{r}{\sigma^2} I_0\left(\frac{rA}{\sigma^2}\right) \exp\left(-\frac{r^2+A^2}{2\sigma^2}\right) dr. \qquad \text{Eq. (11.42)}$$

Assuming that the radar signal is a sinusoid of amplitude A (completely known), then its power is $A^2/2$. Now, by using $SNR = A^2/2\sigma^2$ (single-pulse SNR) and $(v_T^2/2\sigma^2) = \ln(1/P_{fa})$, Eq. (11.42) can be rewritten as

$$P_D = \int_{\sqrt{2\sigma^2\ln(1/p_{fa})}}^{\infty} \frac{r}{\sigma^2} I_0\left(\frac{rA}{\sigma^2}\right) \exp\left(-\frac{r^2+A^2}{2\sigma^2}\right) dr = Q\left[\sqrt{\frac{A^2}{\sigma^2}}, \sqrt{2\ln\left(\frac{1}{P_{fa}}\right)}\right] \qquad \text{Eq. (11.43)}$$

where

$$Q[a, b] = \int_{b}^{\infty} \zeta I_0(a\zeta) e^{-(\zeta^2+a^2)/2}\, d\zeta. \qquad \text{Eq. (11.44)}$$

Q is called Marcum's Q-function. When P_{fa} is small and P_D is relatively large so that the threshold is also large, Eq. (11.43) can be approximated by

$$P_D \approx F\left(\frac{A}{\sigma} - \sqrt{2\ln\left(\frac{1}{P_{fa}}\right)}\right). \qquad \text{Eq. (11.45)}$$

$F(x)$ is given by Eq. (11.34). Many approximations for Eq. (11.45) can be found throughout the literature. One very accurate approximation presented by North (1963) is given by

$$P_D \approx 0.5 \times erfc(\sqrt{-\ln P_{fa}} - \sqrt{SNR + 0.5}) \qquad \text{Eq. (11.46)}$$

where the complementary error function was defined in Eq. (11.13).

The integral given in Eq. (11.44) is complicated and can be computed using numerical integration techniques. Parl[1] developed an excellent algorithm to numerically compute this integral. It is summarized as follows:

$$Q[a, b] = \begin{Bmatrix} \frac{\alpha_n}{2\beta_n}\exp\left(\frac{(a-b)^2}{2}\right) & a < b \\ 1 - \left(\frac{\alpha_n}{2\beta_n}\exp\left(\frac{(a-b)^2}{2}\right)\right. & \left. a \geq b\right) \end{Bmatrix} \qquad \text{Eq. (11.47)}$$

$$\alpha_n = d_n + \frac{2n}{ab}\alpha_{n-1} + \alpha_{n-2} \qquad \text{Eq. (11.48)}$$

$$\beta_n = 1 + \frac{2n}{ab}\beta_{n-1} + \beta_{n-2} \qquad \text{Eq. (11.49)}$$

$$d_{n+1} = d_n d_1 \qquad \text{Eq. (11.50)}$$

$$\alpha_0 = \begin{Bmatrix} 1 & a < b \\ 0 & a \geq b \end{Bmatrix} \qquad \text{Eq. (11.51)}$$

$$d_1 = \begin{Bmatrix} a/b & a < b \\ b/a & a \geq b \end{Bmatrix}. \qquad \text{Eq. (11.52)}$$

$\alpha_{-1} = 0.0$, $\beta_0 = 0.5$, and $\beta_{-1} = 0$. The recursive Eq. (11.47) through Eq. (11.52) are computed continuously until $\beta_n > 10^p$ for values of $p \geq 3$. The accuracy of the algorithm is enhanced as the value of p is increased. Figure 11.6 shows plots of the probability of detection, P_D, versus the single pulse SNR, with the P_{fa} as a parameter.

11.3. Pulse Integration

In the previous two sections, target detection was introduced in the context of single pulse detection with completely known (i.e., deterministic) amplitude and phase in one case, and known amplitude with random phase in another. The underlying assumption was that radar targets were made of non-varying (non-fluctuating) scatterers. However, in practice that it is rarely the case. First, one would expect the radar to receive multiple returns (pulses) from any given target in its field of view. Furthermore, real-world targets will fluctuate over the duration of a single pulse or from pulse to pulse. Hence, the analysis is extended to account for target fluctuation as well as for target detection where multiple returned pulses are taken into consideration.

1. Parl, S. A., New Method of Calculating the Generalized Q Function, *IEEE Trans. Information Theory*, Vol. IT-26, No. 1, January 1980, pp. 121-124.

Multiple returned pulses can be integrated (combined) coherently or non-coherently. The process of combining radar returns from many pulses is called radar pulse integration. Pulse integration can be performed on the quadrature components prior to the envelope detector. This is called coherent integration or predetection integration. Coherent integration preserves the phase relationship between the received pulses. Thus a buildup in the signal amplitude is expected. Alternatively, pulse integration performed after the envelope detector (where the phase relation is lost) is called non-coherent or post-detection integration, and a buildup in the signal amplitude is guaranteed.

Combining the returns from all pulses returned by a given target during a single scan is very likely to increase the radar sensitivity (i.e., SNR). The number of returned pulses from a given target depends on the antenna scan rate, the antenna beamwidth, and the radar PRF. More precisely, the number of pulses returned from a given target is given by

$$n_P = \frac{\theta_a T_{sc} f_r}{2\pi} \qquad \text{Eq. (11.53)}$$

where θ_a is the azimuth antenna beamwidth, T_{sc} is the scan time, and f_r is the radar PRF. The number of reflected pulses may also be expressed as

$$n_P = \frac{\theta_a f_r}{\dot{\theta}_{scan}} \qquad \text{Eq. (11.54)}$$

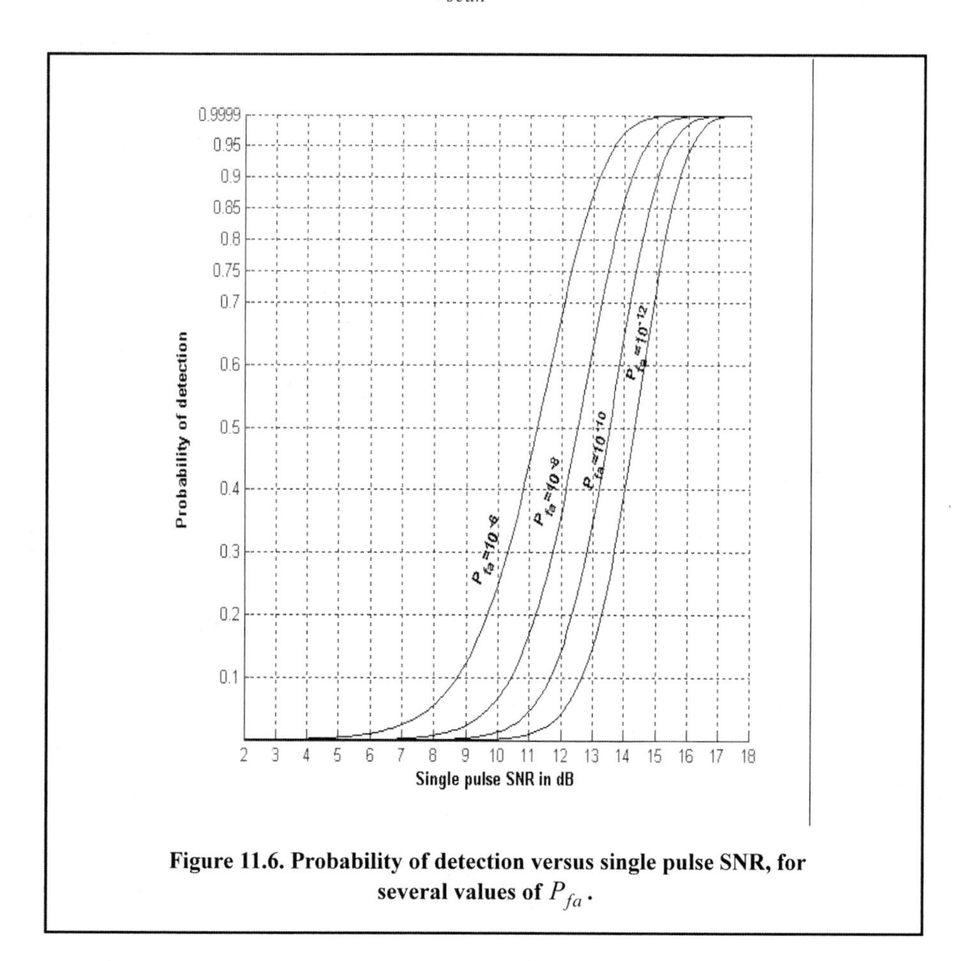

Figure 11.6. Probability of detection versus single pulse SNR, for several values of P_{fa}.

where $\dot{\theta}_{scan}$ is the antenna scan rate in degrees per second. Note that when using Eq. (11.53), θ_a is expressed in radians, while when using Eq. (11.54), it is expressed in degrees. As an example, consider a radar with an azimuth antenna beamwidth $\theta_a = 3°$, antenna scan rate $\dot{\theta}_{scan} = 45°/\text{sec}$ (antenna scan time, $T_{sc} = 8\text{sec}$), and a PRF $f_r = 300Hz$. Using either Eq. (11.53) or Eq. (11.54) yields $n_P = 20$ pulses.

Although, pulse integration will very likely improve the receiver SNR, caution should be exercised when attempting to account for how much SNR is attained through pulse integration. This is because of the following. First, during an antenna scan, a given target will not always be located at the center of the radar beam (i.e., have maximum gain). In fact, during a scan, a target will first enter the antenna beam at the *3* dB point, reach maximum gain, and finally leave the beam at the *3* dB point again. Thus, the returns do not have the same amplitude even though the target RCS may be constant and all other factors that may introduce signal loss remain the same.

Other factors that may introduce further variation to the amplitude of the returned pulses include target RCS and propagation path fluctuations. Additionally, when the radar employs a very fast scan rate, an additional loss term is introduced due to the motion of the beam between transmission and reception. This is referred to as scan loss. A distinction should be made between scan loss due to a rotating antenna (which is described here) and the term scan loss that is normally associated with phased array antennas (which takes on a different meaning in that context).

Finally, since coherent integration utilizes the phase information from all integrated pulses, it is critical that any phase variation between all integrated pulses be known with a great level of confidence. Consequently, target dynamics (such as target range, range rate, tumble rate, RCS fluctuation) must be estimated or computed accurately so that coherent integration can be meaningful. In fact, if a radar coherently integrates pulses from targets without proper knowledge of the target dynamics, it suffers a loss in SNR rather than the expected SNR buildup. Knowledge of target dynamics is not as critical when employing non-coherent integration; nonetheless, target range rate must be estimated so that only the returns from a given target within a specific range bin are integrated. In other words, one must avoid range walk (i.e., having a target cross between adjacent range bins during a single scan).

11.3.1. Coherent Integration

In coherent integration, and when a perfect integrator is used (100% efficiency) to integrate n_P pulses, the SNR is improved by the same factor. Otherwise, integration loss occurs, which is always the case for non-coherent integration. Coherent integration loss occurs when the integration process is not optimum. This could be due to target fluctuation, instability in the radar local oscillator, or propagation path changes.

Denote the single pulse SNR required to produce a given probability of detection as $(SNR)_1$. The SNR resulting from coherently integrating n_P pulses is then given by

$$(SNR)_{CI} = n_P(SNR)_1 . \qquad \text{Eq. (11.55)}$$

Coherent integration cannot be applied over a long interval of time, particularly if the target RCS is varying rapidly. If the target radial velocity is known and no acceleration is assumed, the maximum coherent integration time is limited to

$$t_{CI} = \sqrt{\frac{\lambda}{2a_r}} \qquad \text{Eq. (11.56)}$$

where λ is the radar wavelength and a_r is the target radial acceleration. Coherent integration time can be extended if the target radial acceleration can be compensated for by the radar.

In order to demonstrate the improvement in the SNR using coherent integration, consider the case where the radar return signal contains both signal plus additive noise. The mth pulse is

$$y_m(t) = s(t) + n_m(t) \qquad \text{Eq. (11.57)}$$

where $s(t)$ is the radar signal return of interest and $n_m(t)$ is white uncorrelated additive noise signal with variance σ^2. Coherent integration of n_P pulses yields

$$z(t) = \frac{1}{n_P}\sum_{m=1}^{n_P} y_m(t) = \sum_{m=1}^{n_P}\frac{1}{n_P}[s(t) + n_m(t)] = s(t) + \sum_{m=1}^{n_P}\frac{1}{n_P}n_m(t). \qquad \text{Eq. (11.58)}$$

The total noise power in $z(t)$ is equal to the variance. More precisely,

$$\sigma_{n_P}^2 = E\left[\left(\sum_{m=1}^{n_P}\frac{1}{n_P}n_m(t)\right)\left(\sum_{l=1}^{n_P}\frac{1}{n_P}n_l(t)\right)^*\right] \qquad \text{Eq. (11.59)}$$

where E is the expected value operator. It follows that

$$\sigma_{n_P}^2 = \frac{1}{n_P^2}\sum_{m,l=1}^{n_P} E[n_m(t)n_l^*(t)] = \frac{1}{n_P^2}\sum_{m,l=1}^{n_P}\sigma_{ny}^2\delta_{ml} = \frac{1}{n_P}\sigma_{ny}^2 \qquad \text{Eq. (11.60)}$$

where σ_{ny}^2 is the single pulse noise power and δ_{ml} is equal to zero for $m \neq l$ and unity for $m = l$. Observation of Eqs. (11.58) and (11.60) indicates that the desired signal power after coherent integration is unchanged, while the noise power is reduced by the factor $1/n_P$. Thus, the SNR after coherent integration is improved by n_P.

11.3.2. Non-coherent Integration

When the phase of the integrated pulses is not known, so that coherent integration is no longer possible, another form of pulse integration is done. In this case, pulse integration is performed by adding (integrating) the individual pulses' envelopes or the square of their envelopes. Thus, the term non-coherent integration is adopted. A block diagram of a radar receiver utilizing non-coherent integration is illustrated in Figure 11.7.

The performance difference (measured in SNR) between the linear envelope detector and the quadratic (square law) detector is practically negligible. Robertson (1967) showed that this difference is typically less than $0.2dB$; he showed that the performance difference is higher than $0.2dB$ only for cases where $n_P > 100$ and $P_D < 0.01$. Both of these conditions are of no practical significance in radar applications. It is much easier to analyze and implement the square law detector in real hardware than for the envelope detector. Therefore, most authors make no distinction between the type of detector used when referring to non-coherent integration, and the square law detector is almost always assumed. The analysis presented in this book will always assume, unless indicated otherwise, non-coherent integration using the square law detector.

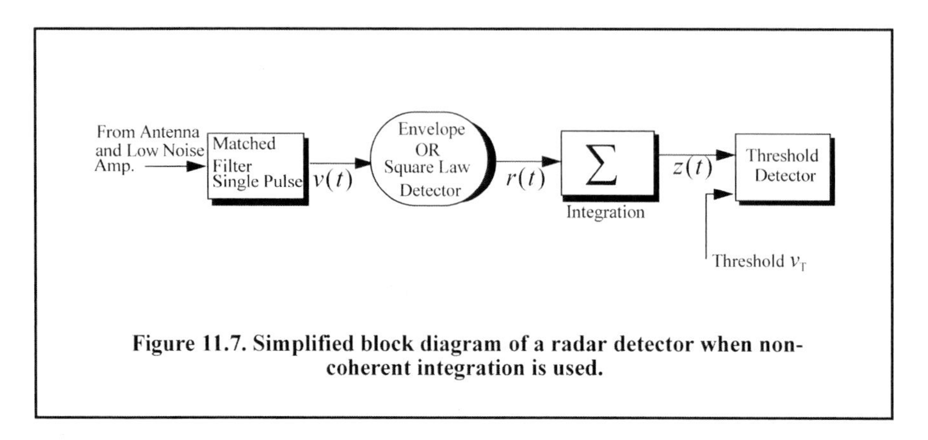

Figure 11.7. Simplified block diagram of a radar detector when non-coherent integration is used.

11.3.3. Improvement Factor and Integration Loss

Non-coherent integration is less efficient than coherent integration. Actually, the non-coherent integration gain is always smaller than the number of non-coherently integrated pulses. This loss in integration is referred to as post-detection or square-law detector loss. Define $(SNR)_{NCI}$ as the SNR required to achieve a specific P_D given a particular P_{fa} when n_P pulses are integrated non-coherently. Also denote the single pulse SNR as $(SNR)_1$. It follows that

$$(SNR)_{NCI} = (SNR)_1 \times I(n_P) \qquad \text{Eq. (11.61)}$$

where $I(n_P)$ is called the integration improvement factor. An empirically derived expression for the improvement factor that is accurate within $0.8dB$ is reported in Peebles (1998) as

$$[I(n_P)]_{dB} = 6.79(1 + 0.253P_D)\left(1 + \frac{\log(1/P_{fa})}{46.6}\right)\log(n_P) \quad . \qquad \text{Eq. (11.62)}$$

$$(1 - 0.140\log(n_P) + 0.018310(\log n_P)^2)$$

The integration loss in *dB* is defined as

$$[L_{NCI}]_{dB} = 10\log n_P - [I(n_P)]_{dB} . \qquad \text{Eq. (11.63)}$$

Tables 11.2 and 11.3 give a few values of the integration improvement factor for $P_{fa} = 10^{-4}$ and $P_{fa} = 10^{-12}$, respectively.

Figure 11.8 shows plots of the improvement factor versus the number of integrated pulses using different combinations of P_D and P_{fa}. The top part of Figure 11.8 shows plots of the integration improvement factor as a function of the number of integrated pulses with P_D and P_{fa} as parameters using Eq. (11.62), while, the lower part of Figure 11.8 shows plots of the corresponding integration loss versus n_P with P_D and P_{fa} as parameters.

Table 11.2. Improvement factor, $P_{fa}=10^{-4}$.

n	1	2	3	4	5	6	7	10	20	100
$P_D=0.5$	1	1.83	2.35	3	3.5	4	4.5	5.8	9.15	27
$P_D=0.9$	1	1.88	2.6	3.25	3.9	4.5	5	6.25	10.5	29.5
$P_D=0.99$	1	1.89	2.75	3.5	4.1	4.8	5.25	6.75	12	32.5

Table 11.3. Improvement factor, $P_{fa}=10^{-12}$.

n	1	2	3	4	5	6	7	10	20	100
$P_D=0.5$	1	2	2.8	3.5	4.2	5	5.8	7	13	35
$P_D=0.9$	1	2	2.85	3.6	4.2	5	5.8	7.25	14	36.5
$P_D=0.99$	1	2	2.9	3.6	4.25	5	5.8	7.5	15	40

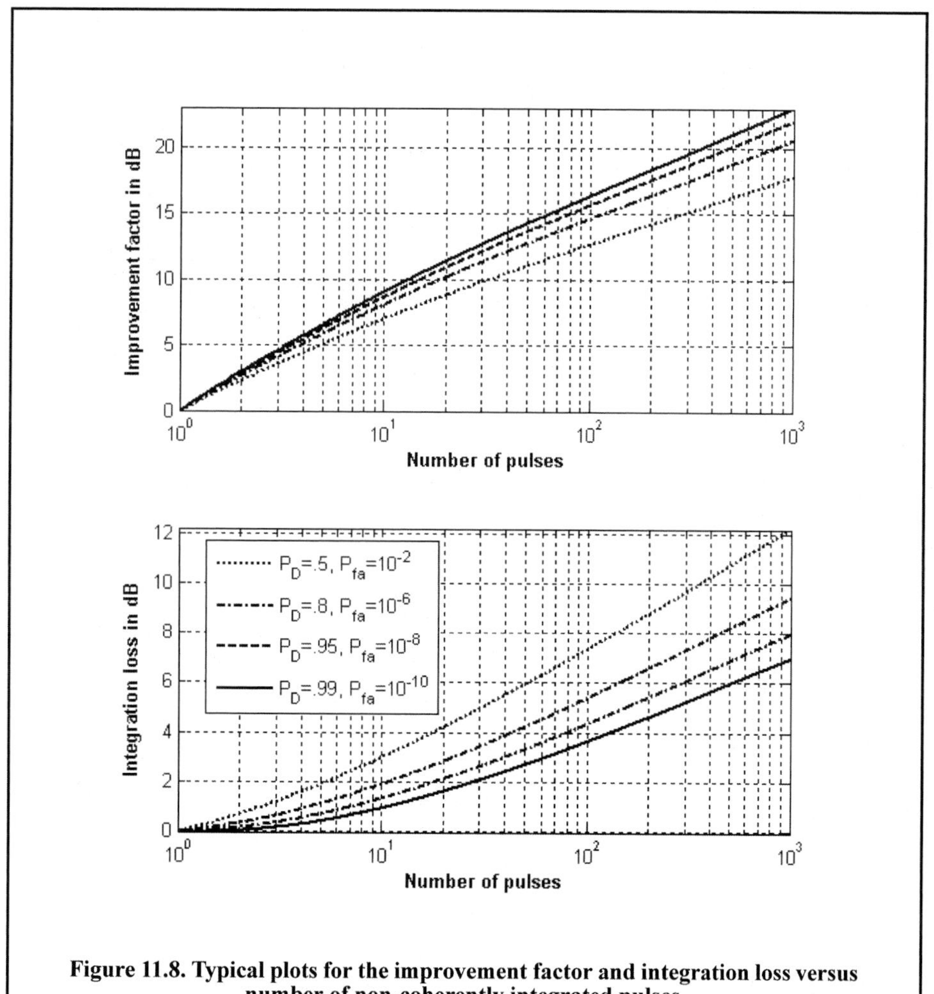

Figure 11.8. Typical plots for the improvement factor and integration loss versus number of non-coherently integrated pulses.

11.4. Target Fluctuation: The Chi-Square Family of Targets

Target detection utilizing the square law detector was first analyzed by Marcum[1], where he assumed a constant RCS (nonfluctuating target). This work was extended by Swerling[2] to four distinct cases of target RCS fluctuation. These cases have come to be known as Swerling models. They are Swerling I, Swerling II, Swerling III, and Swerling IV. The constant RCS case analyzed by Marcum is widely known as Swerling 0 or equivalently as Swerling V. Target fluctuation introduces an additional loss factor in the SNR as compared to the case where fluctuation is not present, given the same P_D and P_{fa}.

Swerling V targets have constant amplitude over one antenna scan or observation interval; however, a Swerling I target amplitude varies independently from scan to scan according to a chi-square probability density function with two degrees of freedom. The amplitude of Swerling II targets fluctuates independently from pulse to pulse according to a chi-square probability density function with two degrees of freedom. Target fluctuation associated with a Swerling III model is from scan to scan according to a chi-square probability density function with four degrees of freedom. Finally, the fluctuation of Swerling IV targets is from pulse to pulse according to a chi-square probability density function with four degrees of freedom.

Swerling showed that the statistics associated with Swerling I and II models apply to targets consisting of many small scatterers of comparable RCS values, while the statistics associated with Swerling III and IV models apply to targets consisting of one large RCS scatterer and many small equal RCS scatterers. Non-coherent integration can be applied to all four Swerling models; however, coherent integration cannot be used when the target fluctuation is either Swerling II or Swerling IV. This is because the target amplitude decorrelates from pulse to pulse (fast fluctuation) for Swerling II and IV models, and thus phase coherency cannot be maintained.

The chi-square *pdf* with $2N$ degrees of freedom can be written as

$$f_X(x) = \frac{N}{(N-1)!\ \sqrt{\sigma_x^2}} \left(\frac{Nx}{\sigma_x}\right)^{N-1} \exp\left(-\frac{Nx}{\sigma_x}\right) \qquad \text{Eq. (11.64)}$$

where σ_x is the standard deviation for the RCS value. Using this equation, the *pdf* associated with Swerling I and II targets can be obtained by letting $N = 1$, which yields a Rayleigh *pdf*. More precisely,

$$f_X(x) = \frac{1}{\sigma_x}\exp\left(-\frac{x}{\sigma_x}\right) \qquad x \ge 0\,. \qquad \text{Eq. (11.65)}$$

Letting $N = 2$ yields the *pdf* for Swerling III and IV type targets,

$$f_X(x) = \frac{4x}{\sigma_x^2}\exp\left(-\frac{2x}{\sigma_x}\right) \qquad x \ge 0\,. \qquad \text{Eq. (11.66)}$$

1. Marcum, J. I., A Statistical Theory of Target Detection by Pulsed Radar, *IRE Transactions on Information Theory*, Vol IT-6, pp. 59-267, April 1960.
2. Swerling, P., Probability of Detection for Fluctuating Targets, *IRE Transactions on Information Theory*, Vol IT-6, pp. 269-308, April 1960.

Figure 11.9 shows a plot for the probability of detection versus SNR for a Swerling V target and $n_p = 1, 10$. Note that it requires less SNR, with ten pulses integrated non-coherently, to achieve the same probability of detection as in the case of a single pulse. Hence, for any given P_D, the SNR improvement can be read from the plot. Figure 11.10 shows a plot of the probability of detection as a function of SNR for $n_p = 1$ and $P_{fa} = 10^{-9}$ for both Swerling I and V (Swerling 0) type fluctuations. Note that it requires more SNR, with fluctuation, to achieve the same P_D as in the case with no fluctuation.

Figure 11.11 is similar to Figure 11.10 except in this case, $n_p = 5$ and $P_{fa} = 10^{-6}$. Figure 11.12 shows a plot of the probability of detection for Swerling 0, Swerling I, and Swerling II with $n_P = 5$, where $P_{fa} = 10^{-7}$. Figure 11.13 is similar to Figure 11.12 except in this case $n_P = 2$ and $P_{fa} = 10^{-6}$.

Figure 11.14 shows a plot of the probability of detection for a Swerling III target as a function of SNR for $n_P = 1, 10, 50, 100$, where $P_{fa} = 10^{-9}$. Figure 11.15 shows a plot of the probability of detection for Swerling 0, Swerling I, Swerling II, and Swerling III with $n_P = 5$ and $P_{fa} = 10^{-7}$. Figure 11.16 shows plots of the probability of detection for a Swerling IV target as a function of SNR for $n_P = 1, 10, 25, 75$, where $P_{fa} = 10^{-6}$.

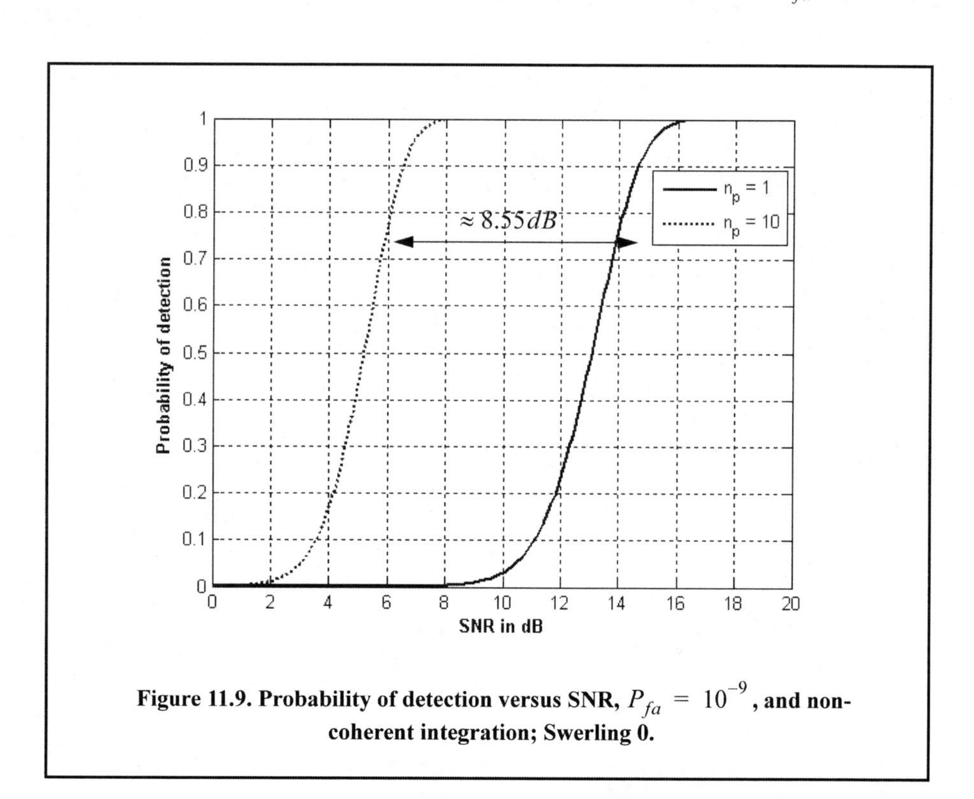

Figure 11.9. Probability of detection versus SNR, $P_{fa} = 10^{-9}$, and non-coherent integration; Swerling 0.

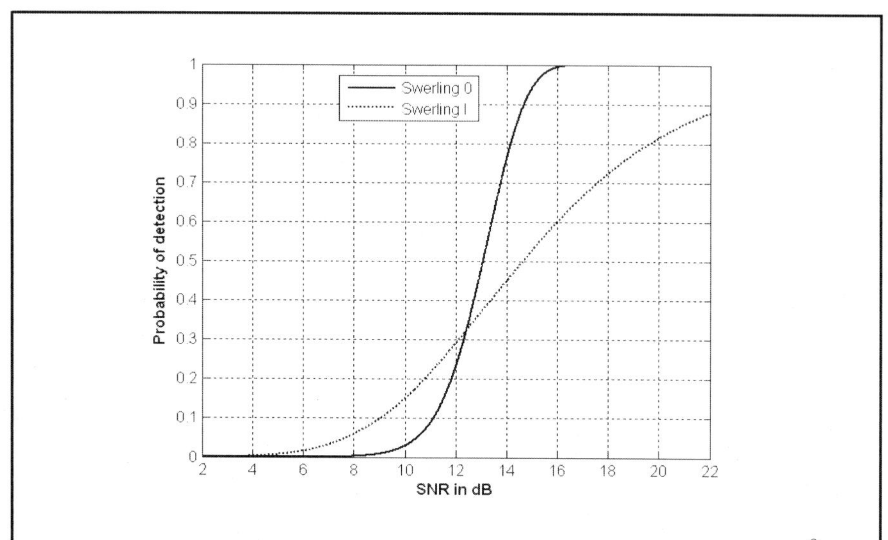

Figure 11.10. Probability of detection versus SNR, single pulse. $P_{fa} = 10^{-9}$**.**

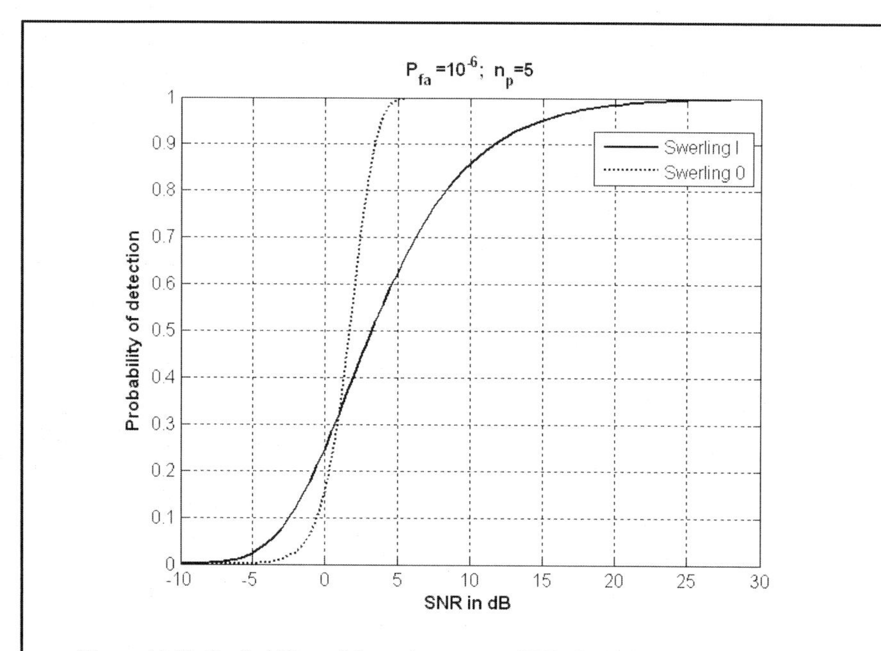

Figure 11.11. Probability of detection versus SNR. Swerling I and Swerling 0.

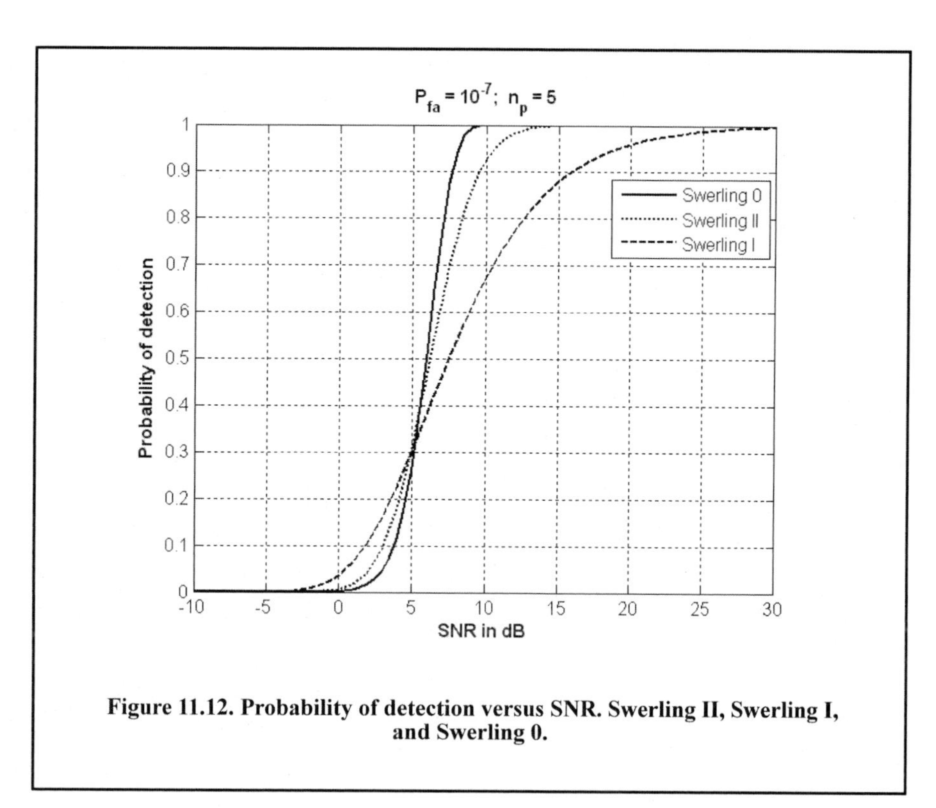

Figure 11.12. Probability of detection versus SNR. Swerling II, Swerling I, and Swerling 0.

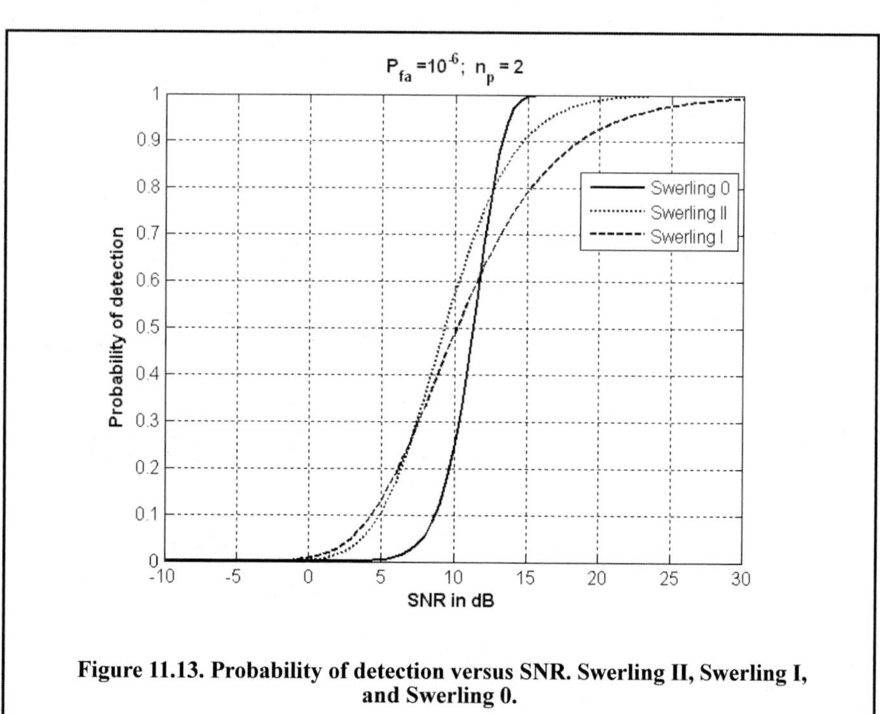

Figure 11.13. Probability of detection versus SNR. Swerling II, Swerling I, and Swerling 0.

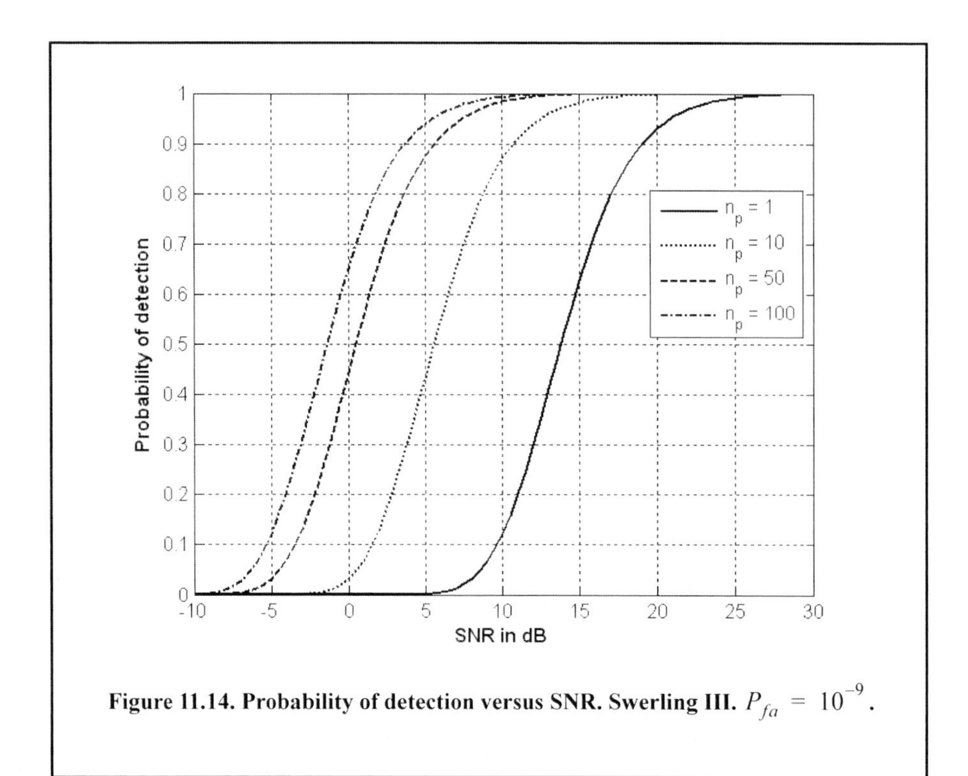

Figure 11.14. Probability of detection versus SNR. Swerling III. $P_{fa} = 10^{-9}$.

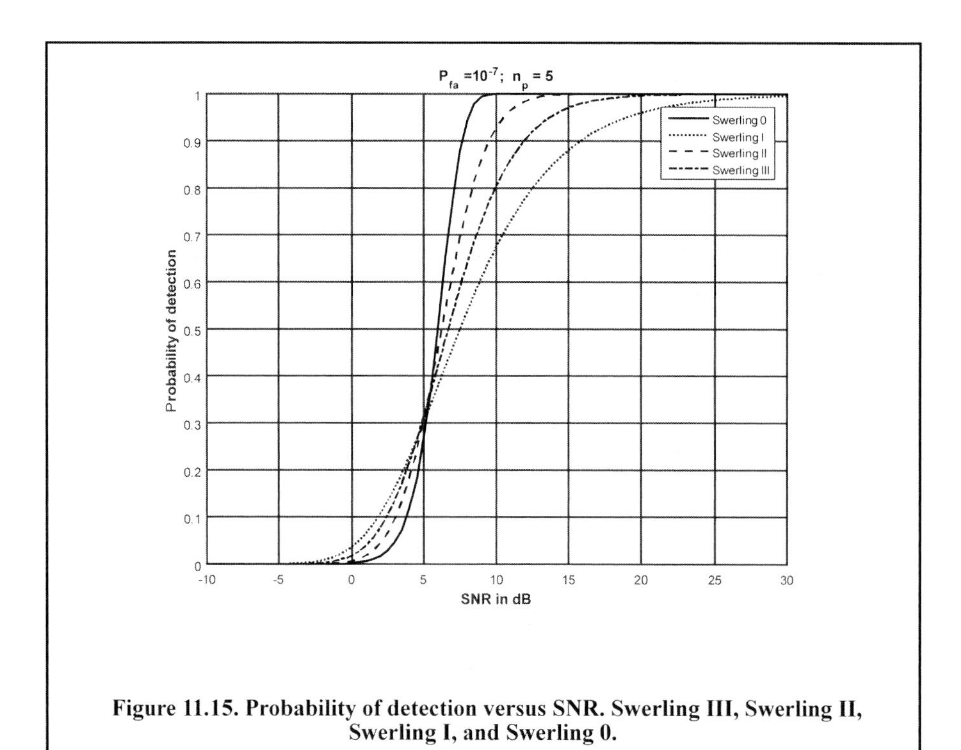

Figure 11.15. Probability of detection versus SNR. Swerling III, Swerling II, Swerling I, and Swerling 0.

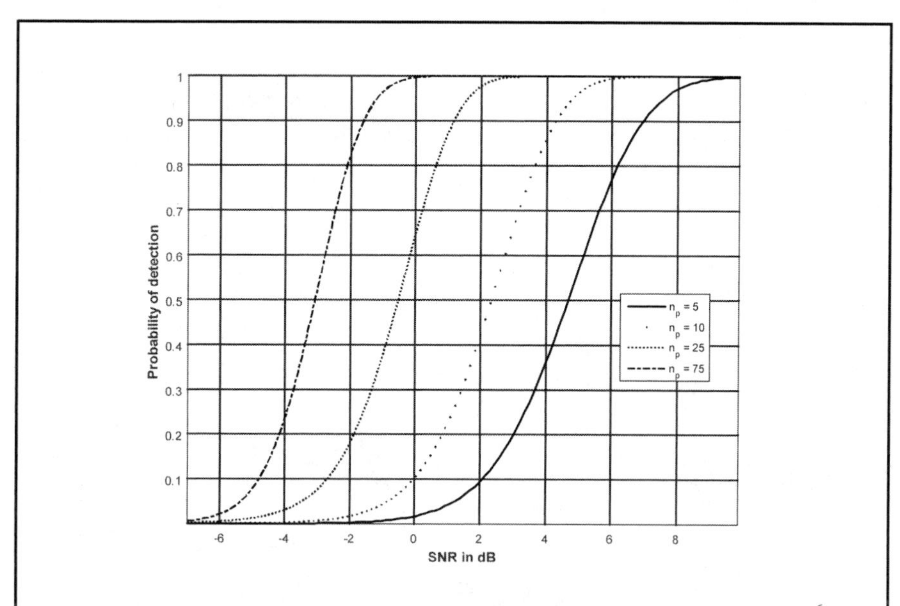

Figure 11.16. Probability of detection versus SNR. Swerling IV. $P_{fa} = 10^{-6}$.

11.5. Cumulative Probability of Detection

Denote the range at which the single pulse SNR is unity (0 dB) as R_0, and refer to it as the reference range. Then, for a specific radar, the single pulse SNR at R_0 is defined by the radar equation and is given by

$$(SNR)_{R_0} = \frac{P_t G^2 \lambda^2 \sigma}{(4\pi)^3 k T_0 B F L R_0^4} = 1. \qquad \text{Eq. (11.67)}$$

The single pulse SNR at any range R is

$$SNR = \frac{P_t G^2 \lambda^2 \sigma}{(4\pi)^3 k T_0 B F L R^4}. \qquad \text{Eq. (11.68)}$$

Dividing Eq. (11.68) by Eq. (11.67) yields

$$\frac{SNR}{(SNR)_{R_0}} = \left(\frac{R_0}{R}\right)^4. \qquad \text{Eq. (11.69)}$$

Therefore, if the range R_0 is known, then the SNR at any other range R is

$$(SNR)_{dB} = 40\log\left(\frac{R_0}{R}\right). \qquad \text{Eq. (11.70)}$$

Also, define the range R_{50} as the range at which $P_D = 0.5 = P_{50}$. Normally, the radar unambiguous range R_u is set equal to $2R_{50}$.

The cumulative probability of detection refers to detecting the target at least once by the time it is at range R. More precisely, consider a target closing on a scanning radar, where the target is illuminated only during a scan (frame). As the target gets closer to the radar, its probability of detection increases since the SNR is increased. Suppose that the probability of detection during the nth frame is P_{D_n}; then, the cumulative probability of detecting the target at least once during the nth frame (see Figure 11.17) is given by

$$P_{C_n} = 1 - \prod_{i=1}^{n} (1 - P_{D_i}).$$ Eq. (11.71)

P_{D_1} is usually selected to be very small. Clearly, the probability of not detecting the target during the nth frame is $1 - P_{C_n}$. The probability of detection for the ith frame, P_{D_i}, is computed as discussed in the previous section.

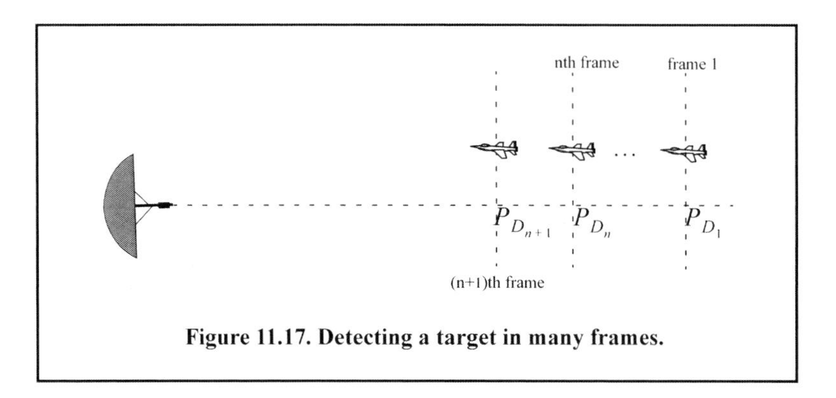

Figure 11.17. Detecting a target in many frames.

11.6. M-out-of-N Detection

A few sources in the literature refer to the *M-out-of-N* detection as *binary integration* and/or as *double threshold* detection; nonetheless, *M-out-of-N* is the most commonly used name. The basic idea behind the *M-out-of-N* detection technique is as follows: in any given resolution cell (range, Doppler, or angle) the detection process is repeated *N* times, where the outcome of each decision cycle is either a "detection" or "no detection". Hence the term *binary* is used in the literature. For each decision cycle, the probability of detection and the probability of false alarm are computed. The final decision criterion declares a target detection if *M* out of *N* decision cycles have resulted in a detection. Clearly, the decision criterion associated with this technique follows a binomial distribution.

To elaborate further on this concept of detection, assume a non-fluctuating target whose single trial probability of detection is P_D and its probability of false alarm is P_{fa}. Denote the total probability of detection resulting from the *M-out-of-N* detection technique as P_{Dmn}. It follows that after *N* independent trials of detection one gets

$$P_{Dmn} = 1 - (1 - P_D)^N.$$ Eq. (11.72)

Similarly, the probability of false alarm after the same number of trials is

$$P_{FA} = 1 - (1 - P_{fa})^N.$$ Eq. (11.73)

For example, if the desired P_{Dmn} is *0.99*, then by using Eq. (11.72), one finds that a $P_D = 0.9$ will accomplish the desired P_{Dmn} after 2 trials (i.e., *N*=2); alternatively, when using a $P_D = 0.2$, it will take 20 trials to reach the desired P_{Dmn}. Furthermore, Eq. (11.72) implicitly indicates that as the number of trials increases so does P_{Dmn}, but this buildup in detection probability is somewhat costly. That is true because as the number of trials is increased, the overall probability of false alarm P_{FA} is also increased. A slight modification to the *M-out-of-N* detection process that guarantees an increase or buildup in P_{Dmn} while simultaneously keeping P_{FA} in check is as follows:

1. A specific P_{fa} value is chosen; typically it is a design constraint.
2. For each value M, compute the corresponding P_{FA} from Eq. (11.73).
3. Using any of the techniques developed in this book to calculate the threshold value V_T so that P_{fa} is maintained, compute its corresponding SNR.
4. Calculate P_D that corresponds to the SNR computed in step 3.
5. Use Eq. (11.82) to compute the probability of detection P_{Dmn}, and from any of the techniques developed in this book, compute the corresponding SNR so that the threshold value computed in step 3 is maintained; therefore, P_{Dmn} is also maintained.
6. Repeat for each M to establish the specific combination of M (i.e., yielding P_{FA}) so that the SNR is minimized for a given P_{Dmn}.

Following this modified approach, P_{Dmn} and P_{FA} are given by

$$P_{Dmn} = \sum_{k=M}^{N} C_k^N \; P_D^k \; (1-P_D)^{N-k} \qquad \text{Eq. (11.74)}$$

$$P_{FA} = \sum_{k=M}^{N} C_k^N \; P_{fa}^k \; (1-P_{fa})^{N-k} \qquad \text{Eq. (11.75)}$$

where

$$C_k^N = \frac{N!}{k!(N-k)!} \,. \qquad \text{Eq. (11.76)}$$

For small values of P_D, Eq. (11.72) keeps the overall detection probability P_{Dmn} to less than or equal to P_D. Alternatively, for larger values of P_D, a quick buildup in the value of P_{Dmn} occurs. Selecting the specific combination of *N* and *M* that yields a desired P_{Dmn} is typically a design constraint. In any case, once the choice is made, one must take target fluctuating into account. In this case, the optimal value for M is

$$M_{opt} = 10^{\alpha} N^{\beta} \qquad \text{Eq. (11.77)}$$

where α and β are constants that vary depending on the target fluctuation type. Table 11.4 shows their values corresponding to different Swerling targets.

Example:

A radar detects a closing target at $R = 10Km$, with probability of detection P_D equal to 0.5. Assume $P_{fa} = 10^{-7}$. Compute and sketch the single look probability of detection as a function of normalized range (with respect to $R = 10Km$), over the interval $(2-20)Km$. If the range between two successive frames is $1Km$, what is the cumulative probability of detection at $R = 8Km$?

Table 11.4. Parameters of Eq. (11.77)

Fluctuation Type	α	β	Range of N
Swerling 0	0.8	-0.02	5-700
Swerling I	0.8	-0.02	6-700
Swerling II	0.91	-0.38	9-700
Swerling III	0.8	-0.02	6-700
Swerling IV	0.873	-0.27	10-700

Solution:

From the function "marcumsq.m," the SNR corresponding to $P_D = 0.5$ *and* $P_{fa} = 10^{-7}$ *is approximately* 12*dB. We can express the SNR at any range* R *as*

$$(SNR)_R = (SNR)_{10} + 40\ \log\frac{10}{R} = 52 - 40\ \log R .$$

By using Eq. (11.46) we can construct the following table:

R *Km*	(SNR) dB	P_D
2	39.09	0.999
4	27.9	0.999
6	20.9	0.999
8	15.9	0.999
9	13.8	0.9
10	12.0	0.5
11	10.3	0.25
12	8.8	0.07
14	6.1	0.01
16	3.8	ε
20	0.01	ε

where ε *is very small. A sketch of* P_D *versus normalized range is shown in Figure 11.18 below. The cumulative probability of detection is given in Eq. (11.71), where the probability of detection of the first frame is selected to be very small. Thus, we can arbitrarily choose frame 1 to be at* $R = 16Km$.

Note that selecting a different starting point for frame 1 would have a negligible effect on the cumulative probability (we only need P_{D_1} *to be very small). Below is a range listing for frames 1 through 9, where frame 9 corresponds to* $R = 8Km$

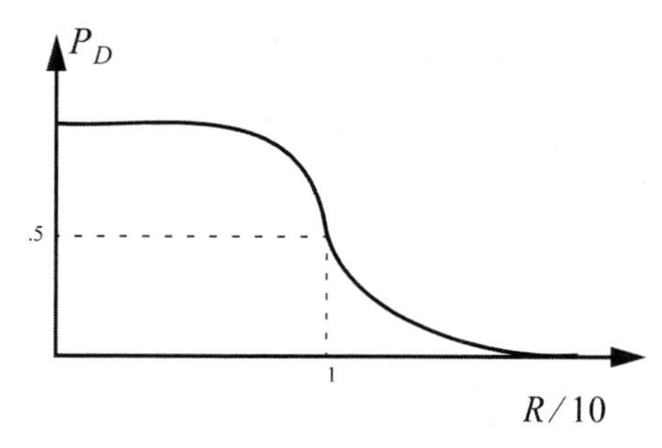

Figure 11.18. Cumulative probability of detection versus normalized range.

frame	1	2	3	4	5	6	7	8	9
range in Km	16	15	14	13	12	11	10	9	8

The cumulative probability of detection at 8Km is then

$$P_{C_9} = 1-(1-0.999)(1-0.9)(1-0.5)(1-0.25)(1-0.07) \quad . \\ (1-0.01)(1-\varepsilon)^2 \approx 0.9998$$

****

11.7. The Radar Equation Revisited

The radar equation developed in Chapter 5 assumed a constant target RCS and did not account for integration loss. In this section, a more comprehensive form of the radar equation is introduced. In this case, the radar equation is given by

$$R^4 = \frac{P_{av} G_t G_r \lambda^2 \sigma I(n_P)}{(4\pi)^3 k T_o F B \tau f_r L_t L_f \ (SNR)_1} \qquad \text{Eq. (11.78)}$$

where $P_{av} = P_t \tau f_r$ is the average transmitted power, P_t is the peak transmitted power, τ is the pulse width, f_r is PRF, G_t is the transmitting antenna gain, G_r is the receiving antenna gain, λ is the wavelength, σ is the target cross section, $I(n_P)$ is the improvement factor, n_P is the number of integrated pulses, k is Boltzman's constant, T_o is 290 degrees Kelvin, F is the system noise figure, B is the receiver bandwidth, L_t is the total system losses including integration loss, L_f is the loss due to target fluctuation, and $(SNR)_1$ is the minimum single pulse SNR required for detection.

Assuming that the radar parameters such as power, antenna gain, wavelength, losses, bandwidth, effective temperature, and noise figure are known, the steps one should follow to solve for range are shown in Figure 11.19. Note that both sides of the bottom half of Figure 11.19 are identical. Nevertheless, two paths are purposely shown so that a distinction between scintillating and nonfluctuating targets is made.

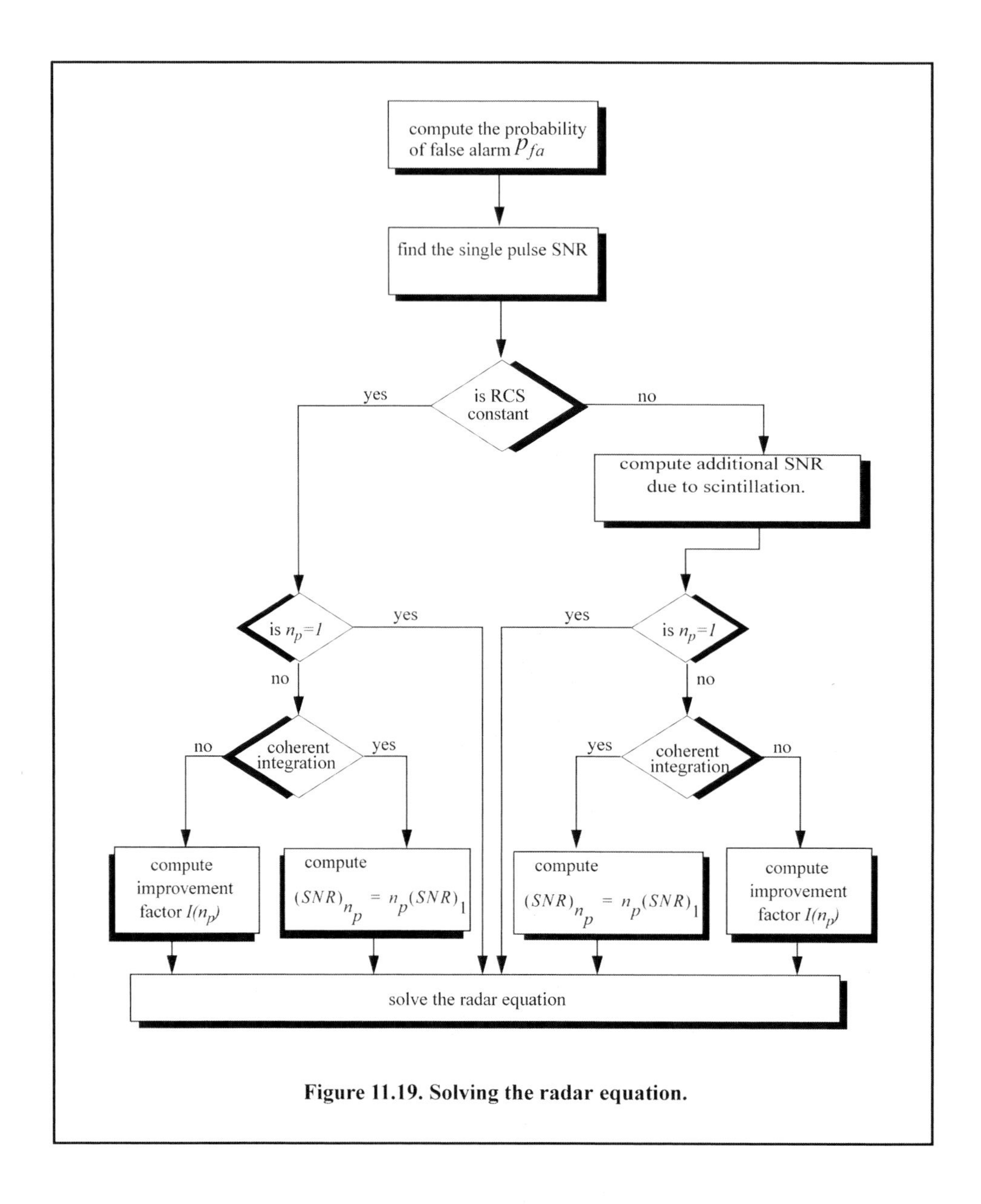

Figure 11.19. Solving the radar equation.

Problems

11.1. Consider the matched filter receiver shown in Figure 11.1. Develop expressions for the single pulse of known parameters probability of detection P_D and probability of false alarm P_{fa}.

11.2. Suppose you want to determine an unknown DC voltage v_{dc} in the presence of additive white Gaussian noise $n(t)$ of zero mean and variance σ_n^2. The measured signal is $x(t) = v_{dc} + n(t)$. An estimate of v_{dc} is computed by making three independent measurements of $x(t)$ and computing the arithmetic mean, $\underset{\sim}{v}_{dc} \approx (x_1 + x_2 + x_3)/3$. (a) Find the mean and variance of the random variable $\underset{\sim}{v}_{dc}$. (b) Does the estimate of v_{dc} get better by using ten measurements instead of three? Why?

11.3. Assume the X and Y miss distances of darts thrown at a bulls-eye dart board are Gaussian with zero mean and variance σ^2. (a) Determine the probability that a dart will fall between 0.8σ and 1.2σ. (b) Determine the radius of a circle about the bull's-eye that contains 80% of the darts thrown. (c) Consider a square with side s in the first quadrant of the board. Determine s so that the probability that a dart will fall within the square is 0.07.

11.4. Derive Eq. (11.15).

11.5. In the case of noise alone, the quadrature components of a radar return are independent Gaussian random variables with zero mean and variance σ^2. Assume that the radar processing consists of envelope detection followed by threshold decision. (a) Write an expression for the *pdf* of the envelope; (b) determine the threshold V_T as a function of σ that ensures a probability of false alarm $P_{fa} \le 10^{-8}$.

11.6. A pulsed radar has the following specifications: time of false alarm $T_{fa} = 10min$, probability of detection $P_D = 0.95$, operating bandwidth $B = 1MHz$. (a) What is the probability of false alarm P_{fa}? (b) What is the single pulse SNR?

11.7. Show that when computing the probability of detection at the output of an envelope detector, it is possible to use Gaussian probability approximation when the SNR is very large.

11.8. A radar system uses a threshold detection criterion. The probability of false alarm is $P_{fa} = 10^{-10}$. (a) What must be the average SNR at the input of a linear detector so that the probability of miss is $P_m = 0.15$? Assume a large SNR approximation. (b) Write an expression for the *pdf* at the output of the envelope detector.

11.9. An X-band radar has the following specifications: received peak power $10^{-10}W$, probability of detection $P_D = 0.95$, time of false alarm $T_{fa} = 8min$, pulse width $\tau = 2\mu s$, operating bandwidth $B = 2MHz$, operating frequency $f_0 = 10GHz$, and detection range $R = 100Km$. Assume single pulse processing. (a) Compute the probability of false alarm P_{fa}. (b) Determine the SNR at the output of the matched filter. (c) At what SNR would the probability of detection drop to 0.9 (P_{fa} does not change)? (d) What is the increase in range that corresponds to this drop in the probability of detection?

11.10. Using the equation

$$P_D = 1 - e^{-SNR} \int_{P_{fa}}^{1} I_0(\sqrt{-4SNR \ln u})du ,$$

calculate P_D when $SNR = 10dB$ and $P_{fa} = 0.01$. Perform the integration numerically.

11.11. A pulsed radar has the following specifications: time of false alarm $T_{fa} = 10\ min$, probability of detection $P_D = 0.95$, operating bandwidth $B = 1MHz$. (a) What is the probability of false alarm P_{fa}? (b) What is the single pulse SNR? (c) Assuming non-coherent integration of 100 pulses, what is the SNR reduction so that P_D and P_{fa} remain unchanged?

11.12. An L-band radar has the following specifications: operating frequency $f_0 = 1.5GHz$, operating bandwidth $B = 2MHz$, noise figure $F = 8dB$, system losses $L = 4dB$, time of false alarm $T_{fa} = 12\ minutes$, detection range $R = 12Km$, probability of detection $P_D = 0.5$, antenna gain $G = 5000$, and target RCS $\sigma = 1m^2$. (a) Determine the PRF f_r, the pulse width τ, the peak power P_t, the probability of false alarm P_{fa}, and the minimum detectable signal level S_{min}. (b) How can you reduce the transmitter power to achieve the same performance when 10 pulses are integrated non-coherently? (c) If the radar operates at a shorter range in the single pulse mode, find the new probability of detection when the range decreases to $9Km$.

11.13. A certain radar utilizes 10 pulses for non-coherent integration. The single pulse SNR is $15dB$ and the probability of miss is $P_m = 0.15$. (a) Compute the probability of false alarm P_{fa}. (b) Find the threshold voltage V_T.

11.14. (a) Show how you can use the radar equation to determine the PRF f_r, the pulse width τ, the peak power P_t, the probability of false alarm P_{fa}, and the minimum detectable signal level S_{min}. Assume the following specifications: operating frequency $f_0 = 1.5MHz$, operating bandwidth $B = 1MHz$, noise figure $F = 10dB$, system losses $L = 5dB$, time of false alarm $T_{fa} = 20\ min$, detection range $R = 12Km$, probability of detection $P_D = 0.5$ (three pulses). (b) If post-detection integration is assumed, determine the SNR.

11.15. Consider a scanning low PRF radar. The antenna half-power beam width is $1.5°$, and the antenna scan rate is $35°$ per second. The pulse width is $\tau = 2\mu s$, and the PRF is $f_r = 400Hz$. (a) Compute the radar operating bandwidth. (b) Calculate the number of returned pulses from each target illumination. (c) Compute the SNR improvement due to post-detection integration (assume 100% efficiency). (d) Find the number of false alarms per minute for a probability of false alarm $P_{fa} = 10^{-6}$.

11.16. Show that the detection probability for a Swerling I & II target is given by the equation

$$P_D = \exp\left\{\frac{\ln(P_{fa})}{1 + SNR}\right\}.$$

11.17. A certain radar has the following specifications: single pulse SNR corresponding to a reference range $R_0 = 200Km$ is $10dB$. The probability of detection at this range is $P_D = 0.95$. Assume a Swerling I type target. Use the radar equation to compute the required pulse widths at ranges $R = 220Km, 250Km$, and $175Km$, so that the probability of detection is maintained.

11.18. A circularly scanning, fan beam radar has a rotation rate of 2 seconds per revolution. The azimuth beamwidth is 1.5 degrees and the radar uses a PRF of $12.5KHz$. The radar uses an unmodulated pulse with a width of $1.2\mu s$ and searches a range window that extends from $15Km$ to $100Km$. The range cells used during search are separated by one pulse width. It is desired that the false alarm probability be set so that the radar experiences only one false alarm every $2min$. What is the required P_{fa} for each range cell? What is the threshold-to-noise ratio, in dB, needed to maintain that P_{fa}?

11.19. The probability of recording a detection in a particular range-angle cell of the scanning radar of Problem 11.18 is 0.7. What is the cumulative detection probability if the cell is checked on three successive scans? If the false alarm probability for a certain range-angle cell of the same radar is 10^{-6} what is the cumulative false alarm probability for that cell over three scans?

11.20. A radar with a phased array antenna conducts a search using a 1500-beam search raster. That is, it steps through 1500 beam positions that span a certain angular area. It transmits one pulse per beam. The radar uses range gates separated by $10m$. The output of each range gate is sent to a bank of Doppler filters with a width of $1000Hz$ each. Thus, the signal processor consists of a set of range gates with a bank of Doppler filters connected to each range gate output. The output of the signal processor consists of a range-Doppler array of signals that consists of MN elements where M is the number of range gates and N is the number of Doppler filter outputs. During the particular search of interest, the detection processor covers a range extent of $10Km$ and a Doppler extent of $25KHz$. The design specifications state that, in this mode, the radar must have less than one false alarm every 10 scans through the search raster. What is the required P_{fa} in each range-Doppler-beam cell needed to support this requirement?

11.21. A certain radar employs a non-coherent integrator that integrates 25 pulses. What are the integrator gains, in dB, for a Swerling 0, a Swerling I, a Sewrling II, a Sewrling III, and a Swerling IV target? Briefly discuss how you arrived at each of your answers. If needed, assume that the radar is to operate with a desired detection probability of 0.9.

11.22. A certain radar has the following parameters: Peak power $P_t = 500KW$, total losses $L = 12dB$, operating frequency $f_o = 5.6GHZ$, PRF $f_r = 2KHz$, pulse width $\tau = 0.5\mu s$, antenna beamwidth $\theta_{az} = 2°$ and $\theta_{el} = 7°$, noise figure $F = 6dB$, and scan time $T_{sc} = 2s$. The radar can experience one false alarm per scan. (a) What is the probability of false alarm? Assume that the radar searches a minimum range of $10Km$ to its maximum unambiguous range. (b) Plot the detection range versus RCS in dBsm. The detection range is defined as the range at which the single scan probability of detection is equal to 0.94. Generate curves for Swerling I, a Sewrling II, a Sewrling III, and a Swerling IV type targets. (c) Repeat part (b) above when non-coherent integration is used.

11.23. A certain circularly scanning radar with a fan beam has a rotation rate of 3 seconds per revolution. The azimuth beamwidth is 3 degrees, and the radar uses a PRI of 600 microseconds. The radar pulse width is 2 microseconds and the radar searches a range window that extends from $15Km$ to $100Km$. It is desired that the false alarm rate not be higher than two false alarms per revolution. What is the required probability of false alarm? What is the minimum SNR so that the minimum probability of false alarm can be maintained?

11.24. Starting with Eq. (11.75), show that as N is increased so is the overall probability of false alarm. More specifically, prove that $P_{FA} \approx NP_{fa}$.

Chapter 12

Target Tracking

Tracking radar systems are used to measure the target's relative position in range, azimuth angle, elevation angle, and velocity. Then, by using and keeping track of these measured parameters, the radar can predict their future values. Target tracking is important to military radars as well as to most civilian radars. In military radars, tracking is responsible for fire control and missile guidance; in fact, missile guidance is almost impossible without proper target tracking. Commercial radar systems, such as civilian airport traffic control radars, may utilize tracking as a means of controlling incoming and departing airplanes. Tracking techniques can be divided into range/velocity tracking and angle tracking. It is also customary to distinguish between continuous single-target tracking radars and multi-target track-while-scan (TWS) radars. Tracking radars utilize pencil beam (very narrow) antenna patterns. It is for this reason that a separate search radar is needed to facilitate target acquisition by the tracker. Still, the tracking radar has to search the volume where the target's presence is suspected. For this purpose, tracking radars use special search patterns.

12.1. Angle Tracking

Angle tracking is concerned with generating continuous measurements of the target's angular position in the azimuth and elevation coordinates. The accuracy of early generation angle tracking radars depended heavily on the size of the pencil beam employed. Most modern radar systems achieve very fine angular measurements by utilizing monopulse tracking techniques.

Tracking radars use the angular deviation from the antenna main axis of the target within the beam to generate an error signal. This deviation is normally measured from the antenna's main axis. The resultant error signal describes how much the target has deviated from the beam main axis. Then, the beam position is continuously changed in an attempt to produce a zero error signal. If the radar beam is normal to the target (maximum gain), then the target angular position would be the same as that of the beam. In practice, this is rarely the case.

In order to be able to quickly change the beam position, the error signal needs to be a linear function of the deviation angle. It can be shown that this condition requires the beam's axis to be squinted by some angle (squint angle) off the antenna's main axis.

12.1.1. Sequential Lobing

Sequential lobing is one of the first tracking techniques that was utilized by the early generation of radar systems. Sequential lobing is often referred to as lobe switching or sequential switching. It has a tracking accuracy that is limited by the pencil beamwidth used and by the noise caused by either mechanical or electronic switching mechanisms. However, it is very simple to implement. The pencil beam used in sequential lobing must be symmetrical (equal azimuth and elevation beamwidths).

Tracking is achieved (in one coordinate) by continuously switching the pencil beam between two pre-determined symmetrical positions around the antenna's Line of Sight (LOS) axis. Hence, the name sequential lobing is adopted. The LOS is called the radar tracking axis, as illustrated in Figure 12.1.

As the beam is switched between the two positions, the radar measures the returned signal levels. The difference between the two measured signal levels is used to compute the angular error signal. For example, when the target is tracked on the tracking axis, as the case in Figure 12.1a, the voltage difference is zero. However, when the target is off the tracking axis, as in Figure 12.1b, a nonzero error signal is produced. The sign of the voltage difference determines the direction in which the antenna must be moved. Keep in mind, the goal here is to make the voltage difference be equal to zero.

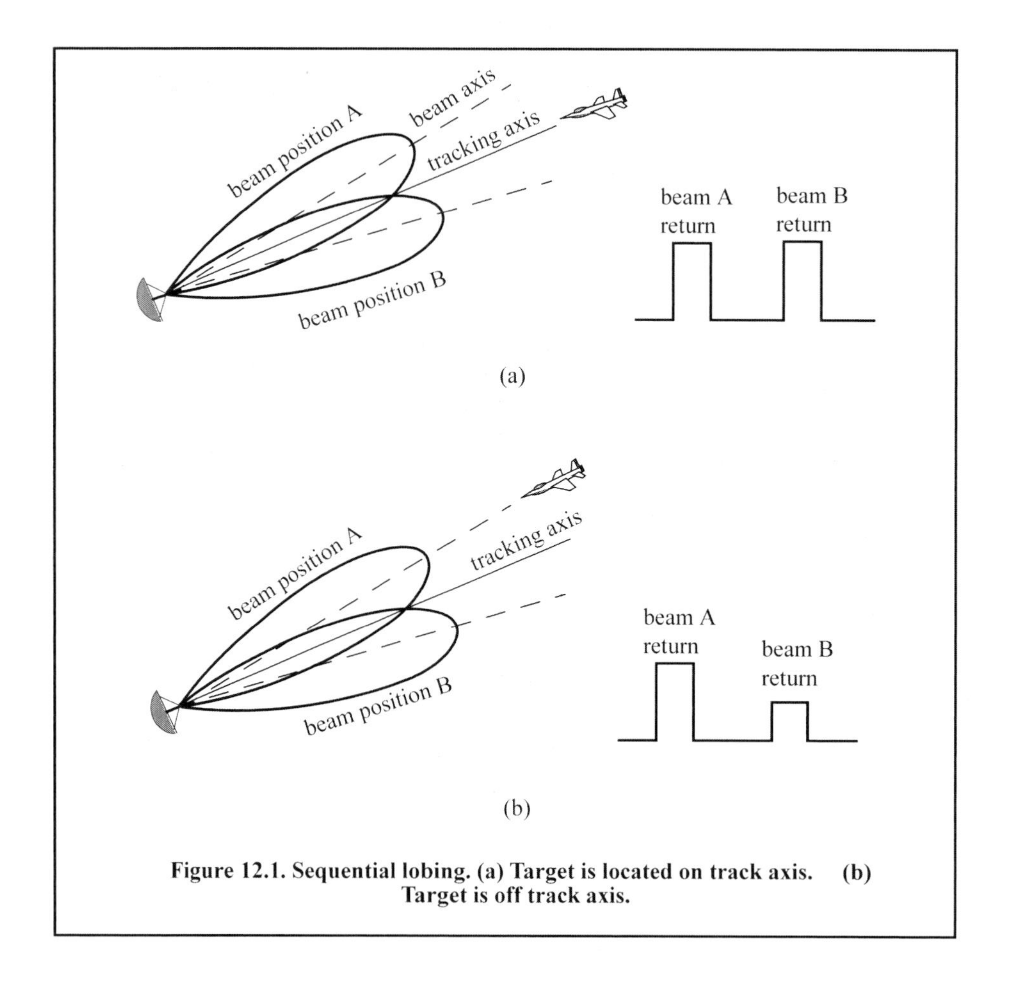

Figure 12.1. Sequential lobing. (a) Target is located on track axis. (b) Target is off track axis.

In order to obtain the angular error in the orthogonal coordinate, two more switching positions are required for that coordinate. Thus, tracking in two coordinates can be accomplished by using a cluster of four antennas (two for each coordinate) or by a cluster of five antennas. In the latter case, the middle antenna is used to transmit, while the other four are used to receive.

12.1.2. Conical Scan

Conical scan is a logical extension of sequential lobing where, in this case, the antenna is continuously rotated at an offset angle, or has a feed that is rotated about the antenna's main axis. Figure 12.2 shows a typical conical scan beam. The beam scan frequency, in radians per second, is denoted as ω_s. The angle between the antenna's LOS and the rotation axis is the squint angle φ. The antenna's beam position is continuously changed so that the target will always be on the tracking axis.

Figure 12.3 shows a simplified conical scan radar system. The envelope detector is used to extract the return signal amplitude and the Automatic Gain Control (AGC) tries to hold the receiver output to a constant value. Since the AGC operates on large time constants, it can hold the average signal level constant and still preserve the signal rapid scan variation. It follows that the tracking error signals (azimuth and elevation) are functions of the target's RCS; they are functions of its angular position off the main beam axis.

In order to illustrate how conical scan tracking is achieved, we will first consider the case shown in Figure 12.4. In this case, as the antenna rotates around the tracking axis, all target returns have the same amplitude (zero error signal). Thus, no further action is required. Next, consider the case depicted by Figure 12.5. Here, when the beam is at position B returns from the target will have maximum amplitude, and when the antenna is at position A, returns from the target have minimum amplitude. Between those two positions, the amplitude of the target returns will vary between the maximum value at position B, and the minimum value at position A. In other words, Amplitude Modulation (AM) exists on top of the returned signal. This AM envelope corresponds to the relative position of the target within the beam. Thus, the extracted AM envelope can be used to derive a servo-control system in order to position the target on the tracking axis.

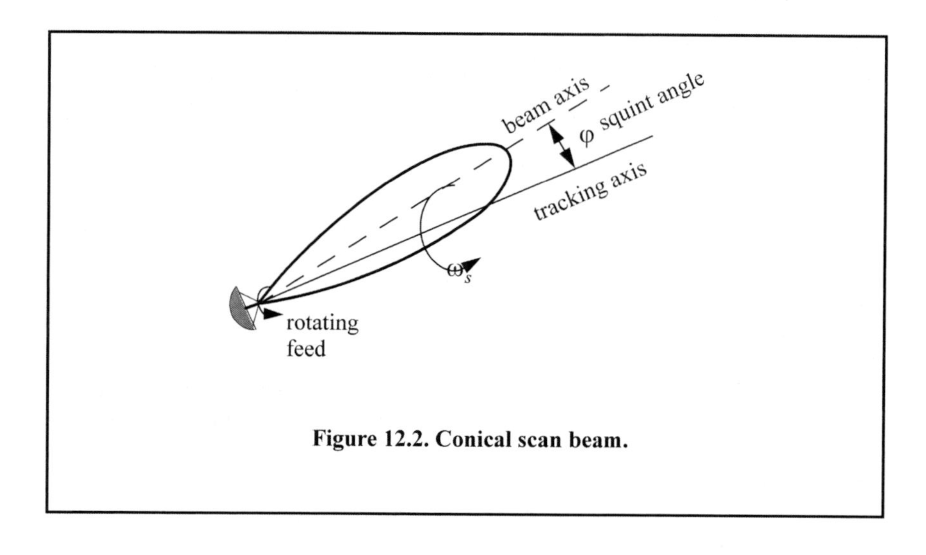

Figure 12.2. Conical scan beam.

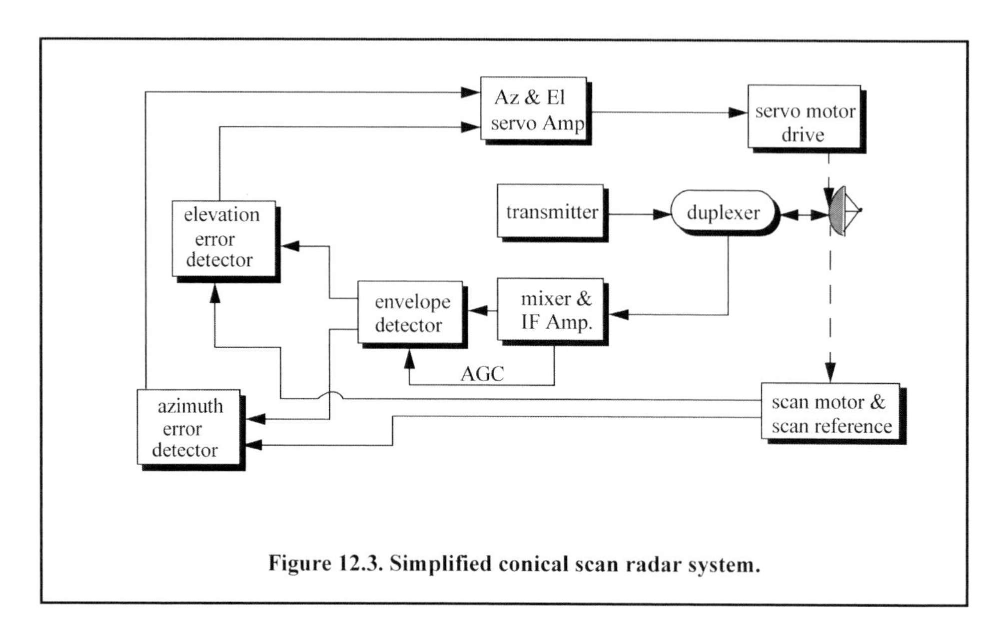

Figure 12.3. Simplified conical scan radar system.

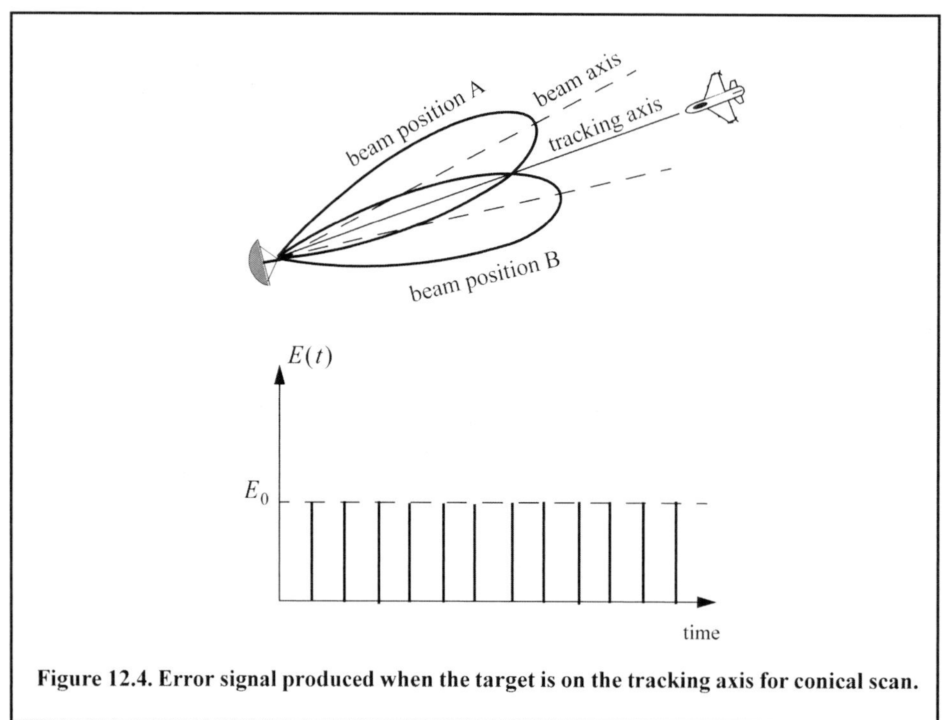

Figure 12.4. Error signal produced when the target is on the tracking axis for conical scan.

Now, let us derive the error signal expression that is used to drive the servo-control system. Consider the top view of the beam axis location shown in Figure 12.6. Assume that $t = 0$ is the starting beam position. The locations for maximum and minimum target returns are also identified. The quantity ε defines the distance between the target location and the antenna's tracking axis. It follows that the azimuth and elevation errors are, respectively, given by

$$\varepsilon_a = \varepsilon \sin\varphi \qquad \text{Eq. (12.1)}$$

$$\varepsilon_e = \varepsilon \cos\varphi \qquad \text{Eq. (12.2)}$$

These are the error signals that the radar uses to align the tracking axis on the target.

The AM signal $E(t)$ can then be written as

$$E(t) = E_0 \cos(\omega_s t - \varphi) = E_0 \varepsilon_e \cos\omega_s t + E_0 \varepsilon_a \sin\omega_s t \qquad \text{Eq. (12.3)}$$

where E_0 is a constant called the error slope, ω_s is the scan frequency in radians per seconds, and φ is the angle already defined. The scan reference is the signal that the radar generates to keep track of the antenna's position around a complete path (scan). The elevation error signal is obtained by mixing the signal $E(t)$ with $\cos\omega_s t$ (the reference signal) followed by low pass filtering. More precisely,

$$E_e(t) = E_0 \cos(\omega_s t - \varphi) \cos\omega_s t = -\frac{1}{2} E_0 \cos\varphi + \frac{1}{2} \cos(2\omega_s t - \varphi) \qquad \text{Eq. (12.4)}$$

and after low pass filtering we get

$$E_e(t) = -\frac{1}{2} E_0 \cos\varphi . \qquad \text{Eq. (12.5)}$$

Negative elevation error drives the antenna beam downward, while positive elevation error drives the antenna beam upward. Similarly, the azimuth error signal is obtained by multiplying $E(t)$ by $\sin\omega_s t$ followed by low pass filtering. It follows that

$$E_a(t) = \frac{1}{2} E_0 \sin\varphi . \qquad \text{Eq. (12.6)}$$

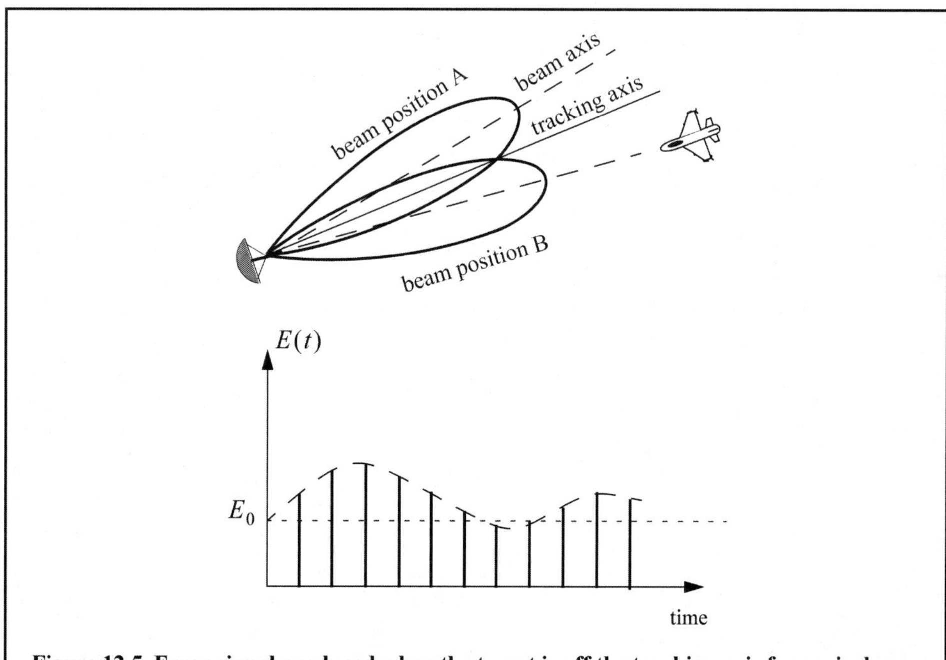

Figure 12.5. Error signal produced when the target is off the tracking axis for conical scan.

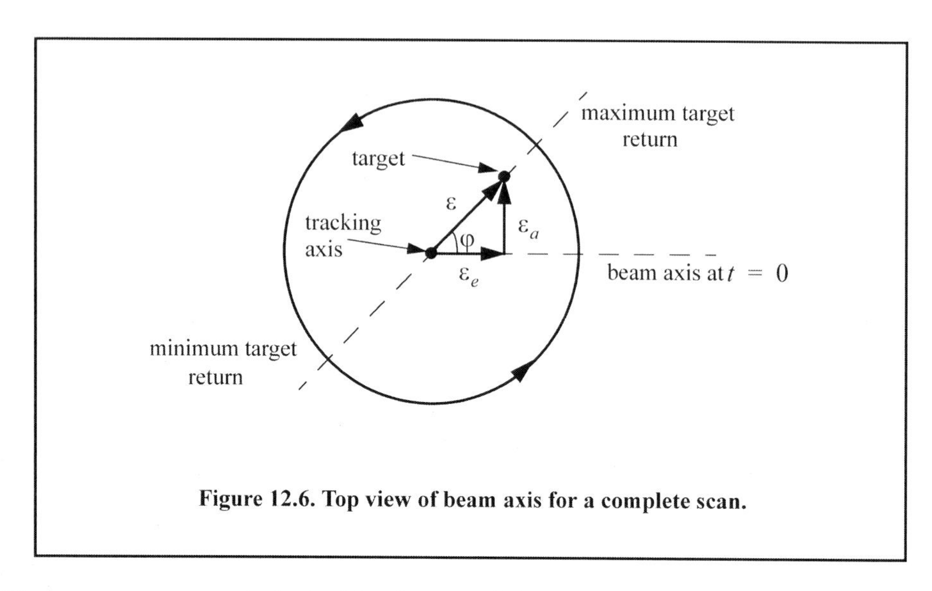

Figure 12.6. Top view of beam axis for a complete scan.

The antenna scan rate is limited by the scanning mechanism (mechanical or electronic), where electronic scanning is much faster and more accurate than mechanical scanning. In either case, the radar needs at least four target returns to be able to determine the target azimuth and elevation coordinates (two returns per coordinate). Therefore, the maximum conical scan rate is equal to one fourth of the PRF. Rates as high as 30 scans per seconds are commonly used.

The conical scan squint angle needs to be large enough so that a good error signal can be measured. However, due to the squint angle, the antenna gain in the direction of the tracking axis is less than maximum. Thus, when the target is in track (located on the tracking axis), the SNR suffers a loss equal to the drop in the antenna gain. This loss is known as the squint or crossover loss. The squint angle is normally chosen such that the two-way (transmit and receive) crossover loss is less than a few decibels.

12.2. Amplitude Comparison Monopulse

Amplitude comparison monopulse tracking is similar to lobing in the sense that four squinted beams are required to measure the target's angular position. The difference is that the four beams are generated simultaneously rather than sequentially. For this purpose, a special antenna feed is utilized such that the four beams are produced using a single pulse, hence the name "monopulse." Additionally, monopulse tracking is more accurate and is not susceptible to lobing anomalies, such as AM jamming and gain inversion ECM. Finally, in sequential and conical lobing, variations in the radar echoes degrade the tracking accuracy; however, this is not a problem for monopulse techniques since a single pulse is used to produce the error signals. Monopulse tracking radars can employ both antenna reflectors as well as phased array antennas.

Figure 12.7 shows a typical monopulse antenna pattern. The four beams A, B, C, and D represent the four conical scan beam positions. Four feeds, mainly horns, are used to produce the monopulse antenna pattern. Amplitude monopulse processing requires that the four signals have the same phase and different amplitudes.

A good way to explain the concept of amplitude monopulse technique is to represent the target echo signal by a circle centered at the antenna's tracking axis, as illustrated by Figure 12.8a, where the four quadrants represent the four beams. In this case, the four horns receive an equal amount of energy, which indicates that the target is located on the antenna's tracking axis. However, when the target is off the tracking axis (Figures 12.8b-d), an imbalance of energy occurs in the different beams. This imbalance of energy is used to generate an error signal that drives the servo-control system. Monopulse processing consists of computing a sum Σ and two difference Δ (azimuth and elevation) antenna patterns. Then by dividing a Δ channel voltage by the Σ channel voltage, the angle of the signal can be determined.

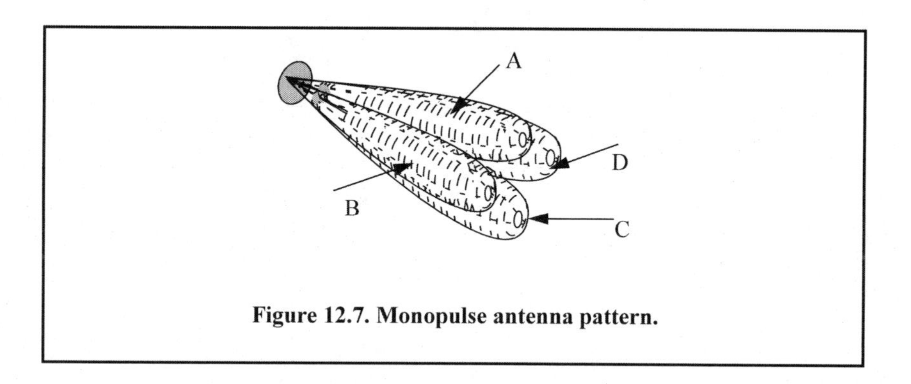

Figure 12.7. Monopulse antenna pattern.

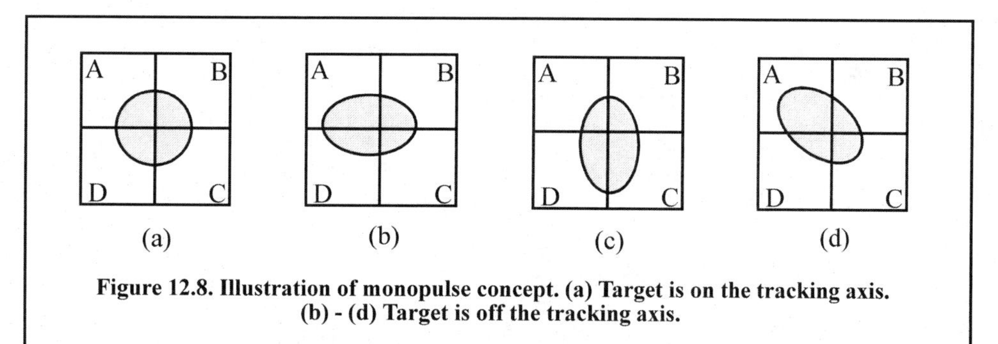

Figure 12.8. Illustration of monopulse concept. (a) Target is on the tracking axis. (b) - (d) Target is off the tracking axis.

The radar continuously compares the amplitudes and phases of all beam returns to sense the amount of target displacement off the tracking axis. It is critical that the phases of the four signals be constant in both transmit and receive modes. For this purpose, either digital networks or microwave comparator circuitry are utilized. Figure 12.9 shows a block diagram for a typical microwave comparator, where the three receiver channels are declared as the sum channel, elevation angle difference channel, and azimuth angle difference channel.

To generate the elevation difference beam, one can use the beam difference (A-D) or (B-C). However, by first forming the sum patterns (A+B) and (D+C) and then computing the difference (A+B)-(D+C), we achieve a stronger elevation difference signal, Δ_{el}. Similarly, by first forming the sum patterns (A+D) and (B+C) and then computing the difference (A+D)-(B+C), a stronger azimuth difference signal, Δ_{az}, is produced. A simplified monopulse radar block diagram is shown in Figure 12.10.

The sum channel is used for both transmit and receive. In the receive mode the sum channel provides the phase reference for the other two difference channels. Range measurements can also be obtained from the sum channel. In order to illustrate how the sum and difference antenna patterns are formed, we will assume a $\sin\varphi/\varphi$ single element antenna pattern and squint angle φ_0. The sum signal in one coordinate (azimuth or elevation) is then given by

$$\Sigma(\varphi) = \frac{\sin(\varphi - \varphi_0)}{(\varphi - \varphi_0)} + \frac{\sin(\varphi + \varphi_0)}{(\varphi + \varphi_0)} \qquad \text{Eq. (12.7)}$$

and a difference signal in the same coordinate is

$$\Delta(\varphi) = \frac{\sin(\varphi - \varphi_0)}{(\varphi - \varphi_0)} - \frac{\sin(\varphi + \varphi_0)}{(\varphi + \varphi_0)}. \qquad \text{Eq. (12.8)}$$

Figure 12.11 (a-c) shows the corresponding plots for the sum and difference patterns for $\varphi_0 = 0.15$ radians. Figure 12.12 (a-c) is similar to Figure 12.11, except in this case $\varphi_0 = 0.75$ radians. Clearly, the sum and difference patterns depend heavily on the squint angle. Using a relatively small squint angle produces a better sum pattern than that resulting from a larger angle. Additionally, the difference pattern slope is steeper for the small squint angle.

The difference channels give us an indication of whether the target is on or off the tracking axis. However, this signal amplitude depends not only on the target angular position, but also on the target's range and RCS. For this reason the ratio Δ/Σ (delta over sum) can be used to accurately estimate the error angle that only depends on the target's angular position.

Let us now address how the error signals are computed. First, consider the azimuth error signal. Define the signals S_1 and S_2 as

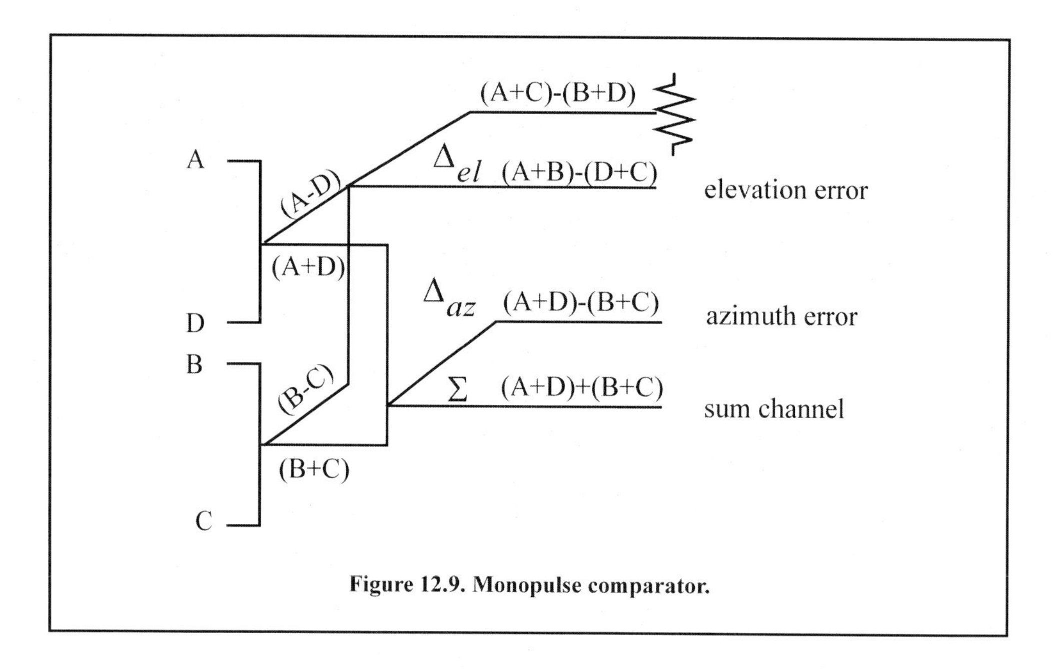

Figure 12.9. Monopulse comparator.

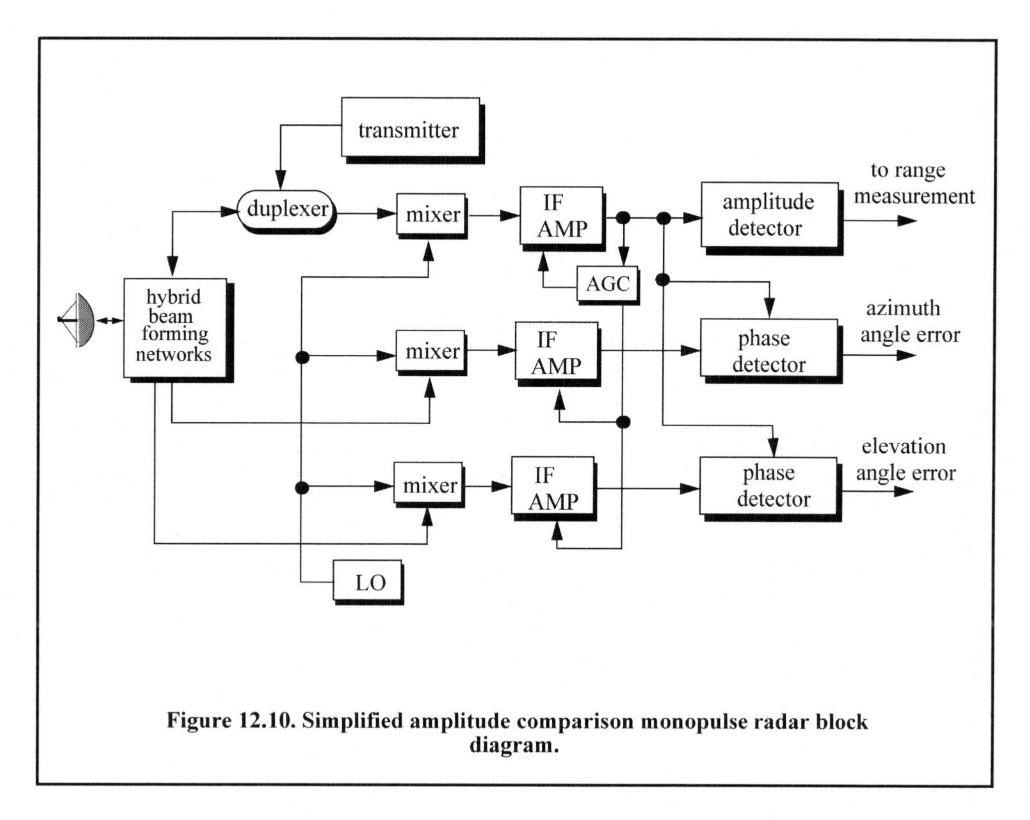

Figure 12.10. Simplified amplitude comparison monopulse radar block diagram.

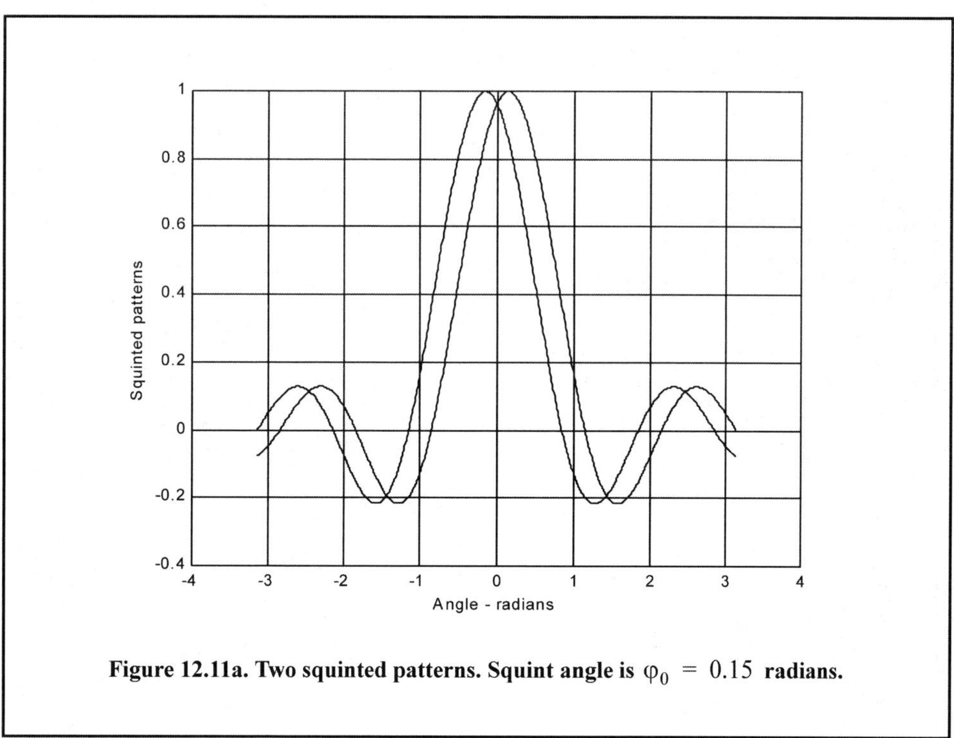

Figure 12.11a. Two squinted patterns. Squint angle is $\varphi_0 = 0.15$ **radians.**

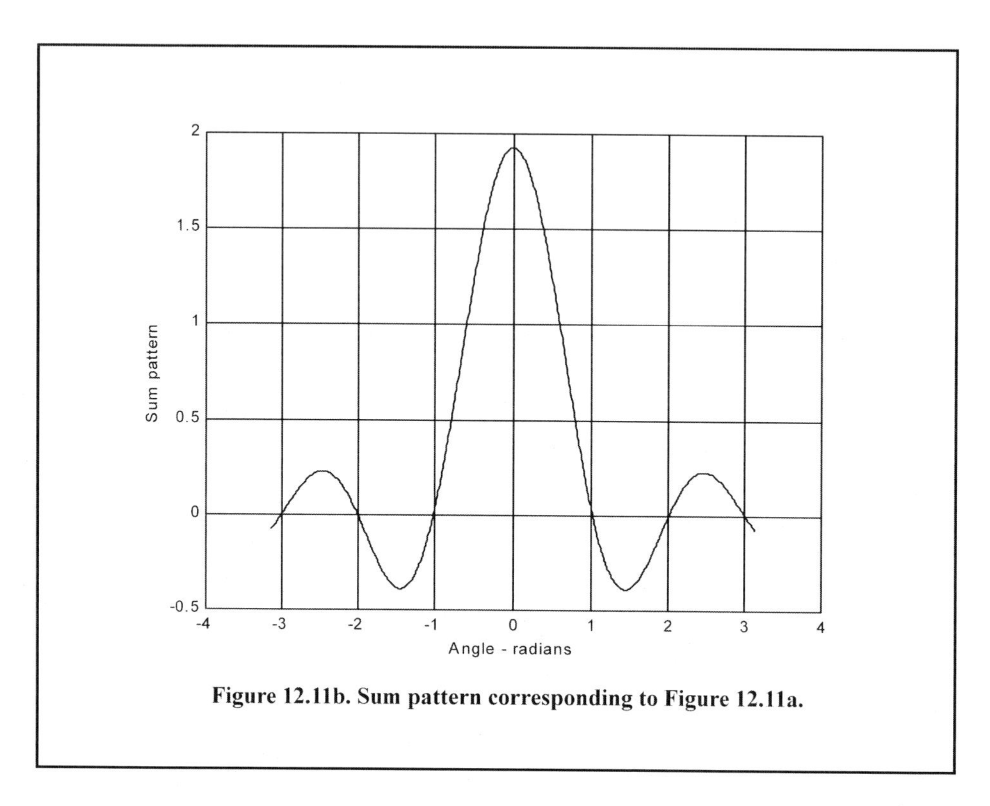

Figure 12.11b. Sum pattern corresponding to Figure 12.11a.

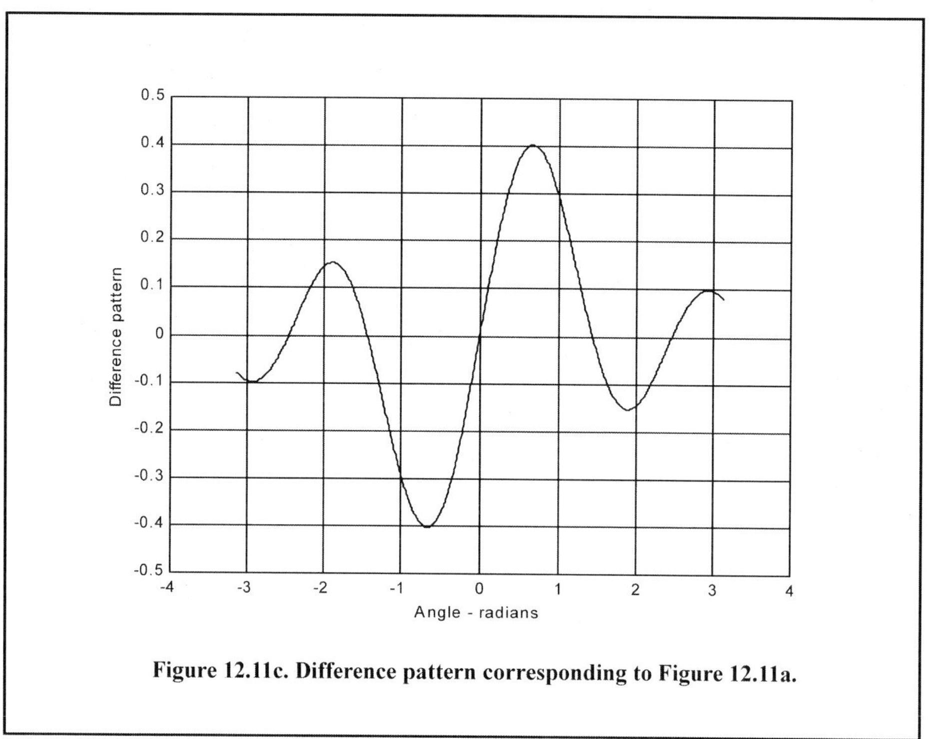

Figure 12.11c. Difference pattern corresponding to Figure 12.11a.

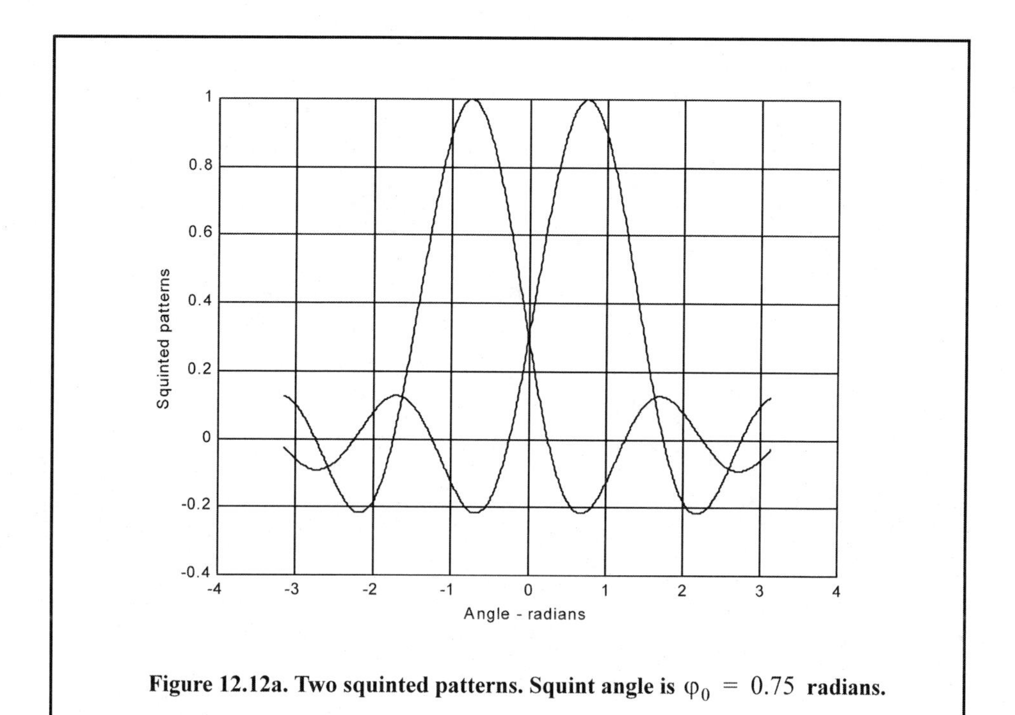

Figure 12.12a. Two squinted patterns. Squint angle is $\varphi_0 = 0.75$ radians.

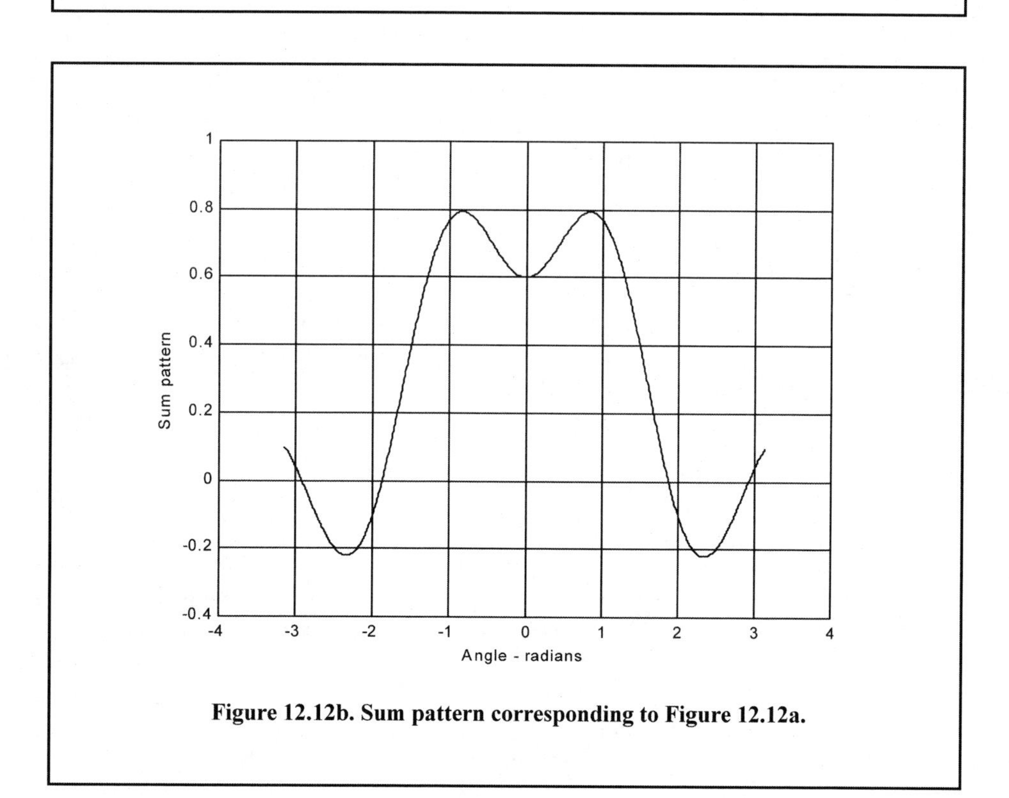

Figure 12.12b. Sum pattern corresponding to Figure 12.12a.

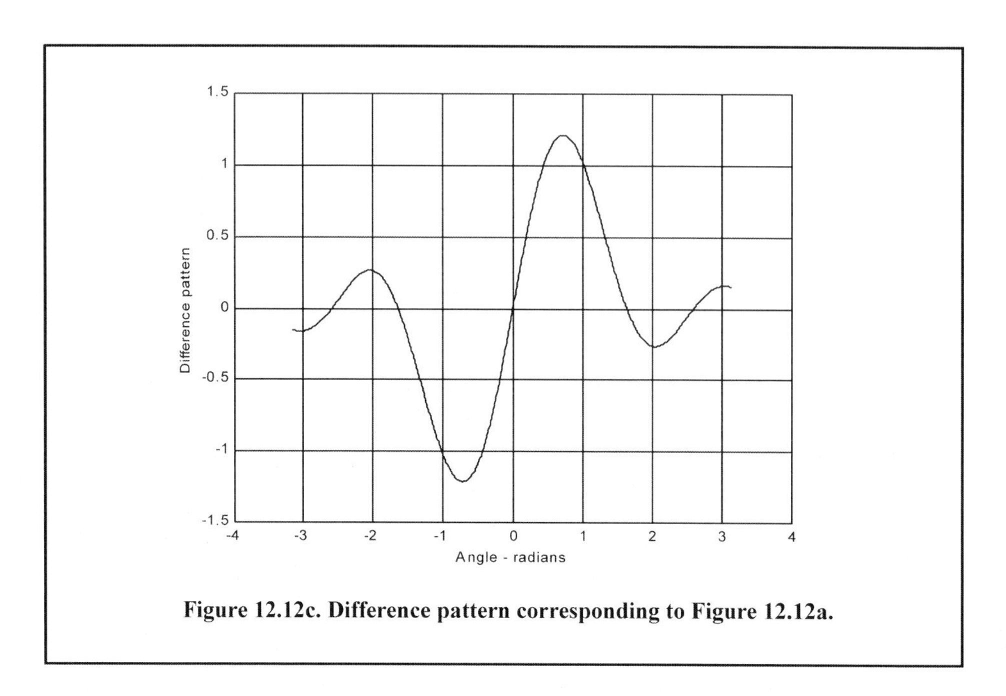

Figure 12.12c. Difference pattern corresponding to Figure 12.12a.

$$S_1 = A + D \qquad \text{Eq. (12.9)}$$

$$S_2 = B + C. \qquad \text{Eq. (12.10)}$$

The sum signal is $\Sigma = S_1 + S_2$, and the azimuth difference signal is $\Delta_{az} = S_1 - S_2$. If $S_1 \geq S_2$, then both channels have the same phase $0°$ (since the sum channel is used for phase reference). Alternatively, if $S_1 < S_2$, then the two channels are $180°$ out of phase. Similar analysis can be done for the elevation channel, where in this case $S_1 = A + B$ and $S_2 = D + C$. Thus, the error signal output is

$$\varepsilon_{\varphi} = \frac{|\Delta|}{|\Sigma|} \cos\xi \qquad \text{Eq. (12.11)}$$

where ξ is the phase angle between the sum and difference channels and it is equal to $0°$ or $180°$. More precisely, if $\xi = 0$, then the target is on the tracking axis; otherwise it is off the tracking axis. Figure 12.13 (a, b) shows a plot for the ratio Δ/Σ for the monopulse radar whose sum and difference patterns are in Figures 12.11 and 12.12.

12.3. Phase Comparison Monopulse

Phase comparison monopulse is similar to amplitude comparison monopulse in the sense that the target angular coordinates are extracted from one sum and two difference channels. The main difference is that the four signals produced in amplitude comparison monopulse will have similar phases but different amplitudes; however, in phase comparison monopulse the signals have the same amplitude and different phases. Phase comparison monopulse tracking radars use a minimum of a two-element array antenna for each coordinate (azimuth and elevation), as illustrated in Figure 12.14. A phase error signal (for each coordinate) is computed from the phase difference between the signals generated in the antenna elements.

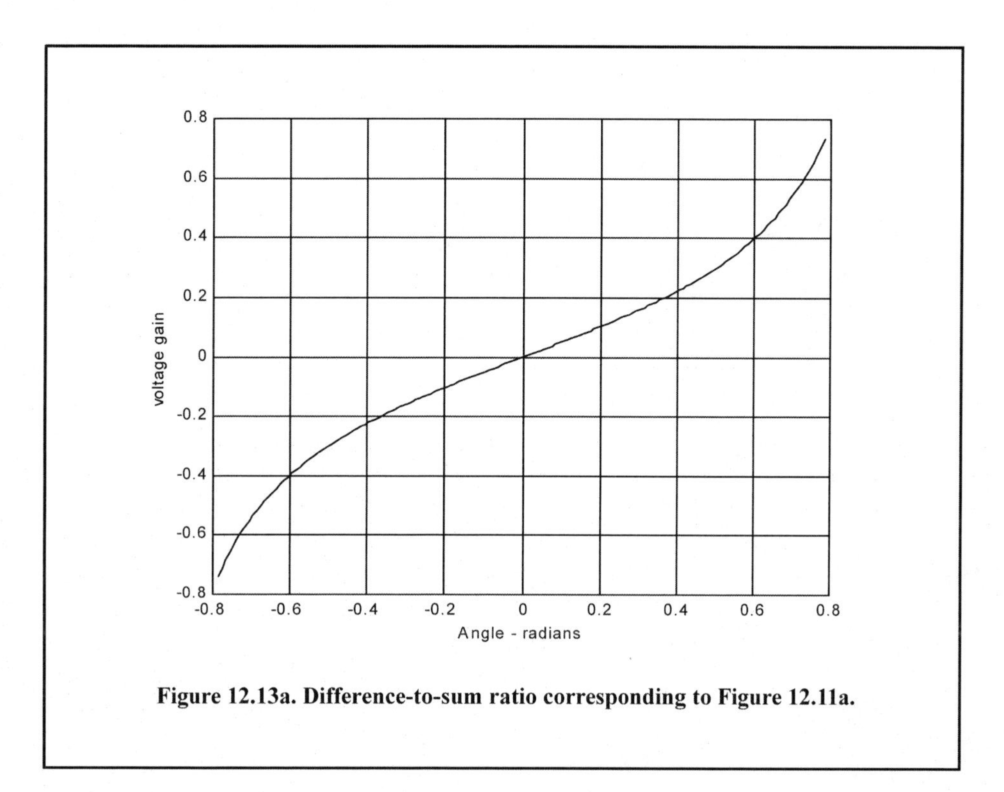

Figure 12.13a. Difference-to-sum ratio corresponding to Figure 12.11a.

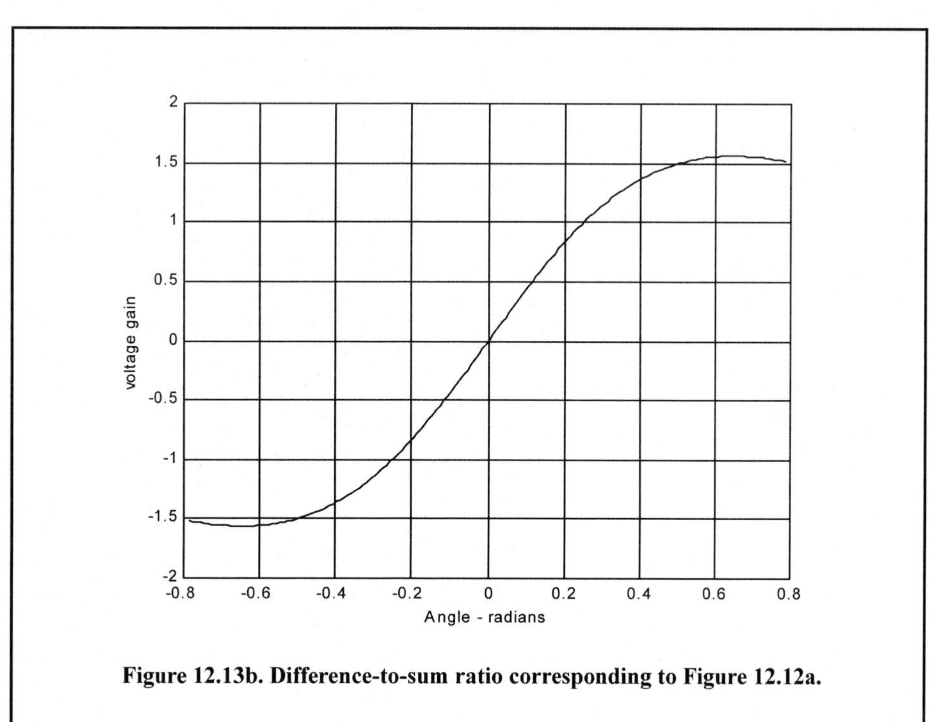

Figure 12.13b. Difference-to-sum ratio corresponding to Figure 12.12a.

Consider Figure 12.14; since the angle α is equal to $\varphi + \pi/2$, it follows that

$$R_1^2 = R^2 + \left(\frac{d}{2}\right)^2 - 2\frac{d}{2}R\cos\left(\varphi + \frac{\pi}{2}\right) = R^2 + \frac{d^2}{4} - dR\sin\varphi \qquad \text{Eq. (12.12)}$$

and since $d \ll R$ we can use the binomial series expansion to get

$$R_1 \approx R\left(1 + \frac{d}{2R}\sin\varphi\right). \qquad \text{Eq. (12.13)}$$

Similarly,

$$R_2 \approx R\left(1 - \frac{d}{2R}\sin\varphi\right). \qquad \text{Eq. (12.14)}$$

The phase difference between the two elements is then given by

$$\phi = \frac{2\pi}{\lambda}(R_1 - R_2) = \frac{2\pi}{\lambda}d\sin\varphi \qquad \text{Eq. (12.15)}$$

where λ is the wavelength. The phase difference ϕ is used to determine the angular target location. Note that if $\phi = 0$, then the target would be on the antenna's main axis. The problem with this phase comparison monopulse technique is that it is quite difficult to maintain a stable measurement of the off boresight angle φ, which causes serious performance degradation. This problem can be overcome by implementing a phase comparison monopulse system as illustrated in Figure 12.15.

The (single coordinate) sum and difference signals are, respectively, given by

$$\Sigma(\varphi) = S_1 + S_2 \qquad \text{Eq. (12.16)}$$

$$\Delta(\varphi) = S_1 - S_2 \qquad \text{Eq. (12.17)}$$

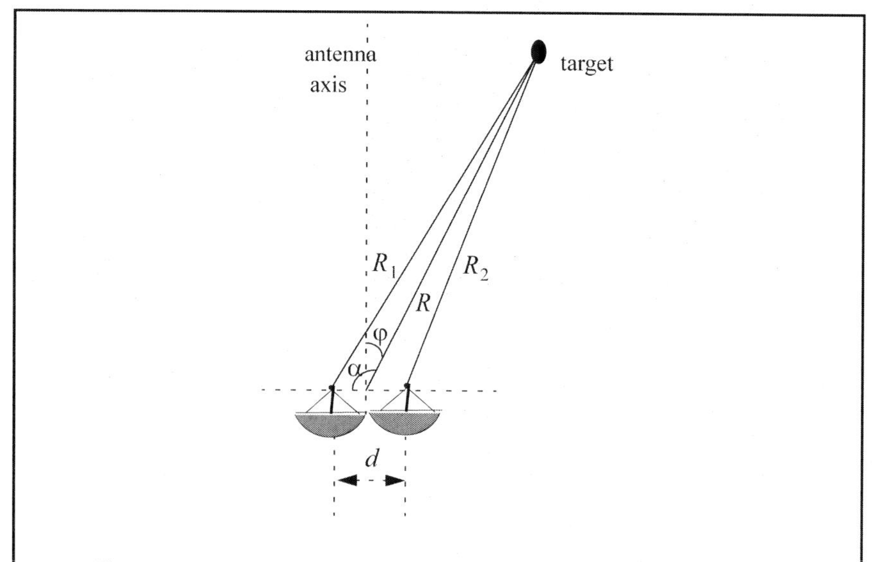

Figure 12.14. Single coordinate phase comparison monopulse antenna.

where the S_1 and S_2 are the signals in the two elements. Now, since S_1 and S_2 have similar amplitude and are different in phase by ϕ, we can write

$$S_1 = S_2 e^{-j\phi} . \qquad \text{Eq. (12.18)}$$

It follows that

$$\Delta(\varphi) = S_2(1 - e^{-j\phi}) \qquad \text{Eq. (12.19)}$$

$$\Sigma(\varphi) = S_2(1 + e^{-j\phi}) . \qquad \text{Eq. (12.20)}$$

The phase error signal is computed from the ratio Δ/Σ. More precisely,

$$\frac{\Delta}{\Sigma} = \frac{1 - e^{-j\phi}}{1 + e^{-j\phi}} = j\tan\left(\frac{\phi}{2}\right) \qquad \text{Eq. (12.21)}$$

which is purely imaginary. The modulus of the error signal is then given by

$$\frac{|\Delta|}{|\Sigma|} = \tan\left(\frac{\phi}{2}\right) . \qquad \text{Eq. (12.22)}$$

This kind of phase comparison monopulse tracker is often called the half-angle tracker.

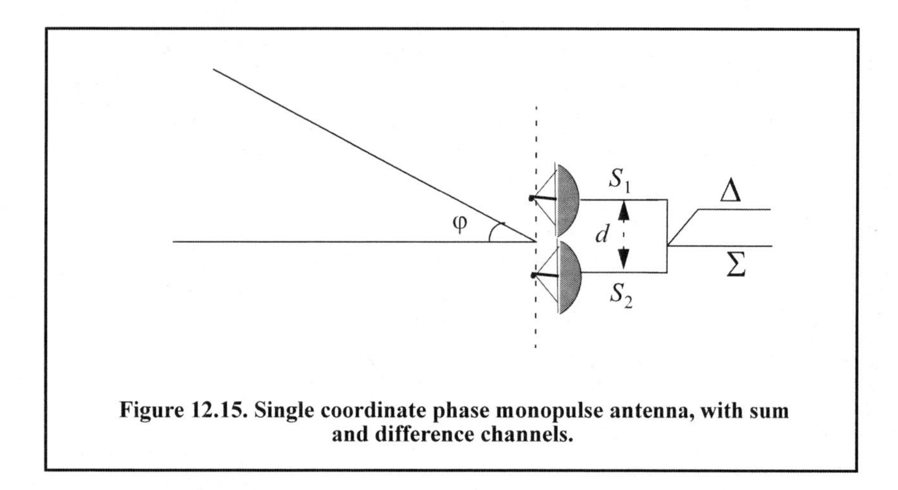

Figure 12.15. Single coordinate phase monopulse antenna, with sum and difference channels.

12.4. Range Tracking

Target range is measured by estimating the round-trip delay of the transmitted pulses. The process of continuously estimating the range of a moving target is known as range tracking. Since the range to a moving target is changing with time, the range tracker must be constantly adjusted to keep the target locked in range. This can be accomplished using a split gate system, where two range gates (early and late) are utilized. The concept of split gate tracking is illustrated in Figure 12.16, where a sketch of a typical pulsed radar echo is shown in the figure. The early gate opens at the anticipated starting time of the radar echo and lasts for half its duration. The late gate opens at the center and closes at the end of the echo signal. For this purpose, good estimates of the echo duration and the pulse center time must be reported to the range tracker so that the early and late gates can be placed properly at the start and center times of the expected echo. This reporting process is widely known as the "designation process."

The early gate produces positive voltage output while the late gate produces negative voltage output. The outputs of the early and late gates are subtracted, and the difference signal is fed into an integrator to generate an error signal. If both gates are placed properly in time, the integrator output will be equal to zero. Alternatively, when the gates are not timed properly, the integrator output is not zero, which gives an indication that the gates must be moved in time, left or right depending on the sign of the integrator output.

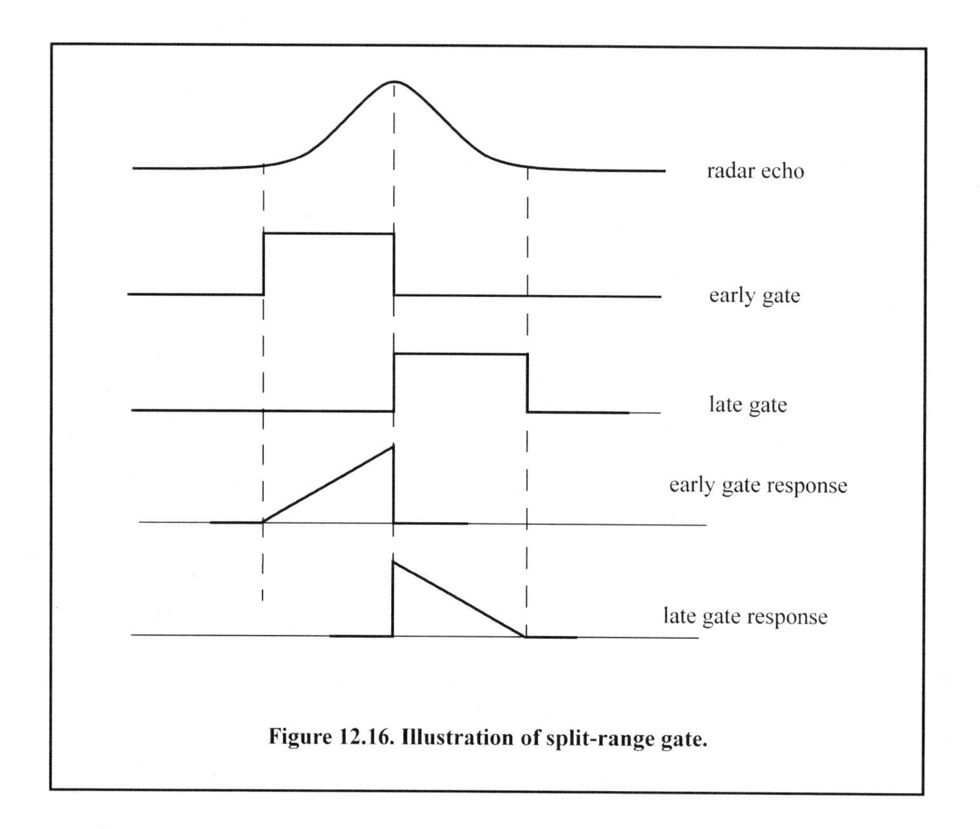

Figure 12.16. Illustration of split-range gate.

12.5. Track-While-Scan (TWS)

Track-while-scan radar systems sample each target once per scan interval, and use sophisticated smoothing and prediction filters to estimate the target parameters between scans. To this end, the Kalman filter and the Alpha-Beta-Gamma ($\alpha\beta\gamma$) filter are commonly used. Once a particular target is detected, the radar may transmit up to a few pulses to verify the target parameters, before it establishes a track file for that target. Target position, velocity, and acceleration comprise the major components of the data maintained by a track file.

The principles of recursive tracking and prediction filters are presented in this part. First, an overview of state representation for Linear Time Invariant (LTI) systems is discussed. Then, second and third order one-dimensional fixed gain polynomial filter trackers are developed. These filters are, respectively, known as the $\alpha\beta$ and $\alpha\beta\gamma$ filters (also known as the g-h and g-h-k filters). Finally, the equations for an n-dimensional multi-state Kalman filter is introduced and analyzed. As a matter of notation, small case letters, with an underbar, are used.

Modern radar systems are designed to perform multi-function operations, such as detection, tracking, and discrimination. With the aid of sophisticated computer systems, multifunction radars are capable of simultaneously tracking many targets. In this case, each target is sampled once (mainly range and angular position) during a dwell interval (scan). Then, by using smoothing and prediction techniques future samples can be estimated. Radar systems that can perform multi-tasking and multi-target tracking are known as Track-While-Scan (TWS) radars.

Once a TWS radar detects a new target, it initiates a separate track file for that detection; this ensures that sequential detections from that target are processed together to estimate the target's future parameters. Position, velocity, and acceleration comprise the main components of the track file. Typically, at least one other confirmation detection (verify detection) is required before the track file is established.

Unlike single target tracking systems, TWS radars must decide whether each detection (observation) belongs to a new target or belongs to a target that has been detected in earlier scans. And in order to accomplish this task, TWS radar systems utilize correlation and association algorithms. In the correlation process each new detection is correlated with all previous detections in order to avoid establishing redundant tracks. If a certain detection correlates with more than one track, then a pre-determined set of association rules are exercised so that the detection is assigned to the proper track. A simplified TWS data processing block diagram is shown in Figure 12.17.

Choosing a suitable tracking coordinate system is the first problem a TWS radar has to confront. It is desirable that a fixed reference of an inertial coordinate system be adopted. The radar measurements consist of target range, velocity, azimuth angle, and elevation angle. The TWS system places a gate around the target position and attempts to track the signal within this gate. The gate dimensions are normally azimuth, elevation, and range. Because of the uncertainty associated with the exact target position during the initial detections, a gate has to be large enough so that targets do not move appreciably from scan to scan; more precisely, targets must stay within the gate boundary during successive scans. After the target has been observed for several scans, the size of the gate is reduced considerably.

Gating is used to decide whether an observation is assigned to an existing track file or to a new track file (new detection). Gating algorithms are normally based on computing a statistical error distance between a measured and an estimated radar observation. For each track file, an upper bound for this error distance is normally set. If the computed difference for a certain radar observation is less than the maximum error distance of a given track file, then the observation is assigned to that track.

All observations that have an error distance less than the maximum distance of a given track are said to correlate with that track. For each observation that does not correlate with any existing tracks, a new track file is established accordingly. Since new detections (measurements) are compared to all existing track files, a track file may then correlate with no observations or with one or more observations. The correlation between observations and all existing track files is identified using a correlation matrix. Rows of the correlation matrix represent radar observations, while columns represent track files. In cases where several observations correlate with more than one track file, a set of pre-determined association rules can be utilized so that a single observation is assigned to a single track file.

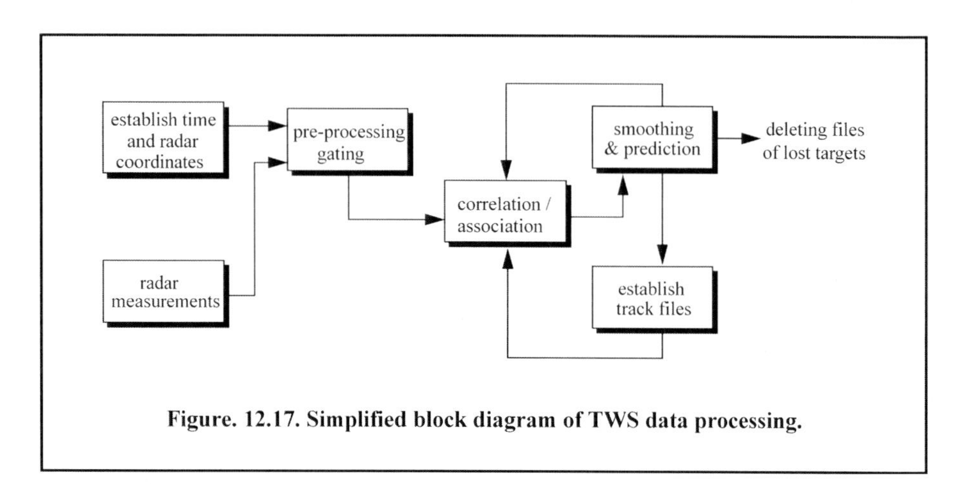

Figure. 12.17. Simplified block diagram of TWS data processing.

12.6. State Variable Representation of an LTI System

A linear time invariant system (continuous or discrete) can be described mathematically using three variables. They are the input, the output, and the state variables. In this representation, any LTI system has observable or measurable objects (abstracts). For example, in the case of a radar system, range may be an object measured or observed by the radar tracking filter. States can be derived in many different ways. For the scope of this book, states of an object or an abstract are the components of the vector that contains the object and its time derivatives. For example, a third-order one-dimensional (in this case range) state vector representing range can be given by

$$\mathbf{x} = \begin{bmatrix} R \\ \dot{R} \\ \ddot{R} \end{bmatrix} \qquad \text{Eq. (12.23)}$$

where R, $\dot{R}$, and $\ddot{R}$ are, respectively, the range measurement, range rate (velocity), and acceleration. The state vector defined in Eq. (12.23) can be representative of continuous or discrete states.

The emphasis is on discrete time representation, since most radar signal processing is executed using digital computers. For this purpose, an n-dimensional state vector has the following form:

$$\mathbf{x} = \begin{bmatrix} x_1 \; \dot{x}_1 \; \dots \; x_2 \; \dot{x}_2 \; \dots \; x_n \; \dot{x}_n \; \dots \end{bmatrix}^t \qquad \text{Eq. (12.24)}$$

where the superscript indicates the transpose operation. The LTI system of interest can be represented using the following state equations:

$$\dot{\mathbf{x}}(t) = \mathbf{A}\mathbf{x}(t) + \mathbf{B}\mathbf{w}(t) \qquad \text{Eq. (12.25)}$$

$$\mathbf{y}(t) = \mathbf{Cx}(t) + \mathbf{Dw}(t). \qquad \text{Eq. (12.26)}$$

$\dot{\mathbf{x}}$ is the value of the $n \times 1$ state vector; $\mathbf{y}$ is the value of the $p \times 1$ output vector; $\mathbf{w}$ is the value of the $m \times 1$ input vector; $\mathbf{A}$ is an $n \times n$ matrix; $\mathbf{B}$ is an $n \times m$ matrix; $\mathbf{C}$ is $p \times n$ matrix; and $\mathbf{D}$ is an $p \times m$ matrix. The homogeneous solution (i.e., $\mathbf{w} = \mathbf{0}$) to this linear system, assuming known initial condition $\mathbf{x}(0)$ at time t_0, has the form

$$\mathbf{x}(t) = \Phi(t - t_0)\mathbf{x}(t - t_0). \qquad \text{Eq. (12.27)}$$

The matrix Φ is known as the state transition matrix, or fundamental matrix, and is equal to

$$\Phi(t - t_0) = e^{\mathbf{A}(t - t_0)}. \qquad \text{Eq. (12.28)}$$

Eq. (12.28) can be expressed in series format as

$$\Phi(t - t_0)\big|_{t_0 = 0} = e^{\mathbf{A}(t)} = \mathbf{I} + \mathbf{A}t + \mathbf{A}^2\frac{t^2}{2!} + \ldots = \sum_{k=0}^{\infty} \mathbf{A}^k \frac{t^k}{k!} \qquad \text{Eq. (12.29)}$$

where $\mathbf{I}$ is the identity matrix.

Example:

Compute the state transition matrix for an LTI system when

$$\mathbf{A} = \begin{bmatrix} 0 & 1 \\ -0.5 & -1 \end{bmatrix}.$$

Solution:

The state transition matrix can be computed using Eq. (12.29). For this purpose, compute $\mathbf{A}^2$ *and* $\mathbf{A}^3$*.... It follows*

$$\mathbf{A}^2 = \begin{bmatrix} -\frac{1}{2} & -1 \\ \frac{1}{2} & \frac{1}{2} \end{bmatrix} \qquad \mathbf{A}^3 = \begin{bmatrix} \frac{1}{2} & \frac{1}{2} \\ -\frac{1}{4} & 0 \end{bmatrix} \qquad \ldots$$

Therefore,

$$\Phi = \begin{bmatrix} 1 + 0t - \dfrac{\frac{1}{2}t^2}{2!} + \dfrac{\frac{1}{2}t^3}{3!} + \ldots & 0 + t - \dfrac{t^2}{2!} + \dfrac{\frac{1}{2}t^3}{3!} + \ldots \\ 0 - \frac{1}{2}t + \dfrac{\frac{1}{2}t^2}{2!} - \dfrac{\frac{1}{4}t^3}{3!} + \ldots & 1 - t + \dfrac{\frac{1}{2}t^2}{2!} + \dfrac{0t^3}{3!} + \ldots \end{bmatrix}.$$

****

The state transition matrix has the following properties:

1. *Derivative property*

$$\frac{\partial}{\partial t}\Phi(t - t_0) = \mathbf{A}\Phi(t - t_0) \qquad \text{Eq. (12.30)}$$

2. *Identity property*

$$\Phi(t_0 - t_0) = \Phi(0) = \mathbf{I} \qquad \text{Eq. (12.31)}$$

3. *Initial value property*

$$\left.\frac{\partial}{\partial t}\Phi(t - t_0)\right|_{t = t_0} = \mathbf{A} \qquad \text{Eq. (12.32)}$$

4. *Transition property*

$$\Phi(t_2 - t_0) = \Phi(t_2 - t_1)\Phi(t_1 - t_0) \qquad ; \; t_0 \le t_1 \le t_2 \qquad \text{Eq. (12.33)}$$

5. *Inverse property*

$$\Phi(t_0 - t_1) = \Phi^{-1}(t_1 - t_0) \qquad \text{Eq. (12.34)}$$

6. *Separation property*

$$\Phi(t_1 - t_0) = \Phi(t_1)\Phi^{-1}(t_0) \qquad \text{Eq. (12.35)}$$

The general solution to the system defined in Eq. (12.25) can be written as

$$\mathbf{x}(t) = \Phi(t - t_0)\mathbf{x}(t_0) + \int_{t_0}^{t} \Phi(t - \tau)\mathbf{B}\mathbf{w}(\tau)d\tau . \qquad \text{Eq. (12.36)}$$

The first term of the right-hand side of Eq. (12.36) represents the contribution from the system response to the initial condition. The second term is the contribution due to the driving force $\mathbf{w}$. By combining Eqs. (12.26) and (12.36), an expression for the output is computed as

$$\mathbf{y}(t) = \mathbf{C}e^{\mathbf{A}(t - t_0)}\mathbf{x}(t_0) + \int_{t_0}^{t} [\mathbf{C}e^{\mathbf{A}(t - \tau)}\mathbf{B} - \mathbf{D}\delta(t - \tau)]\mathbf{w}(\tau)d\tau . \qquad \text{Eq. (12.37)}$$

Note that the system impulse response is equal to $\mathbf{C}e^{\mathbf{A}t}\mathbf{B} - \mathbf{D}\delta(t)$.

The difference equations describing a discrete time system, equivalent to Eqs. (12.25) and (12.26), are

$$\mathbf{x}(n+1) = \mathbf{A}\mathbf{x}(n) + \mathbf{B}\mathbf{w}(n) \qquad \text{Eq. (12.38)}$$

$$\mathbf{y}(n) = \mathbf{C}\ \mathbf{x}(n) + \mathbf{D}\mathbf{w}(n) \qquad \text{Eq. (12.39)}$$

where n defines the discrete time nT and T is the sampling interval. All other vectors and matrices were defined earlier. The homogeneous solution to the system defined in Eq. (12.38), with initial condition $\mathbf{x}(n_0)$, is

$$\mathbf{x}(n) = \mathbf{A}^{n - n_0}\mathbf{x}(n_0) . \qquad \text{Eq. (12.40)}$$

In this case, the state transition matrix is an $n \times n$ matrix given by

$$\Phi(n, n_0) = \Phi(n - n_0) = \mathbf{A}^{n - n_0} . \qquad \text{Eq. (12.41)}$$

The following is the list of properties associated with the discrete transition matrix:

$$\Phi(n+1-n_0) = \mathbf{A}\Phi(n-n_0) \qquad \text{Eq. (12.42)}$$

$$\Phi(n_0-n_0) = \Phi(0) = \mathbf{I} \qquad \text{Eq. (12.43)}$$

$$\Phi(n_0+1-n_0) = \Phi(1) = \mathbf{A} \qquad \text{Eq. (12.44)}$$

$$\Phi(n_2-n_0) = \Phi(n_2-n_1)\Phi(n_1-n_0) \qquad \text{Eq. (12.45)}$$

$$\Phi(n_0-n_1) = \Phi^{-1}(n_1-n_0) \qquad \text{Eq. (12.46)}$$

$$\Phi(n_1-n_0) = \Phi(n_1)\Phi^{-1}(n_0) \qquad \text{Eq. (12.47)}$$

The solution to the general case (i.e., non-homogeneous system) is given by

$$\mathbf{x}(n) = \Phi(n-n_0)\mathbf{x}(n_0) + \sum_{m=n_0}^{n-1} \Phi(n-m-1)\mathbf{B}\mathbf{w}(m). \qquad \text{Eq. (12.48)}$$

It follows that the output is given by

$$\mathbf{y}(n) = \mathbf{C}\Phi(n-n_0)\mathbf{x}(n_0) + \sum_{m=n_0}^{n-1} \mathbf{C}\ \Phi(n-m-1)\mathbf{B}\mathbf{w}(m) + \mathbf{D}\mathbf{w}(n) \qquad \text{Eq. (12.49)}$$

where the system impulse response is given by

$$\mathbf{h}(n) = \sum_{m=n_0}^{n-1} \mathbf{C}\ \Phi(n-m-1)\mathbf{B}\underline{\delta}(m) + \mathbf{D}\underline{\delta}(n) \qquad \text{Eq. (12.50)}$$

where $\underline{\delta}$ is a vector.

Taking the Z-transform for Eqs. (12.38) and (12.39) yields

$$z\mathbf{x}(z) = \mathbf{A}\mathbf{x}(z) + \mathbf{B}\mathbf{w}(z) + z\mathbf{x}(0) \qquad \text{Eq. (12.51)}$$

$$\mathbf{y}(z) = \mathbf{C}\mathbf{x}(z) + \mathbf{D}\mathbf{w}(z). \qquad \text{Eq. (12.52)}$$

Manipulating Eqs. (12.51) and (12.52) yields

$$\mathbf{x}(z) = [z\mathbf{I}-\mathbf{A}]^{-1}\underline{B}\mathbf{w}(z) + [z\mathbf{I}-\mathbf{A}]^{-1}z\mathbf{x}(0) \qquad \text{Eq. (12.53)}$$

$$\mathbf{y}(z) = \{\mathbf{C}[z\mathbf{I}-\mathbf{A}]^{-1}\mathbf{B}+\mathbf{D}\}\mathbf{w}(z) + \mathbf{C}[z\mathbf{I}-\mathbf{A}]^{-1}z\mathbf{x}(0). \qquad \text{Eq. (12.54)}$$

It follows that the state transition matrix is

$$\Phi(z) = z[z\mathbf{I}-\mathbf{A}]^{-1} = [\mathbf{I}-z^{-1}\mathbf{A}]^{-1}, \qquad \text{Eq. (12.55)}$$

and the system impulse response in the z-domain is

$$\mathbf{h}(z) = \mathbf{C}\Phi(z)z^{-1}\mathbf{B}+\mathbf{D}. \qquad \text{Eq. (12.56)}$$

12.7. The LTI System of Interest

For the purpose of establishing the framework necessary for the Kalman filter development, consider the LTI system shown in Figure 12.18. This system (which is a special case of the system described in the previous section) can be described by the following first-order differential vector equations

$$\dot{\mathbf{x}}(t) = \mathbf{A}\ \mathbf{x}(t) + \mathbf{u}(t) \qquad \text{Eq. (12.57)}$$

$$\mathbf{y}(t) = \mathbf{G}\ \mathbf{x}(t) + \mathbf{v}(t) \qquad \text{Eq. (12.58)}$$

where $\mathbf{y}$ is the observable part of the system (i.e., output), $\mathbf{u}$ is a driving force, and $\mathbf{v}$ is the measurement noise. The matrices $\mathbf{A}$ and $\mathbf{G}$ vary depending on the system. The noise observation $\mathbf{v}$ is assumed to be uncorrelated. If the initial condition vector is $\mathbf{x}(t_0)$, then from Eq. (12.36) we get

$$\mathbf{x}(t) = \Phi(t-t_0)\mathbf{x}(t_0) + \int_{t_0}^{t} \Phi(t-\tau)\mathbf{u}(\tau)d\tau . \qquad \text{Eq. (12.59)}$$

The object (abstract) is observed only at discrete times determined by the system. These observation times are declared by discrete time nT where T is the sampling interval. Using the same notation adopted in the previous section, the discrete time representations of Eqs. (12.57) and (12.58) are

$$\mathbf{x}(n) = \mathbf{A}\ \mathbf{x}(n-1) + \mathbf{u}(n) \qquad \text{Eq. (12.60)}$$

$$\mathbf{y}(n) = \mathbf{G}\mathbf{x}(n) + \mathbf{v}(n) \ . \qquad \text{Eq. (12.61)}$$

The homogeneous solution to this system is given in Eq. (12.27) for continuous time, and in Eq. (12.40) for discrete time. The state transition matrix corresponding to this system can be obtained using Taylor series expansion of the vector $\mathbf{x}$. More precisely,

$$\begin{aligned} x &= x + T\dot{x} + \frac{T^2}{2!}\ddot{x} + \dots \\ \dot{x} &= \dot{x} + T\ddot{x} + \dots \\ \ddot{x} &= \ddot{x} + \dots \end{aligned} \qquad \text{Eq. (12.62)}$$

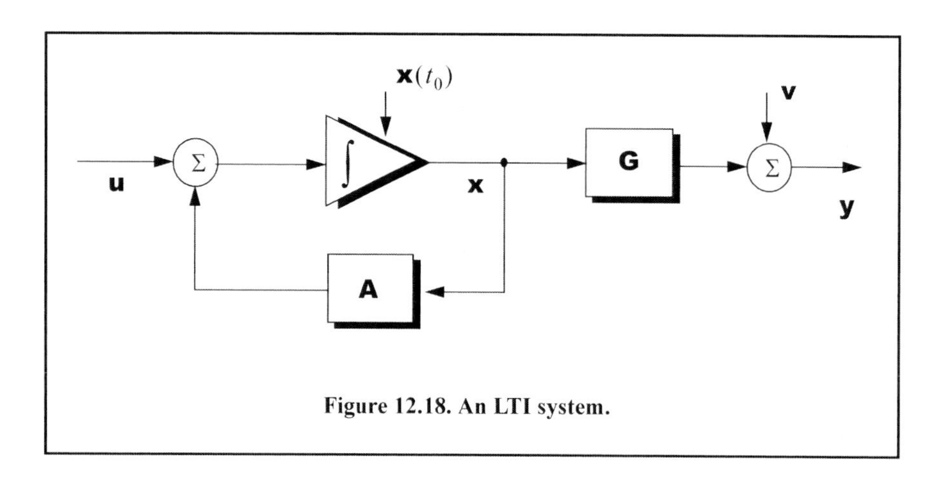

Figure 12.18. An LTI system.

It follows that the elements of the state transition matrix are defined by

$$\Phi[ij] = \begin{cases} T^{j-i} \div (j-i)! & 1 \le i,j \le n \\ 0 & j < i \end{cases}. \qquad \text{Eq. (12.63)}$$

Using matrix notation, the state transition matrix is then given by

$$\Phi = \begin{bmatrix} 1 & T & \frac{T^2}{2!} & \dots \\ 0 & 1 & T & \dots \\ 0 & 0 & 1 & \dots \\ \dots & \dots & \dots & \dots \end{bmatrix}. \qquad \text{Eq. (12.64)}$$

The matrix given in Eq. (12.64) is often called the Newtonian matrix.

12.8. Fixed-Gain Tracking Filters

This class of filters (or estimators) is also known as "Fixed-Coefficient" filters. The most common examples of this class of filters are the $\alpha\beta$ and $\alpha\beta\gamma$ filters and their variations. The $\alpha\beta$ and $\alpha\beta\gamma$ trackers are one-dimensional second- and third-order filters, respectively. They are equivalent to special cases of the one-dimensional Kalman filter. The general structure of this class of estimators is similar to that of the Kalman filter.

The standard $\alpha\beta\gamma$ filter provides smoothed and predicted data for target position, velocity (Doppler), and acceleration. It is a polynomial predictor/corrector linear recursive filter. This filter can reconstruct position, velocity, and constant acceleration based on position measurements. The $\alpha\beta\gamma$ filter can also provide a smoothed (corrected) estimate of the present position, which can be used in guidance and fire control operations.

Notation:

For the purpose of the discussion presented in the remainder of this chapter, the following notation is adopted: $x(n|m)$ represents the estimate during the nth sampling interval, using all data up to and including the mth sampling interval; y_n is the nth measured value; and e_n is the nth residual (error).

The fixed-gain filter equation is given by

$$\mathbf{x}(n|n) = \Phi \underline{x}(n-1|n-1) + \mathbf{K}[y_n - \mathbf{G}\Phi\mathbf{x}(n-1|n-1)]. \qquad \text{Eq. (12.65)}$$

Since the transition matrix assists in predicting the next state,

$$\mathbf{x}(n+1|n) = \Phi \underline{x}(n|n). \qquad \text{Eq. (12.66)}$$

Substituting Eq. (12.66) into Eq. (12.65) yields

$$\mathbf{x}(n|n) = \mathbf{x}(n|n-1) + \mathbf{K}[y_n - \mathbf{G}\mathbf{x}(n|n-1)]. \qquad \text{Eq. (12.67)}$$

The term enclosed within the brackets on the right-hand side of Eq. (12.67) is often called the residual (error), which is the difference between the measured input and predicted output. Eq. (12.67) means that the estimate of $\mathbf{x}(n)$ is the sum of the prediction and the weighted residual. The term $\mathbf{Gx}(n|n-1)$ represents the prediction state. In the case of the $\alpha\beta\gamma$ estimator, $\mathbf{G}$ is the row vector given by

$$\mathbf{G} = \begin{bmatrix} 1 & 0 & 0 & \ldots \end{bmatrix}, \qquad \text{Eq. (12.68)}$$

and the gain matrix $\mathbf{K}$ is given by

$$\mathbf{K} = \begin{bmatrix} \alpha \\ \beta/T \\ \gamma/T^2 \end{bmatrix}. \qquad \text{Eq. (12.69)}$$

One of the main objectives of a tracking filter is to decrease the effect of the noise observation on the measurement. For this purpose, the noise covariance matrix is calculated. More precisely, the noise covariance matrix is

$$\mathbf{C}(n|n) = E[(\mathbf{x}(n|n)\ )\mathbf{x}^t(n|n)] \qquad ;\ y_n = v_n \qquad \text{Eq. (12.70)}$$

where E indicates the expected value operator. Noise is assumed to be a zero mean random process with variance equal to σ_v^2. Additionally, noise measurements are assumed to be uncorrelated,

$$E[v_n v_m] = \begin{cases} \delta\sigma_v^2 & n = m \\ 0 & n \neq m \end{cases}. \qquad \text{Eq. (12.71)}$$

Eq. (12.65) can be written as

$$\mathbf{x}(n|n) = \mathbf{Ax}(n-1|n-1) + \mathbf{K}y_n \qquad \text{Eq. (12.72)}$$

where

$$\mathbf{A} = (\mathbf{I} - \mathbf{KG})\Phi. \qquad \text{Eq. (12.73)}$$

Substituting Eqs. (12.72) and (12.73) into Eq. (12.70) yields

$$\mathbf{C}(n|n) = E[(\mathbf{Ax}(n-1|n-1) + \mathbf{K}y_n)(\mathbf{Ax}(n-1|n-1) + \mathbf{K}y_n)^t]. \qquad \text{Eq. (12.74)}$$

Expanding the right-hand side of Eq. (12.74), and using Eq. (12.71), gives

$$\mathbf{C}(n|n) = \mathbf{AC}(n-1|n-1)\mathbf{A}^t + \mathbf{K}\sigma_v^2\mathbf{K}^t. \qquad \text{Eq. (12.75)}$$

Under the steady-state condition, Eq. (12.75) collapses to

$$\mathbf{C}(n|n) = \mathbf{ACA}^t + \mathbf{K}\sigma_v^2\mathbf{K}^t \qquad \text{Eq. (12.76)}$$

where $\mathbf{C}$ is the steady-state noise covariance matrix. In the steady-state,

$$\mathbf{C}(n|n) = \mathbf{C}(n-1|n-1) = \mathbf{C} \qquad \textit{for any } n. \qquad \text{Eq. (12.77)}$$

Several criteria can be used to establish the performance of the fixed-gain tracking filter. The most commonly used technique is to compute the Variance Reduction Ratio (VRR). The VRR is defined only when the input to the tracker is noise measurements. It follows that in the steady-state case, the VRR is the steady-state ratio of the output variance (autocovariance) to the input measurement variance.

In order to determine the stability of the tracker under consideration, consider the Z-transform for Eq. (12.72),

$$\mathbf{x}(z) = \mathbf{A}z^{-1}\mathbf{x}(z) + \mathbf{K}y_n(z). \qquad \text{Eq. (12.78)}$$

Rearranging Eq. (12.78) yields the following system transfer functions:

$$\mathbf{h}(z) = \frac{\mathbf{x}(z)}{y_n(z)} = (\mathbf{I} - \mathbf{A}z^{-1})^{-1}\mathbf{K} \qquad \text{Eq. (12.79)}$$

where $(\mathbf{I} - \mathbf{A}z^{-1})$ is called the characteristic matrix. Note that the system transfer functions can exist only when the characteristic matrix is a non-singular matrix. Additionally, the system is stable if and only if the roots of the characteristic equation are within the unit circle in the z-plane,

$$|(\mathbf{I} - \mathbf{A}z^{-1})| = 0. \qquad \text{Eq. (12.80)}$$

The filter's steady-state errors can be determined with the help of Figure 12.19. The error transfer function is

$$\mathbf{e}(z) = \frac{\mathbf{y}(z)}{1 + \mathbf{h}(z)}, \qquad \text{Eq. (12.81)}$$

and by using Abel's theorem, the steady-state error is

$$\mathbf{e}_\infty = \lim_{t \to \infty} \mathbf{e}(t) = \lim_{z \to 1}\left(\frac{z-1}{z}\right)\mathbf{e}(z). \qquad \text{Eq. (12.82)}$$

Substituting Eq. (12.82) into (12.81) yields

$$\mathbf{e}_\infty = \lim_{z \to 1} \frac{z-1}{z} \frac{\mathbf{y}(z)}{1 + \mathbf{h}(z)}. \qquad \text{Eq. (12.83)}$$

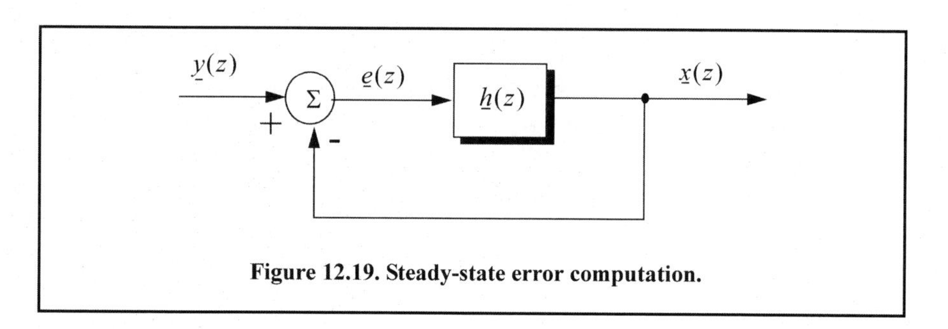

Figure 12.19. Steady-state error computation.

12.8.1. The $\alpha\beta$ Filter

The $\alpha\beta$ tracker produces, on the *nth* observation, smoothed estimates for position and velocity, and a predicted position for the $(n+1)th$ observation. Figure 12.20 shows an implementation of this filter. Note that the subscripts "p" and "s" are used to indicate, respectively, the predicated and smoothed values. The $\alpha\beta$ tracker can follow an input ramp (constant velocity) with no steady-state errors. However, a steady-state error will accumulate when constant acceleration is present in the input. Smoothing is done to reduce errors in the predicted position through adding a weighted difference between the measured and predicted values to the predicted position, as follows:

$$x_s(n) = x(n|n) = x_p(n) + \alpha(x_0(n) - x_p(n)) \qquad \text{Eq. (12.84)}$$

$$\dot{x}_s(n) = x'(n|n) = \dot{x}_s(n-1) + \frac{\beta}{T}(x_0(n) - x_p(n)) \; . \qquad \text{Eq. (12.85)}$$

x_0 is the position input samples. The predicted position is given by

$$x_p(n) = x_s(n|n-1) = x_s(n-1) + T\dot{x}_s(n-1) \; . \qquad \text{Eq. (12.86)}$$

The initialization process is defined by

$$x_s(1) = x_p(2) = x_0(1)$$

$$\dot{x}_s(1) = 0$$

$$\dot{x}_s(2) = \frac{x_0(2) - x_0(1)}{T} \; .$$

A general form for the covariance matrix was developed in the previous section, and is given in Eq. (12.75). In general, a second-order one-dimensional covariance matrix (in the context of the $\alpha\beta$ filter) can be written as

$$\mathbf{C}(n|n) = \begin{bmatrix} C_{xx} & C_{x\dot{x}} \\ C_{\dot{x}x} & C_{\dot{x}\dot{x}} \end{bmatrix} \qquad \text{Eq. (12.87)}$$

where, in general, C_{xy} is

$$C_{xy} = E\{xy^t\} \; . \qquad \text{Eq. (12.88)}$$

By inspection, the $\alpha\beta$ filter has

$$\mathbf{A} = \begin{bmatrix} 1-\alpha & (1-\alpha)T \\ -\beta/T & (1-\beta) \end{bmatrix} \qquad \text{Eq. (12.89)}$$

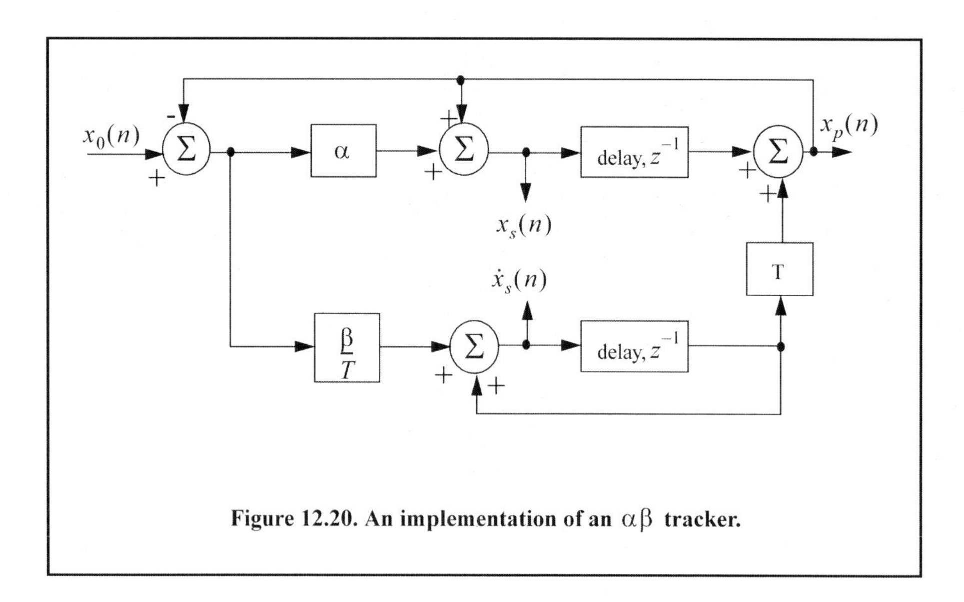

Figure 12.20. An implementation of an $\alpha\beta$ tracker.

$$\mathbf{K} = \begin{bmatrix} \alpha \\ \beta/T \end{bmatrix} \quad \text{Eq. (12.90)}$$

$$\mathbf{G} = \begin{bmatrix} 1 & 0 \end{bmatrix} \quad \text{Eq. (12.91)}$$

$$\Phi = \begin{bmatrix} 1 & T \\ 0 & 1 \end{bmatrix}. \quad \text{Eq. (12.92)}$$

Finally, using Eqs. (12.89) through (12.92) in Eq. (12.72) yields the steady-state noise covariance matrix,

$$\mathbf{C} = \frac{\sigma_v^2}{\alpha(4-2\alpha-\beta)} \begin{bmatrix} 2\alpha^2 - 3\alpha\beta + 2\beta & \dfrac{\beta(2\alpha-\beta)}{T} \\ \dfrac{\beta(2\alpha-\beta)}{T} & \dfrac{2\beta^2}{T^2} \end{bmatrix}. \quad \text{Eq. (12.93)}$$

It follows that the position and velocity VRR ratios are, respectively, given by

$$(VRR)_x = C_{xx}/\sigma_v^2 = \frac{2\alpha^2 - 3\alpha\beta + 2\beta}{\alpha(4-2\alpha-\beta)} \quad \text{Eq. (12.94)}$$

$$(VRR)_{\dot{x}} = C_{\dot{x}\dot{x}}/\sigma_v^2 = \frac{1}{T^2}\,\frac{2\beta^2}{\alpha(4-2\alpha-\beta)}. \quad \text{Eq. (12.95)}$$

The stability of the $\alpha\beta$ filter is determined from its system transfer functions. For this purpose, compute the roots for Eq. (12.80) with $\mathbf{A}$ from Eq. (12.89),

$$\left|\mathbf{I} - \mathbf{A}z^{-1}\right| = 1 - (2-\alpha-\beta)z^{-1} + (1-\alpha)z^{-2} = 0. \quad \text{Eq. (12.96)}$$

Solving Eq. (12.96) for z yields

$$z_{1,2} = 1 - \frac{\alpha+\beta}{2} \pm \frac{1}{2}\sqrt{(\alpha-\beta)^2 - 4\beta}, \quad \text{Eq. (12.97)}$$

and in order to guarantee stability,

$$\left|z_{1,2}\right| < 1. \quad \text{Eq. (12.98)}$$

Two cases are analyzed. First, $z_{1,2}$ are real. In this case,

$$\beta > 0 \quad ; \; \alpha > -\beta. \quad \text{Eq. (12.99)}$$

The second case is when the roots are complex; in this case we find

$$\alpha > 0. \quad \text{Eq. (12.100)}$$

The system transfer functions can be derived by using Eqs. (12.79), (12.89), and (12.90),

$$\begin{bmatrix} h_x(z) \\ h_{\dot{x}}(z) \end{bmatrix} = \frac{1}{z^2 - z(2-\alpha-\beta) + (1-\alpha)} \begin{bmatrix} \alpha z\left(z - \dfrac{(\alpha-\beta)}{\alpha}\right) \\ \dfrac{\beta z(z-1)}{T} \end{bmatrix}. \quad \text{Eq. (12.101)}$$

Up to this point all relevant relations concerning the $\alpha\beta$ filter were made with no regard to how to choose the gain coefficients (α and β). Before considering the methodology of selecting these coefficients, consider the main objective behind using this filter. The two-fold purpose of the $\alpha\beta$ tracker can be described as follows:

1. *The tracker must reduce the measurement noise as much as possible.*
2. *The filter must be able to track maneuvering targets, with as little residual (tracking error) as possible.*

The reduction of measurement noise is normally determined by the VRR ratios. However, the maneuverability performance of the filter depends heavily on the choice of the parameters α and β.

A special variation of the $\alpha\beta$ filter was developed by Benedict and Bordner[1] and is often referred to as the Benedict-Bordner filter. The main advantage of the Benedict-Bordner is reducing the transient errors associated with the $\alpha\beta$ tracker. This filter uses both the position and velocity VRR ratios as measures of performance. It computes the sum of the squared differences between the input (position) and the output when the input has a unit step velocity at time zero. Additionally, it computes the squared differences between the real velocity and the velocity output when the input is as described earlier. Both error differences are minimized when

$$\beta = \frac{\alpha^2}{2-\alpha}. \qquad \text{Eq. (12.102)}$$

In this case, the position and velocity VRR ratios are, respectively, given by

$$(VRR)_x = \frac{\alpha(6-5\alpha)}{\alpha^2 - 8\alpha + 8} \qquad \text{Eq. (12.103)}$$

$$(VRR)_{\dot{x}} = \frac{2}{T^2} \frac{\alpha^3/(2-\alpha)}{\alpha^2 - 8\alpha + 8}. \qquad \text{Eq. (12.104)}$$

Another important sub-class of the $\alpha\beta$ tracker is the critically damped filter, often called the fading memory filter. In this case, the filter coefficients are chosen on the basis of a smoothing factor ξ, where $0 \le \xi \le 1$. The gain coefficients are given by

$$\alpha = 1 - \xi^2 \qquad \text{Eq. (12.105)}$$

$$\beta = (1-\xi)^2. \qquad \text{Eq. (12.106)}$$

Heavy smoothing means $\xi \to 1$ and little smoothing means $\xi \to 0$. The elements of the covariance matrix for a fading memory filter are

$$C_{xx} = \frac{1-\xi}{(1+\xi)^3} (1 + 4\xi + 5\xi^2)\ \sigma_v^2 \qquad \text{Eq. (12.107)}$$

$$C_{x\dot{x}} = C_{\dot{x}x} = \frac{1}{T} \frac{1-\xi}{(1+\xi)^3} (1 + 2\xi + 3\xi^2)\ \sigma_v^2 \qquad \text{Eq. (12.108)}$$

1. Benedict, T. R. and Bordner, G. W., Synthesis of an Optimal Set of Radar Track-While-Scan Smoothing Equations. *IRE Transaction on Automatic Control, AC-7*. July 1962, pp. 27-32.

$$C_{\ddot{x}\ddot{x}} = \frac{2}{T^2}\frac{1-\xi}{(1+\xi)^3}(1-\xi)^2\sigma_v^2 \,.$$ Eq. (12.109)

12.8.2. The $\alpha\beta\gamma$ Filter

The $\alpha\beta\gamma$ tracker produces, for the *nth* observation, smoothed estimates of position, velocity, and acceleration. It also produces the predicted position and velocity for the $(n+1)th$ observation. An implementation of the $\alpha\beta\gamma$ tracker is shown in Figure 12.21. The $\alpha\beta\gamma$ tracker will follow an input whose acceleration is constant with no steady-state errors. Again, in order to reduce the error at the output of the tracker, a weighted difference between the measured and predicted values is used in estimating the smoothed position, velocity, and acceleration as follows:

$$x_s(n) = x_p(n) + \alpha(x_0(n) - x_p(n))$$ Eq. (12.110)

$$\dot{x}_s(n) = \dot{x}_s(n-1) + T\ddot{x}_s(n-1) + \frac{\beta}{T}(x_0(n) - x_p(n))$$ Eq. (12.111)

$$\ddot{x}_s(n) = \ddot{x}_s(n-1) + \frac{2\gamma}{T^2}(x_0(n) - x_p(n))$$ Eq. (12.112)

$$x_p(n+1) = x_s(n) + T\,\dot{x}_s(n) + \frac{T^2}{2}\,\ddot{x}_s(n) \,,$$ Eq. (12.113)

and the initialization process is

$$x_s(1) = x_p(2) = x_0(1)$$

$$\dot{x}_s(1) = \ddot{x}_s(1) = \ddot{x}_s(2) = 0$$

$$\dot{x}_s(2) = \frac{x_0(2) - x_0(1)}{T}$$

$$\ddot{x}_s(3) = \frac{x_0(3) + x_0(1) - 2x_0(2)}{T^2} \,.$$

Using Eq. (12.63), the state transition matrix for the $\alpha\beta\gamma$ filter is

$$\Phi = \begin{bmatrix} 1 & T & \frac{T^2}{2} \\ 0 & 1 & T \\ 0 & 0 & 1 \end{bmatrix} .$$ Eq. (12.114)

The covariance matrix (which is symmetric) can be computed from Eq. (12.76). For this purpose, note that

$$\mathbf{K} = \begin{bmatrix} \alpha \\ \beta/T \\ \gamma/T^2 \end{bmatrix}$$ Eq. (12.115)

$$\mathbf{G} = \begin{bmatrix} 1 & 0 & 0 \end{bmatrix}$$ Eq. (12.116)

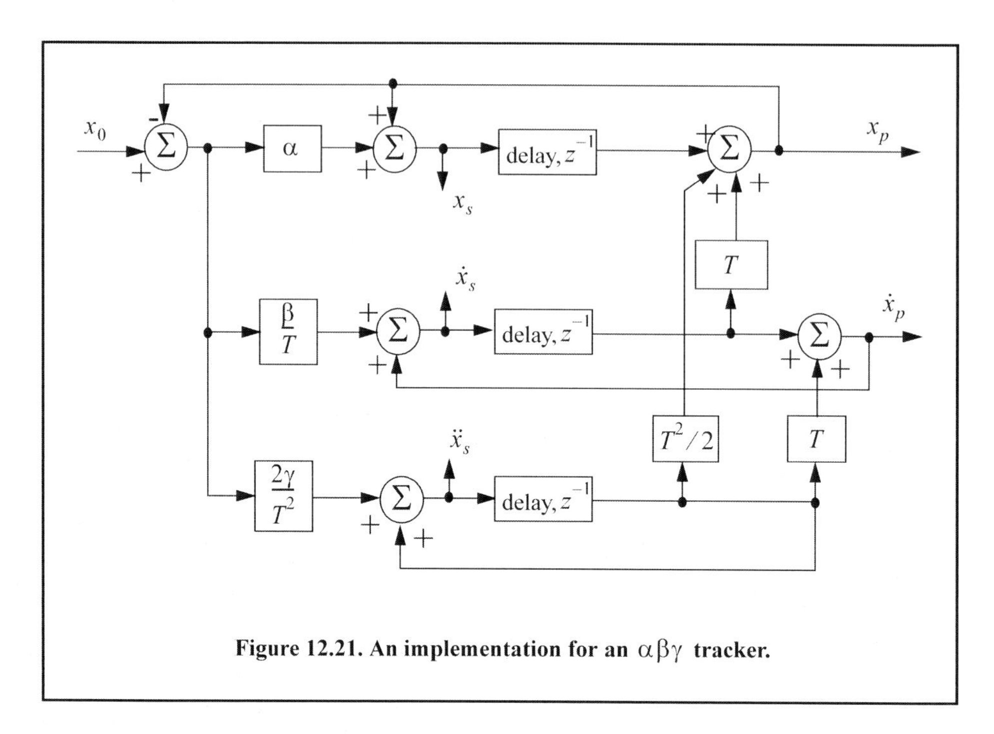

Figure 12.21. An implementation for an $\alpha\beta\gamma$ tracker.

and

$$\mathbf{A} = (\mathbf{I} - \mathbf{KG})\Phi = \begin{bmatrix} 1-\alpha & (1-\alpha)T & (1-\alpha)T^2/2 \\ -\beta/T & -\beta+1 & (1-\beta/2)T \\ -\gamma/T^2 & -\gamma/T & (1-\gamma/2) \end{bmatrix} . \qquad \text{Eq. (12.117)}$$

Substituting Eq. (12.117) into (12.76) and collecting terms, the VRR ratios are computed as

$$(VRR)_x = \frac{2\beta(2\alpha^2 + 2\beta - 3\alpha\beta) - \alpha\gamma(4 - 2\alpha - \beta)}{(4 - 2\alpha - \beta)(2\alpha\beta + \alpha\gamma - 2\gamma)} \qquad \text{Eq. (12.118)}$$

$$(VRR)_{\dot{x}} = \frac{4\beta^3 - 4\beta^2\gamma + 2\gamma^2(2-\alpha)}{T^2(4 - 2\alpha - \beta)(2\alpha\beta + \alpha\gamma - 2\gamma)} \qquad \text{Eq. (12.119)}$$

$$(VRR)_{\ddot{x}} = \frac{4\beta\gamma^2}{T^4(4 - 2\alpha - \beta)(2\alpha\beta + \alpha\gamma - 2\gamma)} . \qquad \text{Eq. (12.120)}$$

As in the case of any discrete time system, this filter will be stable if and only if all of its poles fall within the unit circle in the z-plane.

The $\alpha\beta\gamma$ characteristic equation is computed by setting

$$|\mathbf{I} - \mathbf{A}z^{-1}| = 0 . \qquad \text{Eq. (12.121)}$$

Substituting Eq. (12.117) into (12.121) and collecting terms yield the following characteristic function:

$$f(z) = z^3 + (-3\alpha + \beta + \gamma)z^2 + (3 - \beta - 2\alpha + \gamma)z - (1 - \alpha) . \qquad \text{Eq. (12.122)}$$

The $\alpha\beta\gamma$ becomes a Benedict-Bordner filter when

$$2\beta - \alpha\left(\alpha + \beta + \frac{\gamma}{2}\right) = 0 . \qquad \text{Eq. (12.123)}$$

Note that for $\gamma = 0$, Eq. (12.123) reduces to Eq. (12.102). For a critically damped filter the gain coefficients are

$$\alpha = 1 - \xi^3 \qquad \text{Eq. (12.124)}$$

$$\beta = 1.5(1 - \xi^2)(1 - \xi) = 1.5(1 - \xi)^2(1 + \xi) \qquad \text{Eq. (12.125)}$$

$$\gamma = (1 - \xi)^3 . \qquad \text{Eq. (12.126)}$$

Note that heavy smoothing takes place when $\xi \rightarrow 1$, while $\xi = 0$ means that no smoothing is present.

To illustrate the behavior of the $\alpha\beta\gamma$ filter, consider the input truth signals shown in Figures 12.22 and 12.23. Figure 12.22 assumes an input with lazy maneuvering, while Figure 12.23 assumes an aggressive maneuvering case. Figures 12.24 and 12.25 show the residual error and predicted position corresponding to Figure 12.22 assuming the cases: heavy smoothing and little smoothing with and without noise. The noise is white Gaussian with zero mean and variance of $\sigma_v^2 = 0.05$. Figures 12.26 and 12.27 show the residual error and predicted position corresponding to Figure 12.23 with and without noise.

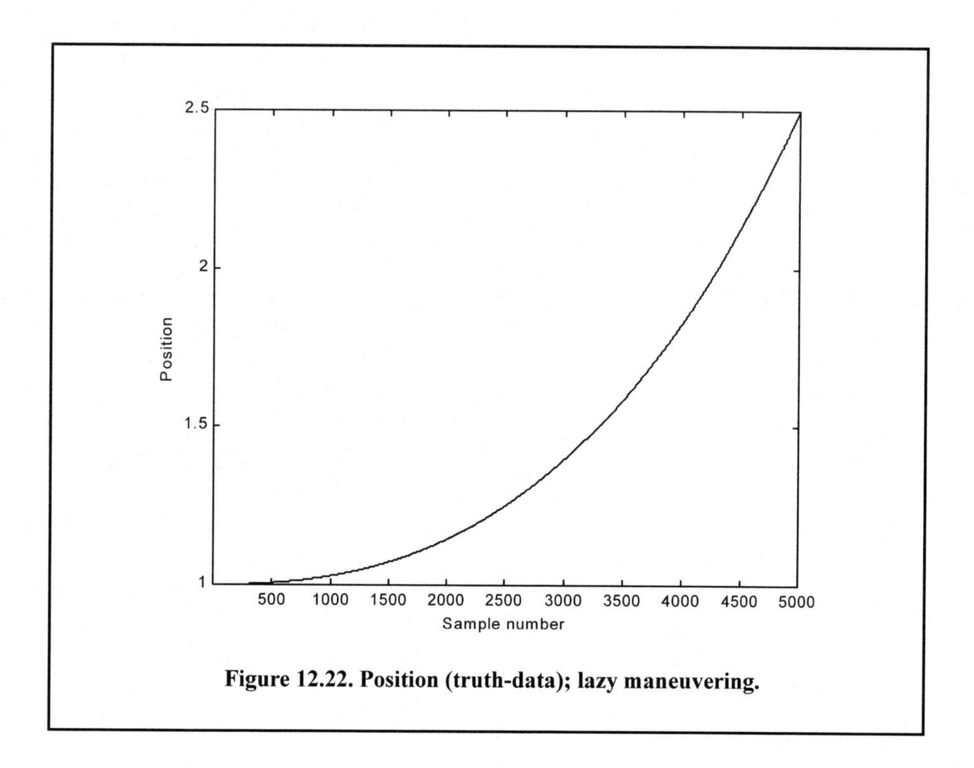

Figure 12.22. Position (truth-data); lazy maneuvering.

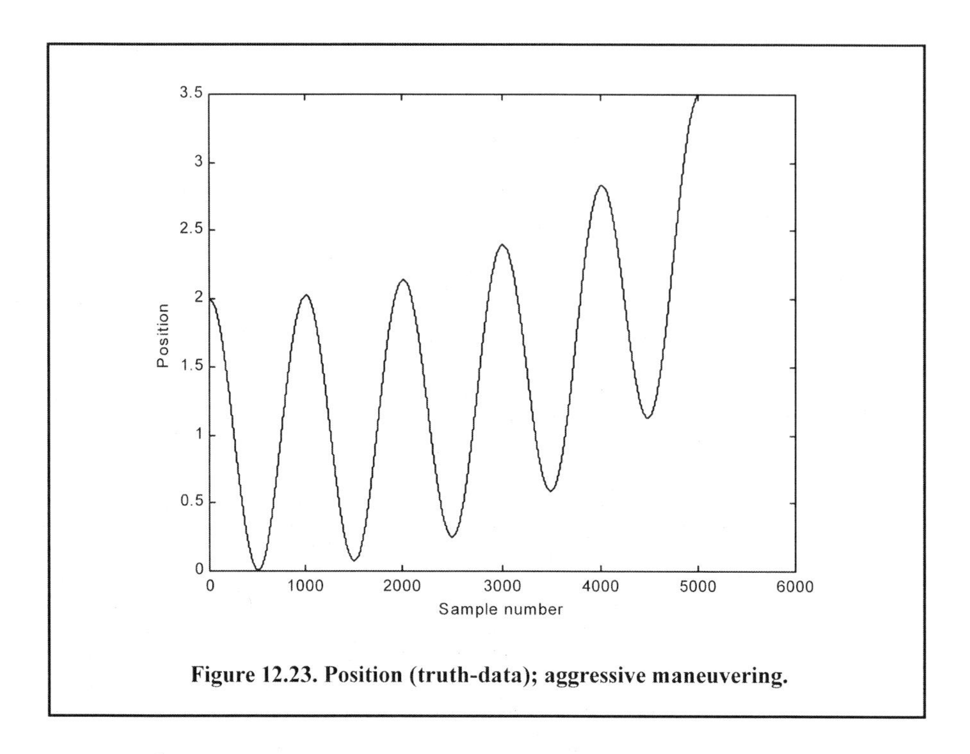

Figure 12.23. Position (truth-data); aggressive maneuvering.

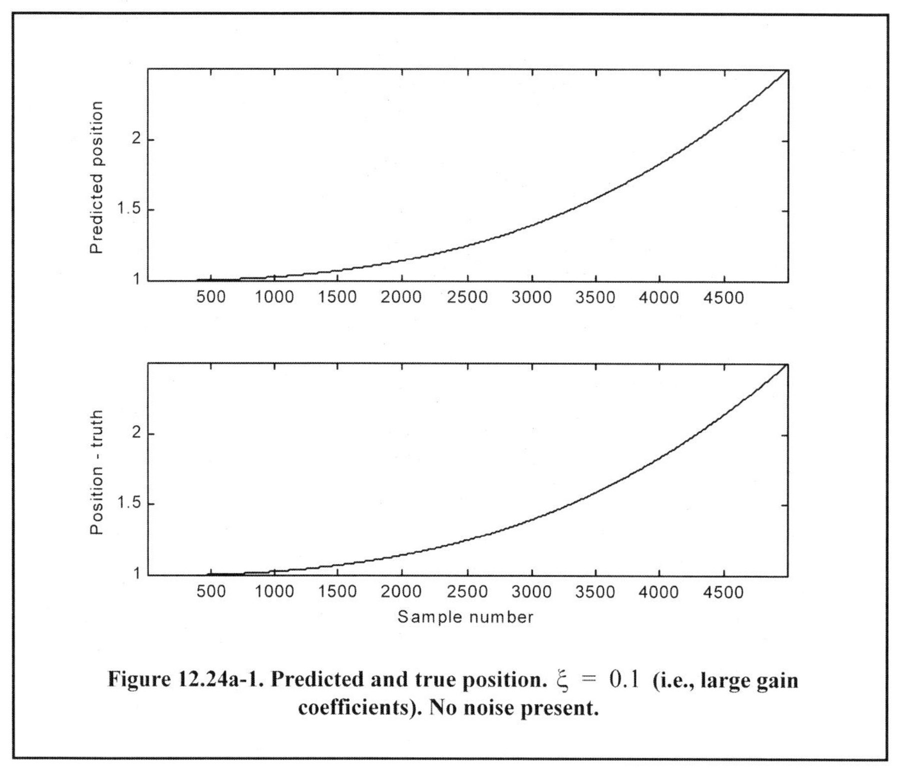

Figure 12.24a-1. Predicted and true position. ξ = 0.1 (i.e., large gain coefficients). No noise present.

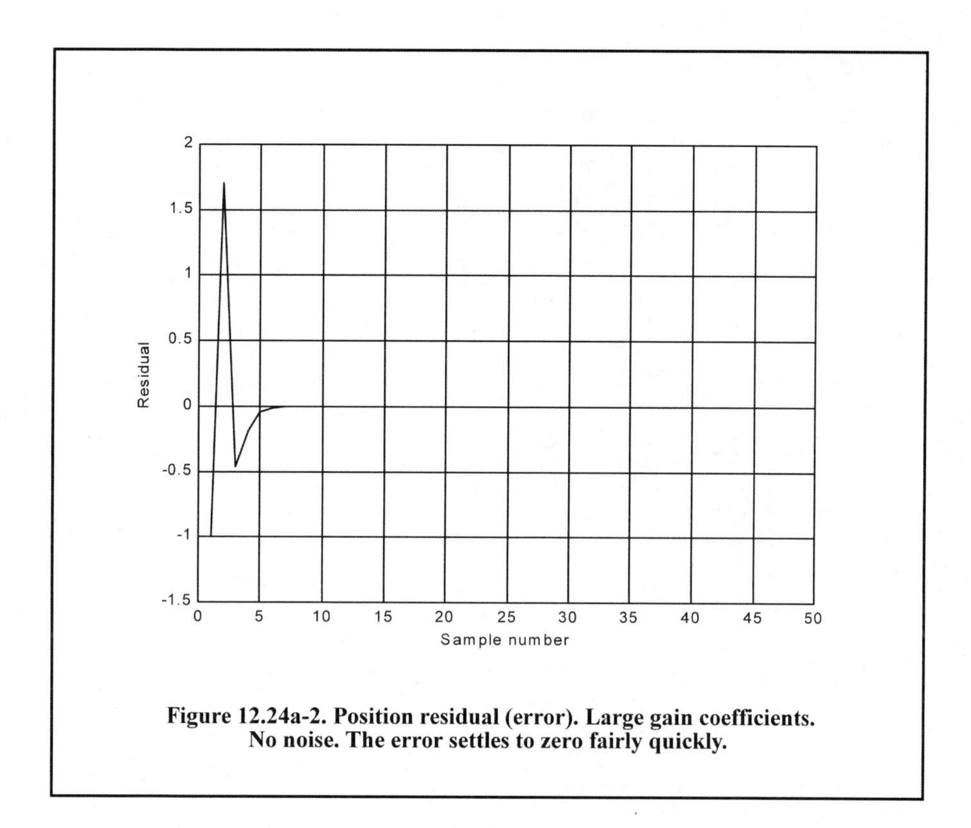

Figure 12.24a-2. Position residual (error). Large gain coefficients. No noise. The error settles to zero fairly quickly.

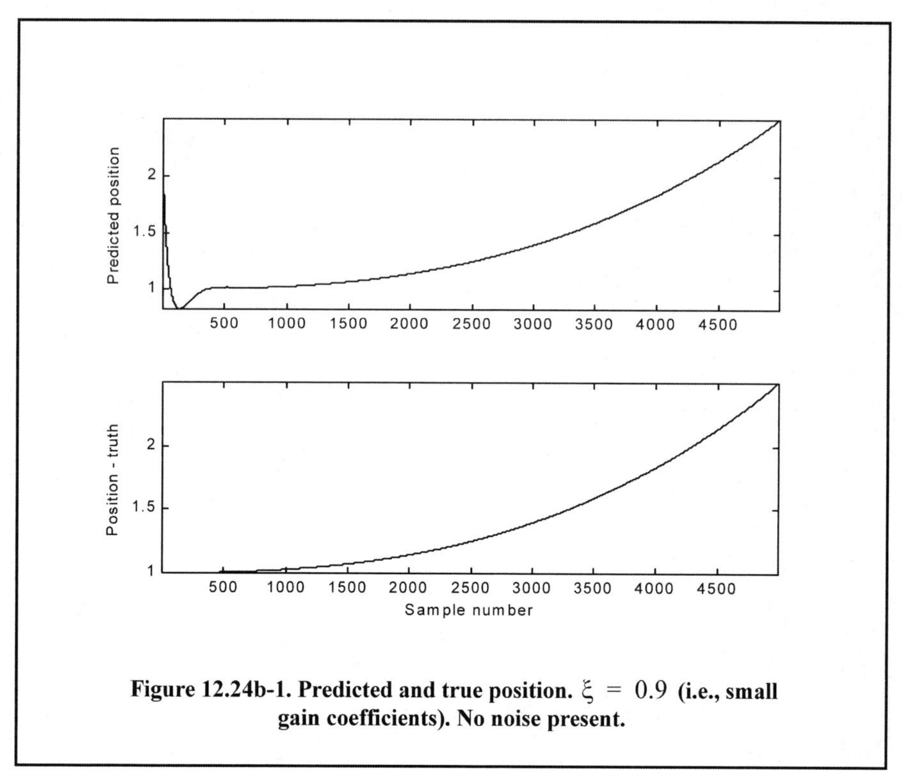

Figure 12.24b-1. Predicted and true position. $\xi = 0.9$ **(i.e., small gain coefficients). No noise present.**

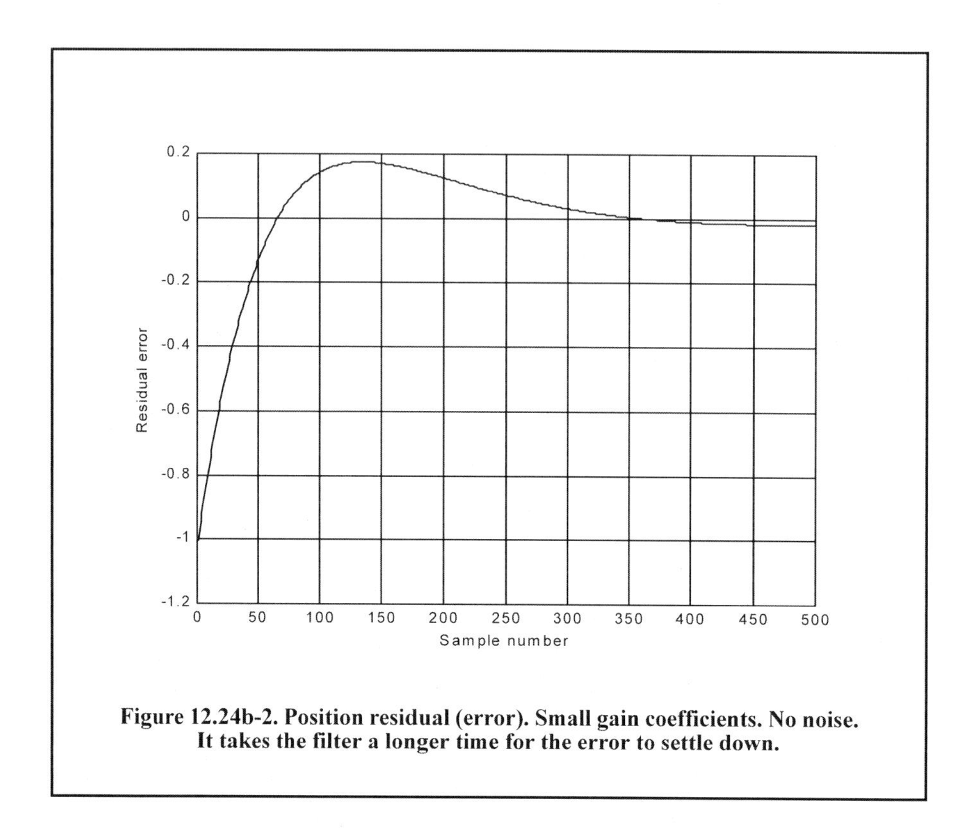

Figure 12.24b-2. Position residual (error). Small gain coefficients. No noise. It takes the filter a longer time for the error to settle down.

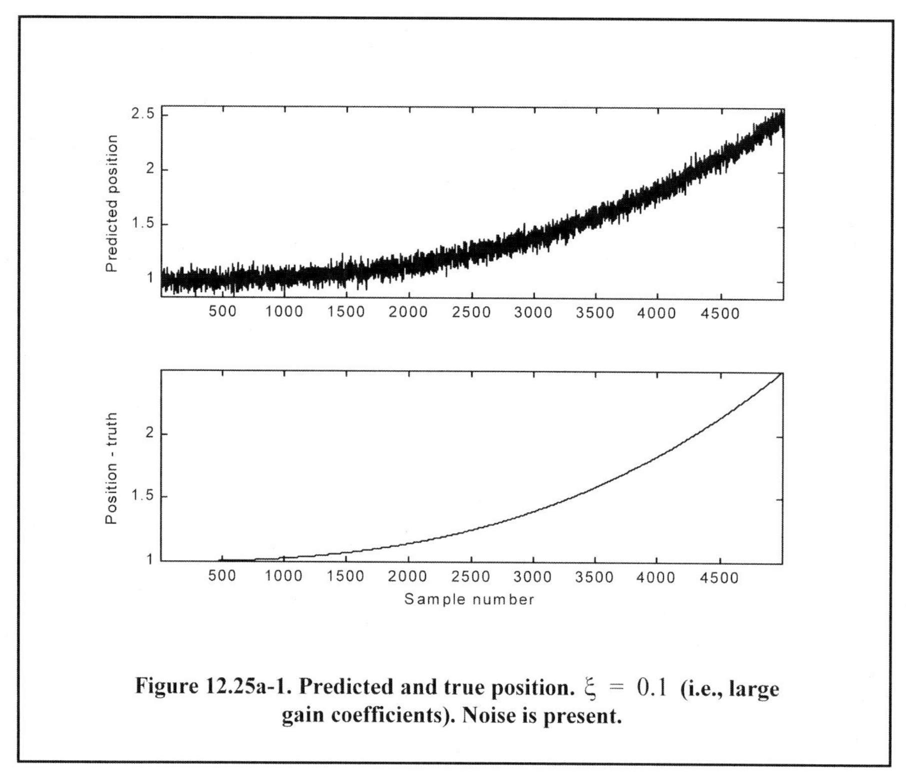

Figure 12.25a-1. Predicted and true position. $\xi = 0.1$ **(i.e., large gain coefficients). Noise is present.**

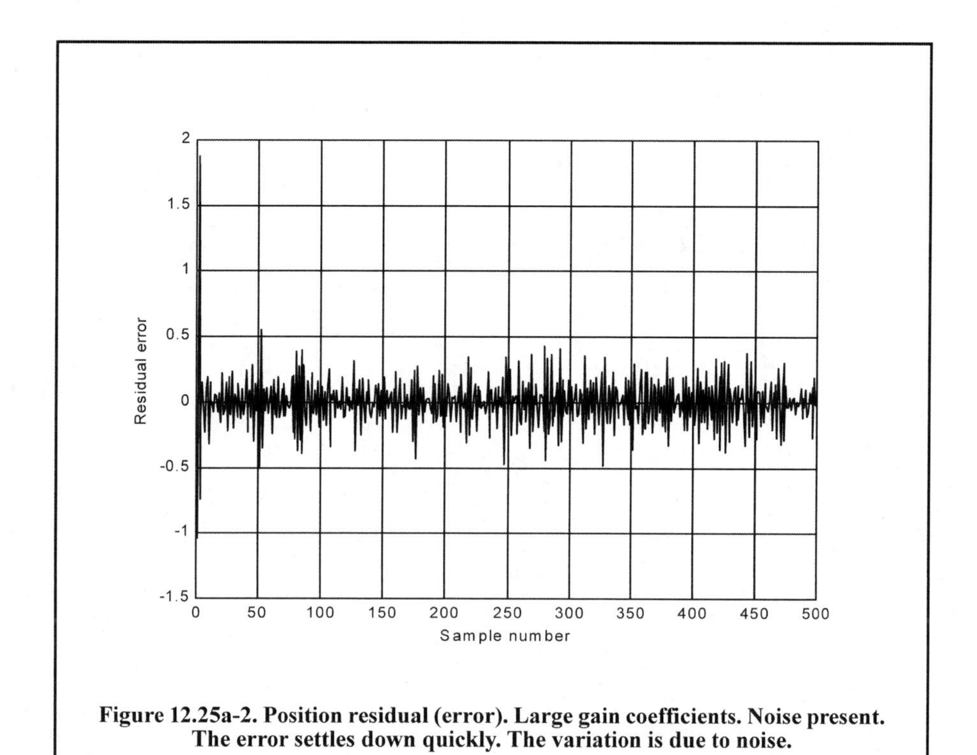

Figure 12.25a-2. Position residual (error). Large gain coefficients. Noise present. The error settles down quickly. The variation is due to noise.

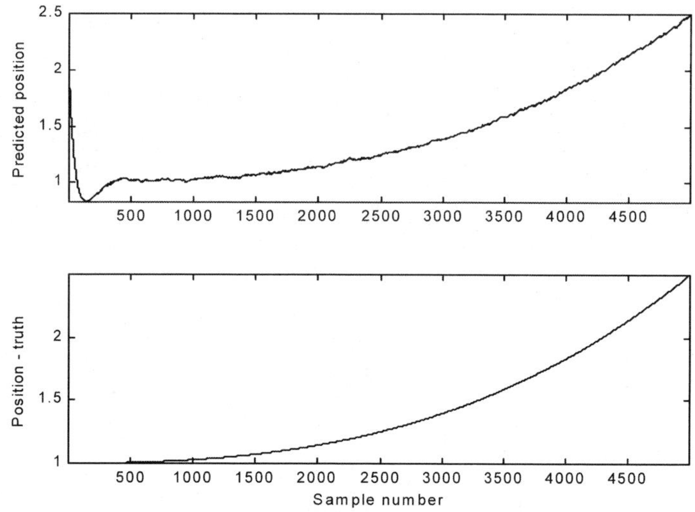

Predicted position
2.5
2
1.5
1
500 1000 1500 2000 2500 3000 3500 4000 4500

Position - truth
2
1.5
1
500 1000 1500 2000 2500 3000 3500 4000 4500
Sample number

Figure 12.25b-1. Predicted and true position. $\xi = 0.9$ (i.e., small gain coefficients). Noise is present.

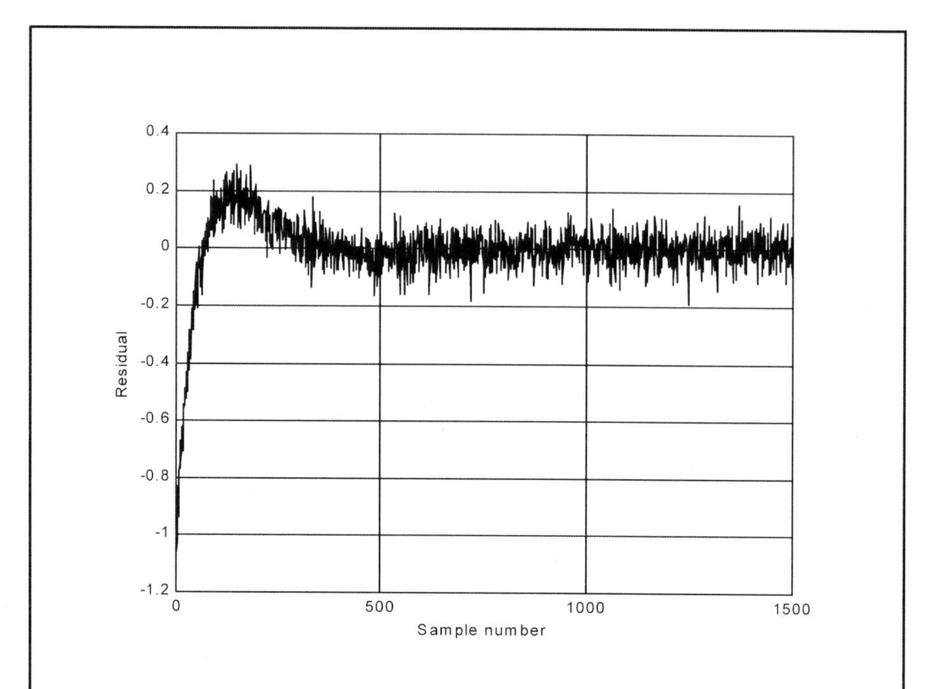

Figure 12.25b-2. Position residual (error). Small gain coefficients. Noise present. The error requires more time before settling down. The variation is due to noise.

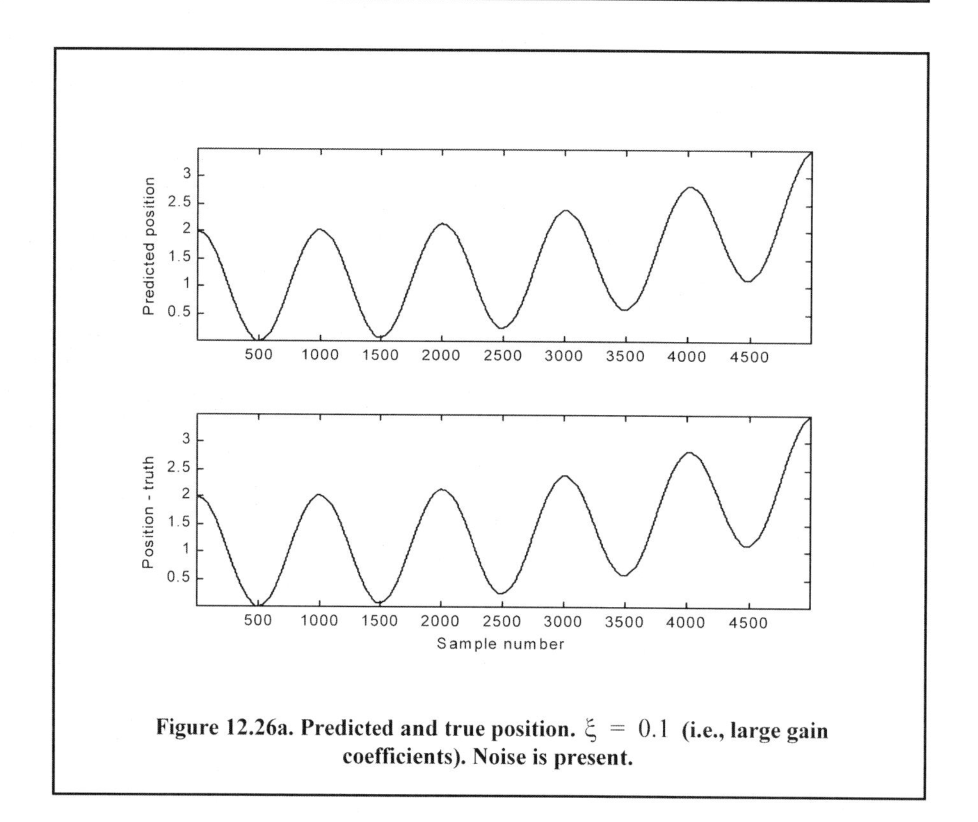

Figure 12.26a. Predicted and true position. $\xi = 0.1$ **(i.e., large gain coefficients). Noise is present.**

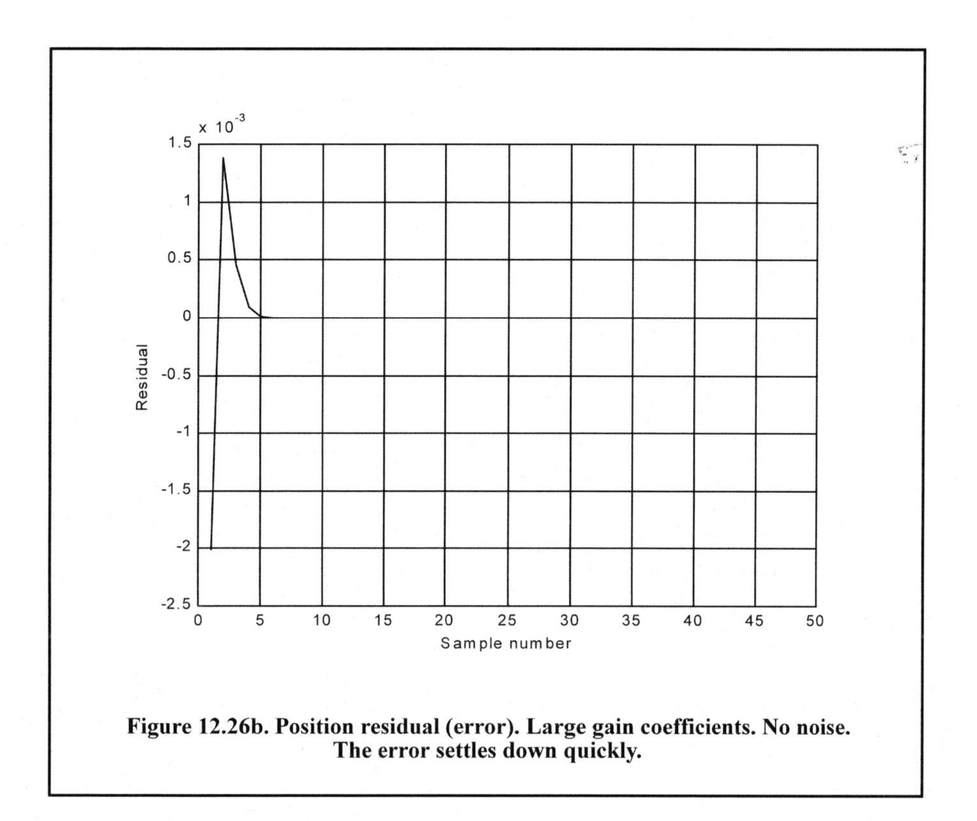

Figure 12.26b. Position residual (error). Large gain coefficients. No noise. The error settles down quickly.

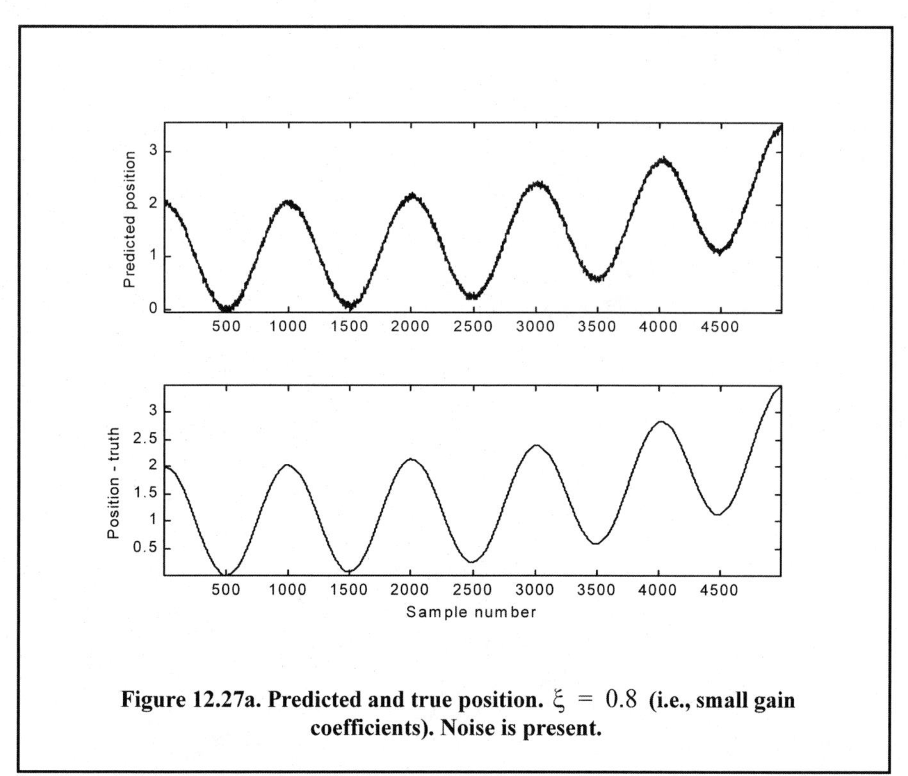

Figure 12.27a. Predicted and true position. $\xi = 0.8$ (i.e., small gain coefficients). Noise is present.

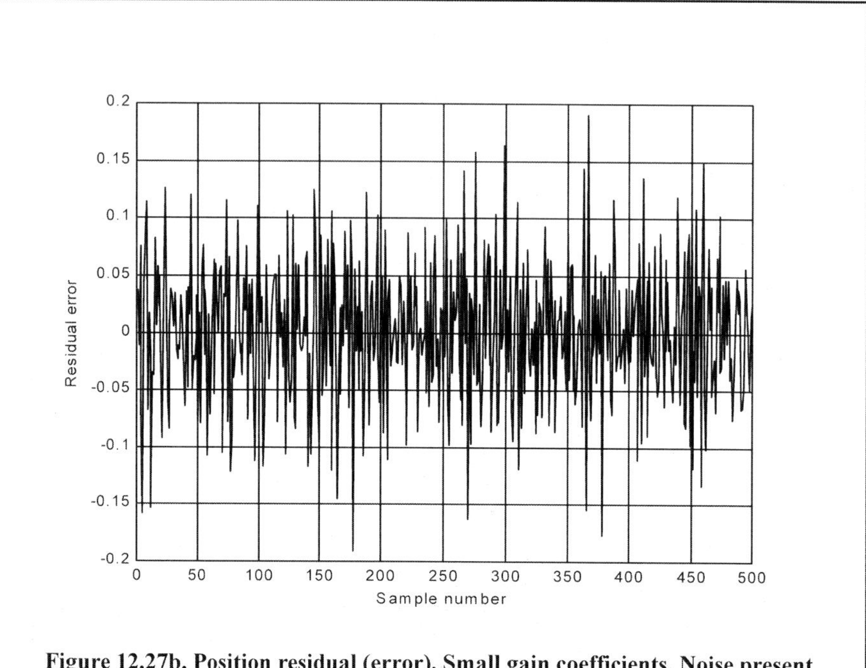

Figure 12.27b. Position residual (error). Small gain coefficients. Noise present. The error stays fairly large; however, its average is around zero. The variation is due to noise.

12.9. The Kalman Filter

The Kalman filter is a linear estimator that minimizes the mean squared error as long as the target dynamics are modeled accurately. All other recursive filters, such as the $\alpha\beta\gamma$ and the Benedict-Bordner filters, are special cases of the general solution provided by the Kalman filter for the mean squared estimation problem. Additionally, the Kalman filter has the following advantages:

1. *The gain coefficients are computed dynamically. This means that the same filter can be used for a variety of maneuvering target environments.*
2. *The Kalman filter gain computation adapts to varying detection histories, including missed detections.*
3. *The Kalman filter provides an accurate measure of the covariance matrix. This allows for better implementation of the gating and association processes.*
4. *The Kalman filter makes it possible to partially compensate for the effects of mis-correlation and mis-association.*

Many derivations of the Kalman filter exist in the literature; only results are provided in this chapter. Figure 12.28 shows a block diagram for the Kalman filter. The Kalman filter equations can be deduced from Figure 12.28. The filtering equation is

$$\mathbf{x}(n|n) = \mathbf{x}_s(n) = \mathbf{x}(n|n-1) + \mathbf{K}(n)[\mathbf{y}(n) - \mathbf{G}\mathbf{x}(n|n-1)] . \qquad \text{Eq. (12.127)}$$

The measurement vector is

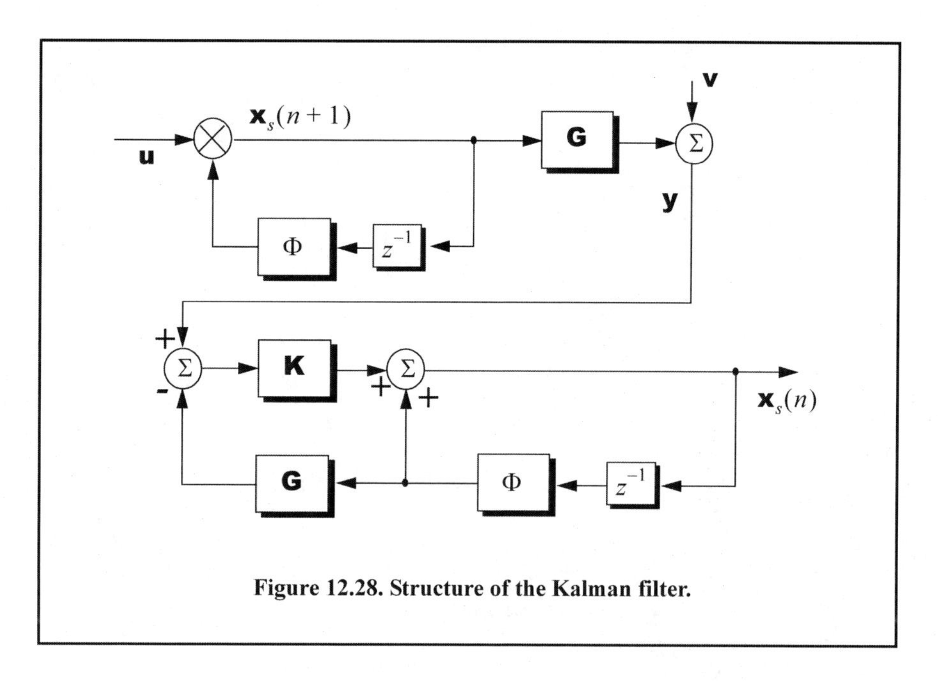

Figure 12.28. Structure of the Kalman filter.

$$\mathbf{y}(n) = \mathbf{G}\mathbf{x}(n) + \mathbf{v}(n) \qquad \text{Eq. (12.128)}$$

where $\mathbf{y}(n)$ is zero mean, white Gaussian noise with covariance $\Re_c$,

$$\Re_c = E\{\mathbf{y}(n)\ \mathbf{y}^t(n)\}\,. \qquad \text{Eq. (12.129)}$$

The gain (weight) vector is dynamically computed as

$$\mathbf{K}(n) = \mathbf{P}(n|n-1)\mathbf{G}^t[\mathbf{G}\mathbf{P}(n|n-1)\mathbf{G}^t + \Re_c]^{-1} \qquad \text{Eq. (12.130)}$$

where the measurement noise matrix $\mathbf{P}$ represents the predictor covariance matrix, and is equal to

$$\mathbf{P}(n+1|n) = E\{\mathbf{x}_s(n+1)\mathbf{x}^*{}_s(n)\} = \Phi\mathbf{P}(n|n)\Phi^t + \mathbf{Q} \qquad \text{Eq. (12.131)}$$

where $\mathbf{Q}$ is the covariance matrix for the input $\mathbf{u}$,

$$\mathbf{Q} = E\{\mathbf{u}(n)\ \mathbf{u}^t(n)\}\,. \qquad \text{Eq. (12.132)}$$

The corrector equation (covariance of the smoothed estimate) is

$$\mathbf{P}(n|n) = [\mathbf{I} - \mathbf{K}(n)\mathbf{G}]\mathbf{P}(n|n-1)\,. \qquad \text{Eq. (12.133)}$$

Finally, the predictor equation is

$$\mathbf{x}(n+1|n) = \Phi\mathbf{x}(n|n)\,. \qquad \text{Eq. (12.134)}$$

12.9.1. Relationship between Kalman and $\alpha\beta\gamma$ Filters

The relationship between the Kalman filter and the $\alpha\beta\gamma$ filters can be easily obtained by using the appropriate state transition matrix Φ, and gain vector $\mathbf{K}$ corresponding to the $\alpha\beta\gamma$. Thus,

$$\begin{bmatrix} x(n|n) \\ \dot{x}(n|n) \\ \ddot{x}(n|n) \end{bmatrix} = \begin{bmatrix} x(n|n-1) \\ \dot{x}(n|n-1) \\ \ddot{x}(n|n-1) \end{bmatrix} + \begin{bmatrix} k_1(n) \\ k_2(n) \\ k_3(n) \end{bmatrix} [x_0(n) - x(n|n-1)] \qquad \text{Eq. (12.135)}$$

with (see Figure 12.21)

$$x(n|n-1) = x_s(n-1) + T\ \dot{x}_s(n-1) + \frac{T^2}{2}\ \ddot{x}_s(n-1) \qquad \text{Eq. (12.136)}$$

$$\dot{x}(n|n-1) = \dot{x}_s(n-1) + T\ \ddot{x}_s(n-1) \qquad \text{Eq. (12.137)}$$

$$\ddot{x}(n|n-1) = \ddot{x}_s(n-1). \qquad \text{Eq. (12.138)}$$

Comparing the previous three equations with the $\alpha\beta\gamma$ filter equations yields,

$$\begin{bmatrix} \alpha \\ \frac{\beta}{T} \\ \frac{\gamma}{T^2} \end{bmatrix} = \begin{bmatrix} k_1 \\ k_2 \\ k_3 \end{bmatrix}. \qquad \text{Eq. (12.139)}$$

Additionally, the covariance matrix elements are related to the gain coefficients by

$$\begin{bmatrix} k_1 \\ k_2 \\ k_3 \end{bmatrix} = \frac{1}{C_{11} + \sigma_v^2} \begin{bmatrix} C_{11} \\ C_{12} \\ C_{13} \end{bmatrix}. \qquad \text{Eq. (12.140)}$$

Eq. (12.140) indicates that the first gain coefficient depends on the estimation error variance of the total residual variance, while the other two gain coefficients are calculated through the covariances between the second and third states and the first observed state. To illustrate the behavior of the Kalman filter, consider the input signals shown in Figures 12.22 and 12.23. Figures 12.29 and 12.30 show the residual error and predicted position corresponding to Figures 12.22 and 12.23.

12.10. Radar Measurement Errors

Earlier, we learned that radar systems extract target information (attributes) such as target RCS, range, range rate, and angular position relative to the radar location by processing the radar echoes. To this end, a radar echo (or simply a radar measurement) passes through three distinct phases. They are, measurement availability, measurement quality, and measurement accuracy. Each phase is typically characterized by the SNR value at the output of the radar receiver (matched filter). Initially, a measurement availability occurs at low SNR values, where the received signal is large enough to surpass a pre-determined threshold value (i.e., detection). Next, when the measurement quality is good enough, the radar data processor establishes distinct track files (i.e, tracking occurs) for each detected target. Finally, when the measured SNR value becomes large the radar is said to have enough SNR to achieve ceratin measurement accuracy. Typically, range, range rate, and angle measurement accuracy values are pre-determined in the radar design process.

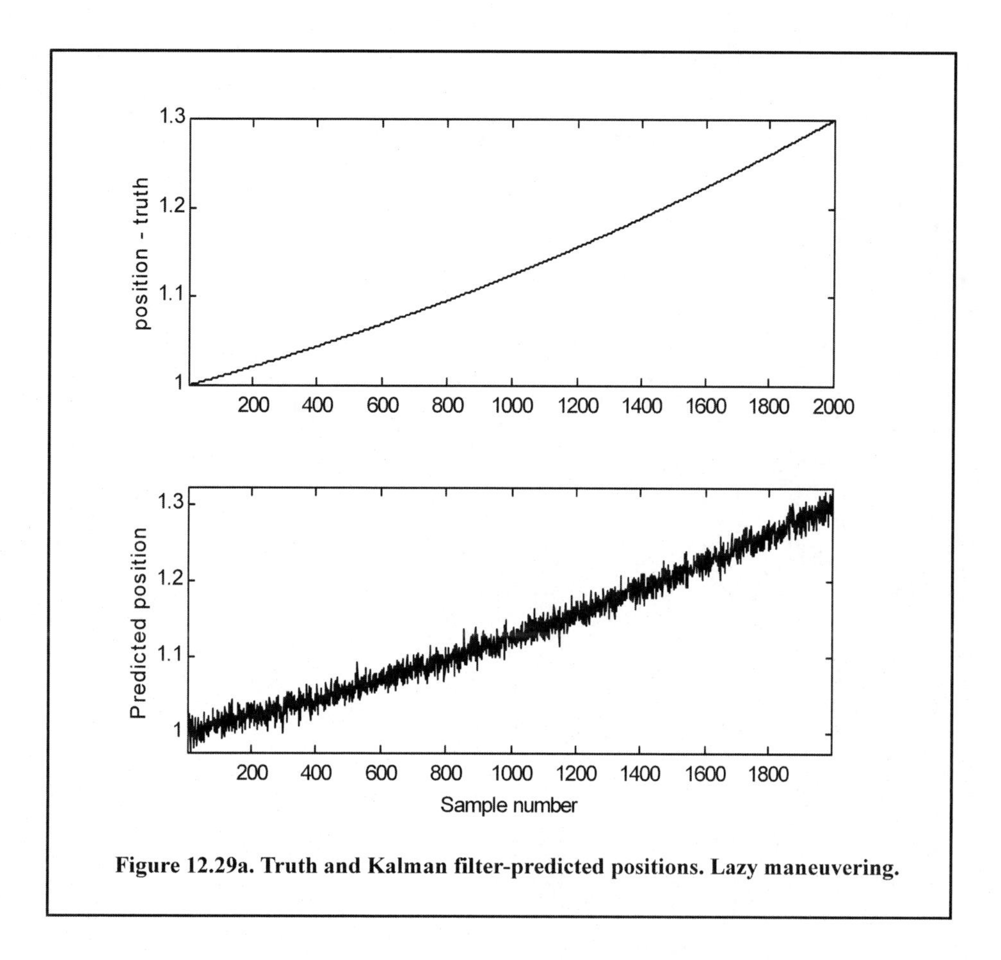

Figure 12.29a. Truth and Kalman filter-predicted positions. Lazy maneuvering.

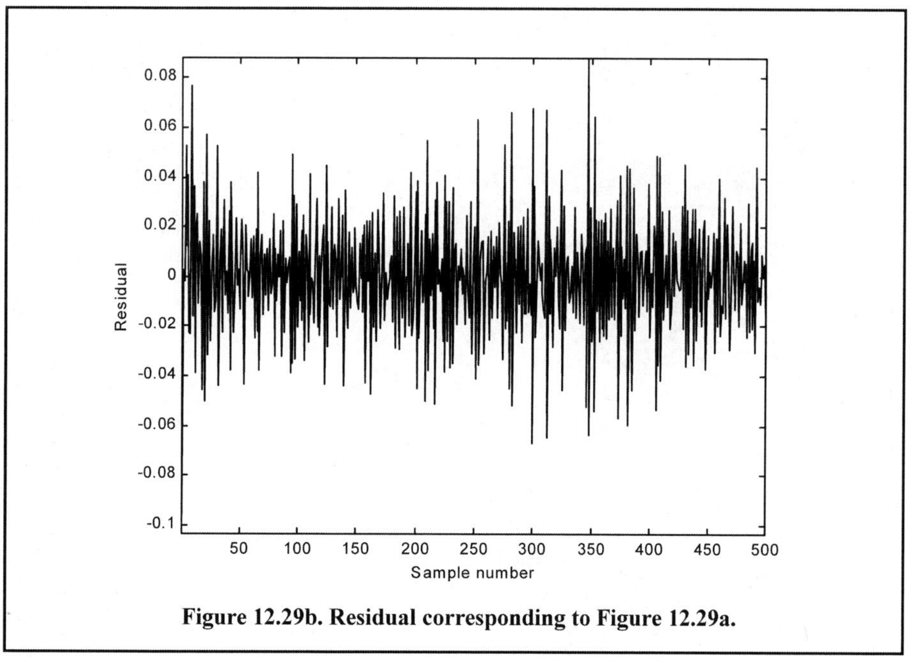

Figure 12.29b. Residual corresponding to Figure 12.29a.

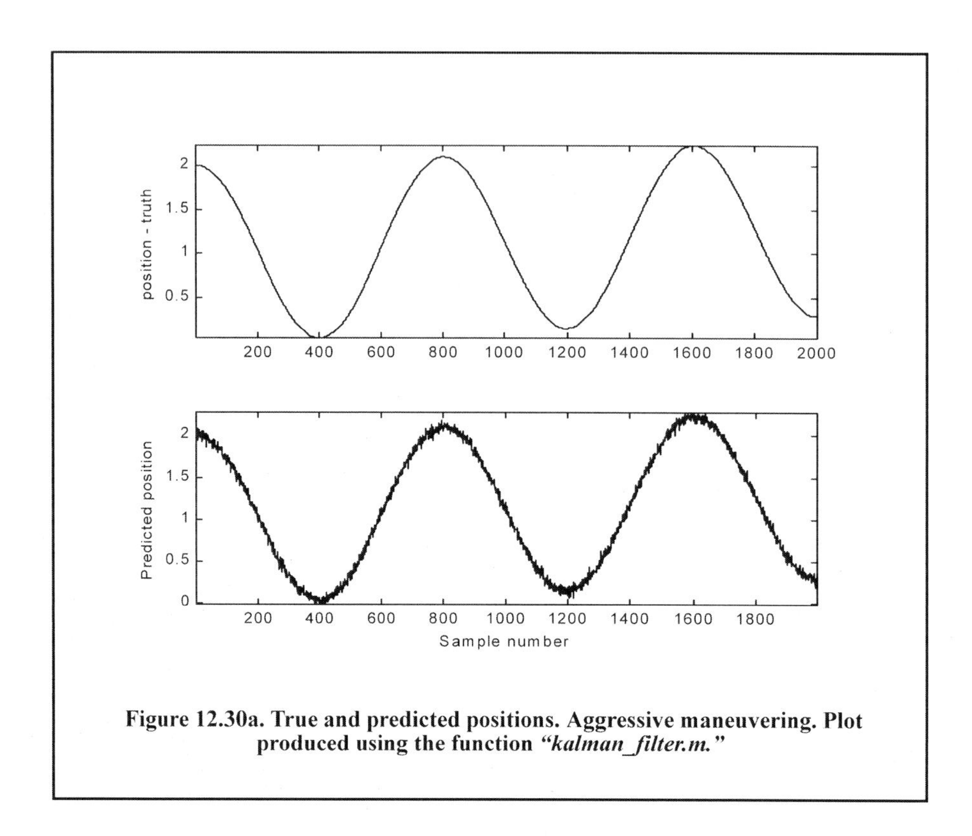

Figure 12.30a. True and predicted positions. Aggressive maneuvering. Plot produced using the function *"kalman_filter.m."*

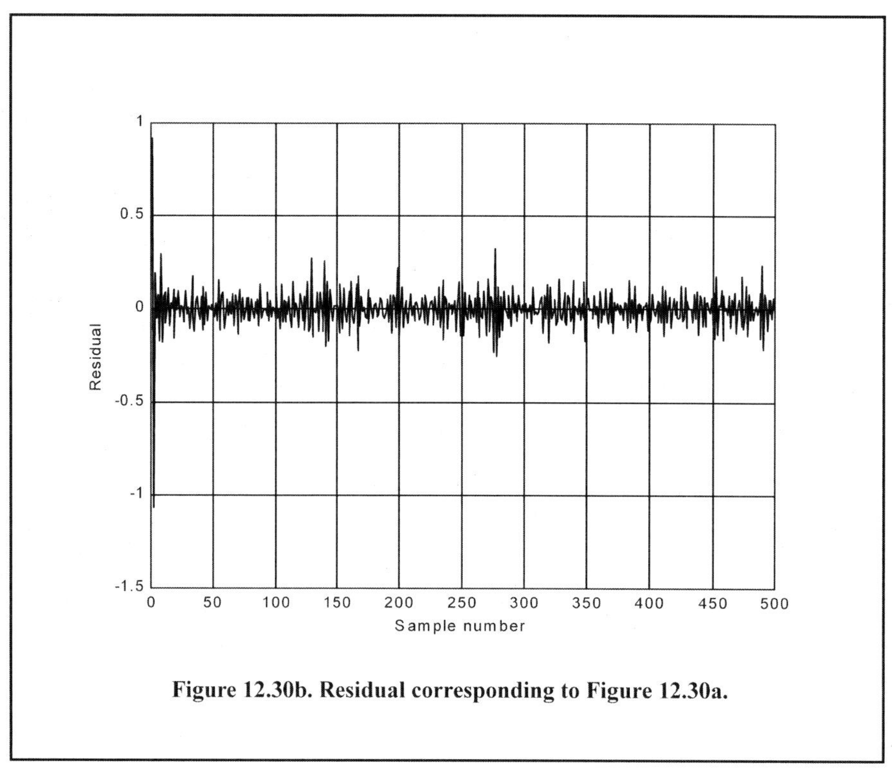

Figure 12.30b. Residual corresponding to Figure 12.30a.

Before presenting the formulas for the radar range measurement, range rate measurement, and angular measurement accuracies, we will first introduce the definition for the term *accuracy.* At the heart of this discussion are the range and velocity (i.e., Doppler) resolutions. The term resolution describes the radar's ability to distinguish close proximity targets (in range or Doppler) as distinct separable targets. In others words, any two adjacent targets separated by this resolution amount in any one coordinate are completely resolved along that coordinate. The radar's ability to quantify a very small change in range or Doppler measurements is referred to as a measurement *precision,* while, measurement accuracy refers to the correctness of the measurement. More precisely, accuracy measures the difference between true and measured values, and precision describes the radar's ability to repeat the same measurements multiple times without any regard to accuracy. In this context, radar systems use measurement accuracy as a means to measure the total measurement error. On the other hand, radar systems use precision to measure the noise error only, since a random noise signal will always interfere with the radar signal. The difference between precision and accuracy is illustrated using the experiment shown in Figure 12.31.

In this experiment, consider three dart players, each with 4 darts, attempting to hit the bull's eye on the dart board. Figure 12.31a shows the results for player 1, while Figures 12.31b and 12.31c show the results for players 2 and 3. Using the definitions introduced above for precision and accuracy, one can conclude the following: player 1 is inaccurate and imprecise; player 2 is inaccurate but is precise; finally player 3 is both accurate and precise.

The mathematical formulations for both range and Doppler resolutions were introduced in earlier chapters of this book. The range resolution, ΔR is

$$\Delta R = \frac{c\tau}{2} = \frac{c}{2B} \qquad \text{Eq. (12.141)}$$

where τ is the total integration time, c the speed of light, and B is the signal bandwidth. The Doppler frequency f_d is

$$f_d = \frac{2v}{\lambda} \qquad \text{Eq. (12.142)}$$

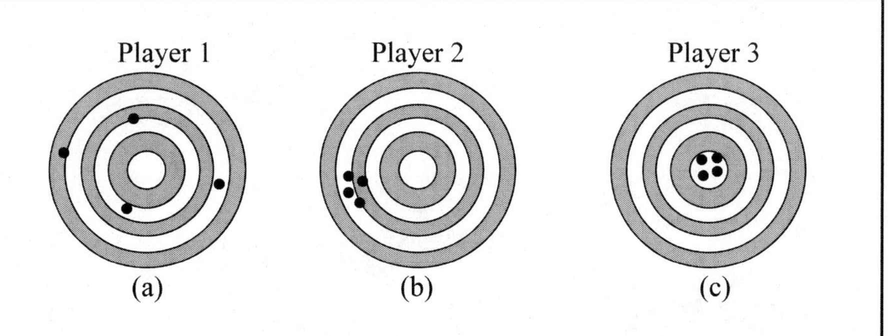

Figure 12.31. (a) Inaccurate and imprecise measurements. (b) Precise but inaccurate measurements. (c) Precise and accurate measurements.

where v is the target radial velocity on the radar line of sight, and λ is the wavelength. It follows that the velocity resolution Δv is given by

$$\Delta v = \frac{\lambda \Delta f_d}{2} = \frac{\lambda}{\tau}. \qquad \text{Eq. (12.143)}$$

Again τ is the total integration time.

The standard deviation describing the integration time error is

$$\sigma_\tau = \frac{1}{B\sqrt{2 \times SNR}}, \qquad \text{Eq. (12.144)}$$

where B is the signal bandwidth and SNR is the measured signal to noise ratio. From Eq. (12.141), the range measurement error is

$$\sigma_R = \frac{c}{2}\sigma_\tau. \qquad \text{Eq. (12.145)}$$

Substituting Eq. (12.144) into Eq. (12.145) yields,

$$\sigma_R = \frac{c}{2B\sqrt{2 \times SNR}}. \qquad \text{Eq. (12.146)}$$

Similarly, the standard deviation in the Doppler measurement can be derived as

$$\sigma_{f_d} = \frac{1}{\tau\sqrt{2 \times SNR}}. \qquad \text{Eq. (12.147)}$$

The velocity measurement error is computed from Eq. (12.143) as

$$\sigma_v = \frac{\lambda}{2}\sigma_{fd}. \qquad \text{Eq. (12.148)}$$

Using Eq. (12.147) into Eq. (12.148) yields

$$\sigma_v = \frac{\lambda}{2\tau\sqrt{2 \times SNR}}. \qquad \text{Eq. (12.149)}$$

The standard deviation of the error in the angle measurement is

$$\sigma_a = \frac{\Theta}{1.56\sqrt{2 \times SNR}} \qquad \text{Eq. (12.150)}$$

where Θ is the antenna beamwidth of the angular coordinate of the measurement (azimuth and elevation).

Problems

12.1. Show that in order to be able to quickly achieve changing the beam position, the error signal needs to be a linear function of the deviation angle.

12.2. Prepare a short report on the vulnerability of conical scan to amplitude modulation jamming. In particular, consider the self-protecting technique called "Gain Inversion."

12.3. Consider a conical scan radar. The pulse repetition interval is $10\mu s$. Calculate the scan rate so that at least ten pulses are emitted within one scan.

12.4. Consider a conical scan antenna whose rotation around the tracking axis is completed in 4 seconds. If during this time 20 pulses are emitted and received, calculate the radar PRF and the unambiguous range.

12.5. Reproduce Figure 12.11 for $\varphi_0 = 0.05, 0.1$ and $\varphi_0 = 0.15$ radians.

12.6. Reproduce Figure 12.13 for the squint angles defined in the previous problem.

12.7. Consider a monopulse radar where the input signal comprises both target return and additive white Gaussian noise. Develop an expression for the complex ratio Σ/Δ.

12.8. To generate the sum and difference patterns for a linear array of size N, follow this algorithm: to form the difference pattern, multiply the first $N/2$ elements by -1 and the second $N/2$ elements by +1. Plot the sum and difference patterns for a linear array of size 60.

12.9. Generate the delta/sum patterns for a 21-element linear array using the form

$$\frac{\Delta}{\Sigma} = j\frac{V_\Delta}{\sqrt{|V_\Delta|^2 + |V_\Sigma|^2}}$$

where V_Δ is the difference voltage pattern and V_Σ is the sum voltage pattern.

12.10. Consider the sum and difference signals defined in Eqs. (12.7) and (12.8). What is the squint angle φ_0 that maximizes $\Sigma(\varphi = 0)$?

12.11. A certain system is defined by the following difference equation:

$$y(n) + 4y(n-1) + 2y(n-2) = w(n).$$

Find the solution to this system for $n > 0$ and $w = \delta$.

12.12. Prove the state transition matrix properties (i.e., Eqs. (12.30) through (12.36)).

12.13. Suppose that the state equations for a certain discrete time LTI system are

$$\begin{bmatrix} x_1(n+1) \\ x_2(n+1) \end{bmatrix} = \begin{bmatrix} 0 & 1 \\ -2 & -3 \end{bmatrix} \begin{bmatrix} x_1(n) \\ x_2(n) \end{bmatrix} + \begin{bmatrix} 0 \\ 1 \end{bmatrix} w(n).$$

If $y(0) = y(1) = 1$, find $y(n)$ when the input is a step function.

12.14. Using Eq. (12.83), compute a general expression (in terms of the transfer function) for the steady-state errors when the input sequence is:

$$u1 = \{0, 1, 1, 1, 1, \ldots\}$$

$$u2 = \{0, 1, 2, 3, \ldots\}$$

$$u3 = \{0, 1^2, 2^2, 3^2, \ldots\}$$

$$u4 = \{0, 1^3, 2^3, 3^3, \ldots\}$$

12.15. Develop an expression for the steady-state error transfer function for an $\alpha\beta$ tracker.

12.16. Design a critically damped $\alpha\beta$, when the measurement noise variance associated with position is $\sigma_v^2 = 50m$ and when the desired standard deviation of the filter prediction error is $5.5m$.

12.17. Consider a $\alpha\beta\gamma$ filter. We can define six transfer functions: $H_1(z)$, $H_2(z)$, $H_3(z)$, $H_4(z)$, $H_5(z)$, and $H_6(z)$ (predicted position, predicted velocity, predicted acceleration, smoothed position, smoothed velocity, and smoothed acceleration). Each transfer function has the form

$$H(z) = \frac{a_3 + a_2 z^{-1} + a_1 z^{-2}}{1 + b_2 z^{-1} + b_1 z^{-2} + b_0 z^{-3}}.$$

12.18. The denominator remains the same for all six transfer functions. Compute all the relevant coefficients for each transfer function.

12.19. Verify the results obtained for the two limiting cases of the Singer-Kalman filter.

12.20. Prepare a short report on association and correlation algorithms for TWS radars.

Chapter 13

Phased Array Antennas

13.1. Directivity, Power Gain, and Effective Aperture

Radar antennas can be characterized by the directive gain G_D, power gain G, and effective aperture A_e. Antenna gain is a term used to describe the ability of an antenna to concentrate the transmitted energy in a certain direction. Directive gain, or simply directivity, is more representative of the antenna radiation pattern, while power gain is normally used in the radar equation. Plots of the power gain and directivity, when normalized to unity, are called the *antenna radiation pattern*. The directivity of a transmitting antenna can be defined by

$$G_D = \frac{maximum\ radiation\ intensity}{average\ radiation\ intensity}. \qquad \text{Eq. (13.1)}$$

The radiation intensity is the power-per-unit solid angle in the direction (θ, ϕ) and denoted by $P(\theta, \phi)$. The average radiation intensity over 4π radians (solid angle) is the total power divided by 4π. Hence, Eq. (13.1) can be written as

$$G_D = \frac{4\pi(maximum\ radiated\ power/unit\ solid\ angle)}{total\ radiated\ power}. \qquad \text{Eq. (13.2)}$$

It follows that

$$G_D = 4\pi \frac{P(\theta, \phi)_{max}}{\int_0^{\pi}\int_0^{2\pi} P(\theta, \phi) d\theta d\phi}. \qquad \text{Eq. (13.3)}$$

As an approximation, it is customary to rewrite Eq. (13.3) as

$$G_D \approx (4\pi)/(\theta_3\phi_3) \qquad \text{Eq. (13.4)}$$

where θ_3 and ϕ_3 are the antenna half-power (3-dB) beamwidths in either direction. The antenna power gain and its directivity are related by

$$G = \rho_r G_D \qquad \text{Eq. (13.5)}$$

where ρ_r is the radiation efficiency factor. In this book, the antenna power gain will be denoted as *gain*. The radiation efficiency factor accounts for the ohmic losses associated with the antenna. Therefore, the definition for the antenna gain is also given in Eq. (13.1). The antenna effective aperture A_e is related to gain by

$$A_e = \frac{G\lambda^2}{4\pi} \qquad \text{Eq. (13.6)}$$

where λ is the wavelength. The relationship between the antenna's effective aperture A_e and the physical aperture A is

$$A_e = \rho A \qquad \text{Eq. (13.7)}$$
$$0 \le \rho \le 1.$$

ρ is referred to as the aperture efficiency, and good antennas require $\rho \to 1$ (in this book $\rho = 1$ is always assumed, i.e., $A_e = A$).

Using simple algebraic manipulations of Eqs. (13.4) through (13.6) (assuming that $\rho_r = 1$) yields

$$G = \frac{4\pi A_e}{\lambda^2} \approx \frac{4\pi}{\theta_3 \phi_3}. \qquad \text{Eq. (13.8)}$$

Consequently, the angular cross section of the beam is

$$\theta_3 \phi_3 \approx \frac{\lambda^2}{A_e}. \qquad \text{Eq. (13.9)}$$

Eq. (13.9) indicates that the antenna beamwidth decreases as $\sqrt{A_e}$ increases. Thus, in surveillance operations, the number of beam positions an antenna will take on to cover a volume V is

$$N_{Beams} > \frac{V}{\theta_3 \phi_3}, \qquad \text{Eq. (13.10)}$$

and when V represents the entire hemisphere, Eq. (13.10) is modified to

$$N_{Beams} > \frac{2\pi}{\theta_3 \phi_3} \approx \frac{2\pi A_e}{\lambda^2} \approx \frac{G}{2}. \qquad \text{Eq. (13.11)}$$

13.2. Near and Far Fields

The electric field intensity generated from the energy emitted by an antenna is a function of the antenna physical aperture shape and the electric current amplitude and phase distribution across the aperture. Plots of the modulus of the electric field intensity of the emitted radiation, $|E(\theta, \phi)|$, are referred to as the *intensity pattern* of the antenna. Alternatively, plots of $|E(\theta, \phi)|^2$ are called the *power radiation pattern* (the same as $P(\theta, \phi)$).

Based on the distance from the face of the antenna, where the radiated electric field is measured, three distinct regions are identified. They are the near field, the Fresnel, and the Fraunhofer regions. In the near field and the Fresnel regions, rays emitted from the antenna have spherical wavefronts (equiphase fronts). In the Fraunhofer region, the wavefronts can be locally represented by plane waves. The near field and the Fresnel regions are normally of little interest to most radar applications. Most radar systems operate in the Fraunhofer region, which is also known as the far field region. In the far field region, the electric field intensity can be computed from the aperture Fourier transform.

Construction of the far criterion can be developed with the help of Figure 13.1. Consider a radiating source at point O that emits spherical waves. A receiving antenna of length d is at distance r away from the source. The phase difference between a spherical wave and a local plane wave at the receiving antenna can be expressed in terms of the distance δr. The distance δr is given by

$$\delta r = \overline{AO} - \overline{OB} = \sqrt{r^2 + \left(\frac{d}{2}\right)^2} - r, \qquad \text{Eq. (13.12)}$$

and since in the far field $r \gg d$, Eq. (13.12) is approximated via binomial expansion by

$$\delta r = r\left(\sqrt{1 + \left(\frac{d}{2r}\right)^2} - 1\right) \approx \frac{d^2}{8r}. \qquad \text{Eq. (13.13)}$$

It is customary to assume far field when the distance δr corresponds to less than $1/16$ of a wavelength (i.e., $22.5°$). More precisely, if

$$\delta r = d^2/8r \leq \lambda/16, \qquad \text{Eq. (13.14)}$$

then a useful expression for far field is

$$r \geq 2d^2/\lambda. \qquad \text{Eq. (13.15)}$$

Note that far field is a function of both the antenna size and the operating wavelength.

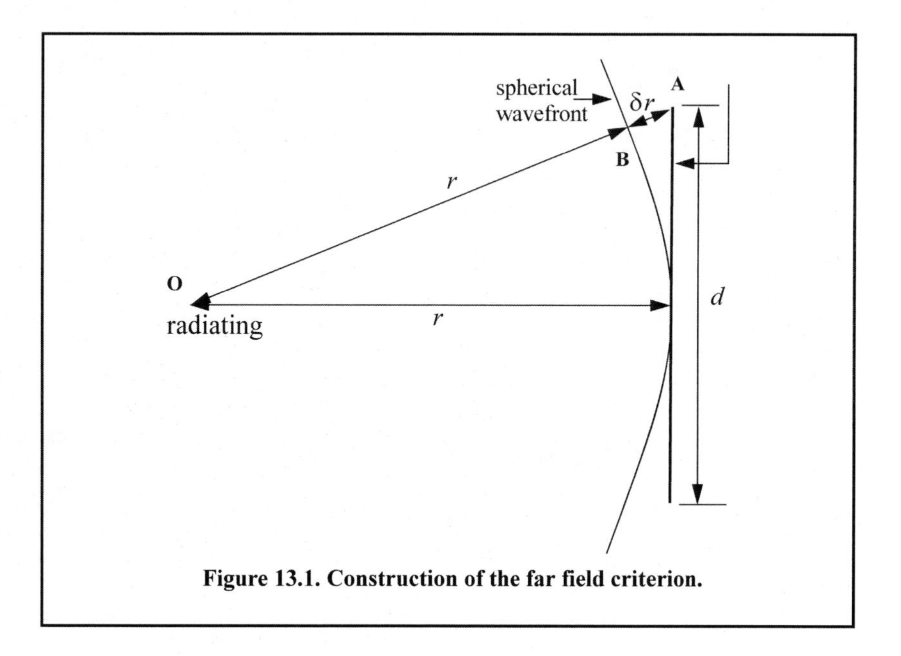

Figure 13.1. Construction of the far field criterion.

13.3. General Arrays

An array is a composite antenna formed from two or more basic radiators. Each radiator is denoted as an element. The elements forming an array could be dipoles, dish reflectors, slots in a wave guide, or any other type of radiator. Array antennas synthesize narrow directive beams that may be steered, mechanically or electronically, in many directions.

Electronic steering is achieved by controlling the phase of the current feeding the array elements. Arrays with electronic beam steering capability are called phased arrays. Phased array antennas, when compared to other simple antennas such as dish reflectors, are costly and complicated to design. However, the inherent flexibility of phased array antennas to steer the beam electronically, and also the need for specialized multifunction radar systems, have made phased array antennas attractive for radar applications.

Figure 13.2 shows the geometrical fundamentals associated with this problem. In general, consider the radiation source located at (x_1, y_1, z_1) with respect to a phase reference at $(0, 0, 0)$. The electric field measured at far field point P is

$$E(\theta, \phi) = I_0 \frac{e^{-jkR_1}}{R_1} f(\theta, \phi) \qquad \text{Eq. (13.16)}$$

where I_0 is the complex amplitude, $k = 2\pi/\lambda$ is the wave number, and $f(\theta, \phi)$ is the radiation pattern.

Now, consider the case where the radiation source is an array made of many elements, as shown in Figure 13.3. The coordinates of each radiator with respect to the phase reference is (x_i, y_i, z_i), and the vector from the origin to the ith element is given by

$$\vec{r}_i = \hat{a}_x x_i + \hat{a}_y y_i + \hat{a}_z z_i \,. \qquad \text{Eq. (13.17)}$$

The far field components that constitute the total electric field are

$$E_i(\theta, \phi) = I_i \frac{e^{-jkR_i}}{R_i} f(\theta_i, \phi_i) \qquad \text{Eq. (13.18)}$$

where

$$\begin{aligned} R_i = \left|\vec{R}_i\right| &= \left|\vec{r} - \vec{r}_i\right| = \sqrt{(x-x_i)^2 + (y-y_i)^2 + (z-z_i)^2} \\ &= r\sqrt{1 + \frac{(x_i^2 + y_i^2 + z_i^2)}{r^2} - 2\frac{(xx_i + yy_i + zz_i)}{r^2}} \end{aligned} \qquad \text{Eq. (13.19)}$$

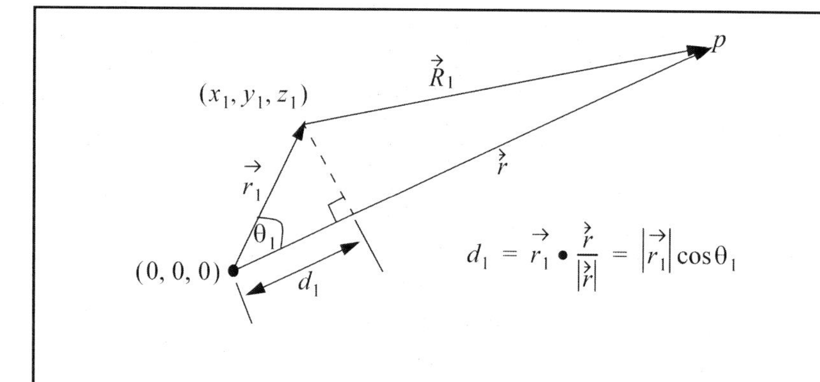

Figure 13.2. Geometry for an array antenna. Single element.

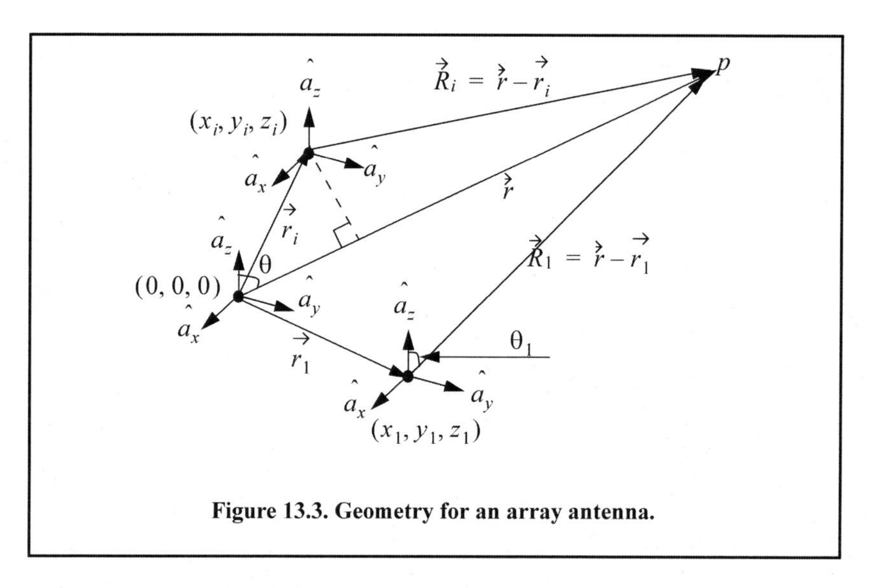

Figure 13.3. Geometry for an array antenna.

Using the spherical coordinates $x = r\sin\theta\cos\varphi$, $y = r\sin\theta\sin\varphi$, and $z = r\cos\theta$, yields

$$\frac{(x_i^2 + y_i^2 + z_i^2)}{r^2} = \frac{|\vec{r}_i|^2}{r^2} \ll 1 . \qquad \text{Eq. (13.20)}$$

Thus, a good approximation (using binomial expansion) for Eq. (13.19) is

$$R_i = r - r(x_i\sin\theta\cos\phi + y_i\sin\theta\sin\phi + z_i\cos\theta) . \qquad \text{Eq. (13.21)}$$

It follows that the phase contribution at the far field point from the *ith* radiator with respect to the phase reference is

$$e^{-jkR_i} = e^{-jkr}\, e^{jk(x_i\sin\theta\cos\phi + y_i\sin\theta\sin\phi + z_i\cos\theta)} . \qquad \text{Eq. (13.22)}$$

Remember, however, that the unit vector $\vec{r}_0$ along the vector $\vec{r}$ is

$$\vec{r}_0 = \frac{\vec{r}}{|\vec{r}|} = \hat{a}_x\sin\theta\cos\phi + \hat{a}_y\sin\theta\sin\phi + \hat{a}_z\cos\theta . \qquad \text{Eq. (13.23)}$$

Hence, we can rewrite Eq. (13.22) as

$$e^{-jkR_i} = e^{-jkr}\, e^{jk(\vec{r}_i \bullet \vec{r}_0)} = e^{-jkr}\, e^{j\Psi_i(\theta,\, \phi)} . \qquad \text{Eq. (13.24)}$$

Finally, by virtue of superposition, the total electric field is

$$E(\theta, \phi) = \sum_{i=1}^{N} I_i e^{j\Psi_i(\theta,\, \phi)} , \qquad \text{Eq. (13.25)}$$

which is known as the array factor for an array antenna where the complex current for the *ith* element is I_i. In general, an array can be fully characterized by its array factor. This is true since knowing the array factor provides the designer with knowledge of the array's (1) 3-dB beamwidth; (2) null-to-null beamwidth; (3) distance from the main peak to the first sidelobe; (4) height of the first sidelobe as compared to the main beam; (5) location of the nulls; (6) rate of decrease of the sidelobes; and (7) grating lobe locations.

13.4. Linear Arrays

Figure 13.4 shows a linear array antenna consisting of N identical elements. The element spacing is d (normally measured in wavelength units). Let element #1 serve as a phase reference for the array. From the geometry, it is clear that an outgoing wave at the nth element leads the phase at the $(n+1)th$ element by $kd\sin\theta$, where $k = 2\pi/\lambda$. The combined phase at the far field observation point P is independent of ϕ and is computed from Eq. (13.24) as

$$\Psi(\theta, \phi) = k(\vec{r}_n \bullet \vec{r}_0) = (n-1)kd\sin\theta . \qquad \text{Eq. (13.26)}$$

Thus, from Eq. (13.25), the electric field at a far field observation point with direction-sine equal to $\sin\theta$ (assuming isotropic elements) is

$$E(\sin\theta) = \sum_{n=1}^{N} e^{j(n-1)(kd\sin\theta)} . \qquad \text{Eq. (13.27)}$$

Expanding the summation in Eq. (13.27) yields

$$E(\sin\theta) = 1 + e^{jkd\sin\theta} + \dots + e^{j(N-1)(kd\sin\theta)} . \qquad \text{Eq. (13.28)}$$

The right-hand side of Eq. (13.28) is a geometric series, which can be expressed as

$$1 + a + a^2 + a^3 + \dots + a^{(N-1)} = \frac{1-a^N}{1-a} . \qquad \text{Eq. (13.29)}$$

Replacing a by $e^{jkd\sin\theta}$ yields

$$E(\sin\theta) = \frac{1-e^{jNkd\sin\theta}}{1-e^{jkd\sin\theta}} = \frac{1-(\cos Nkd\sin\theta)-j(\sin Nkd\sin\theta)}{1-(\cos kd\sin\theta)-j(\sin kd\sin\theta)} . \qquad \text{Eq. (13.30)}$$

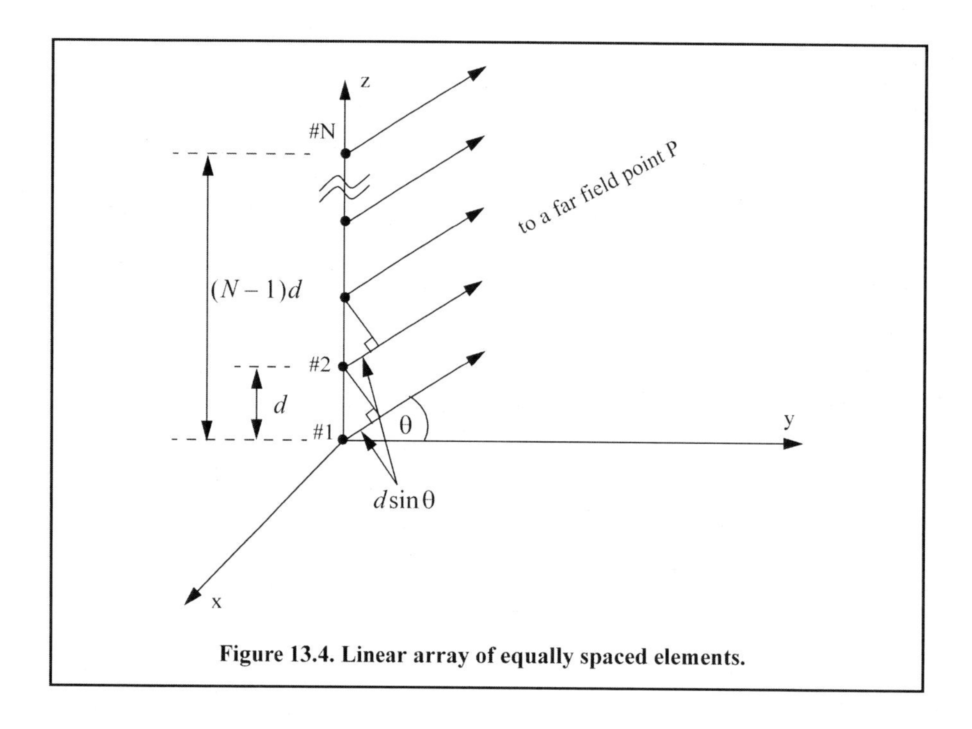

Figure 13.4. Linear array of equally spaced elements.

The far field array intensity pattern is then given by

$$|E(\sin\theta)| = \sqrt{E(\sin\theta)E^*(\sin\theta)} \; . \qquad \text{Eq. (13.31)}$$

Substituting Eq. (13.30) into Eq. (13.31) and collecting terms yield

$$|E(\sin\theta)| = \sqrt{\frac{(1-\cos Nkd\sin\theta)^2 + (\sin Nkd\sin\theta)^2}{(1-\cos kd\sin\theta)^2 + (\sin kd\sin\theta)^2}} = \sqrt{\frac{1-\cos Nkd\sin\theta}{1-\cos kd\sin\theta}} \; , \qquad \text{Eq. (13.32)}$$

and using the trigonometric identity $1 - \cos\theta = 2(\sin\theta/2)^2$ yields

$$|E(\sin\theta)| = \left|\frac{\sin(Nkd\sin\theta/2)}{\sin(kd\sin\theta/2)}\right| \; , \qquad \text{Eq. (13.33)}$$

which is a periodic function of $kd\sin\theta$, with a period equal to 2π. The maximum value of $|E(\sin\theta)|$, which occurs at $\theta = 0$, is equal to N. It follows that the normalized intensity pattern is equal to

$$|E_n(\sin\theta)| = \frac{1}{N}\left|\frac{\sin((Nkd\sin\theta)/2)}{\sin((kd\sin\theta)/2)}\right| \; . \qquad \text{Eq. (13.34)}$$

The normalized two-way array pattern (radiation pattern) is given by

$$G(\sin\theta) = |E_n(\sin\theta)|^2 = \frac{1}{N^2}\left(\frac{\sin((Nkd\sin\theta)/2)}{\sin((kd\sin\theta)/2)}\right)^2 \; . \qquad \text{Eq. (13.35)}$$

Figure 13.5 shows a plot of Eq. (13.35) versus $\sin\theta$ for $N = 8$. The pattern $G(\sin\theta)$ has cylindrical symmetry about its axis ($\sin\theta = 0$), and is independent of the azimuth angle. Thus, it is completely determined by its values within the interval $(0 < \theta < \pi)$. The main beam of an array can be steered electronically by varying the phase of the current applied to each array element. Steering the main beam into the direction-sine $\sin\theta_0$ is accomplished by making the phase difference between any two adjacent elements equal to $kd\sin\theta_0$. In this case, the normalized radiation pattern can be written as

$$G(\sin\theta) = \frac{1}{N^2}\left(\frac{\sin[(Nkd/2)(\sin\theta - \sin\theta_0)]}{\sin[(kd/2)(\sin\theta - \sin\theta_0)]}\right)^2 \; . \qquad \text{Eq. (13.36)}$$

If $\theta_0 = 0$, then the main beam is perpendicular to the array axis, and the array is said to be a broadside array. Alternatively, the array is called an endfire array when the main beam points along the array axis. The radiation pattern maxima are computed using L'Hopital's rule when both the denominator and numerator of Eq. (13.35) are zeros. More precisely,

$$\left(\frac{kd\sin\theta}{2} = \pm m\pi\right); \; m = 0, 1, 2, \ldots \; . \qquad \text{Eq. (13.37)}$$

Solving for θ yields

$$\theta_m = \mathrm{asin}\left(\pm\frac{\lambda m}{d}\right); \; m = 0, 1, 2, \ldots \qquad \text{Eq. (13.38)}$$

where the subscript m is used as a maxima indicator. The first maximum occurs at $\theta_0 = 0$, and is denoted as the main beam (lobe). Other maxima occurring at $|m| \geq 1$ are called grating lobes. Grating lobes are undesirable and must be suppressed.

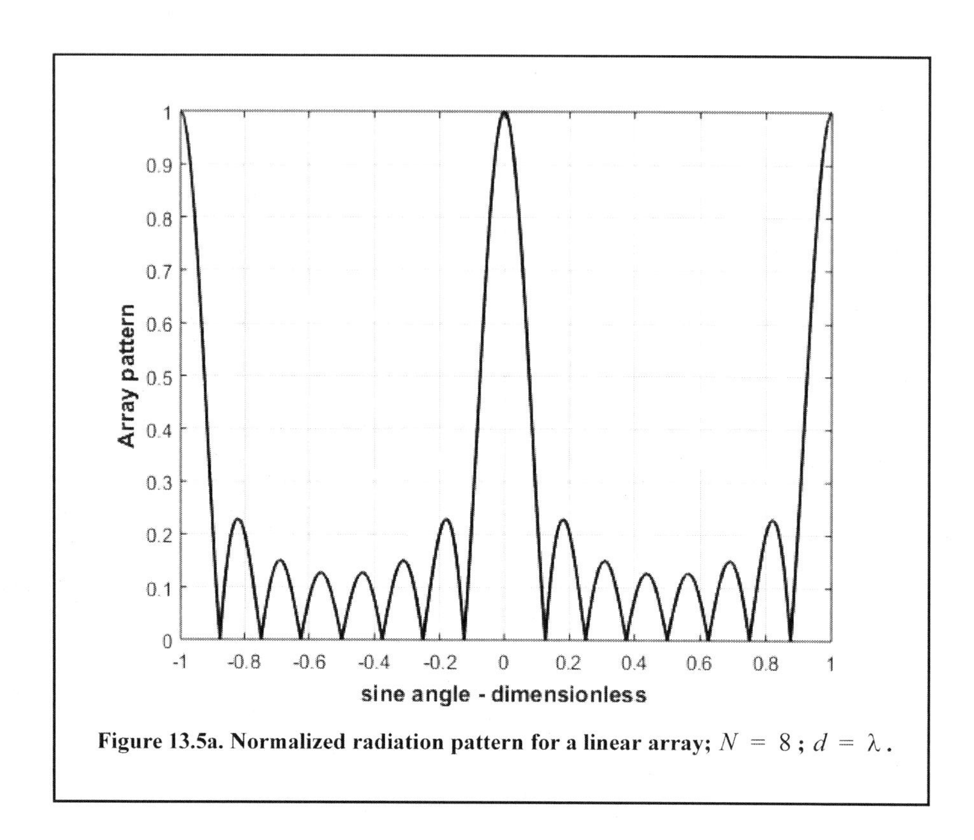

Figure 13.5a. Normalized radiation pattern for a linear array; $N = 8$; $d = \lambda$.

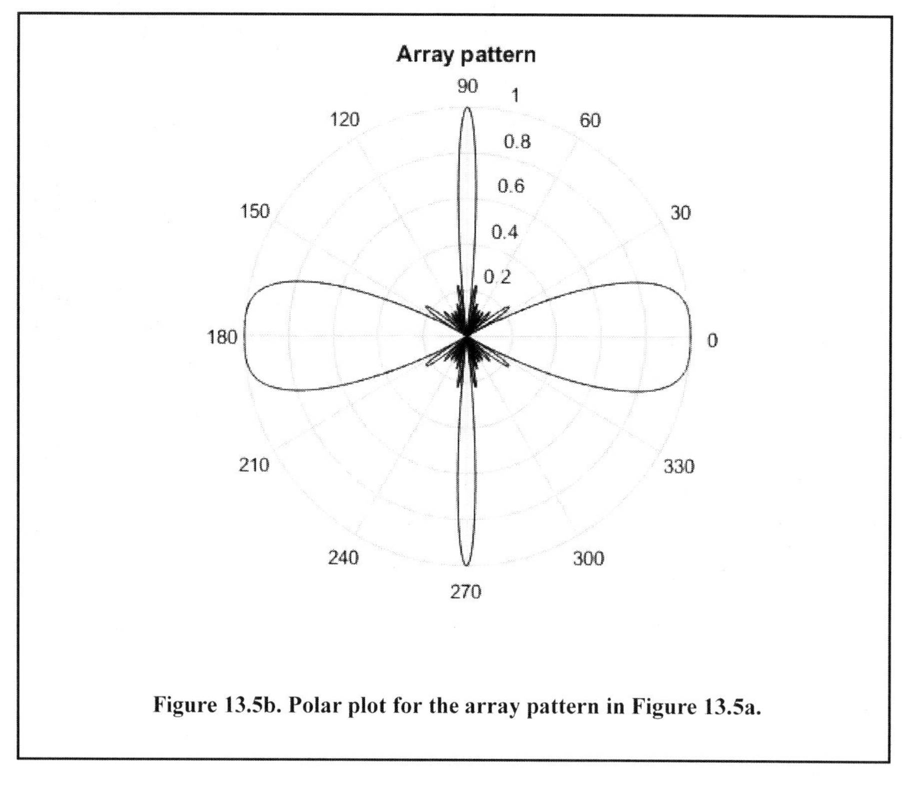

Figure 13.5b. Polar plot for the array pattern in Figure 13.5a.

The grating lobes occur at nonreal angles when the absolute value of the arc-sine argument in Eq. (13.38) is greater than unity; it follows that $d < \lambda$. Under this condition, the main lobe is assumed to be at $\theta = 0$ (broadside array). Alternatively, when electronic beam steering is considered, the grating lobes occur at

$$|\sin\theta - \sin\theta_0| = \pm\frac{\lambda n}{d};\ n = 1, 2, \ldots . \qquad \text{Eq. (13.39)}$$

Thus, in order to prevent the grating lobes from occurring between $\pm 90°$, the element spacing should be $d < \lambda/2$. The radiation pattern attains secondary maxima (sidelobes) when the numerator of Eq. (13.35) is maximum, or equivalently

$$\frac{Nkd\sin\theta}{2} = \pm(2l + 1)\frac{\pi}{2};\ l = 1, 2, \ldots . \qquad \text{Eq. (13.40)}$$

Solving for θ yields

$$\theta_l = \operatorname{asin}\left(\pm\frac{\lambda}{2d}\frac{2l + 1}{N}\right);\ l = 1, 2, \ldots . \qquad \text{Eq. (13.41)}$$

The subscript l is used as an indication of sidelobe maxima. The nulls of the radiation pattern occur when only the numerator of Eq. (13.35) is zero. More precisely,

$$\frac{N}{2}kd\sin\theta = \pm n\pi;\ \begin{matrix} n = 1, 2, \ldots \\ n \neq N, 2N, \ldots \end{matrix} . \qquad \text{Eq. (13.42)}$$

Again solving for θ yields

$$\theta_n = \operatorname{asin}\left(\pm\frac{\lambda}{d}\frac{n}{N}\right);\ \begin{matrix} n = 1, 2, \ldots \\ n \neq N, 2N, \ldots \end{matrix} \qquad \text{Eq. (13.43)}$$

where the subscript n is used as a null indicator. Define the angle that corresponds to the half power point as θ_h. It follows that the half power ($3dB$) beamwidth is $2|\theta_m - \theta_h|$. This occurs when

$$\frac{N}{2}kd\sin\theta_h = 1.391\ radians \Rightarrow \theta_h = \operatorname{asin}\left(\frac{\lambda}{2\pi d}\frac{2.782}{N}\right) . \qquad \text{Eq. (13.44)}$$

13.4.1. Array Tapering

Figure 13.6 shows a normalized two-way radiation pattern of a uniformly excited linear array of size $N = 8$, element spacing $d = \lambda/2$. The first sidelobe is $13.46dB$ below the main lobe, and for most radar applications this may not be sufficient, particularly in the presence of a strong source of jamming or high levels of noise. Under such conditions, target detection in the main beam becomes rather challenging, since the SNR is reduced.

In order to reduce the sidelobe levels, the array must be designed to radiate more power toward the center, and much less at the edges. This can be achieved through tapering (windowing) the current distribution over the face of the array. There are many possible tapering sequences that can be used for this purpose. However, as known from spectral analysis, windowing reduces sidelobe levels at the expense of widening the main beam. Thus, for a given radar application, the choice of the tapering sequence must be based on the trade-off between sidelobe reduction and main-beam widening. The same type of windows discussed earlier in Chapter 2 can be used for array tapering. Table 13.1 summarizes the impact of most common windows on the array pattern in terms of main-beam widening and peak reduction. Note that the rectangular window is used as the baseline. This is also illustrated in Figure 13.7.

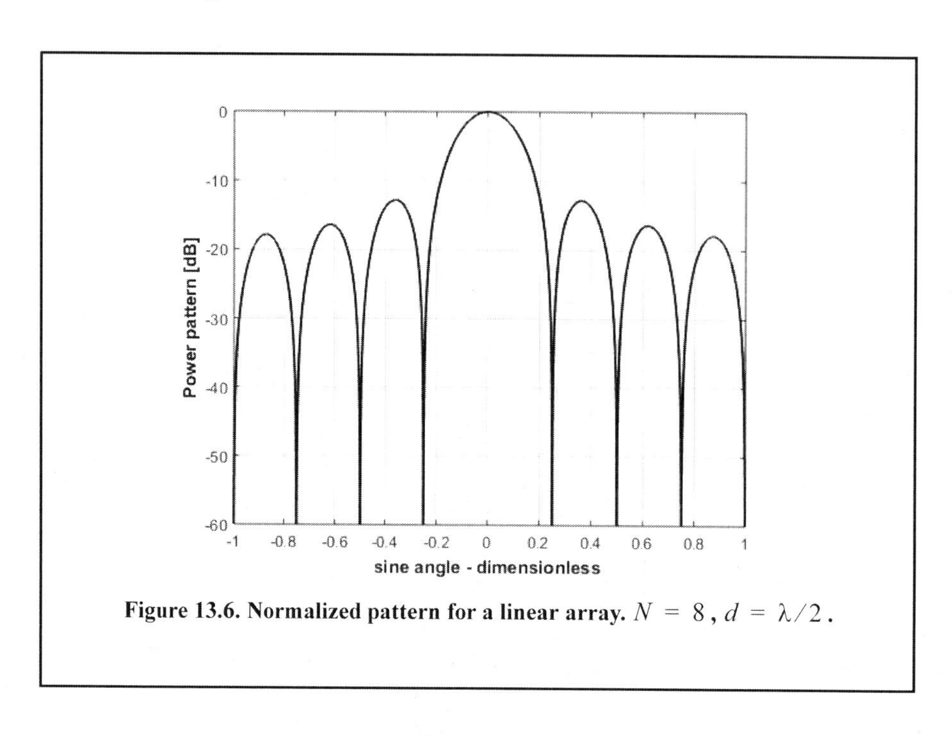

Figure 13.6. Normalized pattern for a linear array. $N = 8$, $d = \lambda/2$.

Table 13.1. Common windows.

Window	Null-to-Null Beamwidth	Peak Reduction
Rectangular	*1*	*1*
Hamming	*2*	*0.73*
Hanning	*2*	*0.664*
Blackman	*6*	*0.577*
Kaiser ($\beta = 6$)	*2.76*	*0.683*
Kaiser ($\beta = 3$)	*1.75*	*0.882*

13.4.2. Computation of the Radiation Pattern via the DFT

Figure 13.8 shows a linear array of size N, element spacing d, and wavelength λ. The radiators are circular dishes of diameter d. Let $w(n)$ and $\Phi(n)$, respectively, denote the tapering and phase shifting sequences. The normalized electric field at a far field point in the direction-sine $\sin\theta$ is

$$E(\sin\theta) = \sum_{n=0}^{N-1} w(n) e^{j\Delta\theta\left(n - \left(\frac{N-1}{2}\right)\right)} \qquad \text{Eq. (13.45)}$$

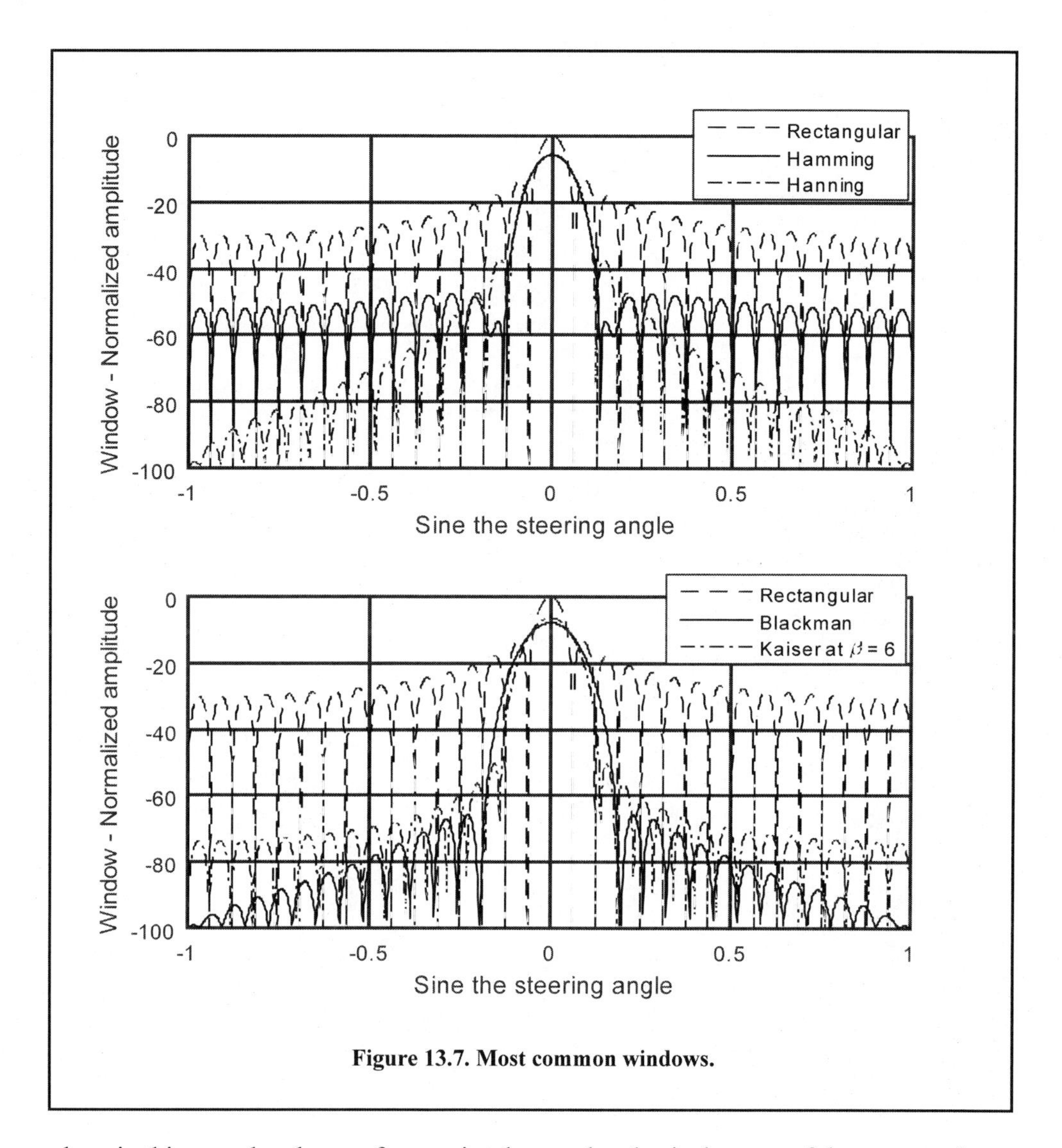

Figure 13.7. Most common windows.

where in this case the phase reference is taken as the physical center of the array, and

$$\Delta\theta = \frac{2\pi d}{\lambda}\sin\theta . \qquad \text{Eq. (13.46)}$$

Expanding Eq. (13.45) and factoring the common phase term $\exp[j(N-1)\Delta\theta/2]$ yields

$$E(\sin\theta) = e^{j(N-1)\Delta\theta/2}\{w(0)e^{-j(N-1)\Delta\theta} + w(1)e^{-j(N-2)\Delta\theta} + \ldots + w(N-1)\} . \qquad \text{Eq. (13.47)}$$

By using the symmetry property of a window sequence (remember that a window must be symmetrical about its central point), we can rewrite Eq. (13.47) as

$$E(\sin\theta) = e^{j\theta_0}\{w(N-1)e^{-j(N-1)\Delta\theta} + w(N-2)e^{-j(N-2)\Delta\theta} + \ldots + w(0)\} \qquad \text{Eq. (13.48)}$$

where $\theta_0 = (N-1)\Delta\theta/2$.

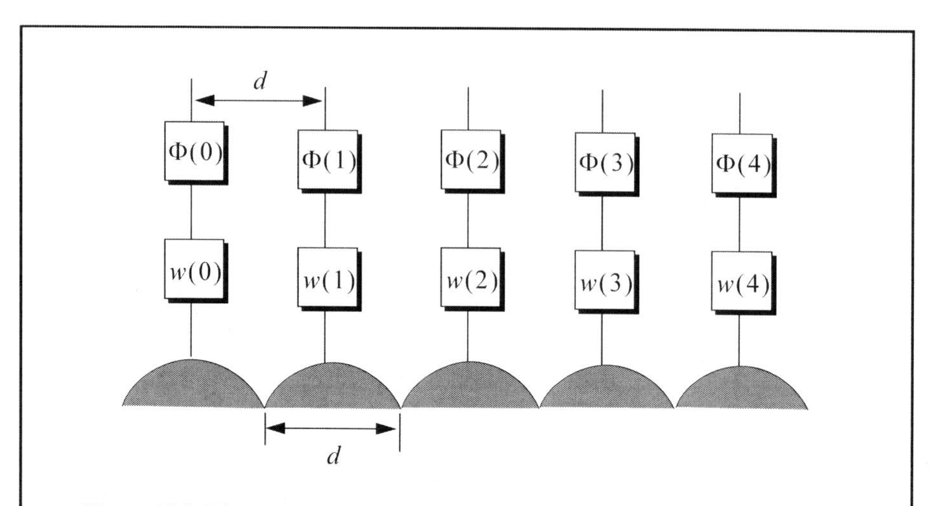

Figure 13.8. Linear array of size 5, with tapering and phase shifting hardware.

Define $\{ V_1^n = \exp(-jn\Delta\theta) ; n = 0, 1, \ldots, N-1 \}$. It follows that

$$E(\sin\theta) = e^{j\theta_0}[w(0) + w(1)V_1^1 + \ldots + w(N-1)V_1^{N-1}] = e^{j\theta_0}\sum_{n=0}^{N-1} w(n)V_1^n . \qquad \text{Eq. (13.49)}$$

The discrete Fourier transform of the sequence $w(n)$ is defined as

$$W(q) = \sum_{n=0}^{N-1} w(n)e^{-\frac{(j2\pi nq)}{N}} \quad ; \ q = 0, 1, \ldots, N-1 . \qquad \text{Eq. (13.50)}$$

The set $\{ \sin\theta_q \}$ that makes V_1 equal to the DFT kernel is

$$\sin\theta_q = \frac{\lambda q}{Nd}; \ q = 0, 1, \ldots, N-1 . \qquad \text{Eq. (13.51)}$$

Then, using Eq. (13.51) in Eq. (13.50) yields

$$E(\sin\theta) = e^{j\phi_0}W(q) . \qquad \text{Eq. (13.52)}$$

The one-way array pattern is computed as the modulus of Eq. (13.52). It follows that the one-way radiation pattern of a tapered linear array of circular dishes is

$$G(\sin\theta) = G_e \ |W(q)| \qquad \text{Eq. (13.53)}$$

where G_e is the element pattern. In practice, phase shifters are normally implemented as part of the Transmit/Receive (TR) modules, using a finite number of bits. Consequently, due to the quantization error (difference between desired phase and actual quantized phase) the sidelobe levels are affected. Figures 13.10 through 13.18 show a few examples for a linear array gain pattern using different windows and different numbers of bits in the phase shifters. Note that the term *"dolr"* refers to the element spacing in wavelength units and the term *"Nr"* refers to the number of elements in the array.

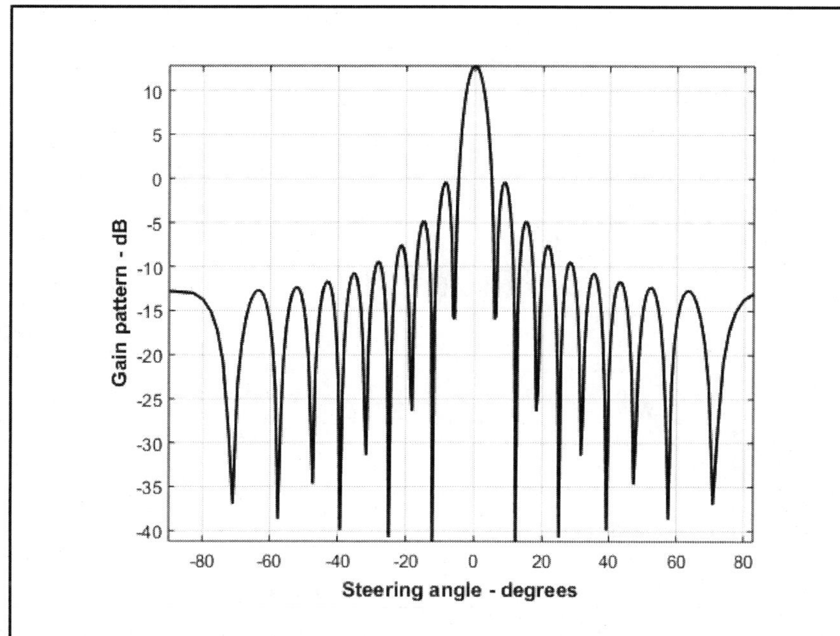

Figure 13.10. Array gain pattern: $Nr = 19$; $dolr = 0.5$; $\theta_0 = 0°$ **; using perfect phase shifters and no window.**

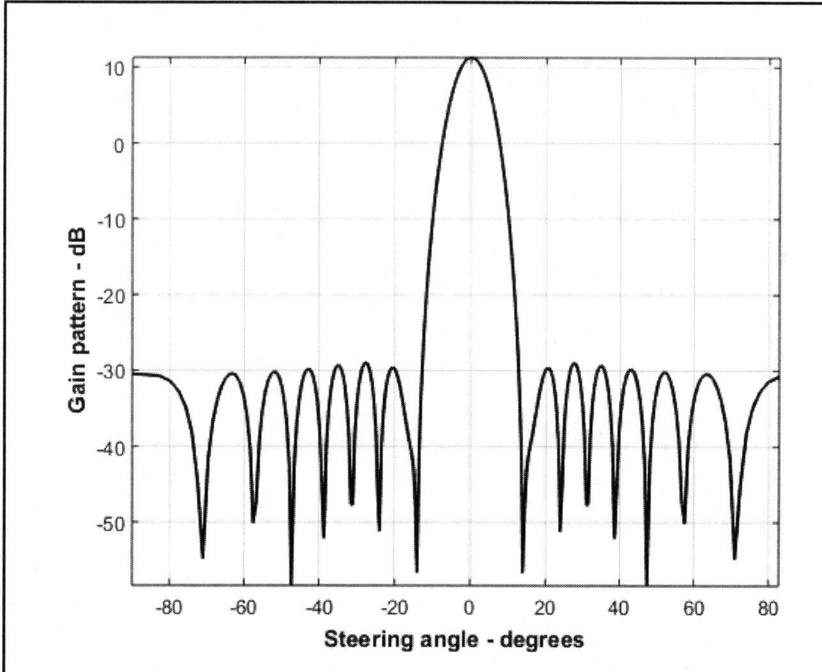

Figure 13.11. Array gain pattern: $Nr = 19$; $dolr = 0.5$; $\theta_0 = 0°$ **; using perfect phase shifters and Hamming window.**

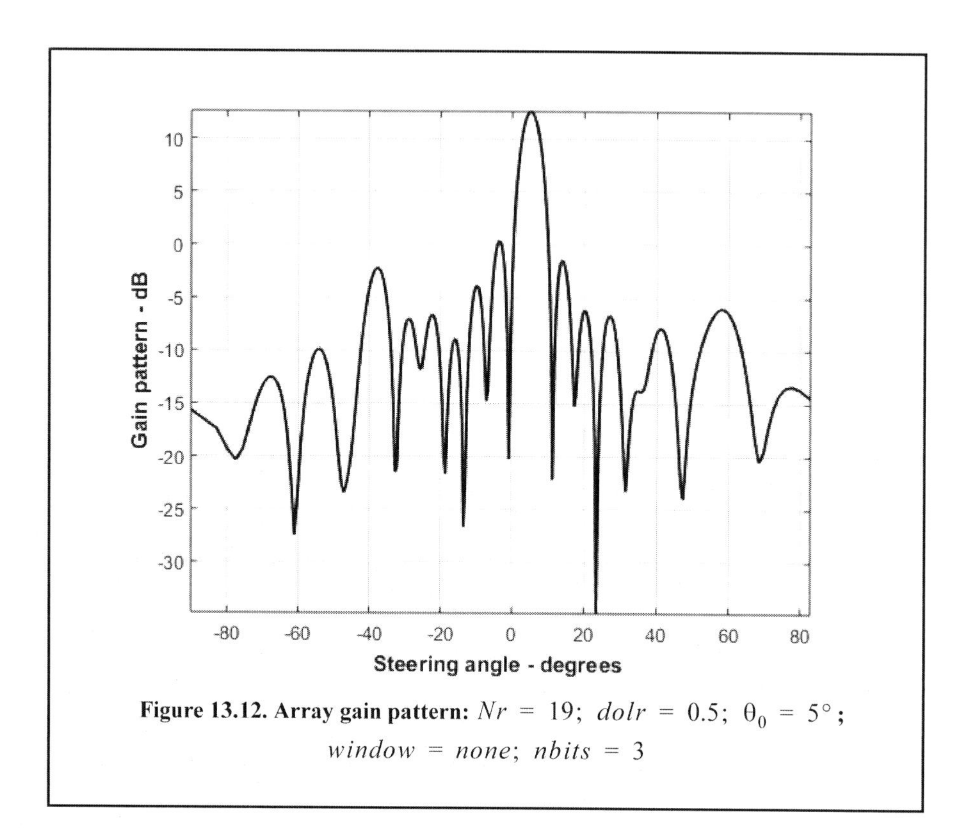

Figure 13.12. Array gain pattern: $Nr = 19$; $dolr = 0.5$; $\theta_0 = 5°$; $window = none$; $nbits = 3$

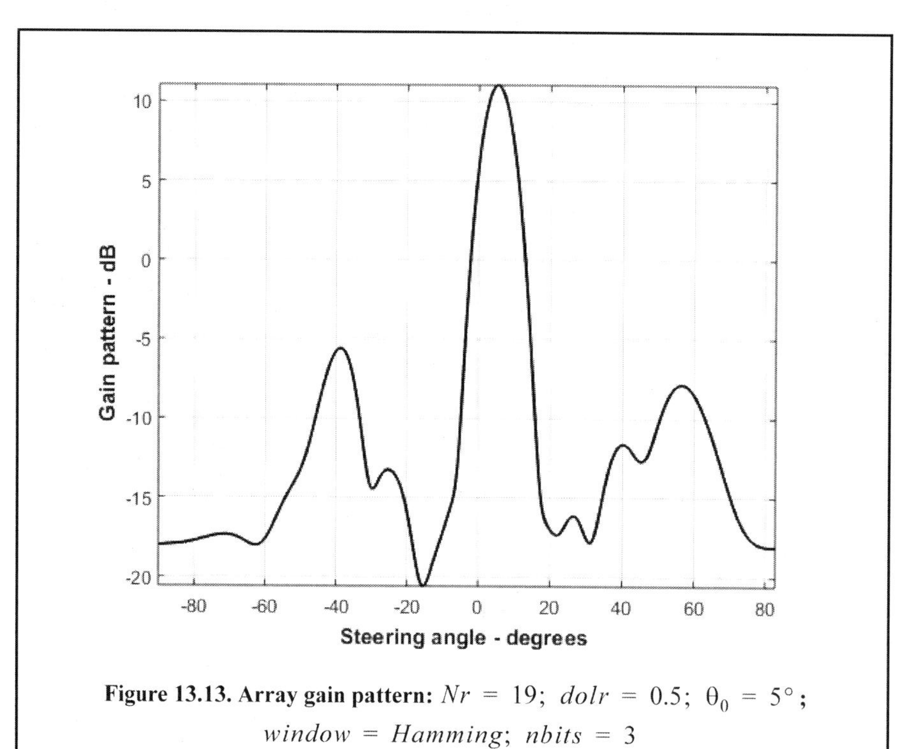

Figure 13.13. Array gain pattern: $Nr = 19$; $dolr = 0.5$; $\theta_0 = 5°$; $window = Hamming$; $nbits = 3$

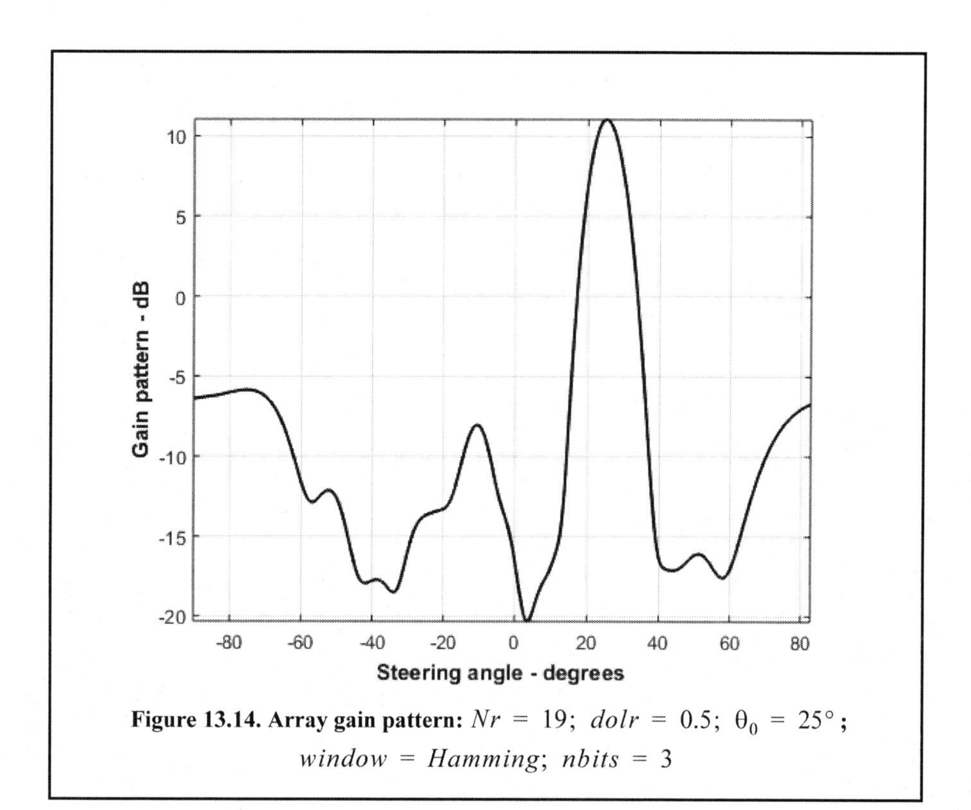

Figure 13.14. Array gain pattern: $Nr = 19$; $dolr = 0.5$; $\theta_0 = 25°$; $window = Hamming$; $nbits = 3$

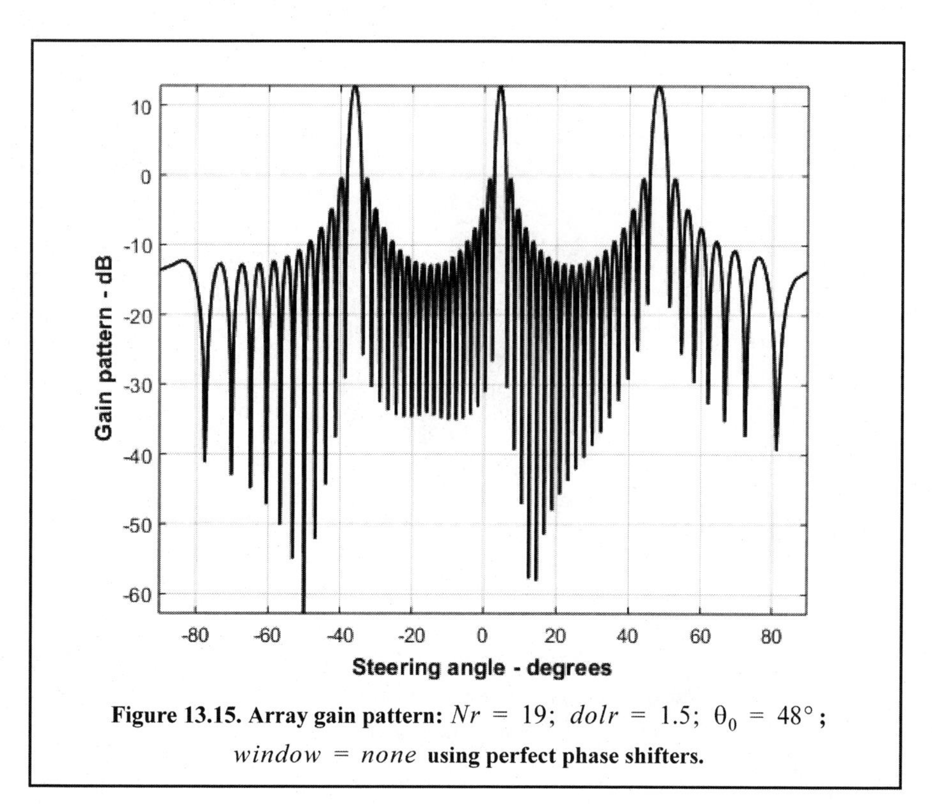

Figure 13.15. Array gain pattern: $Nr = 19$; $dolr = 1.5$; $\theta_0 = 48°$; $window = none$ **using perfect phase shifters.**

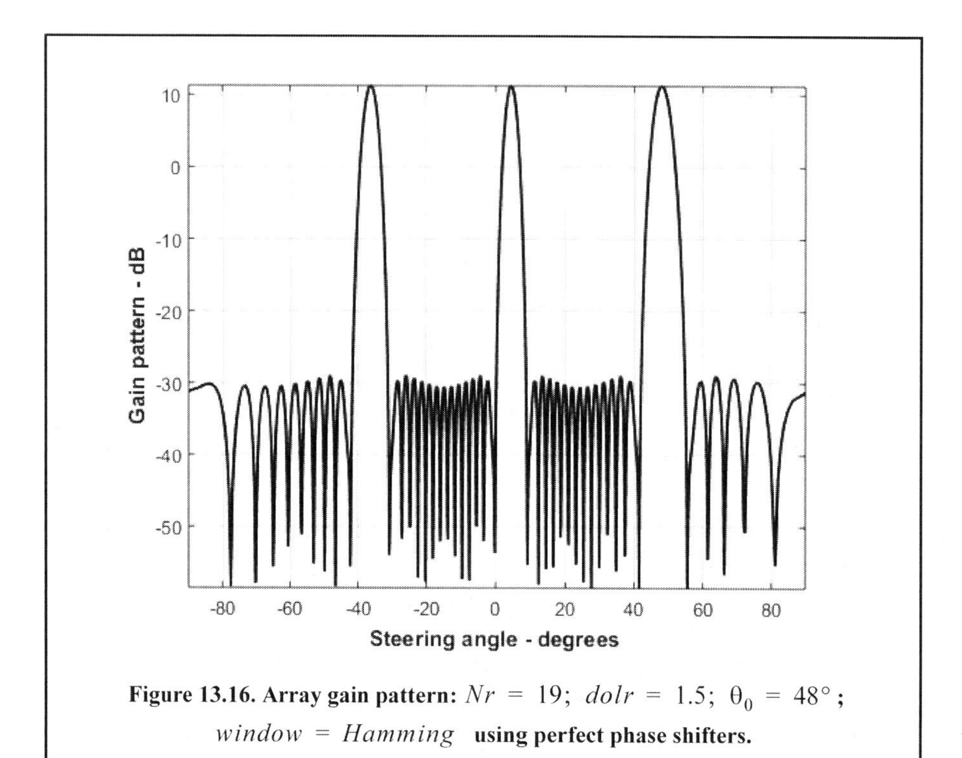

Figure 13.16. Array gain pattern: $Nr = 19$; $dolr = 1.5$; $\theta_0 = 48°$; $window = Hamming$ **using perfect phase shifters.**

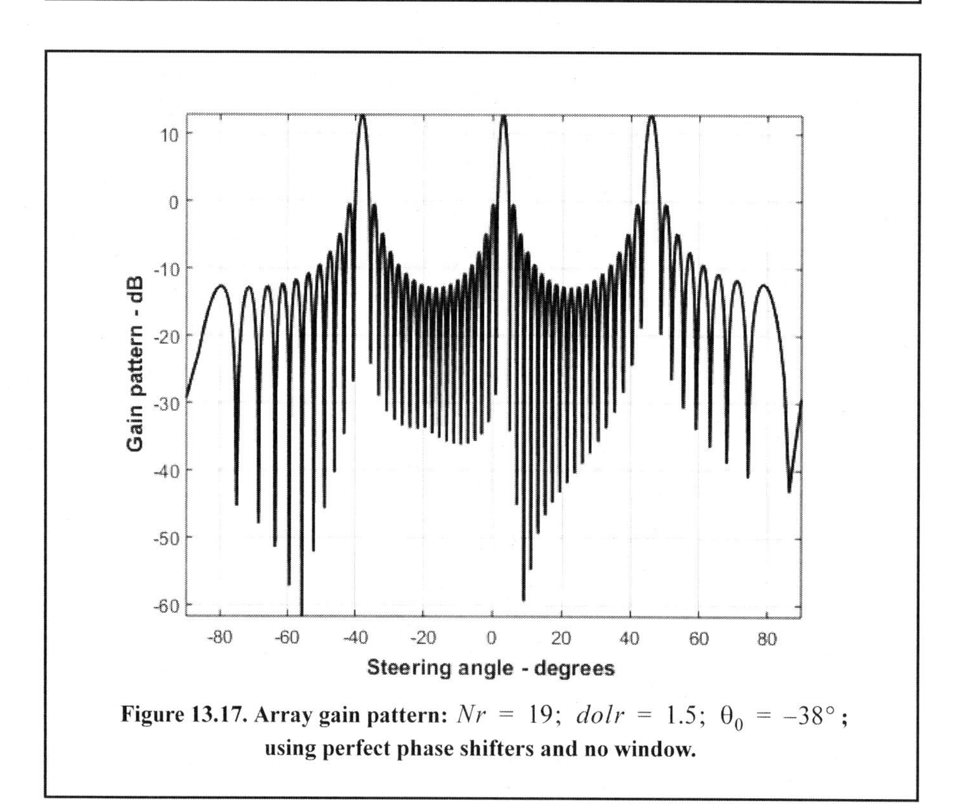

Figure 13.17. Array gain pattern: $Nr = 19$; $dolr = 1.5$; $\theta_0 = -38°$; **using perfect phase shifters and no window.**

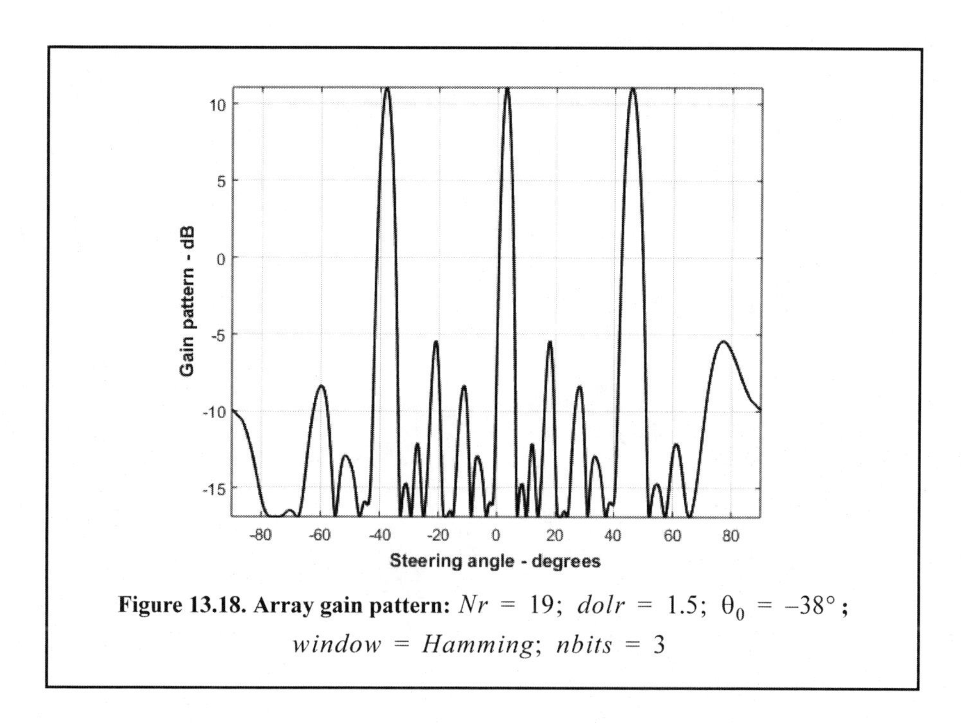

Figure 13.18. Array gain pattern: $Nr = 19$; $dolr = 1.5$; $\theta_0 = -38°$; $window = Hamming$; $nbits = 3$

13.5. Planar Arrays

Planar arrays are a natural extension of linear arrays. Planar arrays can take on many configurations, depending on the element spacing and distribution defined by a "grid." Examples include rectangular, rectangular with circular boundary, hexagonal with circular boundary, circular, and concentric circular grids, as illustrated in Figure 13.19.

Planar arrays can be steered in elevation and azimuth $((\theta, \phi)$, as illustrated in Figure 13.20 for a rectangular grid array. The element spacing along the x- and y-directions are respectively denoted by d_x and d_y. The total electric field at a far field observation point for any planar array can be computed using Eqs. (13.24) and (13.25).

13.5.1. Rectangular Grid Arrays

Consider the $N \times M$ rectangular grid as shown in Figure 13.20. The dot product $\vec{r}_i \bullet \vec{r}_0$, where the vector $\vec{r}_i$ is the vector to the ith element in the array and $\vec{r}_0$ is the unit vector to the far field observation point, can be broken linearly into its x- and y-components. It follows that the electric field components due to the elements distributed along the x- and y-directions are, respectively,

$$E_x(\theta, \phi) = \sum_{n=1}^{N} I_{x_n} e^{j(n-1)kd_x \sin\theta \cos\phi} \quad \text{Eq. (13.54)}$$

$$E_y(\theta, \phi) = \sum_{m=1}^{N} I_{y_m} e^{j(m-1)kd_y \sin\theta \sin\phi} . \quad \text{Eq. (13.55)}$$

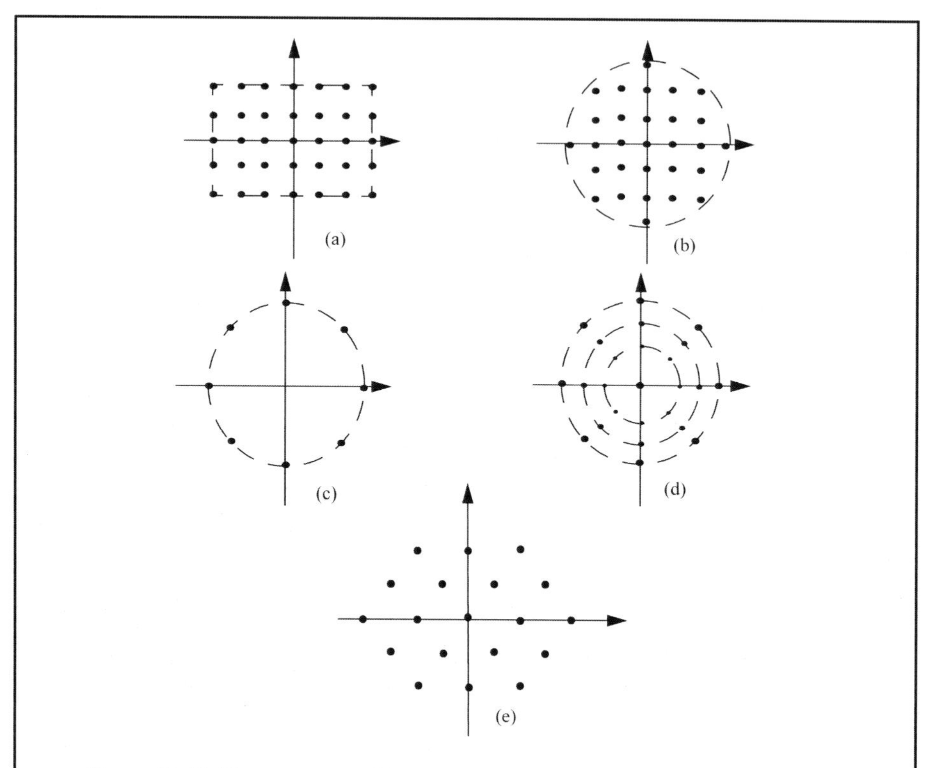

Figure 13.19. Planar array grids. (a) Rectangular, (b) rectangular with circular boundary, (c) circular, (d) concentric circular, and (e) hexagonal.

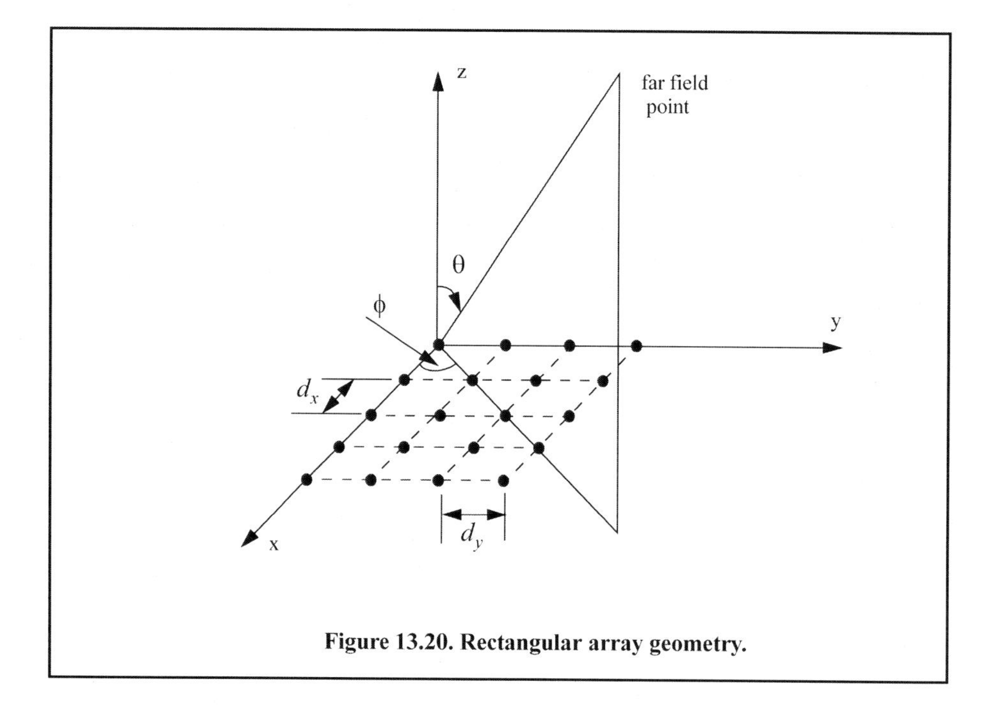

Figure 13.20. Rectangular array geometry.

The total electric field at the far field observation point is then given by

$$E(\theta, \phi) = E_x\ E_y = \left(\sum_{m=1}^{N} I_{y_m} e^{j(m-1)kd_y \sin\theta \sin\phi}\right)\left(\sum_{n=1}^{N} I_{x_n} e^{j(n-1)kd_x \sin\theta \cos\phi}\right). \qquad \text{Eq. (13.56)}$$

Eq. (13.56) can be expressed in terms of the directional cosines

$$u = \sin\theta\cos\phi$$
$$v = \sin\theta\sin\phi \qquad \text{Eq. (13.57)}$$

$$\phi = \operatorname{atan}\left(\frac{u}{v}\right); \qquad \theta = \operatorname{asin}\sqrt{u^2+v^2}. \qquad \text{Eq. (13.58)}$$

The visible region is then defined by

$$\sqrt{u^2+v^2} \le 1. \qquad \text{Eq. (13.59)}$$

It is very common to express a planar array's ability to steer the beam in space in terms of the U, V space instead of the angles θ, ϕ. Figure 13.21 shows how a beam steered in a certain θ, ϕ direction is translated into U, V space.

The rectangular array one-way intensity pattern is then equal to the product of the individual patterns. More precisely for a uniform excitation ($I_{y_m} = I_{x_n} = const$),

$$E(\theta, \phi) = \left|\frac{\sin\left(\frac{Nkd_x \sin\theta\cos\phi}{2}\right)}{\sin\left(\frac{kd_x \sin\theta\cos\phi}{2}\right)}\right| \left|\frac{\sin\left(\frac{Nkd_y \sin\theta\sin\phi}{2}\right)}{\sin\left(\frac{kd_y \sin\theta\sin\phi}{2}\right)}\right|. \qquad \text{Eq. (13.60)}$$

The radiation pattern maxima, nulls, sidelobes, and grating lobes in both the x- and y-axes are computed in a similar fashion to the linear array case. Additionally, the same conditions for grating lobe control are applicable. Note the symmetry is about the angle ϕ.

13.5.2. Circular Grid Arrays

The geometry of interest is shown in Figure 13.19c. In this case, N elements are distributed equally on the outer circle whose radius is a. For this purpose, consider the geometry shown in Figure 13.22. From the geometry

$$\Phi_n = \frac{2\pi}{N} n \qquad ; n = 1, 2, \ldots, N. \qquad \text{Eq. (13.61)}$$

The coordinates of the nth element are

$$x_n = a\ \cos\Phi_n \qquad ; y_n = a\ \sin\Phi_n \qquad ; z_n = 0. \qquad \text{Eq. (13.62)}$$

It follows that

$$k(\vec{r}_n \bullet \vec{r}_0) = \Psi_n = k(a\sin\theta\cos\phi\cos\Phi_n + a\sin\theta\sin\phi\sin\Phi_n + 0). \qquad \text{Eq. (13.63)}$$

Equation (13.63) can be rearranged as

$$\Psi_n = ak\sin\theta(\cos\phi\cos\Phi_n + \sin\phi\sin\Phi_n). \qquad \text{Eq. (13.64)}$$

By using the identity $\cos(A-B) = \cos A\cos B + \sin A\sin B$, Eq.(13.63) collapses to

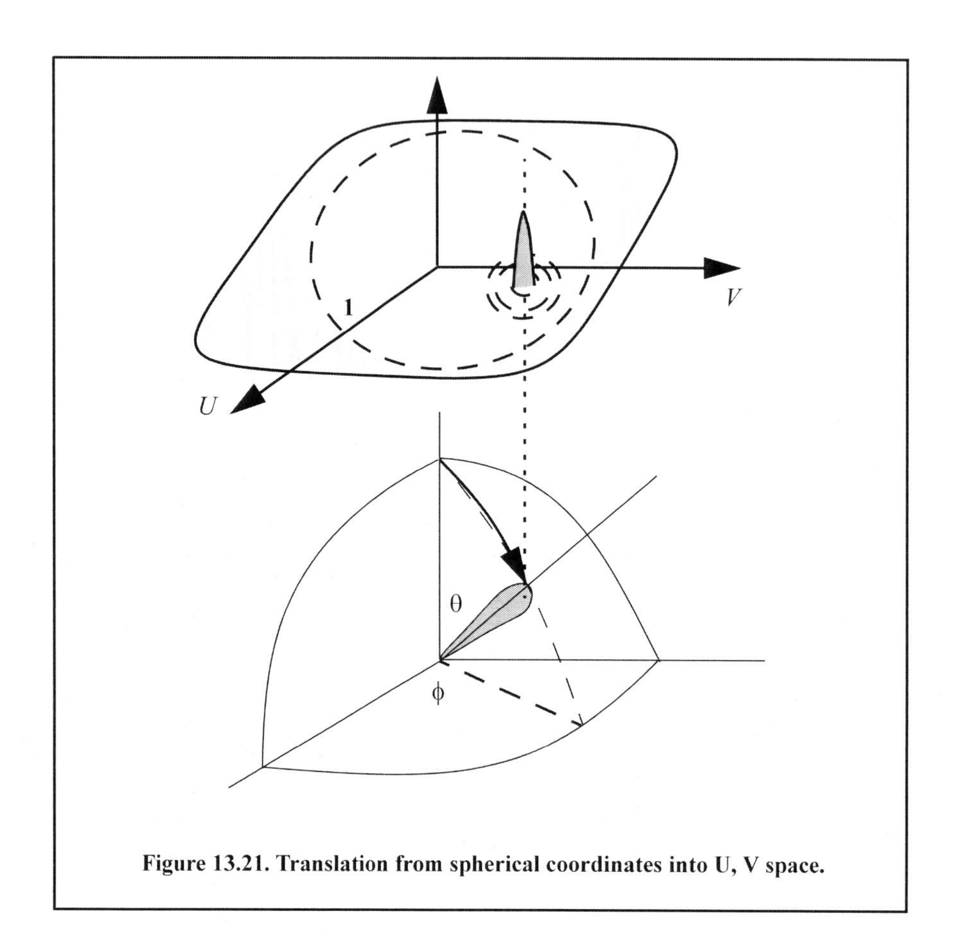

Figure 13.21. Translation from spherical coordinates into U, V space.

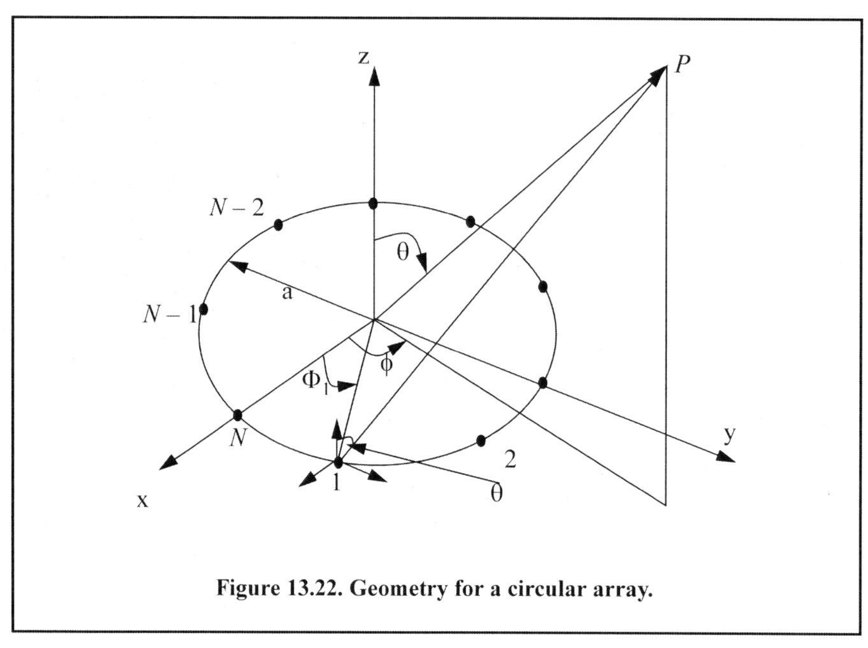

Figure 13.22. Geometry for a circular array.

$$\Psi_n = ak\sin\theta\cos(\Phi_n - \phi) .\qquad \text{Eq. (13.65)}$$

Finally, by using Eq. (13.25), the far field electric field is then given by

$$E(\theta, \phi;a) = \sum_{n=1}^{N} I_n \exp\left\{j\frac{2\pi a}{\lambda}\sin\theta\cos(\Phi_n - \phi)\right\}\qquad \text{Eq. (13.66)}$$

where I_n represents the complex current distribution for the nth element. When the array main beam is directed in the (θ_0, ϕ_0), Eq. (13.65) takes on the following form

$$E(\theta, \phi;a) = \sum_{n=1}^{N} I_n \exp\left\{j\frac{2\pi a}{\lambda}[\sin\theta\cos(\Phi_n - \phi) - \sin\theta_0\cos(\Phi_n - \phi_0)]\right\} .\qquad \text{Eq. (13.67)}$$

Consider the following two cases with inputs defined by

Parameter	Case I	Case II
a	*1.*	*1.5*
N	*10*	*10*
θ_0	*45*	*45*
ϕ_0	*60*	*60*
variation	*'Theta'*	*'Phi'*
ϕ_d	*60*	*60*
θ_d	*45*	*45*

Figures 13.23 and 13.24 respectively show the array patterns in relative amplitude and the power patterns versus the angle θ corresponding to Case I parameters. Figures 13.25 and 13.26 are similar to Figures 13.23 and 13.24, except in this case the patterns are plotted in polar coordinates.

Figure 13.27 shows a plot of the normalized single element pattern (upper left corner), the normalized array factor (upper right corner), the total array pattern (lower left corner). Figures 13.28 through 13.32 are similar to those in Figures 13.23 through 15.27, except in this case the input parameters correspond to Case II.

13.5.3. Concentric Grid Circular Arrays

The geometry of interest is shown in Figure 13.33. In this case, N_2 elements are distributed equally on the outer circle whose radius is a_2, while another N_1 elements are linearly distributed on the inner circle whose radius is a_1. The element located on the center of both circles is used as the phase reference. In this configuration, there are $N_1 + N_2 + 1$ total elements in the array.

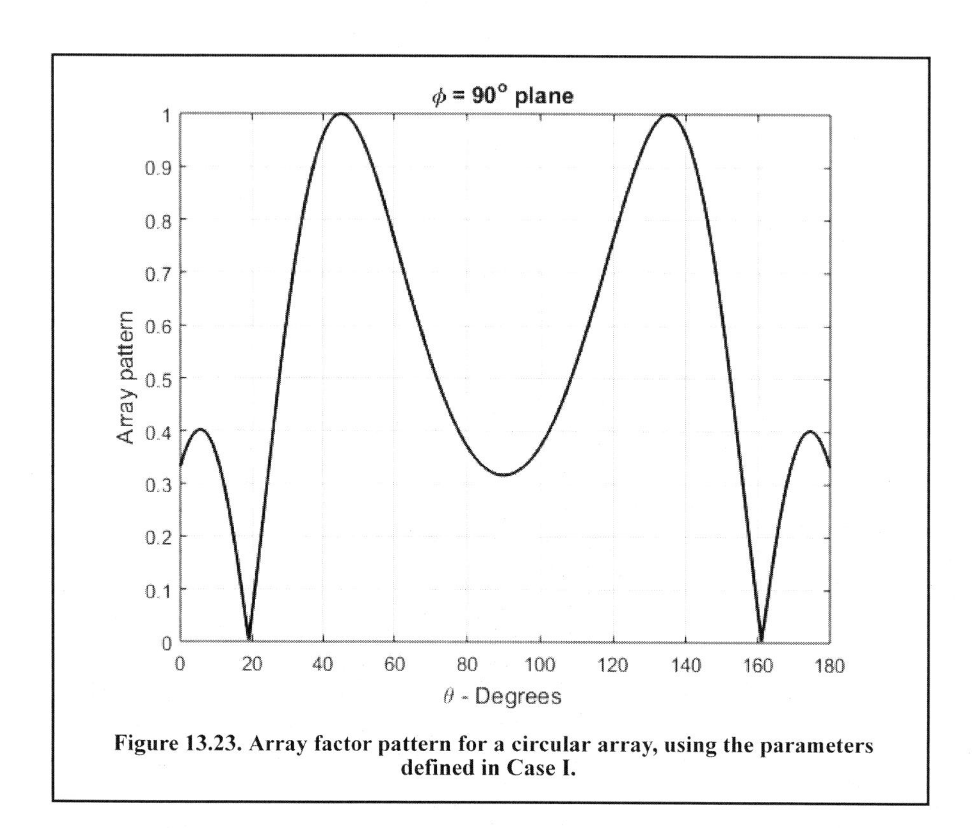

Figure 13.23. Array factor pattern for a circular array, using the parameters defined in Case I.

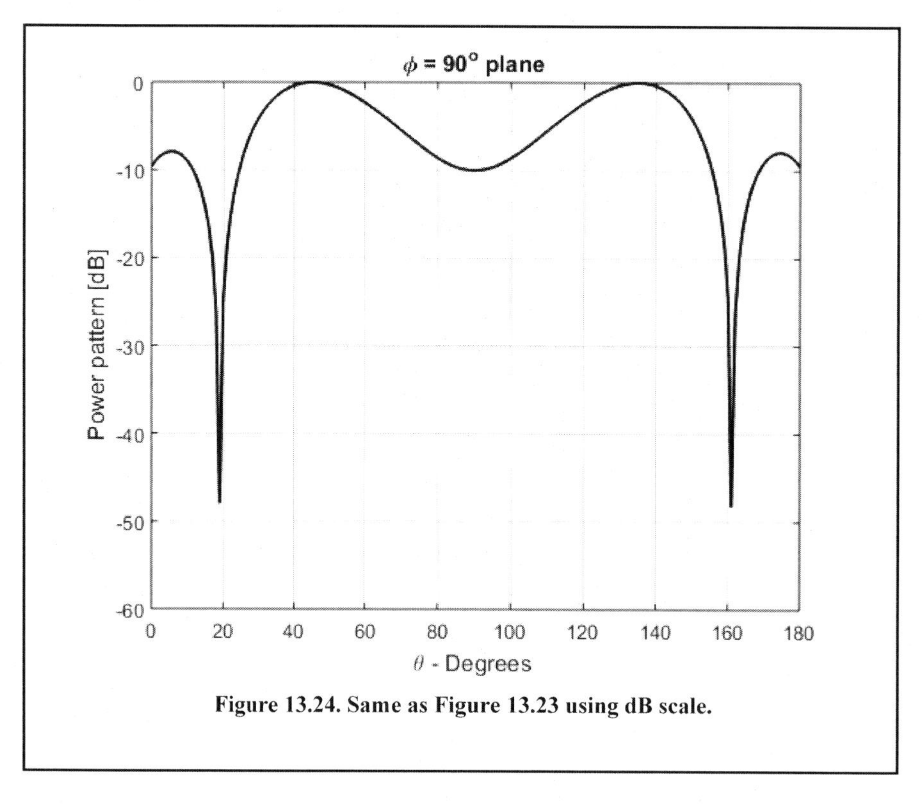

Figure 13.24. Same as Figure 13.23 using dB scale.

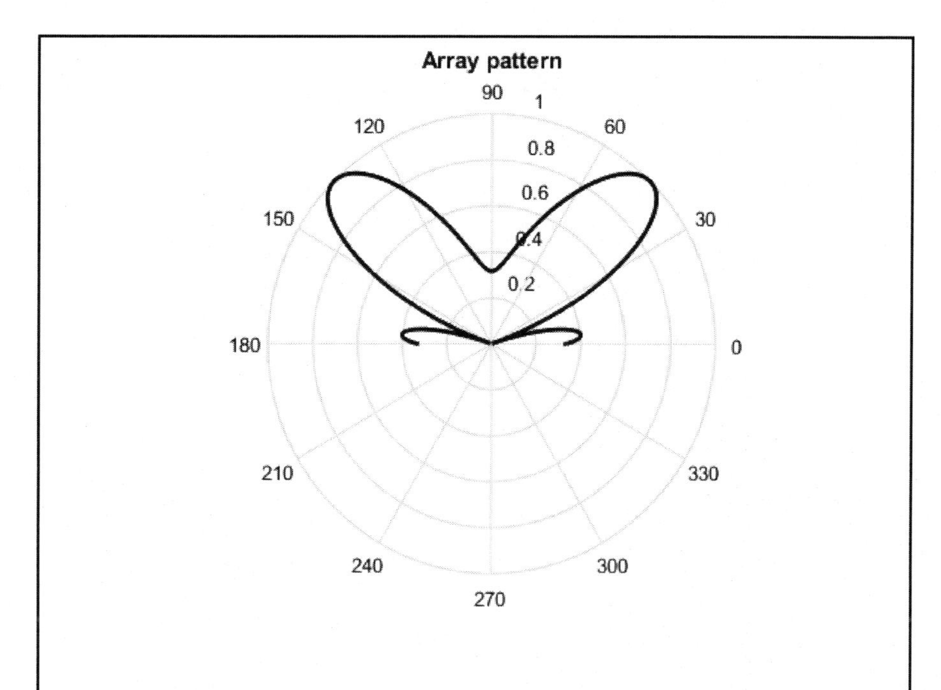

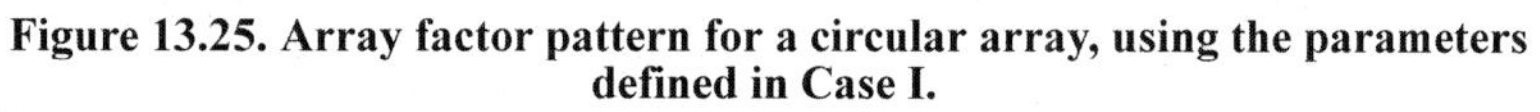

Figure 13.25. Array factor pattern for a circular array, using the parameters defined in Case I.

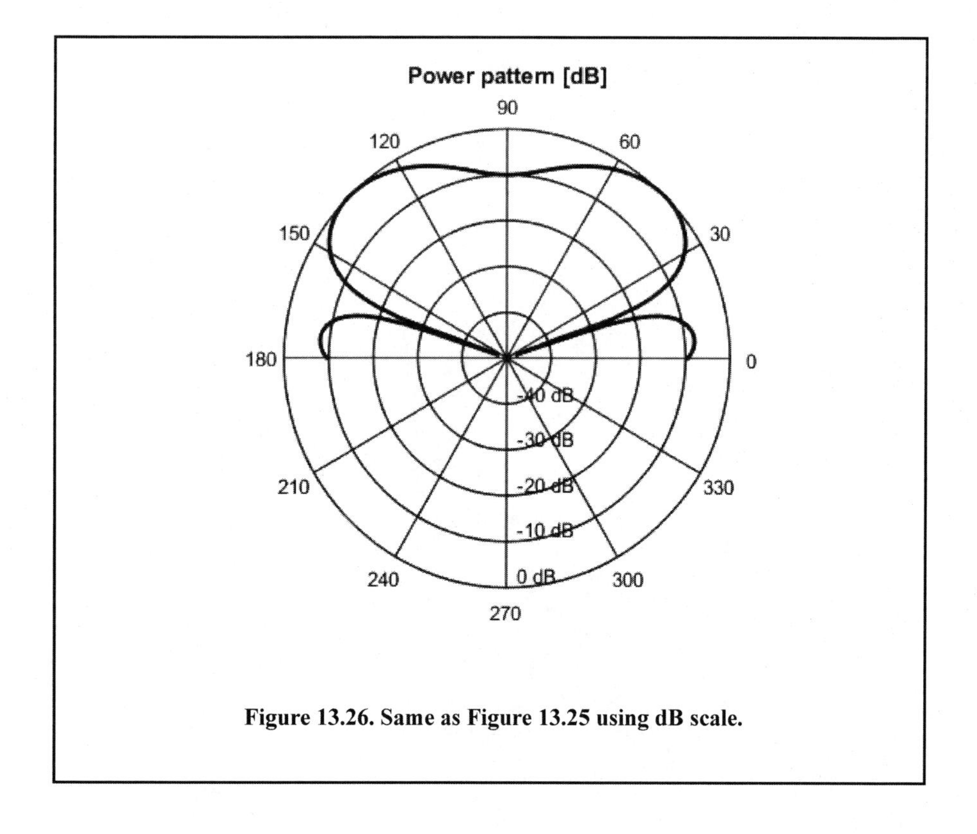

Figure 13.26. Same as Figure 13.25 using dB scale.

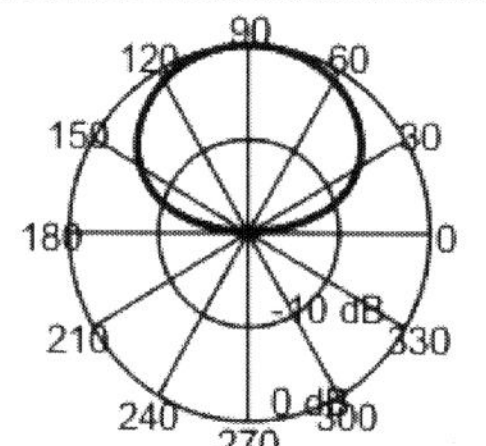

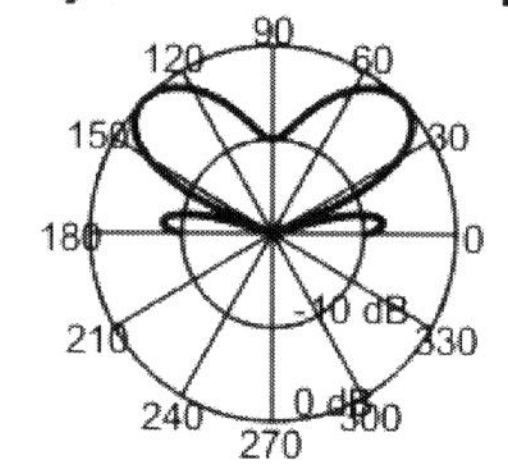

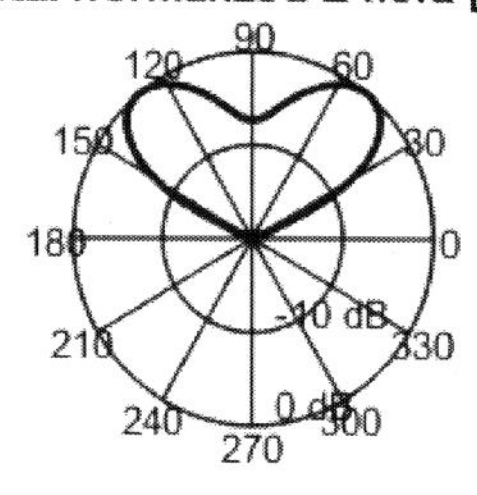

Figure 13.27. Element, array factor and total pattern using the parameters defined in Case I.

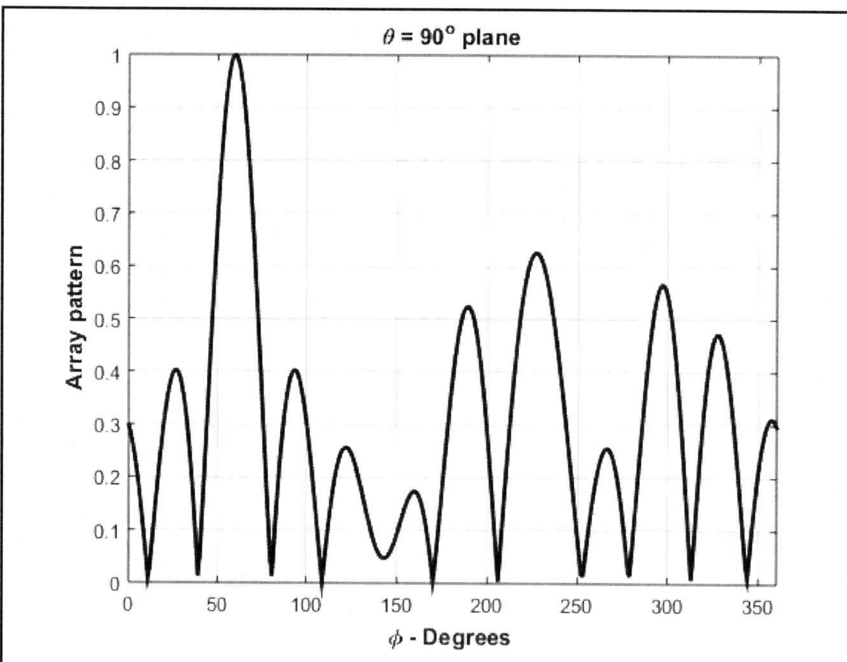

Figure 13.28. Array factor pattern for a circular array, using the parameters defined in Case II.

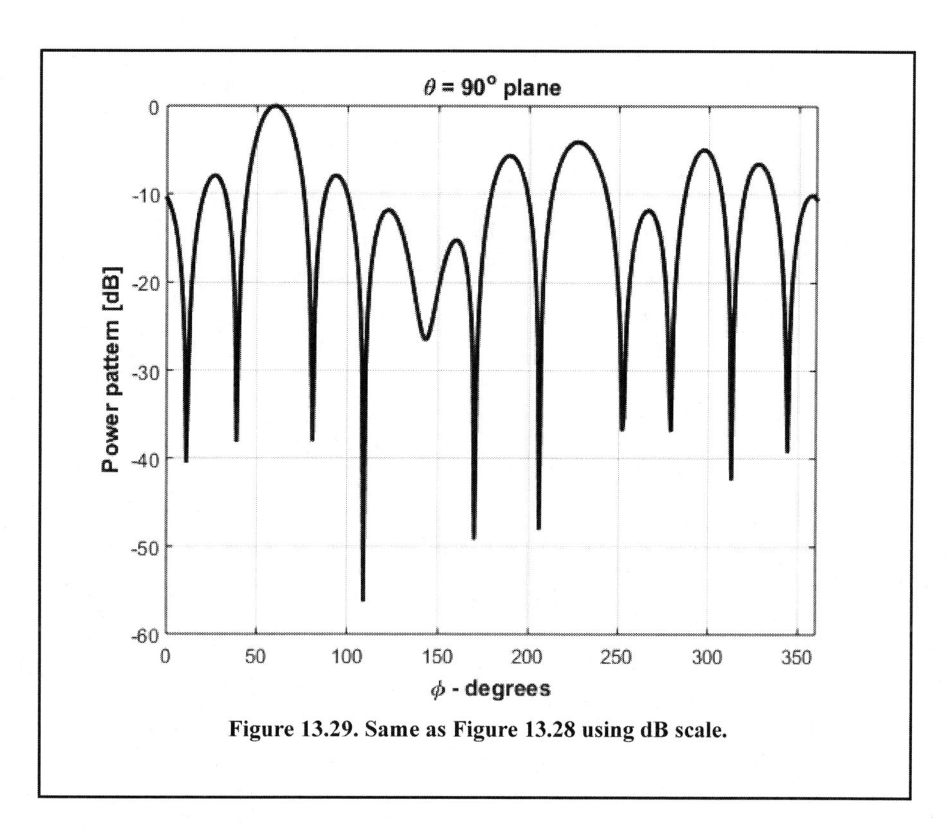

Figure 13.29. Same as Figure 13.28 using dB scale.

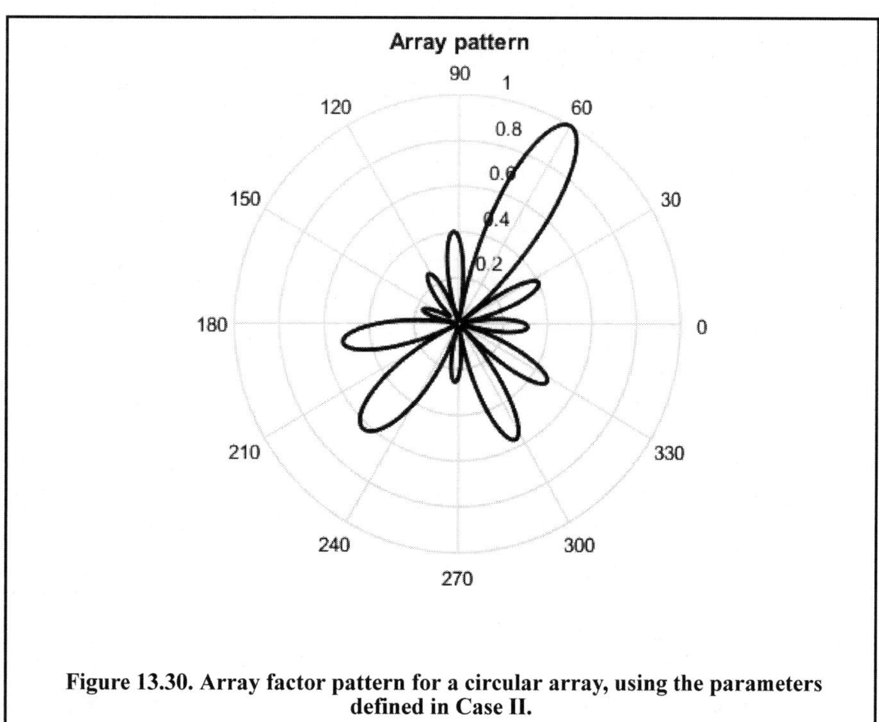

Figure 13.30. Array factor pattern for a circular array, using the parameters defined in Case II.

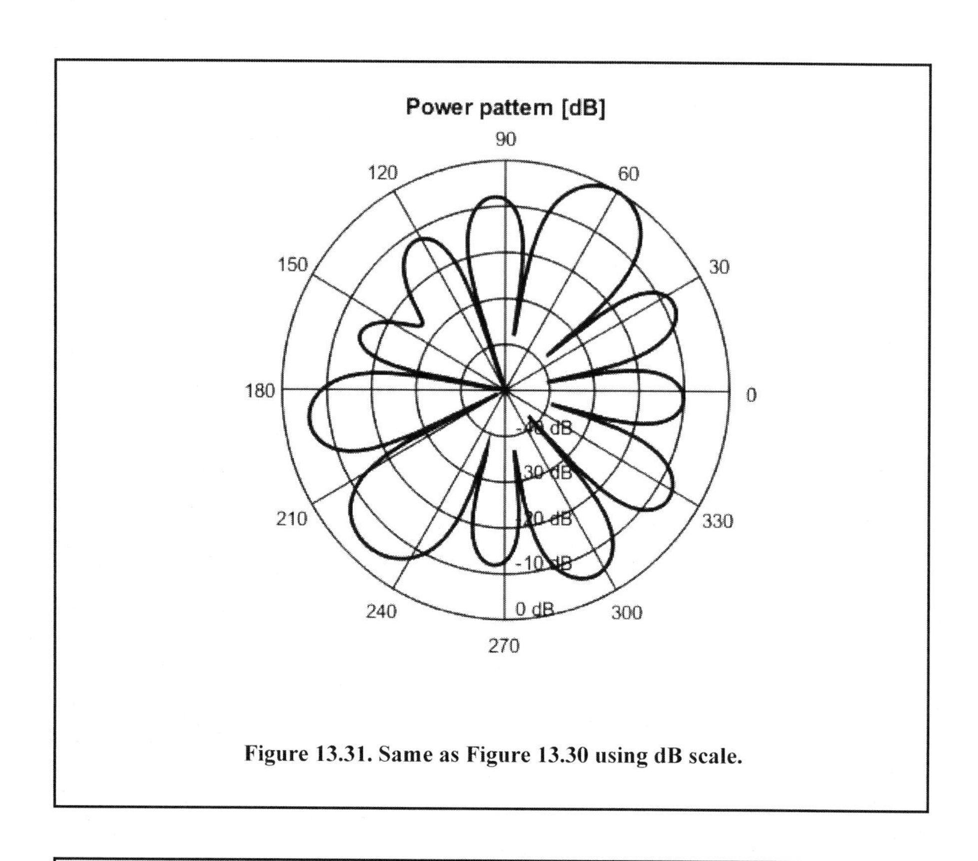

Figure 13.31. Same as Figure 13.30 using dB scale.

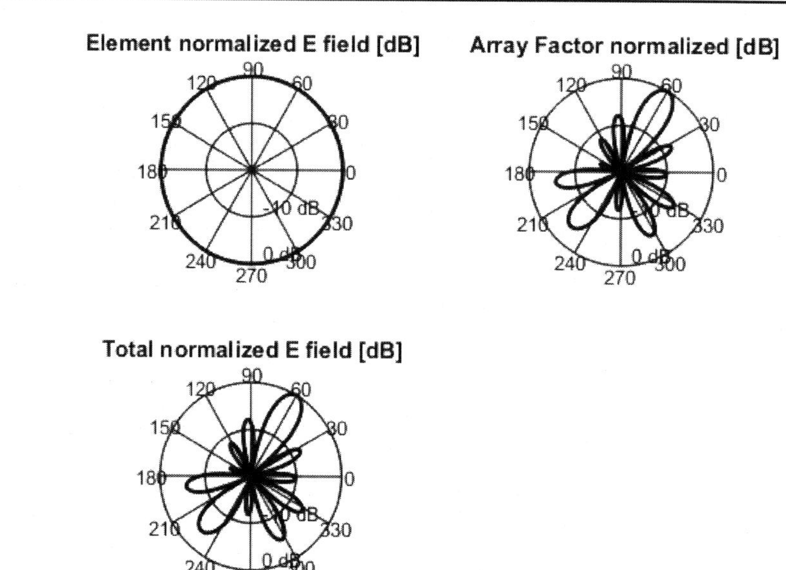

Figure 13.32. Element, array factor and total pattern for the circular using the parameters defined in Case II.

The array factor is derived in two steps. First, the array factor corresponding to a linearly distributed circular array is computed. Second, the overall array factor corresponding to all elements will be the product of each individual circular array times the pattern of the central element. More precisely,

$$E(\theta, \phi) = E_1(\theta, \phi; a_1)E_2(\theta, \phi; a_2)E_0(\theta, \phi) \,. \qquad \text{Eq. (13.68)}$$

Figure 13.34 shows a 3-D plot for a concentric circular array in the θ, ϕ space for the following parameters: $a_1 = 1\lambda$, $N_1 = 8 = N_2$, and $a_2 = 2\lambda$.

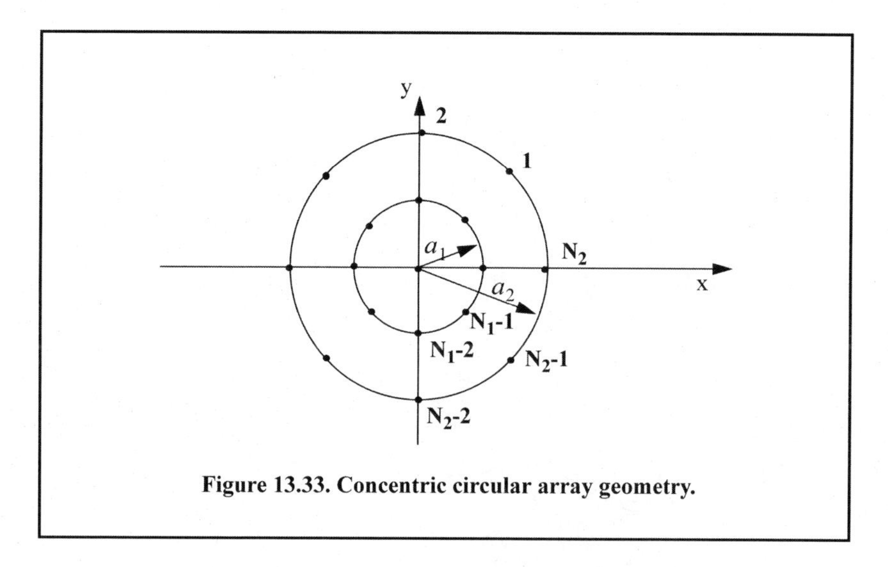

Figure 13.33. Concentric circular array geometry.

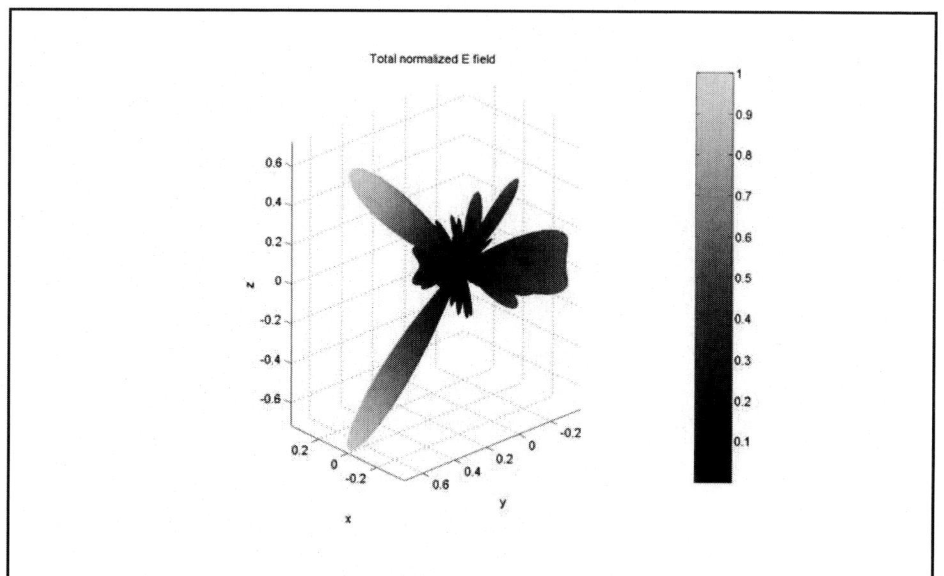

Figure 13.34. A 3-D array pattern - concentric circular array; $\theta = 45°$ and $\phi = 90°$.

13.5.4. Rectangular Grid with Circular Boundary Arrays

The far field electric field associated with this configuration can be easily obtained from that corresponding to a rectangular grid. In order to accomplish this task, follow these steps: first, select the desired maximum number of elements along the diameter of the circle and denote it by N_d. Also select the associated element spacings d_x, d_y. Define a rectangular array of size $N_d \times N_d$. Draw a circle centered at $(x, y) = (0, 0)$ with radius r_d where

$$r_d = \frac{N_d - 1}{2} + \Delta x \qquad \text{Eq. (13.69)}$$

and $\Delta x \le d_x/4$. Finally, modify the weighting function across the rectangular array by multiplying it with the two-dimensional sequence $a(m, n)$, where

$$a(m, n) = \begin{Bmatrix} 1 & , \; if \; dis \; to \; (m, n)th \; element < r_d \\ 0 & ; \; elsewhere \end{Bmatrix} \qquad \text{Eq. (13.70)}$$

where distance, dis, is measured from the center of the circle. This is illustrated in Figure 13.35.

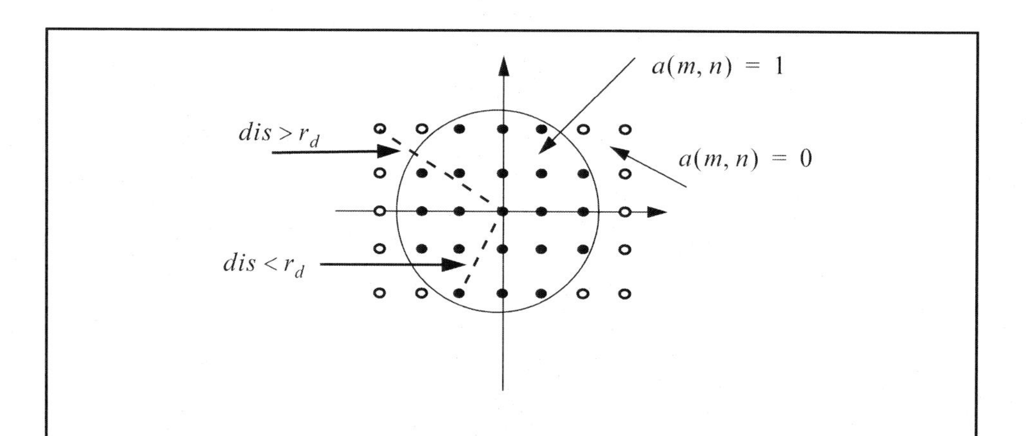

Figure 13.35. Elements with solid dots have $a(m, n) = 1$; others have $a(m, n) = 0$.

13.5.5. Hexagonal Grid Arrays

The analysis provided in this section is limited to hexagonal arrays with circular boundaries. The horizontal element spacing is denoted as d_x, and the vertical element spacing is

$$d_y = (\sqrt{3}\, d_x)/2. \qquad \text{Eq. (13.71)}$$

The array is assumed to have the maximum number of identical elements along the x-axis ($y = 0$). This number is denoted by N_x, where N_x is an odd number, in order to obtain a symmetric array, where an element is present at $(x, y) = (0, 0)$. The number of rows in the array is denoted by M. The horizontal rows are indexed by m, which varies from $-(N_x - 1)/2$ to $(N_x - 1)/2$. The number of elements in the mth row is denoted by N_r and is defined by

$$N_r = N_x - |m| .\qquad \text{Eq. (13.72)}$$

The electric field at a far field observation point is computed using Eq. (13.24) and (13.25). The phase associated with $(m, n)th$ location is

$$\psi_{m,n} = \frac{2\pi d_x}{\lambda}\sin\theta\left[\left(m+\frac{n}{2}\right)\cos\phi + n\frac{\sqrt{3}}{2}\sin\phi\right] .\qquad \text{Eq. (13.73)}$$

Figures 13.37 through 13.42, respectively, show plots of the array gain pattern in the *U*-*V* space, for the following cases:

Syntax: (N_x, N_y, d_x, d_y, azimuth steering angle, elevation steering angle, window type (negative means no window), # of bits (negative means perfect phase shifters)).

Case I: (15, 15, 0.5, 0.5, 0, 0, -1, -1, -3)
Case II: (15, 15, 0.5, 0.5, 20, 30, -1, -1, -3)
Case III: (15, 15, 0.5, 0.5, 45, 45, 1, 'Hamming', -3)
Case IV: (15, 15, 0.5, 0.5, 10, 20, -1, -1, 3)
Case V: (15, 15, 1, 0.5, 20, 25, -1, -1, -3)
Case VI: (15, 15, 1.25, 1.25, 0, 0, -1, -1, -3)

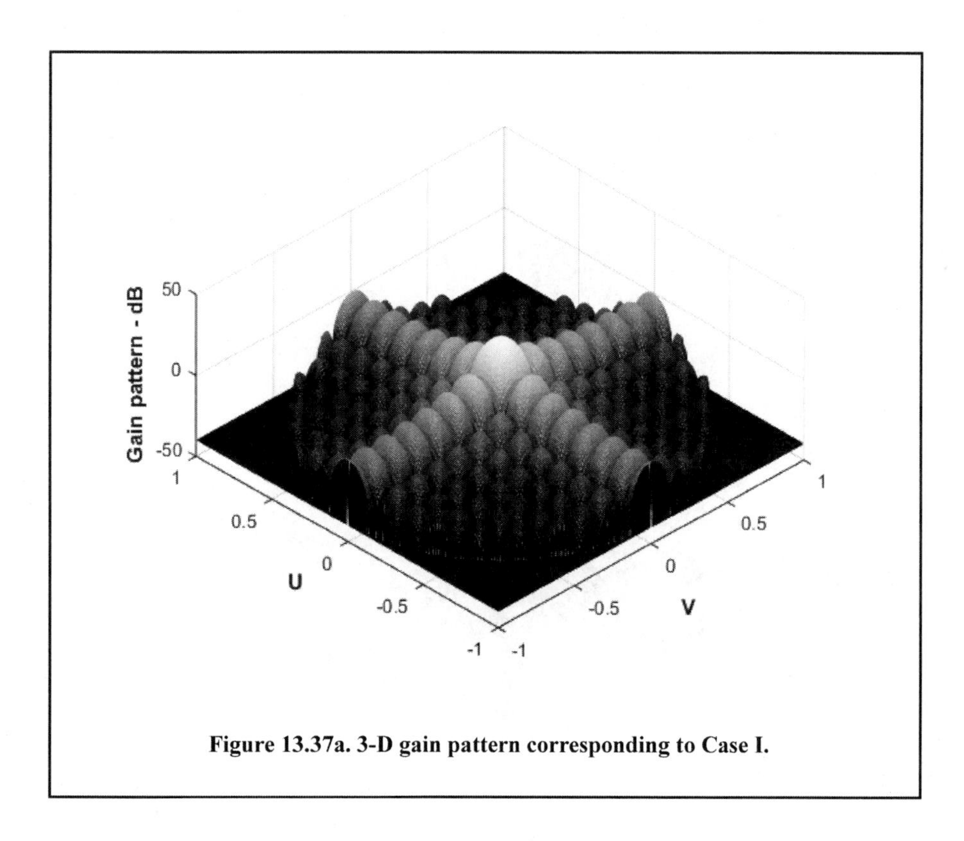

Figure 13.37a. 3-D gain pattern corresponding to Case I.

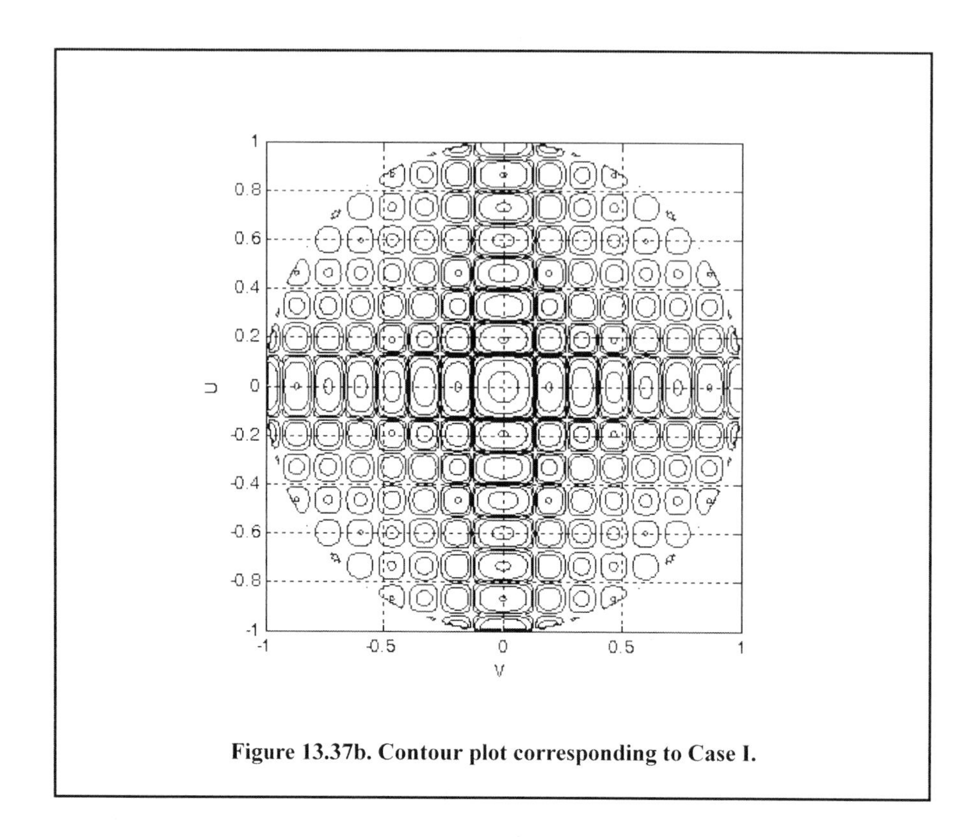

Figure 13.37b. Contour plot corresponding to Case I.

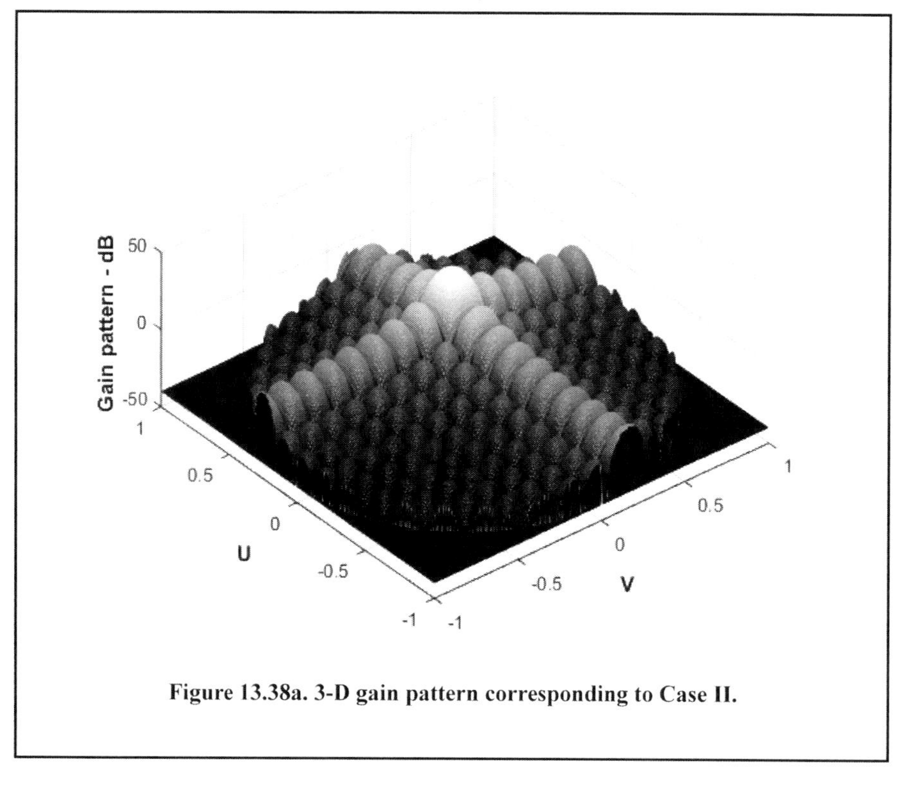

Figure 13.38a. 3-D gain pattern corresponding to Case II.

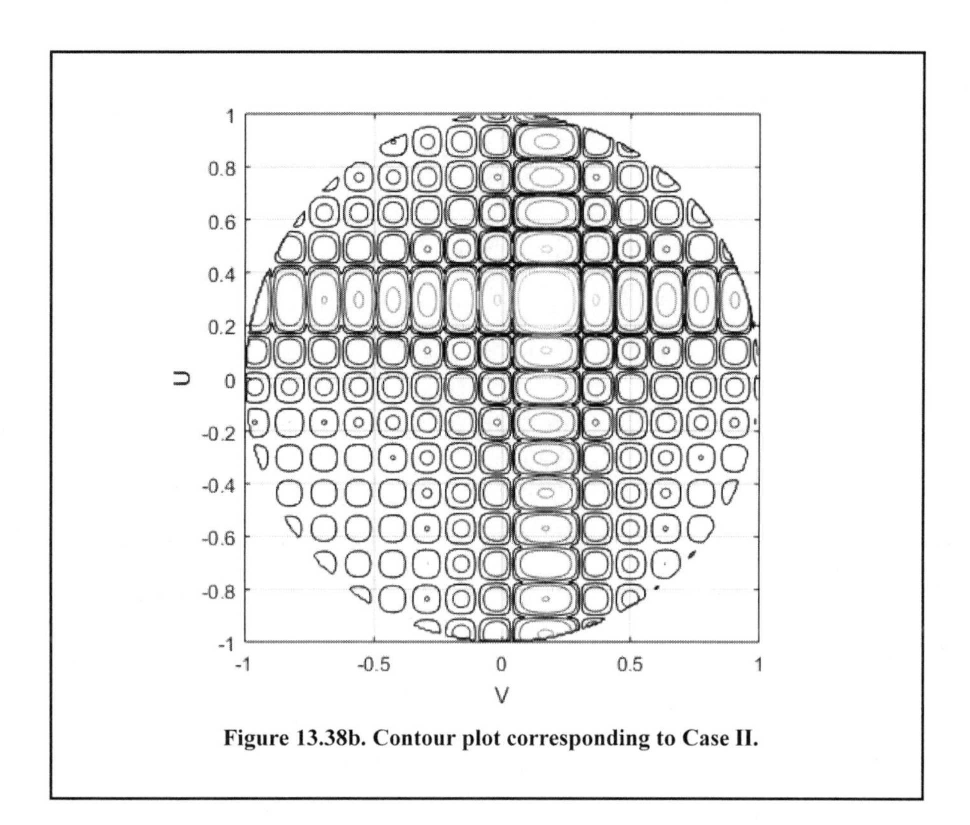

Figure 13.38b. Contour plot corresponding to Case II.

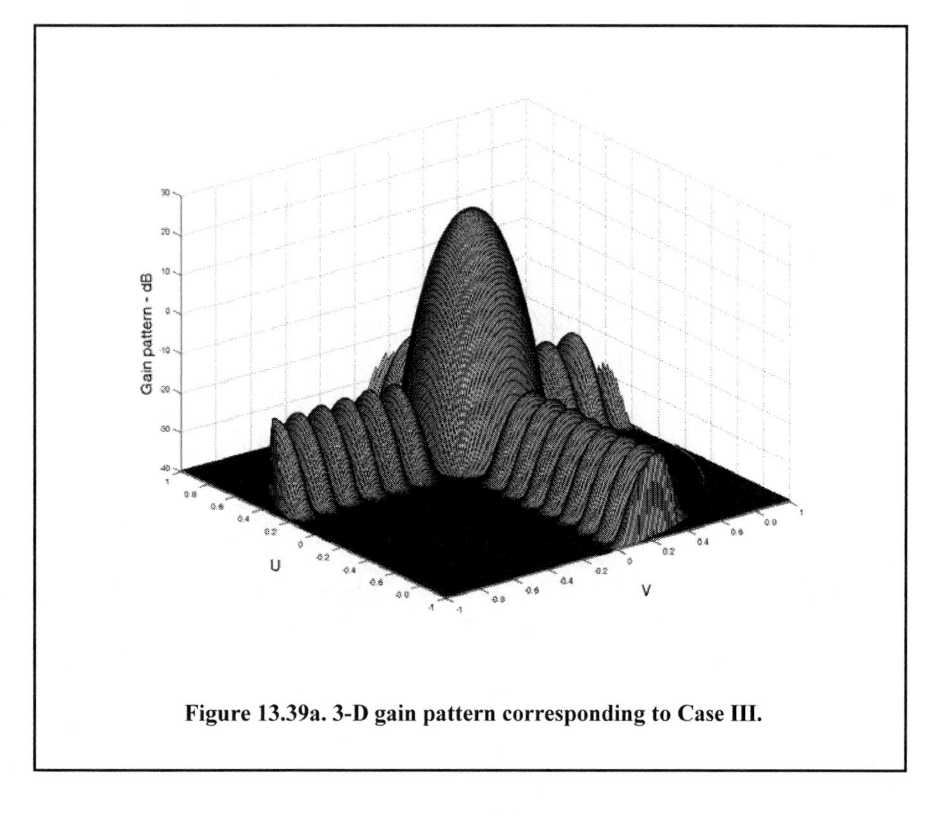

Figure 13.39a. 3-D gain pattern corresponding to Case III.

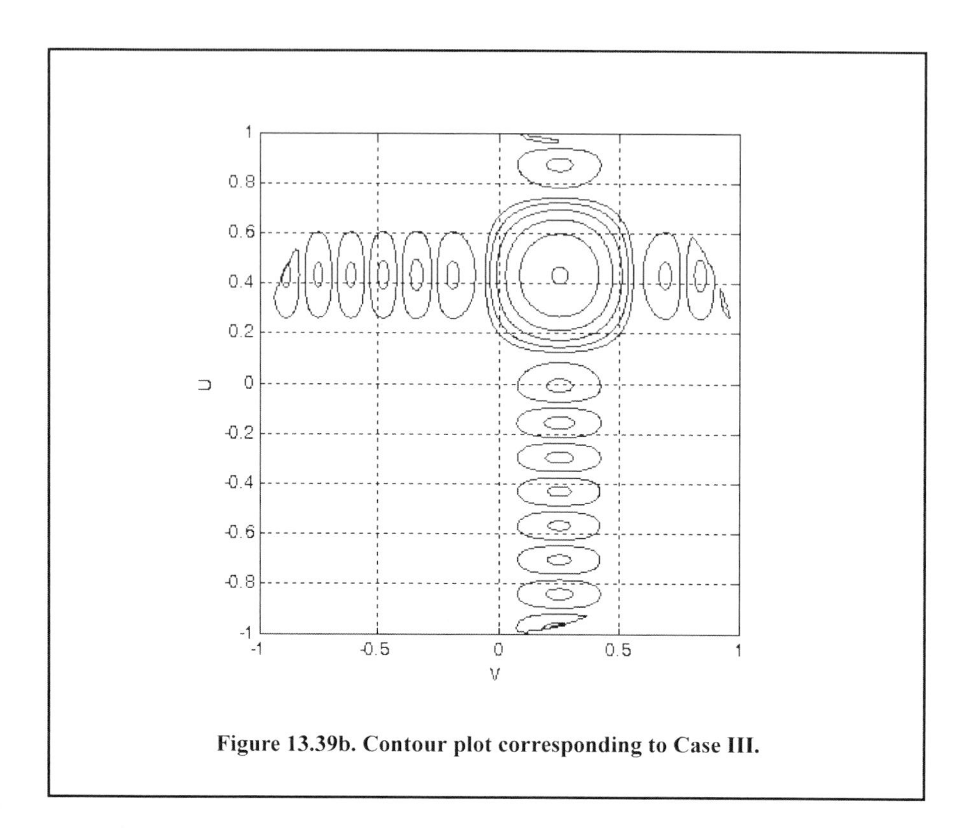

Figure 13.39b. Contour plot corresponding to Case III.

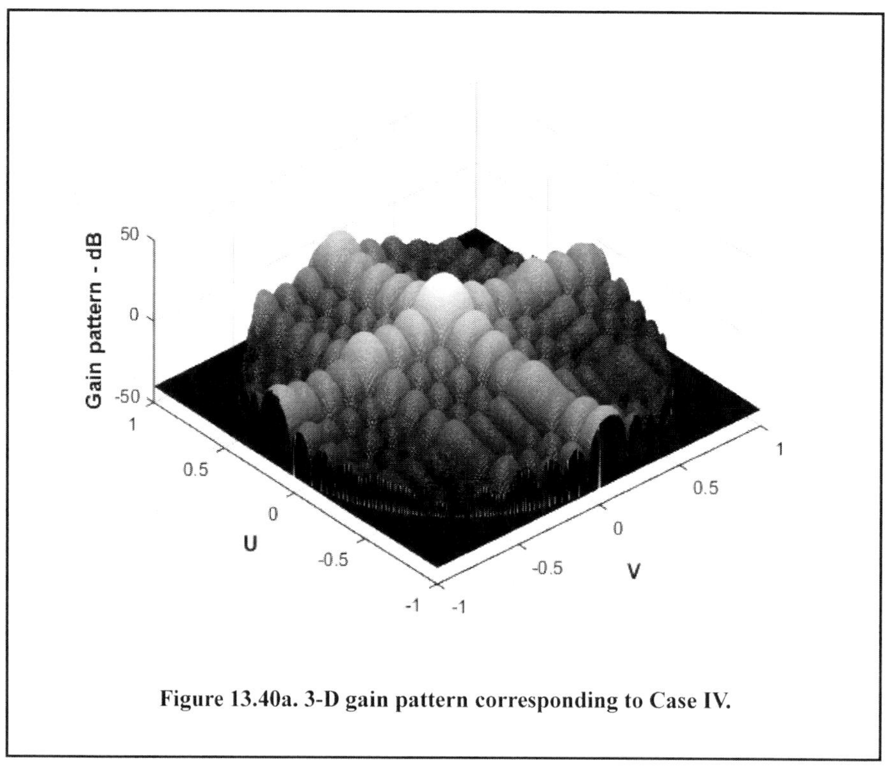

Figure 13.40a. 3-D gain pattern corresponding to Case IV.

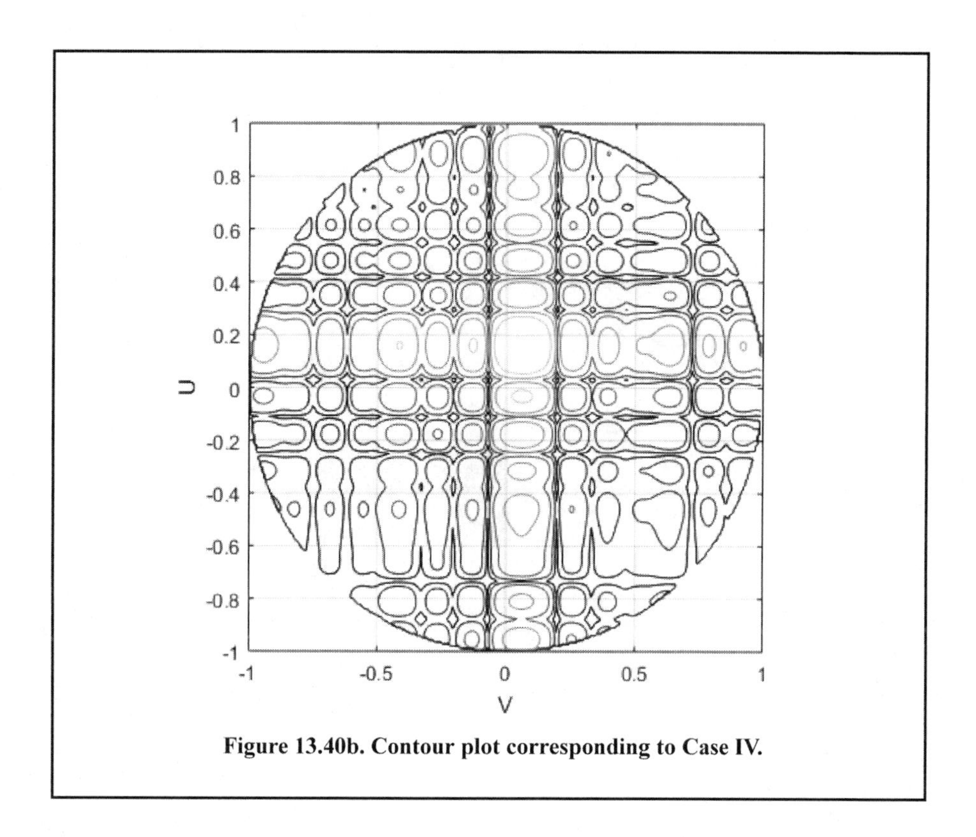

Figure 13.40b. Contour plot corresponding to Case IV.

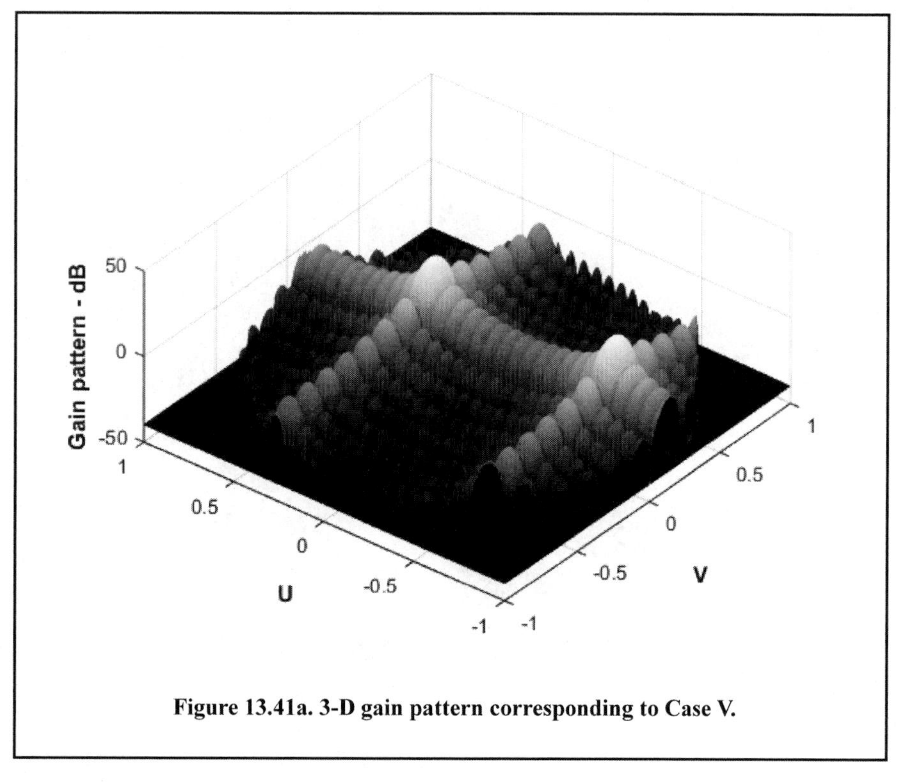

Figure 13.41a. 3-D gain pattern corresponding to Case V.

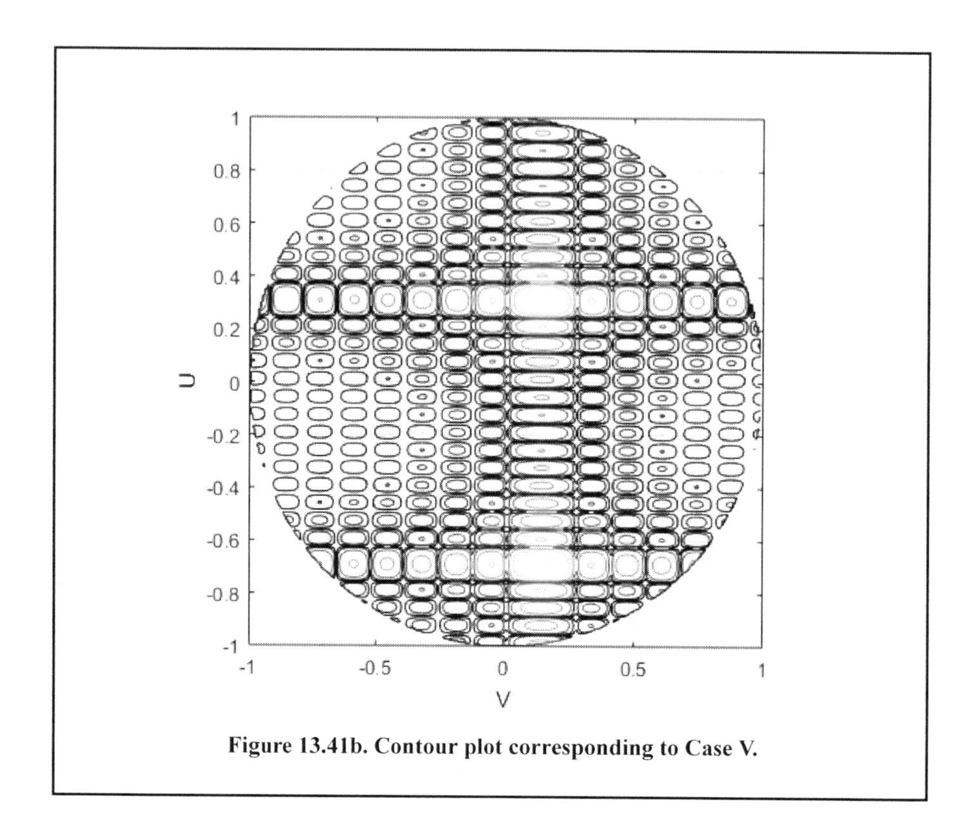

Figure 13.41b. Contour plot corresponding to Case V.

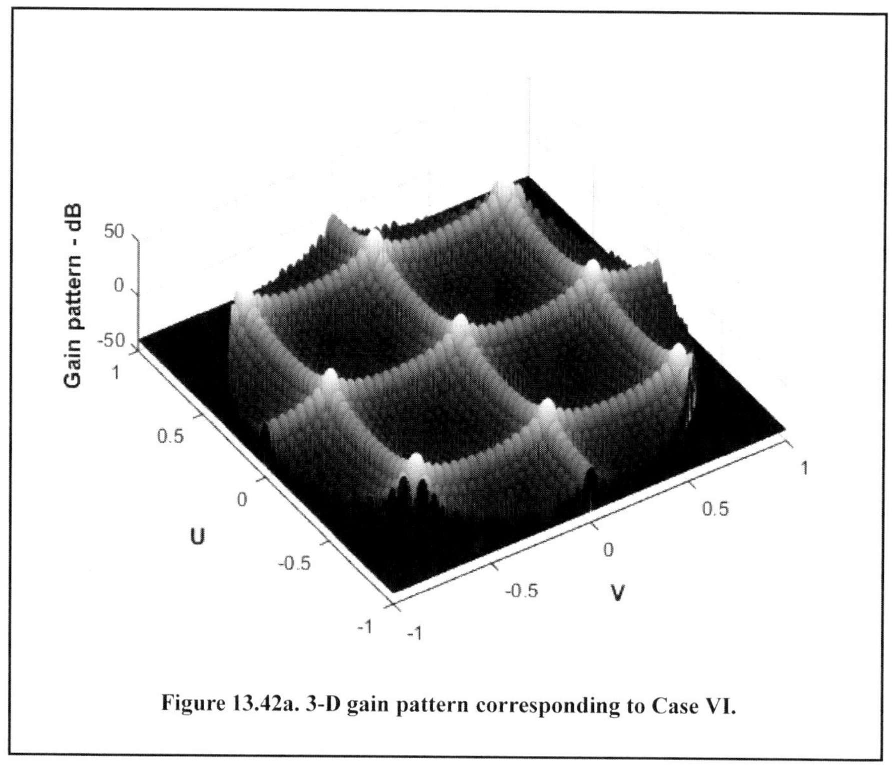

Figure 13.42a. 3-D gain pattern corresponding to Case VI.

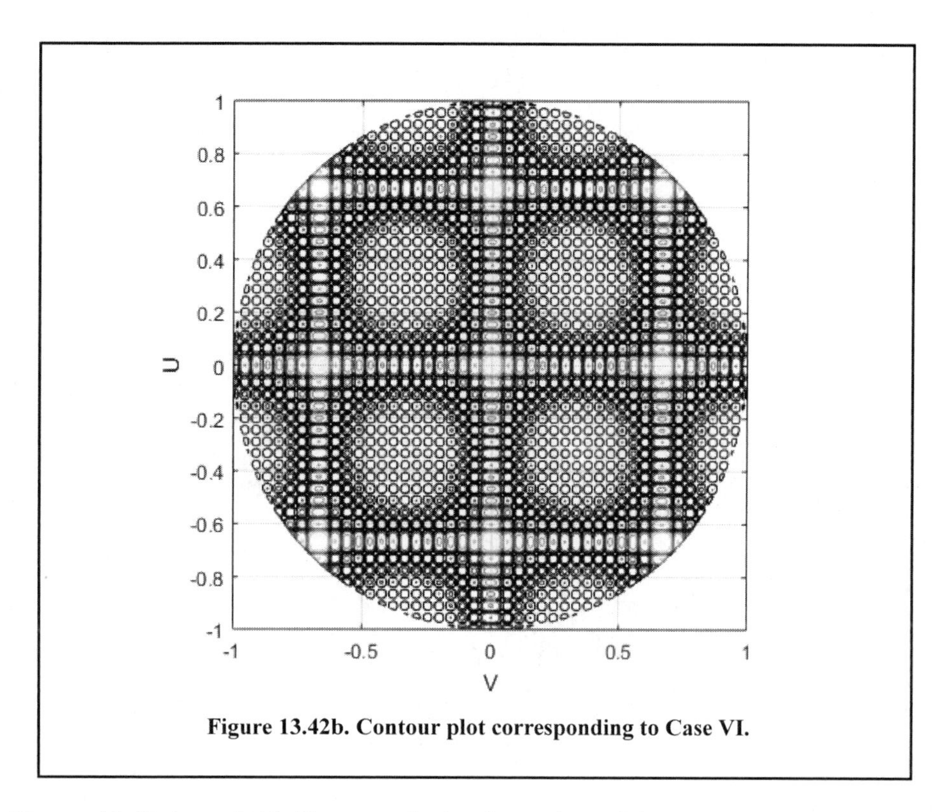

Figure 13.42b. Contour plot corresponding to Case VI.

Figures 13.43 through 13.48, respectively, show plots of the array gain pattern in the *U-V* space, for the following cases:

Syntax: (N, d_x, d_y, azimuth steering angle, elevation steering angle, window type (negative means no window), # of bits (negative means perfect phase shifters)).

Case I: (15, 0.5, 0.5, 0, 0, -1, -1, -3)
Case II: (15, 0.5, 0.5, 20, 30, -1, -1, -3)
Case III: (15, 0.5, 0.5, 30, 30, 1, 'Hamming', -3)
Case IV: (15, 0.5, 0.5, 30, 30, -1, -1, 3)
Case V: (15, 1, 0.5, 30, 30, -1, -1, -3)
Case VI: (15, 1, 1, 0, 0, -1, -1, -3)

13.6. Array Scan Loss

Phased arrays experience gain loss when the beam is steered away from the array boresight, or zenith (normal to the face of the array). This loss is due to the fact that the array effective aperture becomes smaller, and consequently the array beamwidth is broadened, as illustrated in Figure 13.49. This loss in antenna gain is called scan loss, L_{scan}, where

$$L_{scan} = \left(\frac{A}{A_\theta}\right)^2 = \left(\frac{G}{G_\theta}\right)^2. \quad \text{Eq. (13.74)}$$

A_θ is the effective aperture area at scan angle θ, and G_θ is the effective array gain at the same angle.

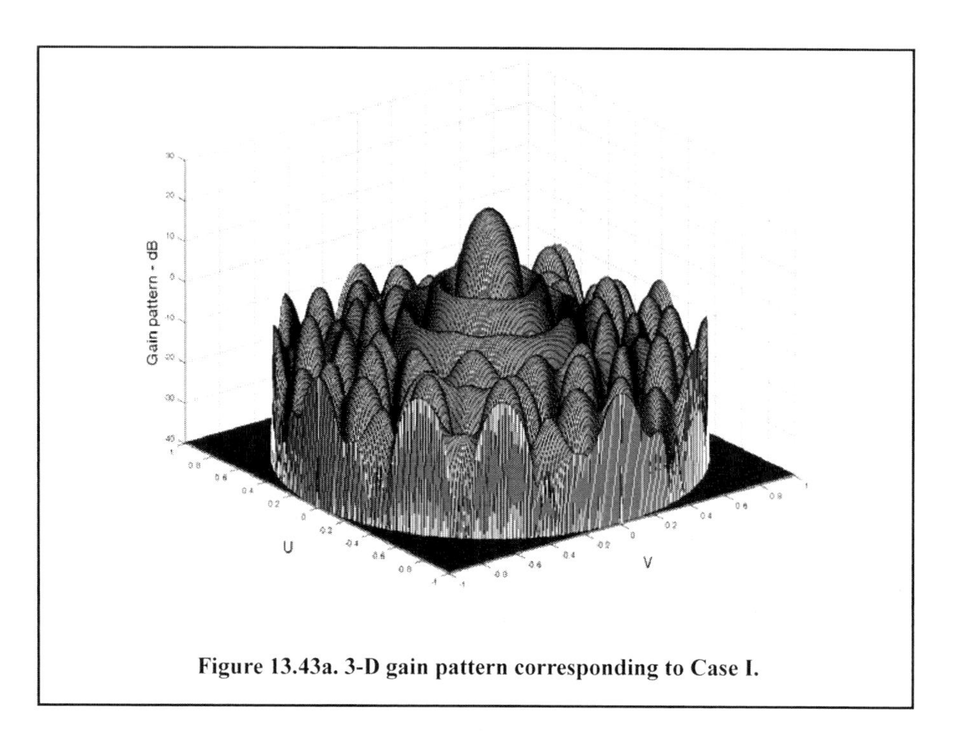

Figure 13.43a. 3-D gain pattern corresponding to Case I.

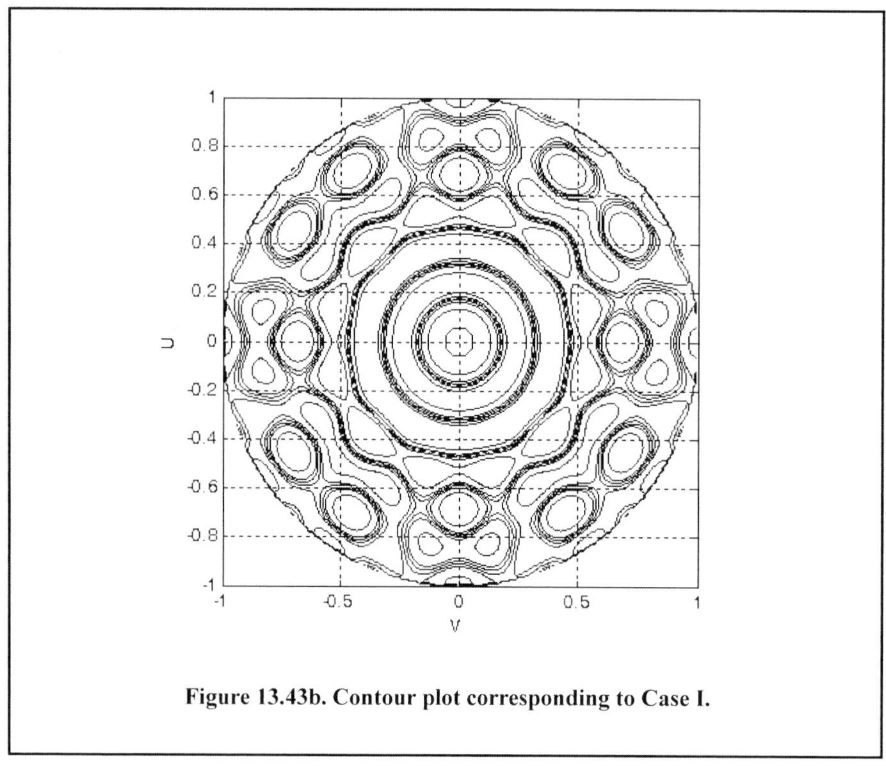

Figure 13.43b. Contour plot corresponding to Case I.

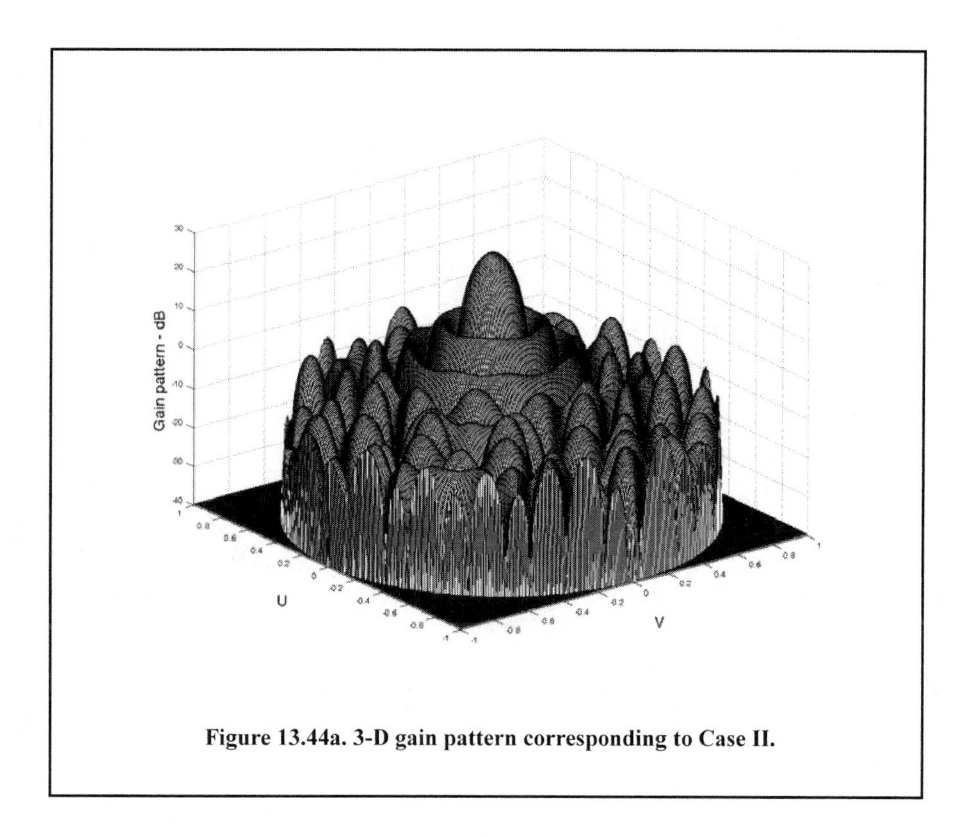

Figure 13.44a. 3-D gain pattern corresponding to Case II.

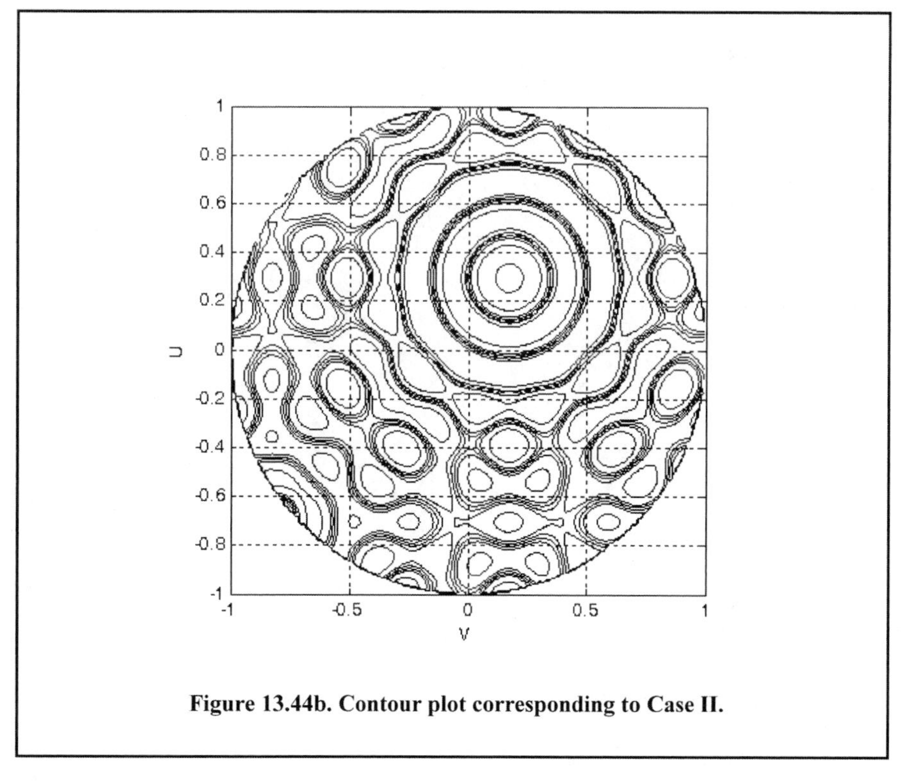

Figure 13.44b. Contour plot corresponding to Case II.

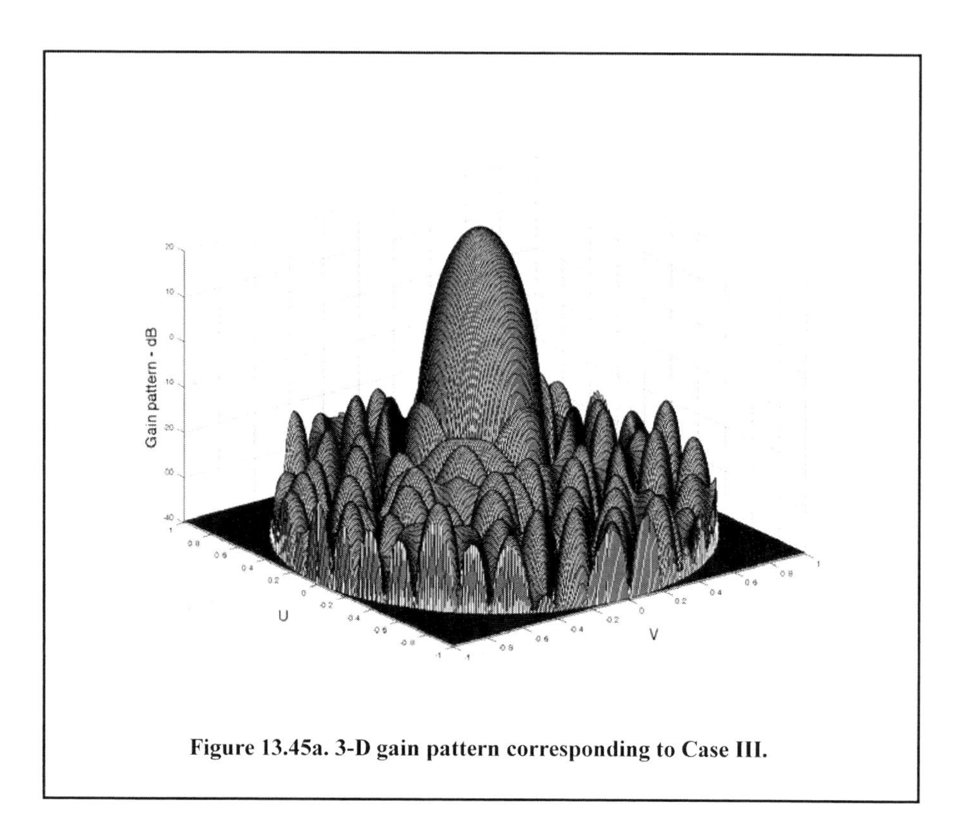

Figure 13.45a. 3-D gain pattern corresponding to Case III.

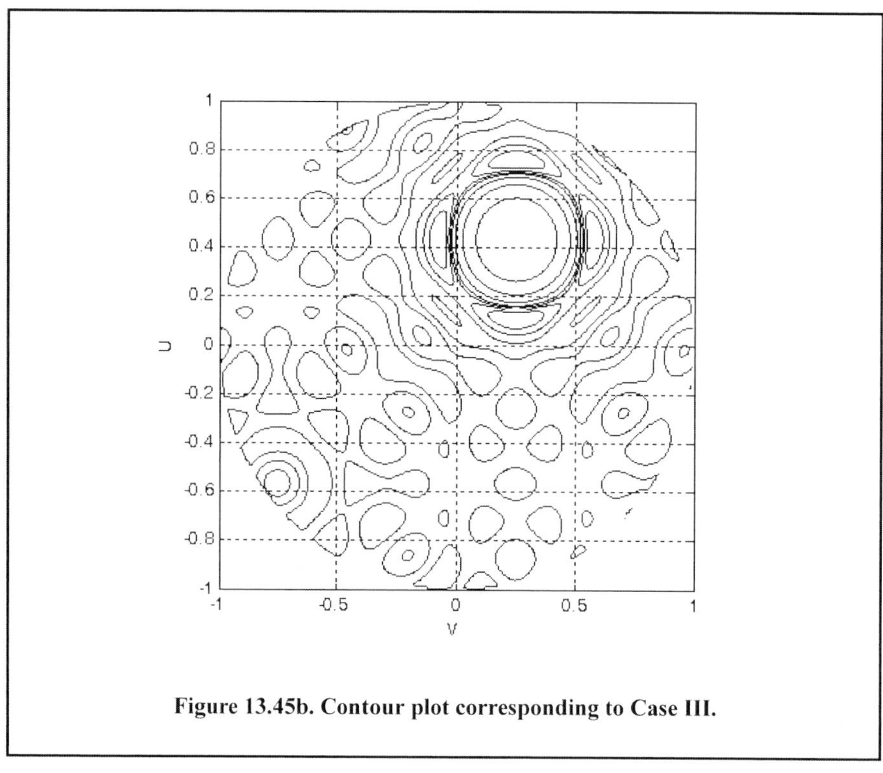

Figure 13.45b. Contour plot corresponding to Case III.

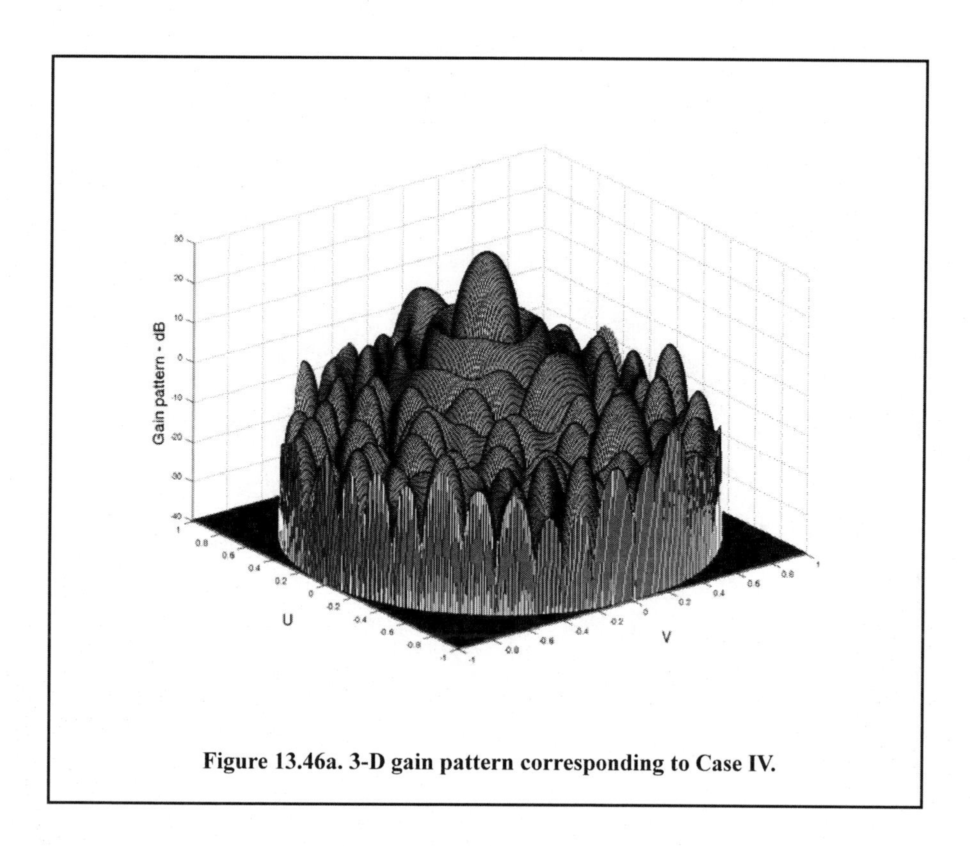

Figure 13.46a. 3-D gain pattern corresponding to Case IV.

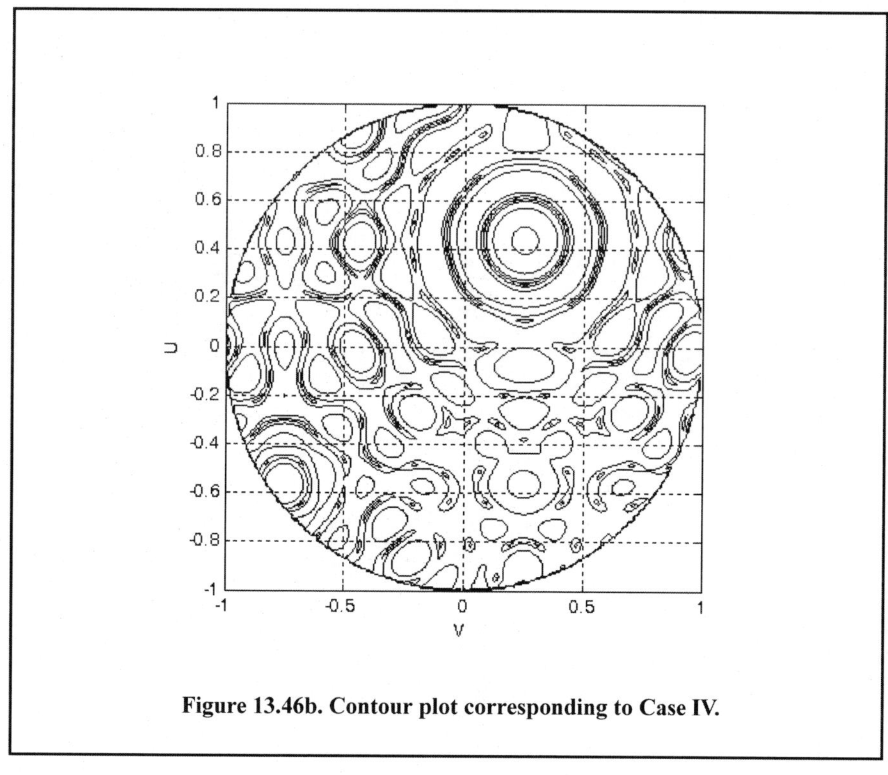

Figure 13.46b. Contour plot corresponding to Case IV.

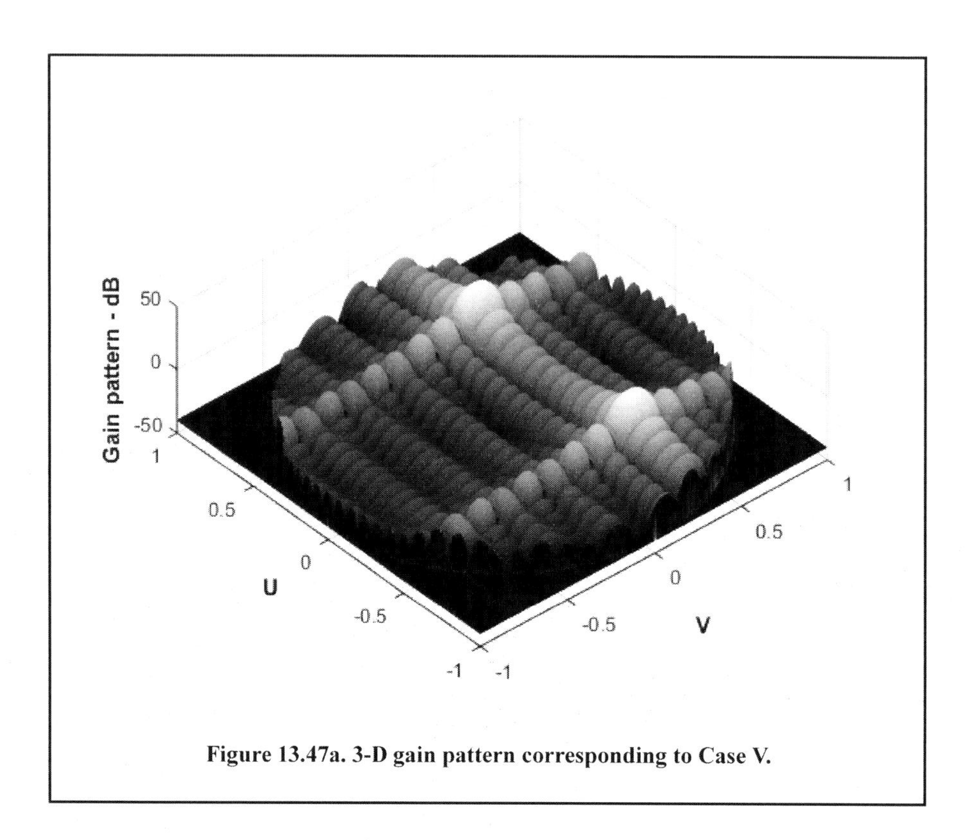

Figure 13.47a. 3-D gain pattern corresponding to Case V.

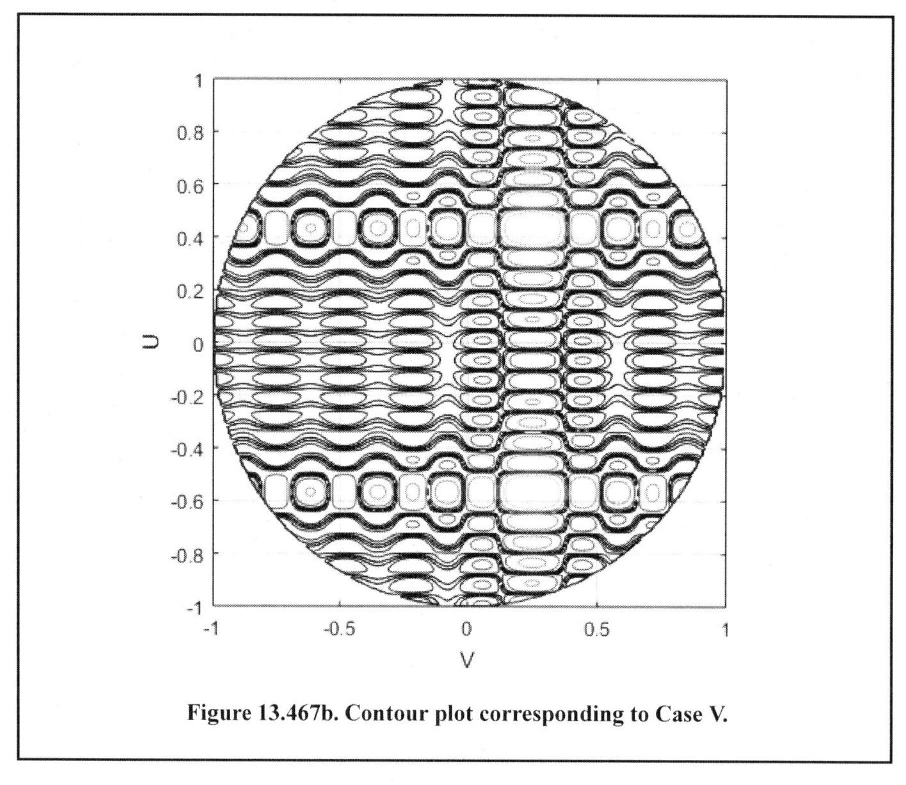

Figure 13.467b. Contour plot corresponding to Case V.

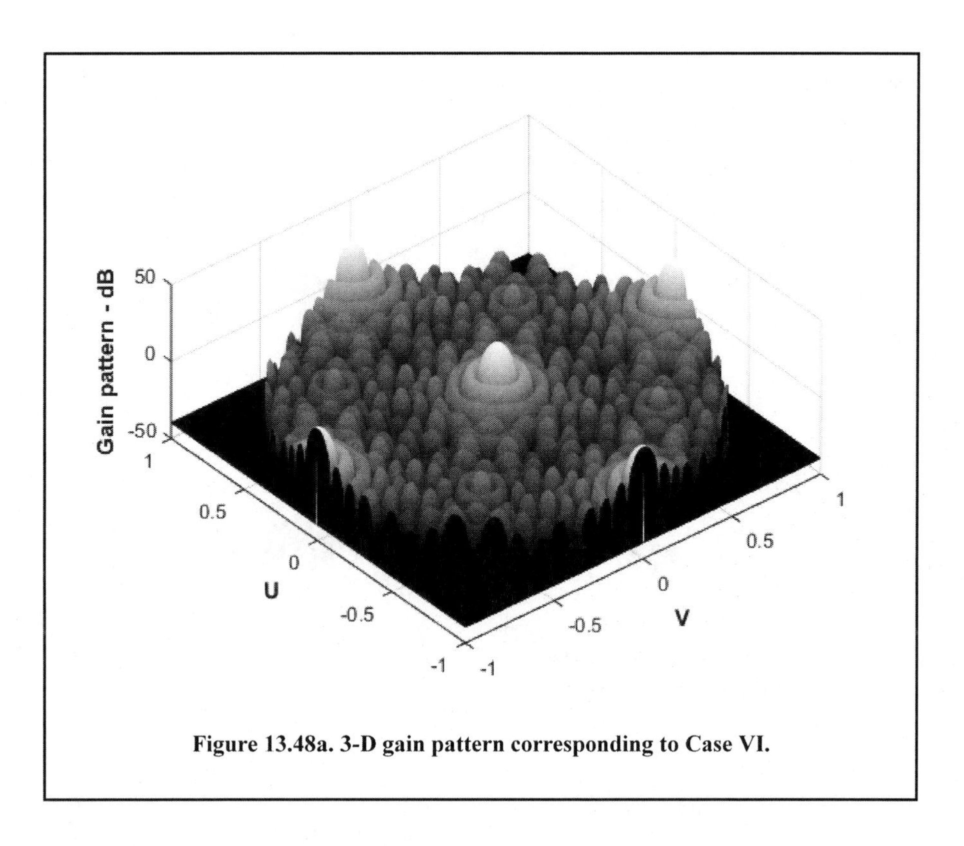

Figure 13.48a. 3-D gain pattern corresponding to Case VI.

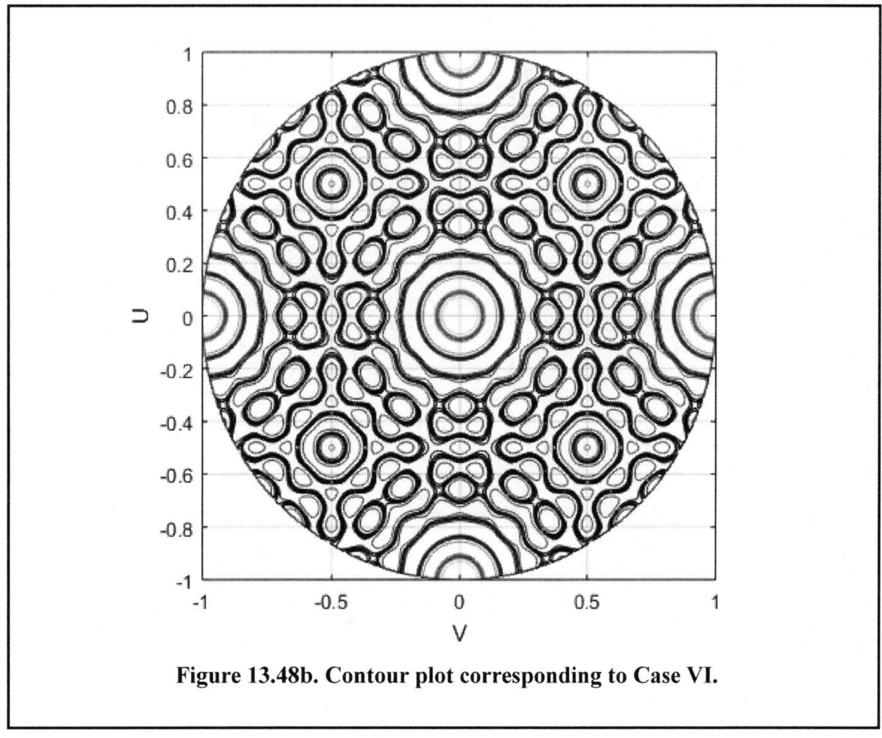

Figure 13.48b. Contour plot corresponding to Case VI.

The beamwidth at scan angle θ is

$$\Theta_{\theta} = \frac{\Theta_{broadside}}{\cos\theta} \qquad \text{Eq. (13.75)}$$

due to the increased scan loss at large scanning angles. In order to limit the scan loss to under some acceptable practical values, most arrays do not scan electronically beyond about $\theta = 60°$. Such arrays are called Full Field of View (FFOV) arrays. FFOV arrays employ element spacing of 0.6λ or less to avoid grating lobes. FFOV array scan loss is approximated by

$$L_{scan} \approx (\cos\theta)^{2.5}. \qquad \text{Eq. (13.76)}$$

Arrays that limit electronic scanning to under $\theta = 60°$ are referred to as Limited Field of View (LFOV) arrays. In this case the scan loss is

$$L_{scan} = \left[\frac{\sin\left(\frac{\pi d}{\lambda}\sin\theta\right)}{\frac{\pi d}{\lambda}\sin\theta}\right]^{-4}. \qquad \text{Eq. (13.77)}$$

Figure 13.50 shows a plot for scan loss versus scan angle.

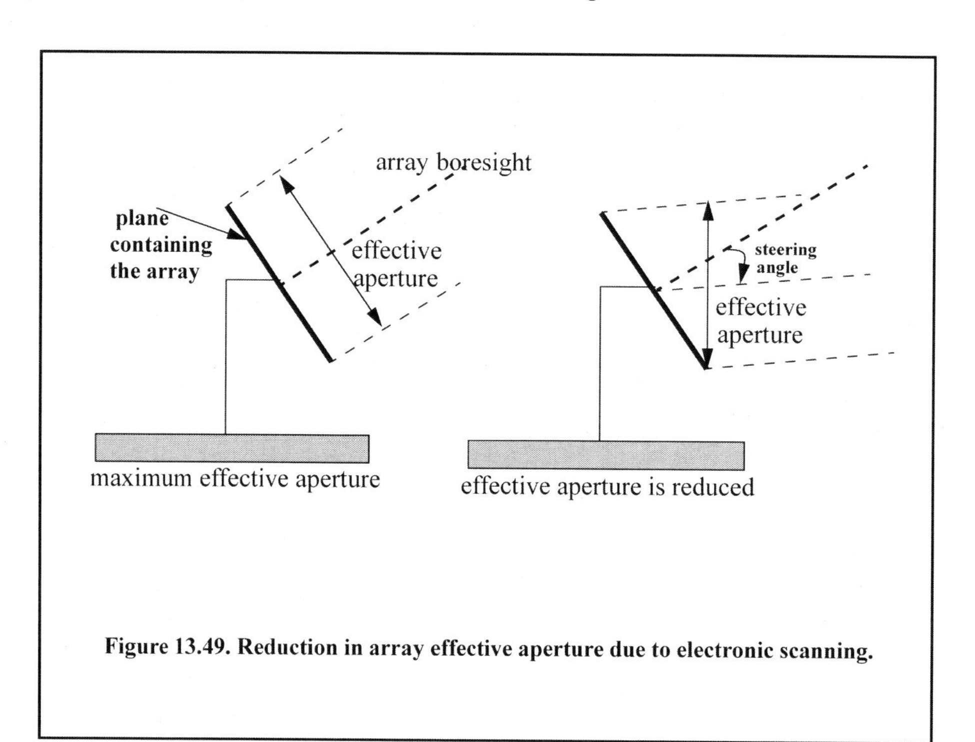

Figure 13.49. Reduction in array effective aperture due to electronic scanning.

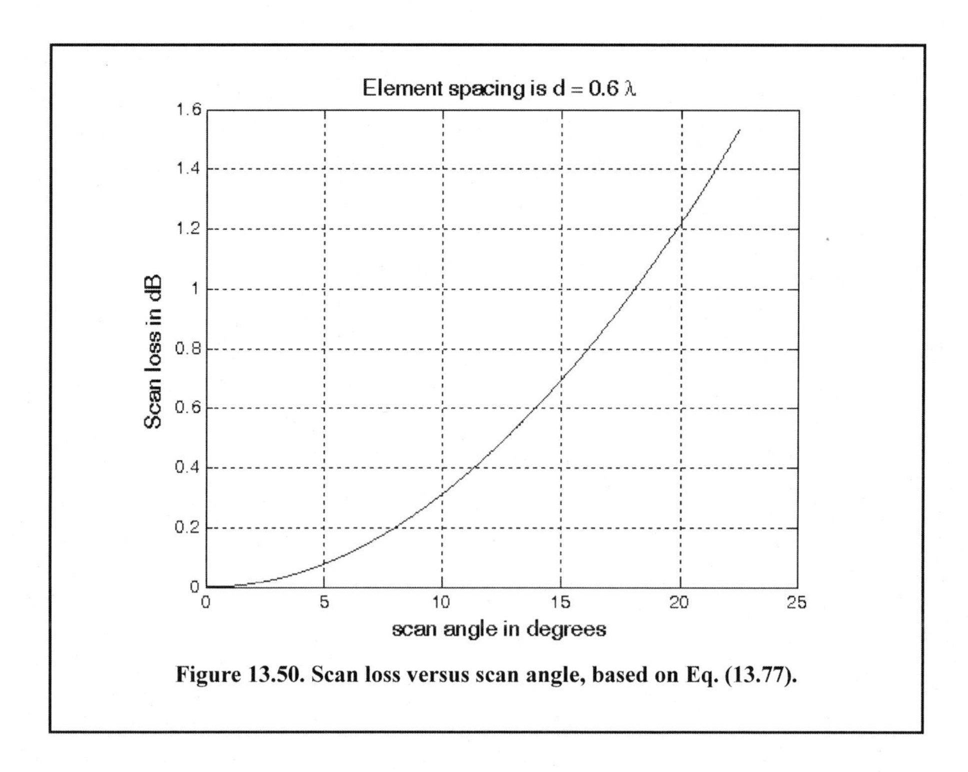

Figure 13.50. Scan loss versus scan angle, based on Eq. (13.77).

Problems

13.1. Consider an antenna whose diameter is $d = 3m$. What is the far field requirement for an X-band or an L-band radar that is using this antenna?

13.2. Consider an antenna with electric field intensity in the xy-plane $E(\varsigma)$. This electric field is generated by a current distribution $D(y)$ in the yz-plane. The electric field intensity is computed using the integral

$$E(\varsigma) = \int_{-r/2}^{r/2} D(y)\exp\left(2\pi j\frac{y}{\lambda}\sin\varsigma\right)dy$$

where λ is the wavelength and r is the aperture. (a) Write an expression for $E(\varsigma)$ when $D(Y) = d_0$ (a constant). (b) Write an expression for the normalized power radiation pattern and plot it in dB.

13.3. A linear phased array consists of 50 elements with $\lambda/2$ element spacing. (a) Compute the $3dB$ beamwidth when the main-beam steering angle is $0°$ and $45°$. (b) Compute the electronic phase difference for any two consecutive elements for steering angle $60°$.

13.4. A linear phased array antenna consists of eight elements spaced with $d = \lambda$ element spacing. (a) Give an expression for the antenna gain pattern (assume no steering and uniform aperture weighting). (b) Sketch the gain pattern versus the sine of the off-boresight angle β. What problems do you see in using $d = \lambda$ rather than $d = \lambda/2$?

13.5. In Section 13.4 we showed how a DFT can be used to compute the radiation pattern of a linear phased array. Consider a linear phased array of 64 elements at half wavelength spacing, where an FFT of size 512 is used to compute the pattern. What are the FFT bins that correspond to steering angles $\beta = 30°, 45°$?

13.6. Derive Eq. (13.73).

13.7. Consider the two-element array shown in the figure below. If the composite array electric field is $E(\theta) = a_1E_1(\theta) + a_2E_2(\theta)$, where a_1, a_2 are constants (can be complex) and E_1, E_2 are the individual elements fields, determine a_1, a_2 so that the electric field is maximum at θ_o. Plot the resulting array pattern.

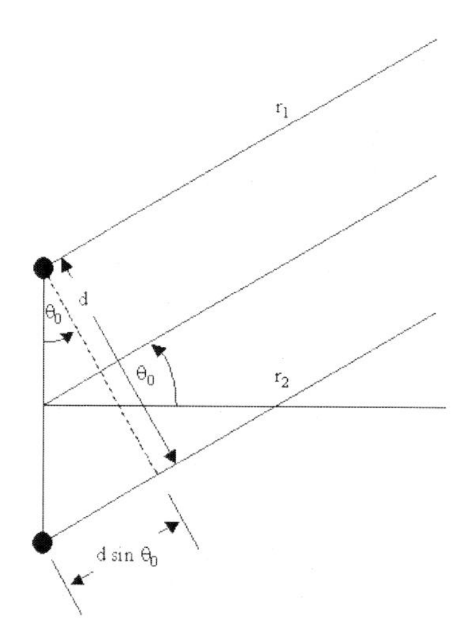

13.8. Use the FFT to compute the radiation pattern for an array of size 21 elements and element spacing (a) $d = 0.5\lambda$, and (b) $d = 0.8\lambda$. In each case, compute and plot the array pattern with and without using Hamming weights.

13.9. Modify the FFT routine developed in the previous problem to compute and plot the power gain pattern.

13.10. Repeat Problems 13.8 and 13.9, where in this case the array pattern can be steered in any off-boresight direction.

13.11. Why do the grating lobes appear when the array beam is steered to angles other than the boresight? Include reasonable plots to back up your argument.

Chapter 14

Synthetic Aperture Radar

14.1. Introduction

Modern airborne radar systems are designed to perform a large number of functions which range from detection and discrimination of targets to mapping large areas of ground terrain. This mapping can be performed by the Synthetic Aperture Radar (SAR). Through illuminating the ground with coherent radiation and measuring the echo signals, SAR can produce high resolution two-dimensional (and in some cases three-dimensional) imagery of the ground surface. The quality of ground maps generated by SAR is determined by the size of the resolution cell. A resolution cell is specified by range and azimuth resolutions of the system. Other factors affecting the size of the resolution cells are (1) size of the processed map and the amount of signal processing involved; (2) cost consideration; and (3) size of the objects that need to be resolved in the map. For example, mapping gross features of cities and coastlines does not require as much resolution when compared to resolving houses, vehicles, and streets.

SAR systems can produce maps of reflectivity versus range and Doppler (cross range). Range resolution is accomplished through range gating. Fine range resolution can be accomplished by using pulse compression techniques. The azimuth resolution depends on antenna size and radar wavelength. Fine azimuth resolution is enhanced by taking advantage of the radar motion in order to synthesize a larger antenna aperture. Let N_r denote the number of range bins and let N_a denote the number of azimuth cells. It follows that the total number of resolution cells in the map is N_rN_a. SAR systems that are generally concerned with improving azimuth resolution are often referred to as Doppler Beam-Sharpening (DBS) SARs. In this case, each range bin is processed to resolve targets in Doppler which corresponds to azimuth. This chapter is presented in the context of DBS.

Due to the large amount of signal processing required in SAR imagery, the early SAR designs implemented optical processing techniques. Although such optical processors can produce high quality radar images, they have several shortcomings. They can be very costly and are, in general, limited to making strip maps. Motion compensation is not easy to implement for radars that utilize optical processors. With the recent advances in solid state electronics and Very Large Scale Integration (VLSI) technologies, digital signal processing in real time has been made possible in SAR systems.

14.2. Real Versus Synthetic Arrays

A linear array of size N, element spacing d, isotropic elements, and wavelength λ is shown in Figure 14.1. A synthetic linear array is formed by linear motion of a single element, transmitting and receiving from distinct positions that correspond to the element locations in a real array. Thus, synthetic array geometry is similar to that of a real array, with the exception that the array exists only at a single element position at a time.

The two-way radiation pattern (in the direction-sine $\sin\beta$) for a real linear array was developed in Chapter 13; it is repeated here as Eq. (14.1)

$$G(\sin\beta) = \left(\frac{\sin((Nkd\sin\beta)/2)}{\sin((kd\sin\beta)/2)}\right)^2 . \qquad \text{Eq. (14.1)}$$

Since a synthetic array exists only at a single location at a time, the array transmission is sequential with only one element receiving. Therefore, the returns received by the successive array positions differ in phase by $\delta = k\Delta r$, where $k = 2\pi/\lambda$, and $\Delta r = 2d\sin\beta$ is the round-trip path difference between contiguous element positions. The two-way array pattern for a synthetic array is the coherent sum of the returns at all the array positions.

Thus, the overall two-way electric field for the synthetic array is

$$E(\sin\beta) = 1 + e^{-j2\delta} + e^{-j4\delta} + \ldots + e^{-j2(N-1)\delta} = \sum_{n=1}^{N} e^{-j2(N-1)kd\sin\beta} . \qquad \text{Eq. (14.2)}$$

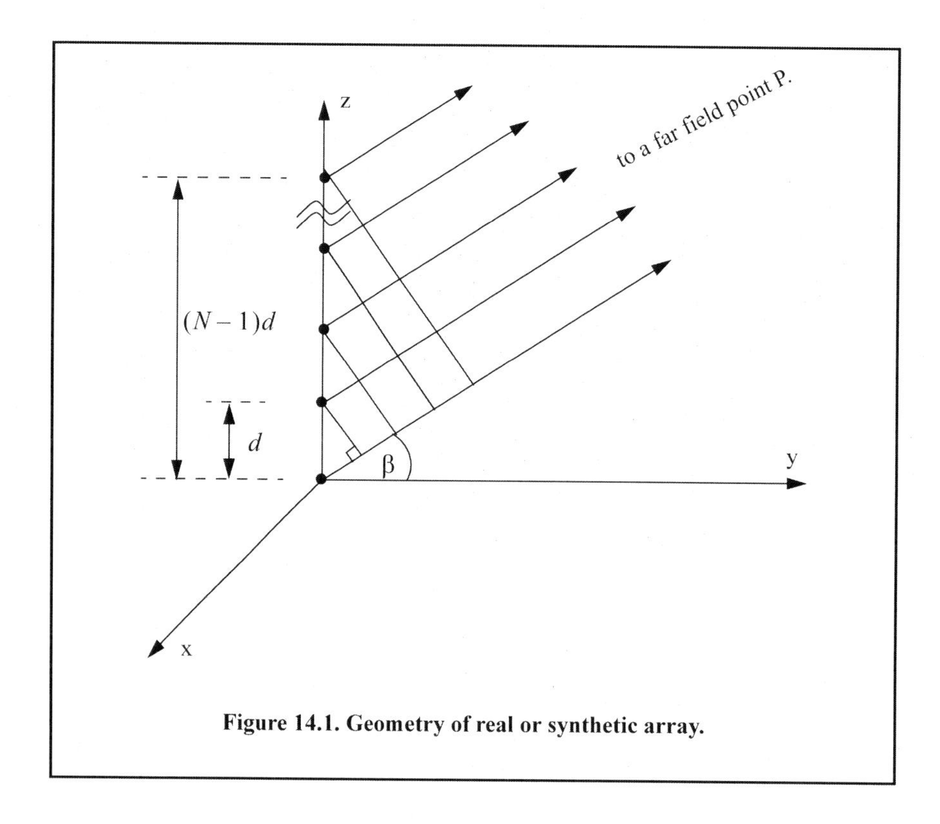

Figure 14.1. Geometry of real or synthetic array.

The two-way electric field for a synthetic array can be expressed as

$$E(\sin\beta) = \frac{\sin(Nkd\sin\beta)}{\sin(kd\sin\beta)}, \qquad \text{Eq. (14.3)}$$

and the two-way radiation pattern is

$$G(\sin\beta) = |E(\sin\beta)| = \left|\frac{\sin(Nkd\sin\beta)}{\sin(kd\sin\beta)}\right|. \qquad \text{Eq. (14.4)}$$

Comparison of Eq. (14.4) and Eq. (14.1) indicates that the two-way radiation pattern for a real array is of the form $(\sin\theta/\theta)^2$ while it is of the form $\sin 2\theta/2\theta$ for the synthetic array. Consequently, for the same size aperture, the main beam of the synthetic array is twice as narrow as that for the real array. Or equivalently, the resolution of a synthetic array of length L (aperture size) is equal to that of a real array with twice the aperture size $(2L)$, as illustrated in Figure 14.2.

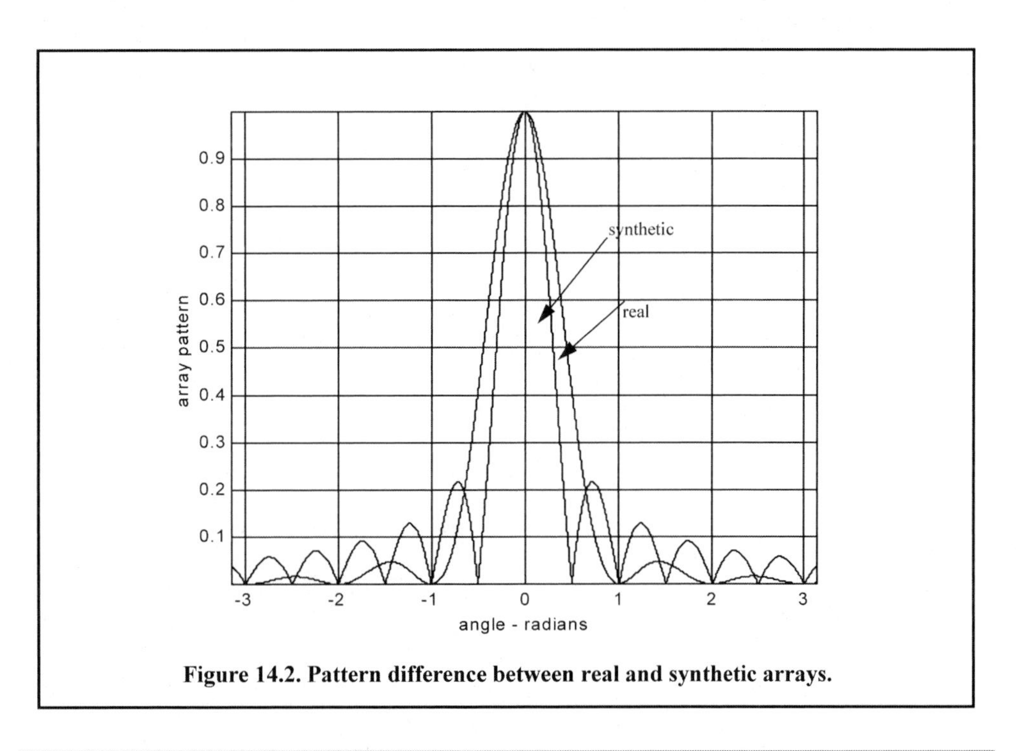

Figure 14.2. Pattern difference between real and synthetic arrays.

14.3. Side Looking SAR Geometry

Figure 14.3 shows the geometry for the standard side looking SAR. We will assume that the platform carrying the radar maintains both fixed altitude h, and velocity v. The antenna $3dB$ beam width is θ, and the elevation angle (measured from the z-axis to the antenna axis) is β. The intersection of the antenna beam with the ground defines a footprint. As the platform moves, the footprint scans a swath on the ground.

The radar position with respect to the absolute origin $\vec{O} = (0, 0, 0)$ at any time is the vector $\vec{a}(t)$. The velocity vector $\vec{a}'(t)$ is

$$\vec{a}'(t) = 0 \times \hat{a}_x + v \times \hat{a}_y + 0 \times \hat{a}_z . \qquad \text{Eq. (14.5)}$$

The Line Of Sight (LOS) for the current footprint centered at $\vec{q}(t_c)$, is defined by the vector $\vec{R}(t_c)$, where t_c denotes the central time of the observation interval T_{ob} (coherent integration interval). More precisely,

$$(t = t_a + t_c) \; ; \; -\frac{T_{ob}}{2} \le t \le \frac{T_{ob}}{2} \qquad \text{Eq. (14.6)}$$

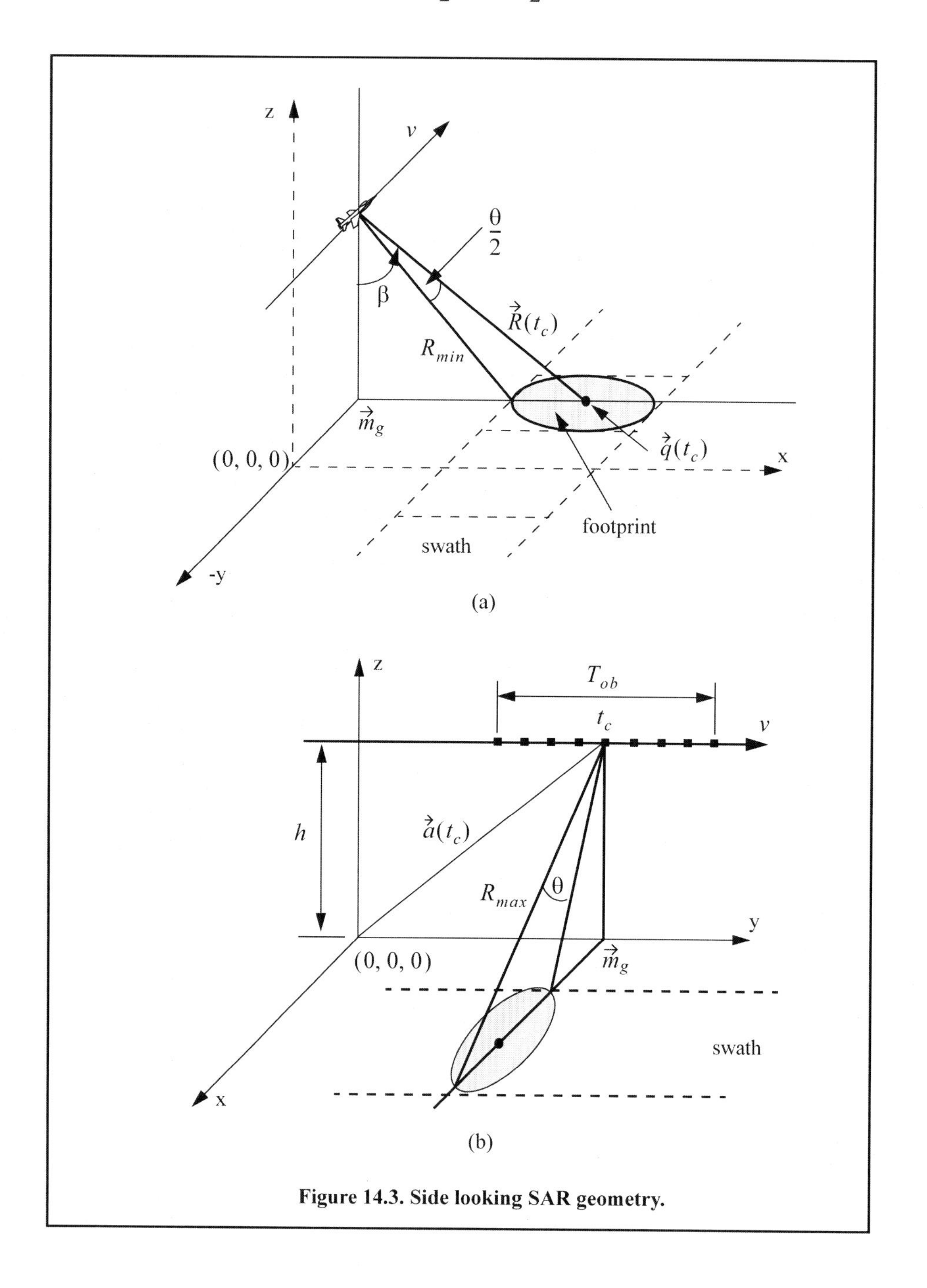

Figure 14.3. Side looking SAR geometry.

where t_a and t are the absolute and relative times, respectively. The vector $\vec{m}_g$ defines the ground projection of the antenna at central time. The minimum slant range to the swath is R_{min}, and the maximum range is denoted R_{max}, as illustrated by Figure 14.4. It follows that,

$$R_{min} = h/\cos(\beta - \theta/2)$$
$$R_{max} = h/\cos(\beta + \theta/2) \quad \text{Eq. (14.7)}$$
$$|\vec{R}(t_c)| = h/\cos\beta$$

Notice that the elevation angle β is equal to

$$\beta = 90 - \psi_g \quad \text{Eq. (14.8)}$$

where ψ_g is the grazing angle. The size of the footprint is a function of the grazing angle and the antenna beam width, as illustrated in Figure 14.5.

The SAR geometry described in this section is referred to as SAR "strip mode" of operation. Another SAR mode of operation, which will not be discussed in this chapter, is called "spot-light mode", where the antenna is steered (mechanically or electronically) to continuously illuminate one spot (footprint) on the ground. In this case, one high resolution image of the current footprint is generated during an observation interval.

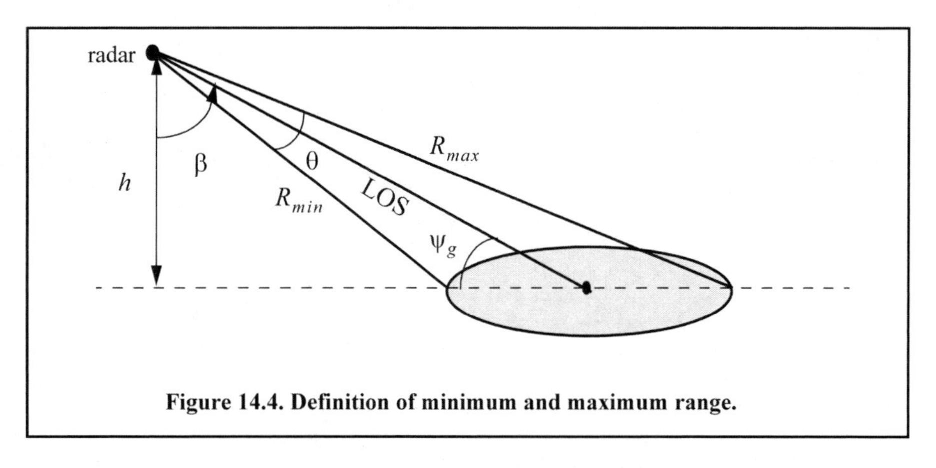

Figure 14.4. Definition of minimum and maximum range.

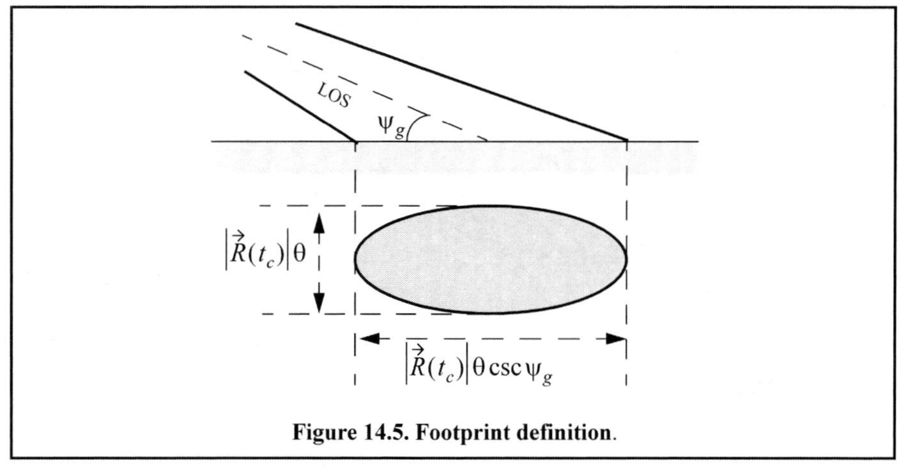

Figure 14.5. Footprint definition.

14.4. SAR Design Considerations

The quality of SAR images is heavily dependent on the size of the map resolution cell shown in Figure 14.6. The range resolution, ΔR, is computed on the beam LOS, and is given by

$$\Delta R = \frac{c\tau}{2} \qquad \text{Eq. (14.9)}$$

where τ is the pulse width. From the geometry in Figure 14.7 the extent of the range cell ground projection ΔR_g, is computed as

$$\Delta R_g = \frac{c\tau}{2}\sec\psi_g . \qquad \text{Eq. (14.10)}$$

The azimuth or cross range resolution for a real antenna with a $3dB$ beam width θ (radians) at range R is

$$\Delta A_r = \theta R . \qquad \text{Eq. (14.11)}$$

However, the antenna beam width is proportional to the aperture size,

$$\theta \approx \frac{\lambda}{L} \qquad \text{Eq. (14.12)}$$

where λ is the wavelength and L is the aperture length. It follows that,

$$\Delta A_r = \frac{\lambda R}{L} . \qquad \text{Eq. (14.13)}$$

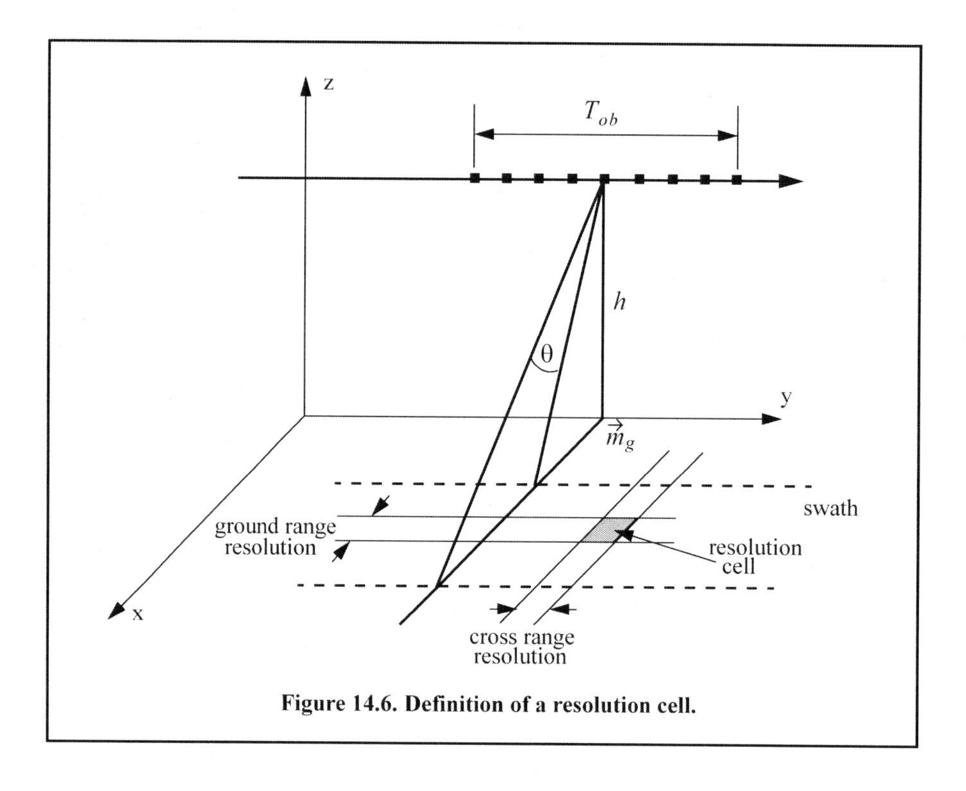

Figure 14.6. Definition of a resolution cell.

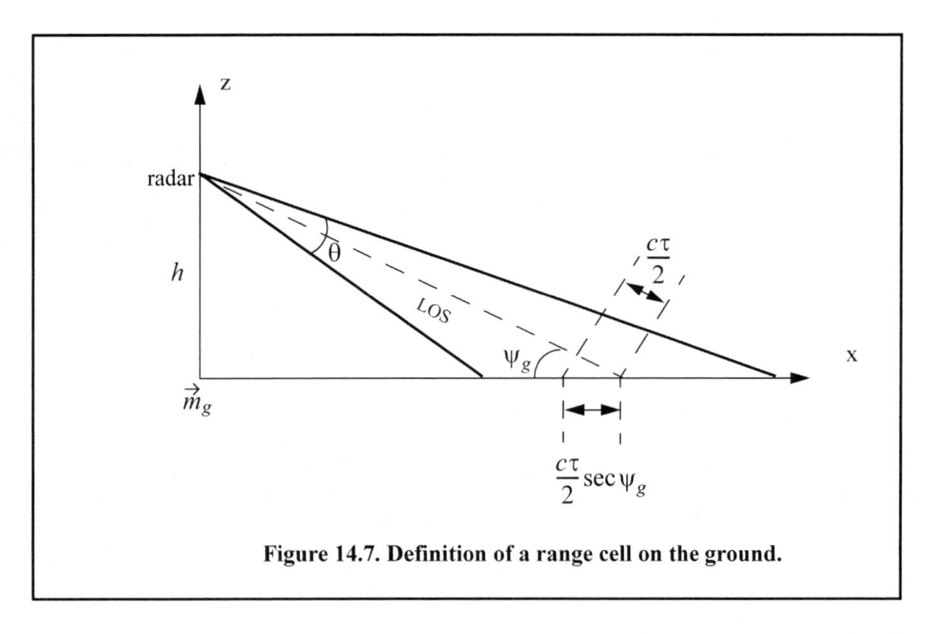

Figure 14.7. Definition of a range cell on the ground.

And since the effective synthetic aperture size is twice that of a real array, the azimuth resolution for a synthetic array is then given by

$$\Delta A = \frac{\lambda R}{2L}. \qquad \text{Eq. (14.14)}$$

Furthermore, since the synthetic aperture length L is equal to vT_{ob}, Eq. (14.14) can be rewritten as

$$\Delta A = \frac{\lambda R}{2vT_{ob}}. \qquad \text{Eq. (14.15)}$$

The azimuth resolution can be greatly improved by taking advantage of the Doppler variation within a footprint (or a beam). As the radar travels along its flight path, the radial velocity to a ground scatterer (point target) within a footprint varies as a function of the radar radial velocity in the direction of that scatterer. The variation of Doppler frequency for a certain scatterer is called the "Doppler history."

Let $R(t)$ denote range to a scatterer at time t, and v_r be the corresponding radial velocity; thus the Doppler shift is

$$f_d = -\frac{2R'(t)}{\lambda} = \frac{2v_r}{\lambda} \qquad \text{Eq. (14.16)}$$

where $R'(t)$ is the range rate to the scatterer. Let t_1 and t_2 be the times when the scatterer enters and leaves the radar beam, respectively, and let t_c be the time that corresponds to minimum range. Figure 14.8 shows a sketch of the corresponding $R(t)$ (see Eq. (14.16)). Since the radial velocity can be computed as the derivative of $R(t)$ with respect to time, one can clearly see that Doppler frequency is maximum at t_1, zero at t_c and minimum at t_2, as illustrated in Figure 14.9.

In general, the radar maximum PRF, $f_{r_{max}}$, must be low enough to avoid range ambiguity. Alternatively, the minimum PRF, $f_{r_{min}}$, must be high enough to avoid Doppler ambiguity. SAR unambiguous range must be at least as wide as the extent of a footprint. More precisely, since target returns from maximum range due to the current pulse must be received by the radar before the next pulse is transmitted, it follows that SAR unambiguous range is given by

$$R_u = R_{max} - R_{min}. \qquad \text{Eq. (14.17)}$$

An expression for the unambiguous range was derived earlier; it is repeated here as Eq. (14.18),

$$R_u = \frac{c}{2f_r}. \qquad \text{Eq. (14.18)}$$

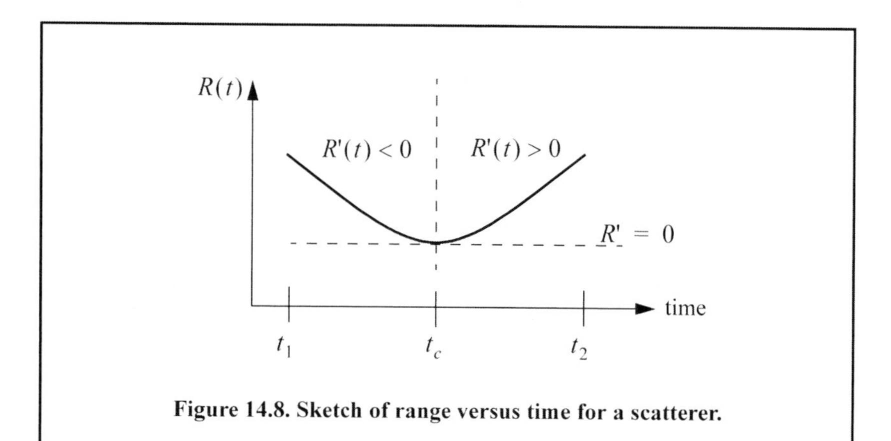

Figure 14.8. Sketch of range versus time for a scatterer.

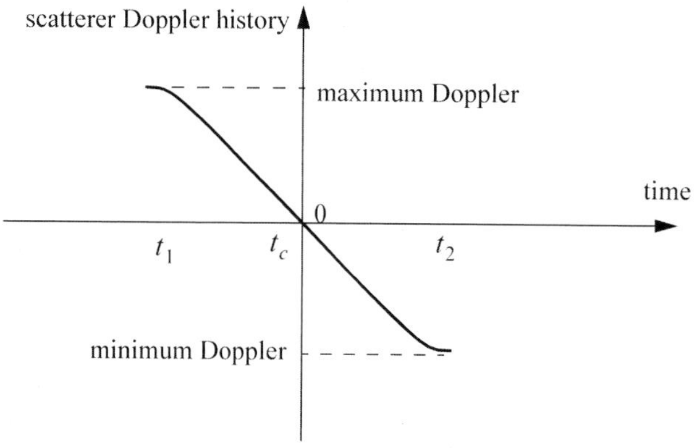

Figure 14.9. Point scatterer Doppler history.

Combining Eq. (14.18) and Eq. (14.17) yields,

$$f_{r_{max}} \le \frac{c}{2(R_{max} - R_{min})} . \qquad \text{Eq. (14.19)}$$

SAR minimum PRF, $f_{r_{min}}$, is selected so that Doppler ambiguity is avoided. In other words, $f_{r_{min}}$ must be greater than the maximum expected Doppler spread within a footprint. From the geometry of Figure 14.10, the maximum and minimum Doppler frequencies are respectively given by,

$$\left(f_{d_{max}} = \frac{2v}{\lambda} \cos\left(90 - \frac{\theta}{2}\right) \sin\beta\right) \quad ; \ at \ t_1 \qquad \text{Eq. (14.20)}$$

$$\left(f_{d_{min}} = \frac{2v}{\lambda} \cos\left(90 + \frac{\theta}{2}\right) \sin\beta\right) \quad ; \ at \ t_2 \ . \qquad \text{Eq. (14.21)}$$

It follows that the maximum Doppler spread is

$$\Delta f_d = f_{d_{max}} - f_{d_{min}} . \qquad \text{Eq. (14.22)}$$

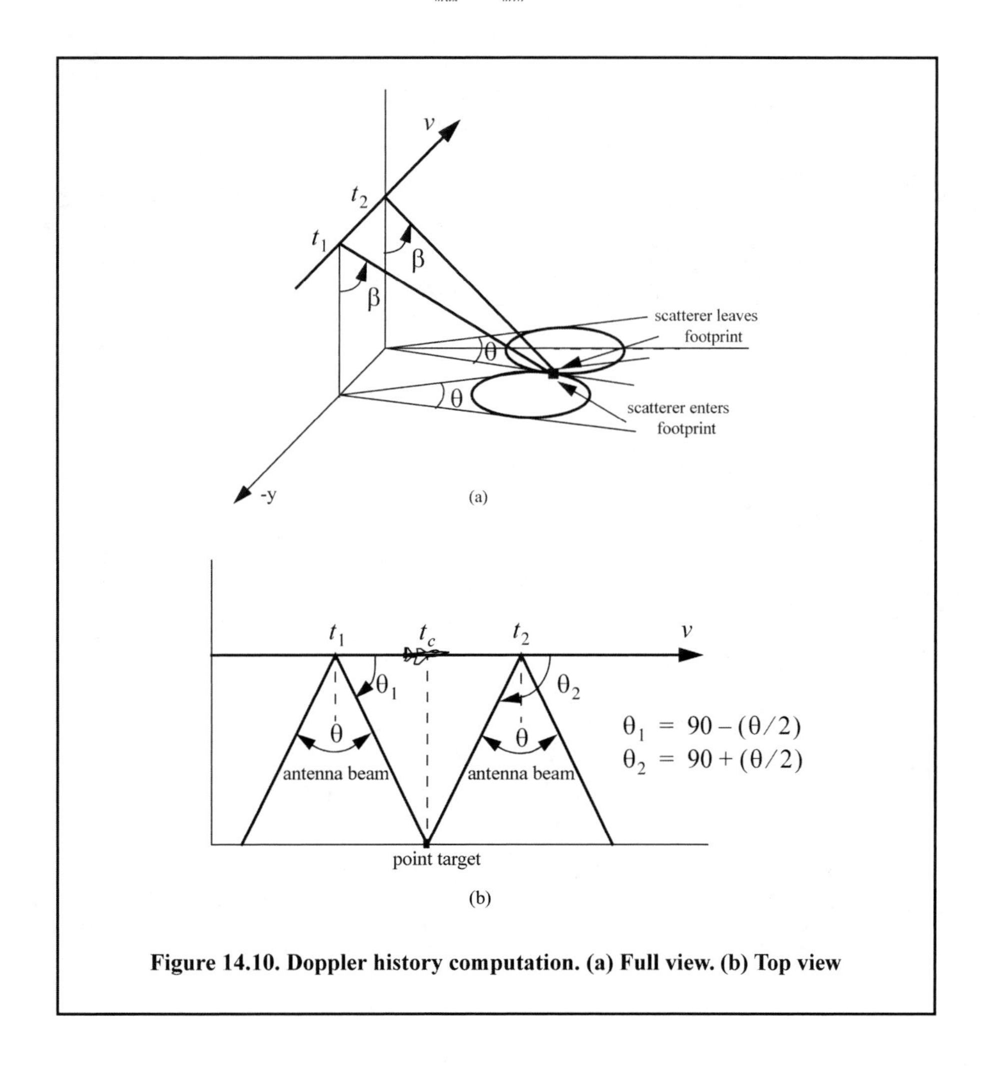

Figure 14.10. Doppler history computation. (a) Full view. (b) Top view

Substituting Eqs. (14.20) and (14.21) into Eq. (14.22) and applying the proper trigonometric identities yield

$$\Delta f_d = \frac{4v}{\lambda} \sin\frac{\theta}{2} \sin\beta . \qquad \text{Eq. (14.23)}$$

Finally, by using the small angle approximation, we get

$$\Delta f_d \approx \frac{4v}{\lambda} \frac{\theta}{2} \sin\beta = \frac{2v}{\lambda} \theta \sin\beta . \qquad \text{Eq. (14.24)}$$

Therefore, the minimum PRF is

$$f_{r_{min}} \ge \frac{2v}{\lambda} \theta \sin\beta . \qquad \text{Eq. (14.25)}$$

Combining Eqs. (14.19) and (14.25) we get

$$\frac{c}{2(R_{max} - R_{min})} \ge f_r \ge \frac{2v}{\lambda} \theta \sin\beta . \qquad \text{Eq. (14.26)}$$

It is possible to resolve adjacent scatterers at the same range within a footprint based only on the difference of their Doppler histories. For this purpose, assume that the two scatterers are within the kth range bin. Denote their angular displacement as $\Delta\theta$, and let $\Delta f_{d_{min}}$ be the minimum Doppler spread between the two scatterers such that they will appear in two distinct Doppler filters. Using the same methodology that led to Eq. (14.24) we get

$$\Delta f_{d_{min}} = \frac{2v}{\lambda} \Delta\theta \sin\beta_k \qquad \text{Eq. (14.27)}$$

where β_k is the elevation angle corresponding to the kth range bin.

The bandwidth of the individual Doppler filters must be equal to the inverse of the coherent integration interval T_{ob} (i.e., $\Delta f_{d_{min}} = 1/T_{ob}$). It follows that

$$\Delta\theta = \frac{\lambda}{2vT_{ob}\sin\beta_k} . \qquad \text{Eq. (14.28)}$$

Substituting L for vT_{ob} yields

$$\Delta\theta = \frac{\lambda}{2L\sin\beta_k} . \qquad \text{Eq. (14.29)}$$

Therefore, the SAR azimuth resolution (within the kth range bin) is

$$\Delta A_g = \Delta\theta R_k = R_k \frac{\lambda}{2L\sin\beta_k} . \qquad \text{Eq. (14.30)}$$

Note that when $\beta_k = 90°$, Eq. (14.30) is identical to Eq. (14.14).

14.5. SAR Radar Equation

The single pulse radar equation was derived in an earlier chapter, and is repeated here as Eq. (14.31),

$$SNR = \frac{P_t G^2 \lambda^2 \sigma}{(4\pi)^3 R_k^4 k T_0 B L_{Loss}} \qquad \text{Eq. (14.31)}$$

where P_t is peak power; G is antenna gain; λ is wavelength; σ is radar cross section; R_k is radar slant range to the kth range bin; k is Boltzman's constant; T_0 is receiver noise temperature; B is receiver bandwidth; and L_{Loss} is radar losses. The radar cross section is a function of the radar resolution cell and terrain reflectivity. More precisely,

$$\sigma = \sigma^0 \Delta R_g \Delta A_g = \sigma^0 \Delta A_g \frac{c\tau}{2} \sec\psi_g \qquad \text{Eq. (14.32)}$$

where σ^0 is the clutter scattering coefficient, ΔA_g is the azimuth resolution, and Eq. (14.10) was used to replace the ground range resolution. The number of coherently integrated pulses within an observation interval is

$$n = f_r T_{ob} = (f_r L)/v \qquad \text{Eq. (14.33)}$$

where L is the synthetic aperture size. Using Eq. (14.30) in Eq. (14.33) and rearranging terms yield

$$n = \frac{\lambda R f_r}{2 \Delta A_g v} \csc\beta_k . \qquad \text{Eq. (14.34)}$$

The radar average power over the observation interval is

$$P_{av} = (P_t / B) f_r . \qquad \text{Eq. (14.35)}$$

The SNR for n coherently integrated pulses is then

$$(SNR)_n = nSNR = n \frac{P_t G^2 \lambda^2 \sigma}{(4\pi)^3 R_k^4 k T_0 B L_{Loss}} . \qquad \text{Eq. (14.36)}$$

Substituting Eqs. (14.35), (14.34), and (14.32) into Eq. (14.36) and performing some algebraic manipulations give the SAR radar equation,

$$(SNR)_n = \frac{P_{av} G^2 \lambda^3 \sigma^0}{(4\pi)^3 R_k^3 k T_0 L_{Loss}} \frac{\Delta R_g}{2v} \csc\beta_k . \qquad \text{Eq. (14.37)}$$

Equation (14.37) leads to the conclusion that in SAR systems the SNR is

1. Inversely proportional to the third power of range
2. Independent of azimuth resolution
3. Function of the ground range resolution inversely proportional to the velocity v, and
4. Proportional to the third power of wavelength.

14.6. SAR Signal Processing

There are two signal processing techniques to sequentially produce a SAR map or image; they are line-by-line processing and Doppler processing. The concept of SAR line-by-line processing is as follows. Through the radar linear motion a synthetic array is formed, where the elements of the current synthetic array correspond to the position of the antenna transmissions during the last observation interval. Azimuth resolution is obtained by forming narrow synthetic beams through combination of the last observation interval returns. Fine range resolution is accomplished in real time by utilizing range gating and pulse compression. For each range bin and each of the transmitted pulses during the last observation interval, the returns are recorded in a two-dimensional array of data that is updated for every pulse. Denote the two-dimensional array of data as MAP.

To further illustrate the concept of line-by-line processing, consider the case where a map of size $N_a \times N_r$ is to be produced; N_a is the number of azimuth cells, and N_r is the number of range bins. Hence, MAP is of size $N_a \times N_r$, where the columns refer to range bins, and the rows refer to azimuth cells. For each transmitted pulse, the echoes from consecutive range bins are recorded sequentially in the first row of MAP. Once the first row is completely filled (i.e., returns from all range bins have been received), all data (in all rows) are shifted downward one row before the next pulse is transmitted. Thus, one row of MAP is generated for every transmitted pulse. Consequently, for the current observation interval, returns from the first transmitted pulse will be located in the bottom row of MAP. And returns from the last transmitted pulse will be in the first row of MAP.

In SAR Doppler processing, the array MAP is updated once every N pulses so that a block of N columns is generated simultaneously. In this case, N refers to the number of transmissions during an observation interval (i.e., size of the synthetic array). From an antenna point of view, this is equivalent to having N adjacent synthetic beams formed in parallel through electronic steering.

14.7. Side Looking SAR Doppler Processing

Consider the geometry shown in Figure 14.11, and assume that the scatterer C_i is located within the kth range bin. The scatterer azimuth and elevation angles are μ_i and β_i, respectively. The scatterer elevation angle β_i is assumed to be equal to β_k, the range bin elevation angle. This assumption is true if the ground range resolution, ΔR_g, is small; otherwise, $\beta_i = \beta_k + \varepsilon_i$ for some small ε_i; in this chapter $\varepsilon_i = 0$.

The normalized transmitted signal can be represented by

$$s(t) = \cos(2\pi f_0 t - \xi_0) \qquad \text{Eq. (14.38)}$$

where f_0 is the radar operating frequency, and ξ_0 denotes the transmitter phase. The returned radar signal from C_i is then equal to

$$s_i(t, \mu_i) = A_i \cos[2\pi f_0(t - \tau_i(t, \mu_i)) - \xi_0] \qquad \text{Eq. (14.39)}$$

where $\tau_i(t, \mu_i)$ is the round-trip delay to the scatterer, and A_i includes scatterer strength, range attenuation, and antenna gain. The round-trip delay is

$$\tau_i(t, \mu_i) = \frac{2r_i(t, \mu_i)}{c} \qquad \text{Eq. (14.40)}$$

where c is the speed of light and $r_i(t, \mu_i)$ is the scatterer slant range. From the geometry in Figure 14.11, one can write the expression for the slant range to the ith scatterer within the kth range bin as

$$r_i(t, \mu_i) = \frac{h}{\cos\beta_i}\sqrt{1 - \frac{2vt}{h}\cos\beta_i \cos\mu_i \sin\beta_i + \left(\frac{vt}{h}\cos\beta_i\right)^2} . \qquad \text{Eq. (14.41)}$$

And by using Eq. (14.40) the round-trip delay can be written as

$$\tau_i(t, \mu_i) = \frac{2}{c}\frac{h}{\cos\beta_i}\sqrt{1 - \frac{2vt}{h}\cos\beta_i\cos\mu_i\sin\beta_i + \left(\frac{vt}{h}\cos\beta_i\right)^2}. \quad \text{Eq. (14.42)}$$

The round-trip delay can be approximated using a two-dimensional second order Taylor series expansion about the reference state $(t, \mu) = (0, 0)$. Performing this Taylor series expansion yields

$$\tau_i(t, \mu_i) \approx \bar{\tau} + \bar{\tau}_{t\mu} \ \mu_i t + \bar{\tau}_{tt} \ \frac{t^2}{2} \quad \text{Eq. (14.43)}$$

where the over-bar indicates evaluation at the state $(0, 0)$, and the subscripts denote partial derivatives. For example, $\bar{\tau}_{t\mu}$ means

$$\bar{\tau}_{t\mu} = \frac{\partial^2}{\partial t \partial \mu}\tau_i(t, \mu_i)\big|_{(t, \mu) = (0, 0)}. \quad \text{Eq. (14.44)}$$

The Taylor series coefficients are (see problem 11.6)

$$\bar{\tau} = \left(\frac{2h}{c}\right)\frac{1}{\cos\beta_i} \quad \text{Eq. (14.45)}$$

$$\bar{\tau}_{t\mu} = \left(\frac{2v}{c}\right)\sin\beta_i \quad \text{Eq. (14.46)}$$

$$\bar{\tau}_{tt} = \left(\frac{2v^2}{hc}\right)\cos\beta_i. \quad \text{Eq. (14.47)}$$

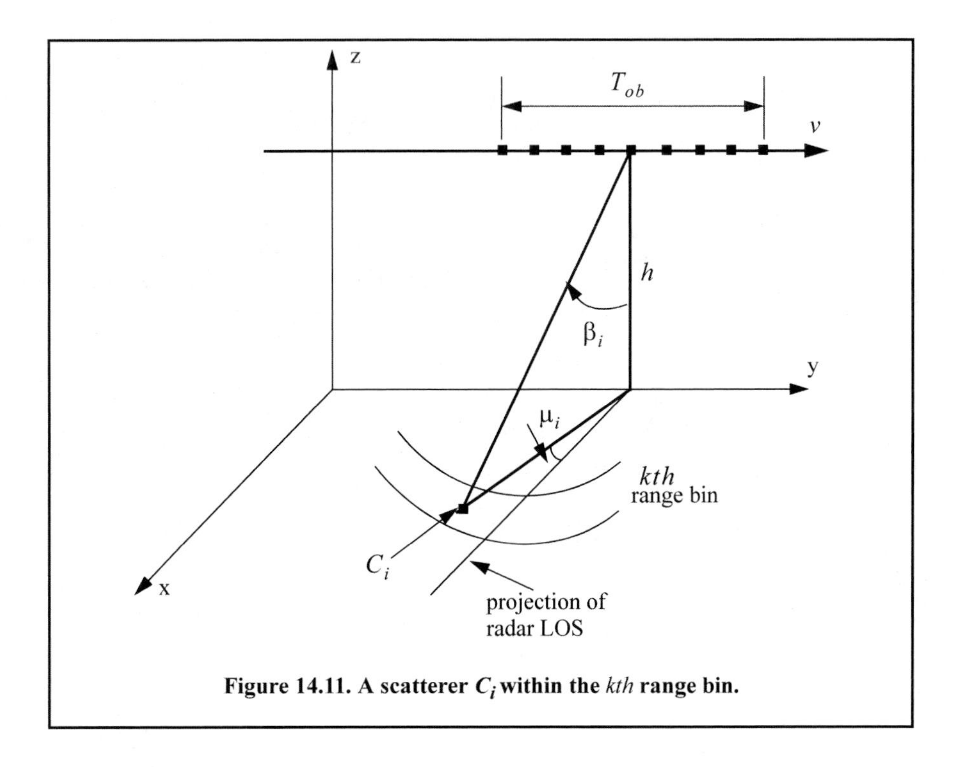

Figure 14.11. A scatterer C_i within the *kth* range bin.

Note that other Taylor series coefficients are either zeros or very small; hence they are neglected. Finally, by substituting Eqs. (14.45) through (14.47) into Eq. (14.43), we can rewrite the returned radar signal as

$$s_i(t, \mu_i) = A_i \cos[\hat{\psi}_i(t, \mu_i) - \xi_0]$$
$$\hat{\psi}_i(t, \mu_i) = 2\pi f_0\left[(1 - \bar{\tau}_{t\mu}\mu_i)t - \bar{\tau} - \bar{\tau}_{tt}\frac{t^2}{2}\right] \cdot \qquad \text{Eq. (14.48)}$$

Observation of Eq. (14.48) indicates that the instantaneous frequency for the ith scatterer varies as a linear function of time due to the second order phase term $2\pi f_0(\bar{\tau}_{tt}t^2/2)$ (this confirms the result we concluded about a scatterer Doppler history). Furthermore, since this phase term is range-bin dependent and not scatterer dependent, all scatterers within the same range bin produce this exact second order phase term. It follows that scatterers within a range bin have identical Doppler histories. These Doppler histories are separated by the time delay required to fly between them, as illustrated in Figure 14.12.

Suppose that there are I scatterers within the kth range bin. In this case, the combined returns for this cell are the sum of the individual returns due to each scatterer as defined by Eq. (14.48). In other words, superposition holds, and the overall echo signal is

$$s_r(t) = \sum_{i=1}^{I} s_i(t, \mu_i) \, . \qquad \text{Eq. (14.49)}$$

A signal processing block diagram for the kth range bin is illustrated in Figure 14.13. It consists of the following steps. First, heterodyning with carrier frequency is performed to extract the quadrature components.

This is followed by LP filtering and A/D conversion. Next, deramping or focusing to remove the second order phase term of the quadrature components is carried out using a phase rotation matrix. The last stage of the processing includes windowing, performing FFT on the windowed quadrature components, and scaling of the amplitude spectrum to account for range attenuation and antenna gain.

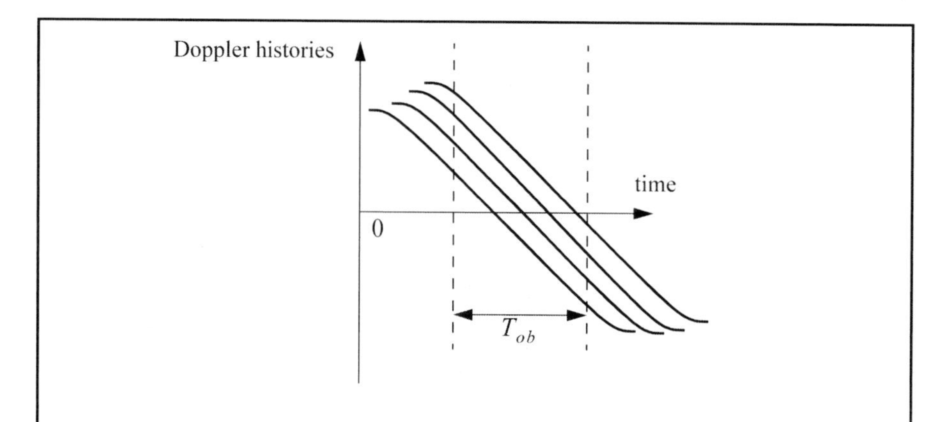

Figure 14.12. Doppler histories for several scatterers within the same range bin.

The discrete quadrature components are

$$\tilde{x}_I(t_n) = \tilde{x}_I(n) = A_i \cos[\tilde{\psi}_i(t_n, \mu_i) - \xi_0]$$
$$\tilde{x}_Q(t_n) = \tilde{x}_Q(n) = A_i \sin[\tilde{\psi}_i(t_n, \mu_i) - \xi_0] \qquad \text{Eq. (14.50)}$$

$$\tilde{\psi}_i(t_n, \mu_i) = \hat{\psi}_i(t_n, \mu_i) - 2\pi f_0 t_n \qquad \text{Eq. (14.51)}$$

and t_n denotes the *nth* sampling time (remember that $-T_{ob}/2 \le t_n \le T_{ob}/2$). The quadrature components after deramping (i.e., removal of the phase $\psi = -\pi f_0 \tau_{tt} t_n^2$) are given by

$$\begin{bmatrix} x_I(n) \\ x_Q(n) \end{bmatrix} = \begin{bmatrix} \cos\psi & -\sin\psi \\ \sin\psi & \cos\psi \end{bmatrix} \begin{bmatrix} \tilde{x}_I(n) \\ \tilde{x}_Q(n) \end{bmatrix}. \qquad \text{Eq. (14.52)}$$

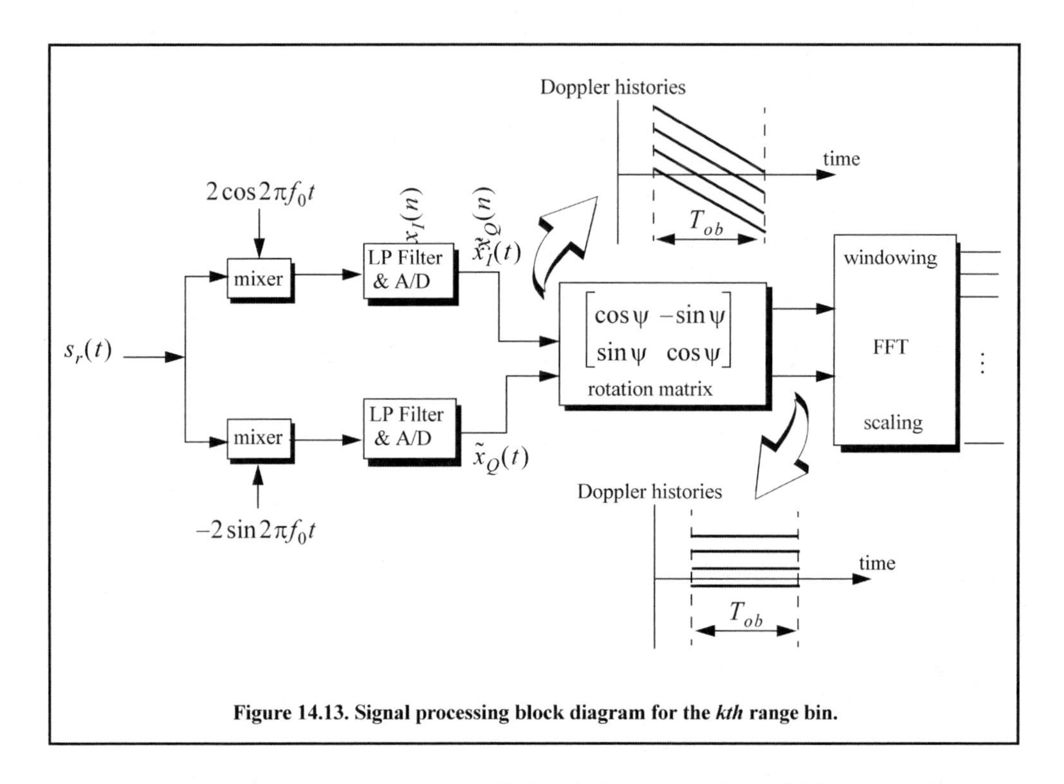

Figure 14.13. Signal processing block diagram for the *kth* range bin.

14.8. SAR Imaging Using Doppler Processing

It was mentioned earlier that SAR imaging is performed using two orthogonal dimensions (range and azimuth). Range resolution is controlled by the receiver bandwidth and pulse compression. Azimuth resolution is limited by the antenna beam width. A one-to-one correspondence between the FFT bins and the azimuth resolution cells can be established by utilizing the signal model described in the previous section. Therefore, the problem of target detection is transformed into a spectral analysis problem, where detection is based on the amplitude spectrum of the returned signal. The FFT frequency resolution Δf is equal to the inverse of the observation interval T_{ob}. It follows that a peak in the amplitude spectrum at $k_1 \Delta f$ indicates the presence of a scatterer at frequency $f_{d1} = k_1 \Delta f$.

For an example, consider the scatterer C_i within the kth range bin. The instantaneous frequency f_{di} corresponding to this scatterer is

$$f_{di} = \frac{1}{2\pi}\frac{d\psi}{dt} = f_0\bar{\tau}_{t\mu}\mu_i = \frac{2v}{\lambda}\sin\beta_i\mu_i, \qquad \text{Eq. (14.53)}$$

which is the same result derived in Eq. (14.27), where $\mu_i = \Delta\theta$. Therefore, the scatterers separated in Doppler by a frequency greater than Δf can then be resolved. An example is shown in Figure 14.14 for five point targets within one footprint.

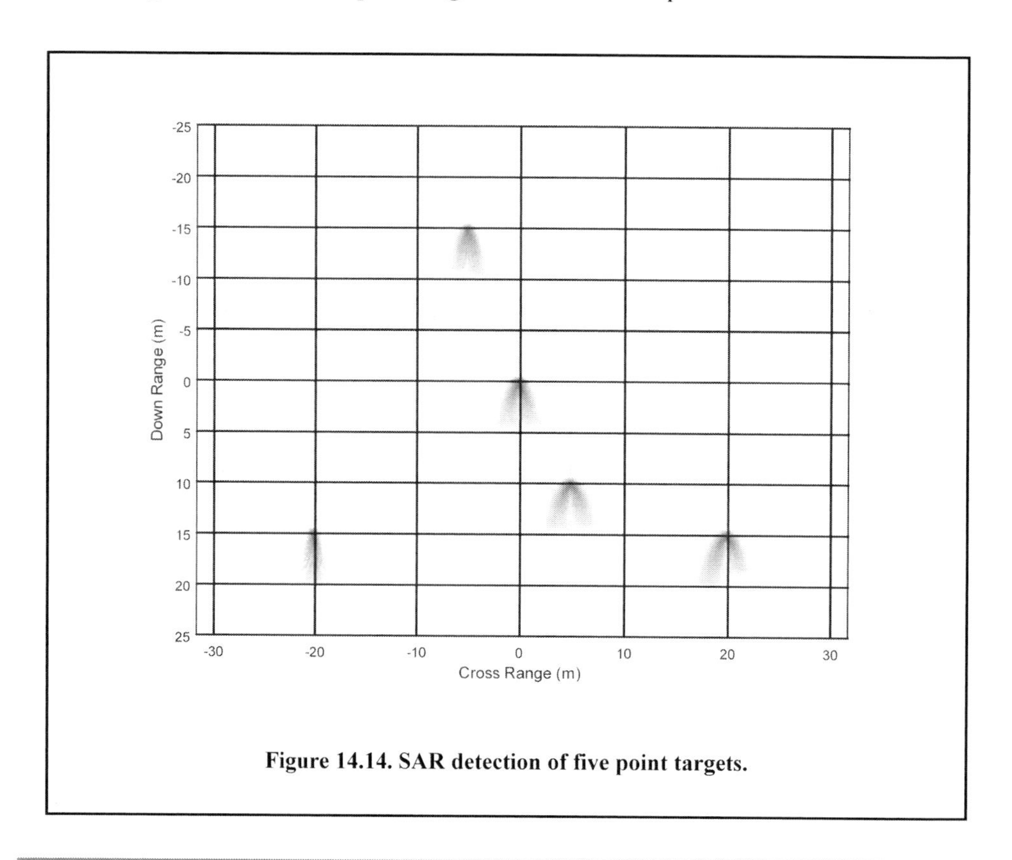

Figure 14.14. SAR detection of five point targets.

14.9. Range Walk

As shown earlier SAR Doppler processing is achieved in two steps, first, range gating and second, azimuth compression within each bin at the end of the observation interval. For this purpose, azimuth compression assumes that each scatterer remains within the same range bin during the observation interval. However, since the range gates are defined with respect to a radar that is moving, the range gate grid is also moving relative to the ground. As a result a scatterer appears to be moving within its range bin. This phenomenon is known as range walk. A small amount of range walk does not bother Doppler processing as long as the scatterer remains within the same range bin. However, range walk over several range bins can constitute serious problems, where in this case Doppler processing is meaningless.

Problems

14.1. A side looking SAR is traveling at an altitude of $15Km$, and the elevation angle is $\beta = 15°$. If the aperture length is $L = 5m$, the pulse width is $\tau = 20\mu s$, and the wavelength is $\lambda = 3.5cm$. (a) Calculate the azimuth resolution; (b) calculate the range and ground range resolutions.

14.2. A MMW side looking SAR has the following specifications: radar velocity $v = 70m/s$, elevation angle $\beta = 35°$, operating frequency $f_0 = 94GHz$, and antenna 3dB beam width $\theta_{3dB} = 65mrad$. (a) Calculate the footprint dimensions; (b) compute the minimum and maximum ranges; (c) compute the Doppler frequency span across the footprint; (d) calculate the minimum and maximum PRFs.

14.3. A side looking SAR takes on eight positions within an observation interval. In each position the radar transmits and receives one pulse. Let the distance between any two consecutive antenna positions be d, and define $\delta = 2\pi\frac{d}{\lambda}(\sin\beta - \sin\beta_0)$ to be the one-way phase difference for a beam steered at angle β_0. (a) In each of the eight positions a sample of the phase pattern is obtained after heterodyning. List the phase samples. (b) How will you process the sequence of samples using an FFT (do not forget windowing)? (c) Give a formula for the angle between the grating lobes.

14.4. Consider a synthetic aperture radar. You are given the following Doppler history for a scatterer: $\{1000Hz, 0, -1000HZ\}$ which correspond to times $\{-10ms, 0, 10ms\}$. Assume that the observation interval is $T_{ob} = 20ms$, and a platform velocity $v = 200m/s$. (a) Show the Doppler history for another scatterer which is identical to the first one except that it is located in azimuth $1m$ earlier. (b) How will you perform deramping on the quadrature components (show only the general approach)? (c) Show the Doppler history for both scatterers after deramping.

14.5. You want to design a side looking synthetic aperture ultrasonic radar operating at $f_0 = 60KHz$ and peak power $P_t = 2W$. The antenna beam is conical with 3dB beam width $\theta_{3dB} = 5°$. The maximum gain is 16. The radar is at a constant altitude $h = 15m$ and is moving at a velocity of $10m/s$. The elevation angle defining the footprint is $\beta = 45°$. (a) Give an expression for the antenna gain assuming a Gaussian pattern. (b) Compute the pulse width corresponding to range resolution of $10mm$. (c) What are the footprint dimensions? (d) Compute and plot the Doppler history for a scatterer located on the central range bin. (e) Calculate the minimum and maximum PRFs; do you need to use more than one PRF? (f) How will you design the system in order to achieve an azimuth resolution of $10mm$?

14.6. Derive Eq. (14.45) through Eq. (14.47).

14.7. In Section 14.7 we assumed the elevation angle increment ε is equal to zero. Develop an equivalent to Eq. (14.43) for the case when $\varepsilon \neq 0$. You need to use a third order three-dimensional Taylor series expansion about the state $(t, \mu, \varepsilon) = (0, 0, 0)$ in order to compute the new round-trip delay expression.

Chapter 15

Textbook Radar Design Case Study

In this chapter, a design case study, referred to as *"Textbook Radar"* design case study is introduced. For this purpose, we will build up the design in steps to mirror the chapters sequence of this book. The radar design case study *"Textbook Radar"* is a ground-based air defense radar based on Brookner's[1] open literature source. However, the design process takes on a different flavor than that introduced by Brookner. Additionally, any and all design alternatives presented in this book are based on and are completely traceable to open literature sources.

The design presented in this book is intended to be tutorial and academic in nature and does not adhere to any other conditions; thus, it is not intended for any other uses. More specifically, the design approach adopted in this book is based on modeling many of the radar system components with no regards to any hardware constraints nor to any practical limitations.

15.1. Textbook Radar Design Case Study - Visit #1

You are to design a ground based radar to fulfill the following mission: Search and Detection. The threat consists of aircraft and missiles, as follows

Threat type	RCS	Velocity	Altitude
Aircraft	$\sigma_a = 6dBsm$	$400m/\sec$	$7Km$
Missile	$\sigma_m = -3dBsm$	$150m/\sec$	$2Km$

Assume,

- A mechanically scanning radar with 360 degrees azimuth coverage
- Scan rate less than or equal to 1 revolution every 2 seconds
- Operating frequency L- to X-band
- Range resolution of 150 meters

1. Brookner, Eli - Editor, *Practical Phased Array Antenna Systems*, Artech House, 1991. Chapter 7.

- No angular resolution specified at this time
- Only one missile and one aircraft constitute the entire threat
- Noise figure F = 6 dB
- Radar total losses L = 8 dB
- For now use a fan beam with azimuth beamwidth of less than 3 degrees
- 13 dB SNR is a reasonable detection threshold
- Only flat earth is considered

A Design - Visit #1

The range resolution is $\Delta R = 150m$. Thus, by using $\Delta R = (c\tau)/2 = c/(2B)$, one calculates the required pulsewidth as $\tau = 1\mu second$, or equivalently the required bandwidth $B = 1MHz$. At this point a few preliminary decisions must be made. This includes the selection of the radar operating frequency, the aperture size, and the single pulse peak power.

The choice of an operating frequency that can fulfill the design requirements is driven by many factors, such as aperture size, antenna gain, clutter, atmospheric attenuation and the maximum peak power, to name a few. In this design, an operating frequency $f_0 = 3GHz$ is selected. This choice is somewhat arbitrary at this point; however, as we proceed with the design process, this choice will be better clarified.

Second, the transportability (mobility) of the radar drives the designer in the direction of a smaller aperture type. A good choice would be less than 5 meters squared. For now choose $A_e = 2.25m^2$. One must consider the energy required per pulse. Note that this design approach assumes that the minimum detection SNR (13 dB) requirement is based on pulse integration. This condition is true because the target is illuminated with several pulses during a single scan, provided that the antenna azimuth beamwidth and the PRF choice satisfy Eq. (5.91) (see Chapter 5).

The single pulse energy is

$$E = P_t\tau . \qquad \text{Eq. (15.1)}$$

Typically, a given radar must be designed such that it has a handful of pulsewidths (waveforms) to choose from. Different waveforms (pulsewidths) are used for distinct modes of operations (search, track, etc.). However, for now only a single pulse which satisfies the range resolution requirement is considered. The radar equation can be modified to compute the pulsewidth that can achieve a certain SNR, and that is,

$$E = P_t\tau = \frac{(4\pi)^3 kT_s FLR^4 SNR_1}{G^2\lambda^2\sigma} . \qquad \text{Eq. (15.2)}$$

All parameters in Eq. (15.1) are known, except for the antenna gain, the detection range, and the single pulse SNR. The antenna gain is calculated from

$$\left(G = \frac{4\pi A_e}{\lambda^2} = \frac{4\pi \times 2.25}{(0.1)^2} = 2827.4\right) \Rightarrow G = 34.5dB \qquad \text{Eq. (15.3)}$$

where the relation ($\lambda = c/f_0$) was used.

In order to estimate the detection range, consider the following argument. Since an aircraft has a larger RCS than a missile, one would expect an aircraft to be detected at a much longer range than that of a missile. This is depicted in Figure 15.1, where R_a refers to the aircraft detection range and R_m denotes the missile detection range. As illustrated in this figure, the minimum search elevation angle θ_1 is driven by the missile detection range, assuming that the missiles are detected, with the proper SNR, as soon as they enter the radar beam. Alternatively, the maximum search elevation angle θ_2 is driven the aircraft's position along with the range that corresponds to the defense last chance to intercept the threat (both aircraft and missile). This range is often called "keep-out minimum range" and is denoted by R_{min}. In this design approach, $R_{min} = 30Km$ is selected.

The determination of R_a and R_m is dictated by how fast a defense interceptor can reach the keep-out minimum range and kill the threat. For example, assume that an interceptor average velocity is $250m/s$. It follows that the interceptor time of flight, based on $R_{min} = 30Km$, is

$$T_{interceptor} = \frac{30 \times 10^3}{250} = 120\,\text{sec} \quad . \qquad \text{Eq. (15.4)}$$

Therefore, an aircraft and a missile must be detected by the radar at

$$\begin{aligned} R_a &= 30Km + 120 \times 400 = 78Km \\ R_m &= 30Km + 120 \times 150 = 48Km \end{aligned} \quad . \qquad \text{Eq. (15.5)}$$

Note that these values should be used only as a guide. The actual detection range must also include a few more kilometers, in order to allow the defense better reaction time. In this design, choose $R_m = 55Km$ and $R_a = 90Km$. Therefore, the maximum PRF that guarantees an unambiguous range of at least 90 Km is calculated from Eq. (4.5) (see Chapter 4). More precisely,

$$f_r \le \frac{c}{2R_u} = \frac{3 \times 10^8}{2 \times 90 \times 10^3} = 1.67KHz \, . \qquad \text{Eq. (15.6)}$$

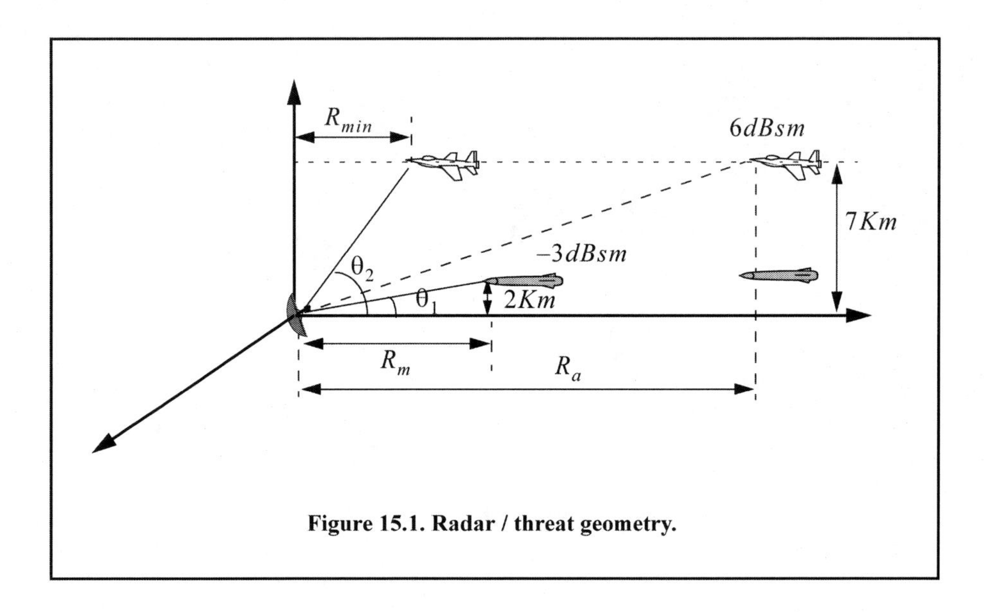

Figure 15.1. Radar / threat geometry.

Since there are no angular resolution requirements imposed on the design at this point, then $1.67KHz$ is the only criterion that will be used to determine the radar operating PRF. Select,

$$f_r = 1000Hz. \qquad \text{Eq. (15.7)}$$

The minimum and maximum elevation angles are, respectively calculated as

$$\theta_1 = \operatorname{atan}\left(\frac{2}{55}\right) = 2.08° \qquad \text{Eq. (15.8)}$$

$$\theta_2 = \operatorname{atan}\left(\frac{7}{30}\right) = 13.13°. \qquad \text{Eq. (15.9)}$$

These angles are then used to compute the elevation search extent (remember that the azimuth search extent is equal to $360°$). More precisely, the search volume Ω (in steradians) is given by

$$\Omega = \frac{\theta_2 - \theta_1}{(57.296)^2} \times 360. \qquad \text{Eq. (15.10)}$$

Consequently, the search volume is

$$\Omega = 360 \times \frac{\theta_2 - \theta_1}{(57.296)^2} = 360 \times \frac{13.13 - 2.08}{(57.296)^2} = 1.212 \ steradians. \qquad \text{Eq. (15.11)}$$

The desired antenna must have a fan beam; thus using a parabolic rectangular antenna will meet the design requirements. Select $A_e = 2.25m^2$; the corresponding antenna 3-dB elevation and azimuth beamwidths are denoted as θ_e, θ_a, respectively. Select

$$\theta_e = \theta_2 - \theta_1 = 13.13 - 2.08 = 11.05°. \qquad \text{Eq. (15.12)}$$

The azimuth 3-dB antenna beamwidth is calculated as

$$\theta_a = \frac{4\pi}{G\theta_e} = \frac{4 \times \pi \times 180^2}{2827.4 \times \pi^2 \times 11} = 1.33°. \qquad \text{Eq. (15.13)}$$

It follows that the number of pulses that strikes a target during a single scan is calculated using

$$n_p \le \frac{\theta_a f_r}{\dot{\theta}_{scan}} = \frac{1.33 \times 1000}{180} = 7.39 \Rightarrow n_p = 7. \qquad \text{Eq. (15.14)}$$

The design approach presented in this book will only assume non-coherent integration (the reader is advised to re-calculate all results by assuming coherent integration, instead). The design requirement mandates a 13 dB SNR for detection. By using Eq. (5.97) (see Chapter 5), one calculates the required single pulse SNR,

$$(SNR)_1 = \frac{10^{1.3}}{2 \times 7} + \sqrt{\frac{(10^{1.3})^2}{4 \times 7^2} + \frac{10^{1.3}}{7}} = 3.635 \Rightarrow (SNR)_1 = 5.6dB. \qquad \text{Eq. (15.15)}$$

Furthermore the non-coherent integration loss associated with this case is computed from Eq. (5.95),

$$L_{NCI} = \frac{1 + 3.635}{3.635} = 1.27 \Rightarrow L_{NCI} = 1.056dB. \qquad \text{Eq. (15.16)}$$

It follows that the corresponding **single pulse** energy for the missile and the aircraft cases are respectively given by

$$E_m = \frac{(4\pi)^3 kT_eFLR_m^4(SNR)_1}{G^2\lambda^2\sigma_m} \Rightarrow$$

$$E_m = \frac{(4\pi)^3(1.38\times10^{-23})(290)(10^{0.8})(10^{0.6})(55\times10^3)^4 10^{0.56}}{(2827.4)^2(0.1)^2(0.5)} = 0.1658 \ J$$ Eq. (15.17)

$$E_a = \frac{(4\pi)^3 kT_eFLR_a^4(SNR)_1}{G^2\lambda^2\sigma_a} \Rightarrow$$

$$E_a = \frac{(4\pi)^3(1.38\times10^{-23})(290)(10^{0.8})(10^{0.6})(90\times10^3)^4 10^{0.56}}{(2827.4)^2(0.1)^2(4)} = 0.1487 \ J \cdot$$ Eq. (15.18)

Hence, the peak power that satisfies the single pulse detection requirement for both target types is

$$P_t = \frac{E}{\tau} = \frac{0.1658}{1\times10^{-6}} = 165.8KW.$$ Eq. (15.19)

The radar equation with pulse integration is

$$SNR = \frac{P_t^1 G^2\lambda^2\sigma}{(4\pi)^3 kT_eBFLR^4}\frac{n_p}{L_{NCI}}.$$ Eq. (15.20)

Figure 15.2 shows the SNR versus detection range for both target-types with and without integration.

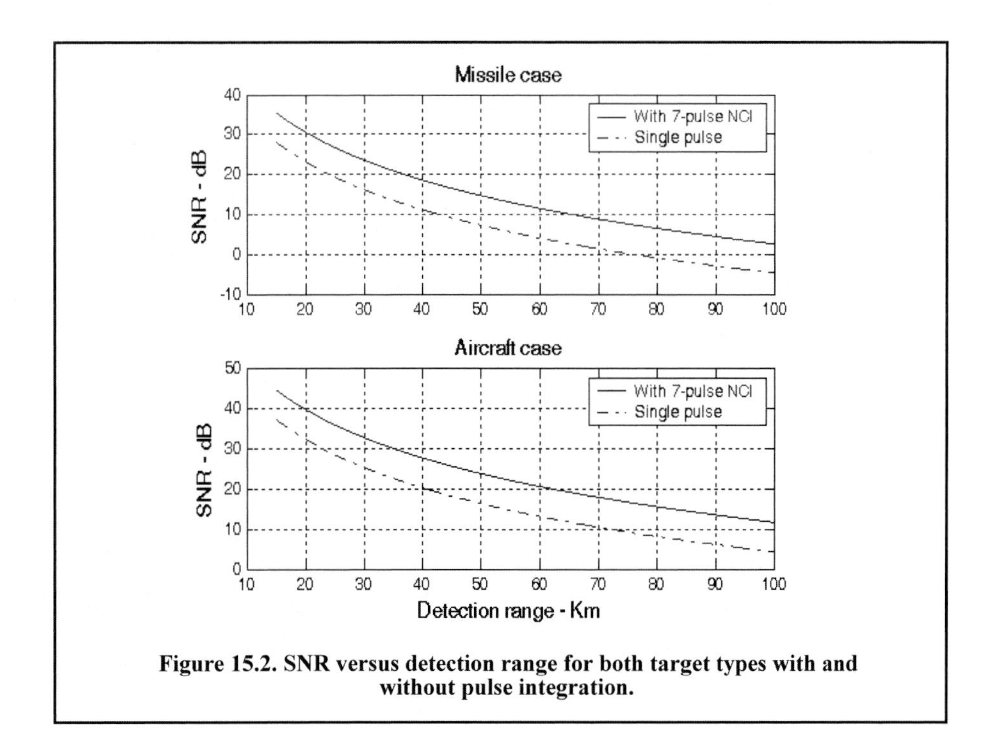

Figure 15.2. SNR versus detection range for both target types with and without pulse integration.

A Design Alternative

One could have elected not to reduce the single pulse peak power, but rather keep the single pulse peak power as computed in Eq. (15.19) and increase the radar detection range. For example, integrating 7 pulses coherently would improve the radar detection range by a factor of

$$R_{imp} = (7)^{0.25} = 1.63 . \qquad \text{Eq. (15.21)}$$

It follows that the new missile and aircraft detection ranges are

$$\begin{aligned} R_a &= 78 \times 1.63 = 126.9Km \\ R_m &= 48 \times 1.63 = 78.08Km \end{aligned} . \qquad \text{Eq. (15.22)}$$

Note that extending the minimum detection range for a missile to $R_m = 78Km$ would increase the size of the extent of the elevation search volume. More precisely,

$$\theta_1 = \operatorname{atan}\left(\frac{2}{78}\right) = 1.47° . \qquad \text{Eq. (15.23)}$$

It follows that the search volume Ω (in steradians) is now

$$\Omega = 360 \times \frac{\theta_2 - \theta_1}{(57.296)^2} = 360 \times \frac{13.13 - 1.47}{(57.296)^2} = 1.279 \ steradians . \qquad \text{Eq. (15.24)}$$

Alternatively, integrating 7 pulses non-coherently with $(SNR)_{NCI} = 13dB$, yields

$$(SNR)_1 = 5.6dB , \qquad \text{Eq. (15.25)}$$

and the integration loss is

$$L_{NCI} = 1.057dB . \qquad \text{Eq. (15.26)}$$

Then, the net non-coherent integration gain is

$$NCI_{gain} = 10 \times \log(7) - 1.057 = 7.394dB \Rightarrow NCI_{gain} = 5.488 . \qquad \text{Eq. (15.27)}$$

Thus, the radar detection range is now improved due to a 7-pulse non-coherent integration to

$$\begin{aligned} R_a &= 78 \times (5.488)^{0.25} = 119.38Km \\ R_m &= 48 \times (5.488)^{0.25} = 73.467Km \end{aligned} . \qquad \text{Eq. (15.28)}$$

Again, the extent of the elevation search volume is changed to,

$$\theta_1 = \operatorname{atan}\left(\frac{2}{73.467}\right) = 1.56° . \qquad \text{Eq. (15.29)}$$

It follows that the search volume Ω (in steradians) is now

$$\Omega = 360 \times \frac{\theta_2 - \theta_1}{(57.296)^2} = 360 \times \frac{13.13 - 1.56}{(57.296)^2} = 1.269 \ steradians . \qquad \text{Eq. (15.30)}$$

15.2. Textbook Radar Design Case Study- Visit #2

Assuming a matched filter receiver, select a set of waveforms that can meet the design requirements as stated in the previous section. Assume,

- Linear frequency modulation
- Do not use more than a total of 5 waveforms
- Modify the design so that the range resolution $\Delta R = 30m$ during the search mode, and $\Delta R = 7.5m$ during tracking

A Design - Visit #2

The major characteristics of radar waveforms include the waveform's energy, range resolution, and Doppler (or velocity) resolution. Close attention should be paid to the selection process of the pulsewidth. In this design we will assume that the pulse energy is the same as that computed in the previous section. The radar operating bandwidth during search and track are calculated from the relation ($\Delta R = c/2B$), where ΔR is the range resolution, c is the speed of light, and B is the bandwidth. It follows that

$$\begin{Bmatrix} B_{search} \\ B_{track} \end{Bmatrix} = \begin{Bmatrix} 3 \times 10^8/(2 \times 30) = 5MHz \\ 3 \times 10^8/(2 \times 7.5) = 20MHz \end{Bmatrix}. \qquad \text{Eq. (15.31)}$$

Since the design calls for a pulsed radar, then for each pulse transmitted (one PRI) the radar should not be allowed to receive any signal until that pulse has been completely transmitted. This limits the radar to a minimum operating range defined by

$$R_{min} = \frac{c\tau}{2}. \qquad \text{Eq. (15.32)}$$

In this design choose $R_{min} \geq 15Km$. It follows that the minimum acceptable pulsewidth is $\tau_{max} \leq 100\mu s$.

For this design select 5 waveforms, one for search and four for track. Typically search waveforms are longer than track waveforms.; alternatively, tracking waveforms require wider bandwidths than search waveforms. However, in the context of range, more energy is required at longer ranges (for both track and search waveforms), since one would expect the SNR to get larger as range becomes smaller.

To illustrate this concept, consider the following radar parameters: peak power $P_t = 1.5MW$; operating frequency $f_0 = 5.6GHz$; antenna gain $G = 45dB$; effective temperature $T_e = 290K$; radar losses $L = 6dB$; noise figure $F = 3dB$; bandwidth $B = 5MHz$; target cross section $\sigma = 0.1m^2$.

One can easily compute the pulsewidth required to achieve a certain SNR for a given detection range. In this case the radar equation can be written as

$$\tau = \frac{(4\pi)^3 kT_0FLR^4SNR}{P_tG^2\lambda^2\sigma}. \qquad \text{Eq. (15.33)}$$

Figure 15.3 shows the corresponding plot for the pulse width versus three different detection range values.

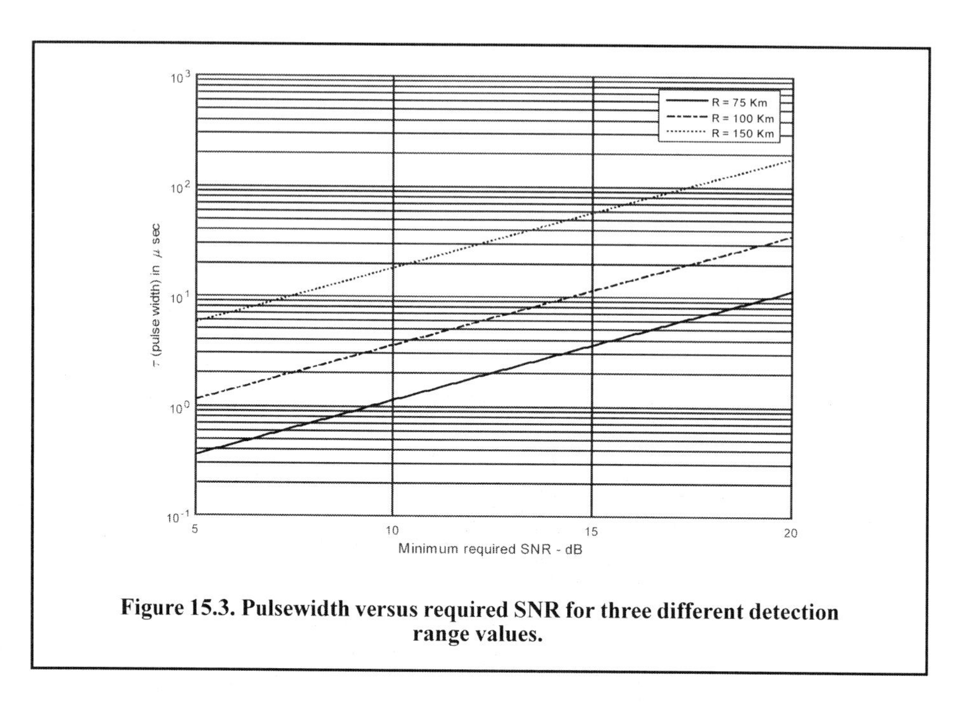

Figure 15.3. Pulsewidth versus required SNR for three different detection range values.

Now, back to our design problem: assume that during search and initial detection the single pulse peak power is to be kept under 10 KW (i.e. $P_t \leq 20KW$). Then by using the single pulse energy calculated using $E_m = 0.1147 Joules$ (this choice will become clearer later on in this chapter), one can compute the minimum required pulsewidth as

$$\tau_{min} \geq \frac{0.1147}{20 \times 10^3} = 5.735 \mu s . \qquad \text{Eq. (15.34)}$$

Choose $\tau_{search} = 20 \mu s$, with bandwidth $B = 5MHz$ and use LFM modulation. Figure 15.4 shows plots of the real part, imaginary part, and the spectrum of this search waveform. As far as the track waveforms, choose four waveforms of the same bandwidth ($B_{track} = 20MHz$) and with the following pulsewidths, as illustrated in Table 15.1.

Note that R_{max} refers to the initial range at which track has been initiated. Figure 15.5 is similar to Figure 15.4 except it is for τ_{t3}. For the waveform set selected in this design option, the radar duty cycle varies from 1.25% to 2.0%. Remember that the PRF was calculated as $f_r = 1KHz$, thus the PRI is $T = 1ms$. Figures 15.6 and 15.7 show the search waveform ambiguity plots.

Table 15.1. "*Textbook Radar*" design case study track waveforms.

Pulsewidth	Range window
$\tau_{t1} = 20 \mu s$	$R_{max} \rightarrow 0.75 R_{max}$
$\tau_{t2} = 17.5 \mu s$	$0.75 R_{max} \rightarrow 0.5 R_{max}$
$\tau_{t3} = 15 \mu s$	$0.5 R_{max} \rightarrow 0.25 R_{max}$
$\tau_{t4} = 12.5 \mu s$	$R \leq 0.25 R_{max}$

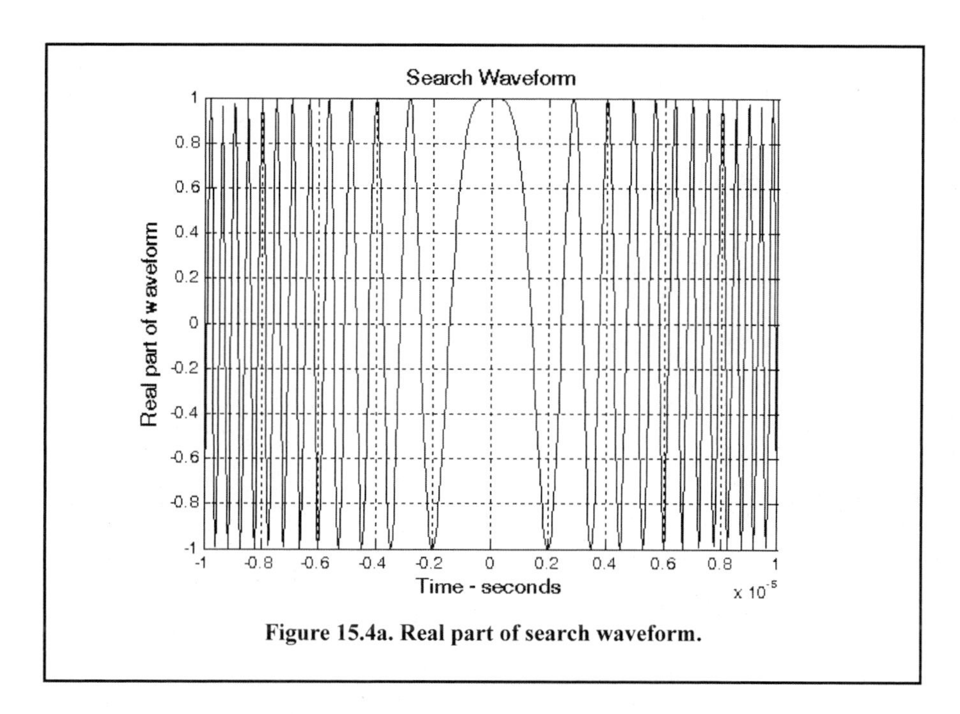

Figure 15.4a. Real part of search waveform.

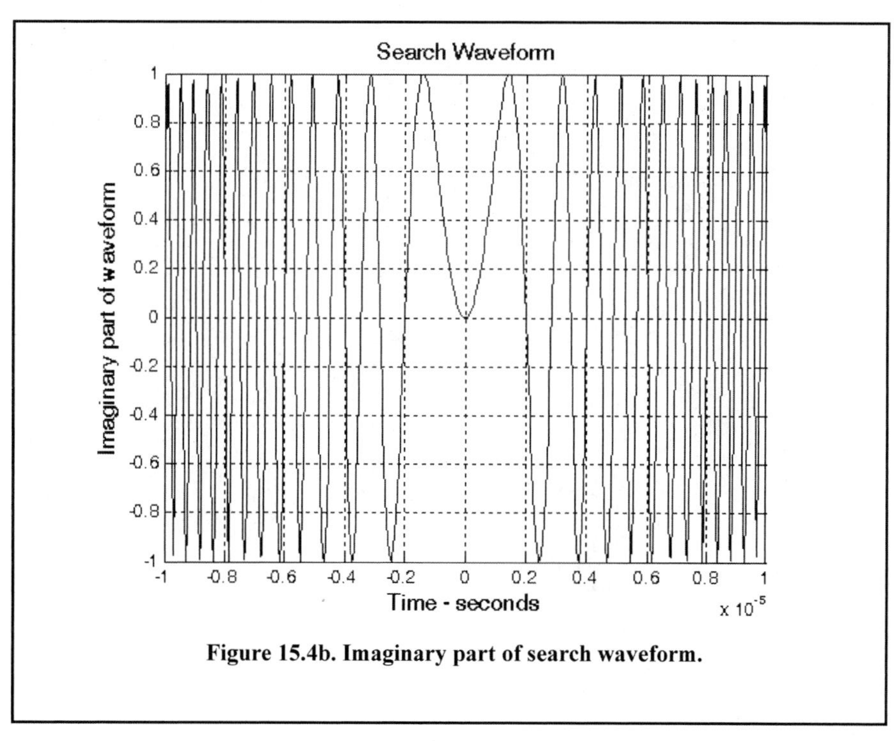

Figure 15.4b. Imaginary part of search waveform.

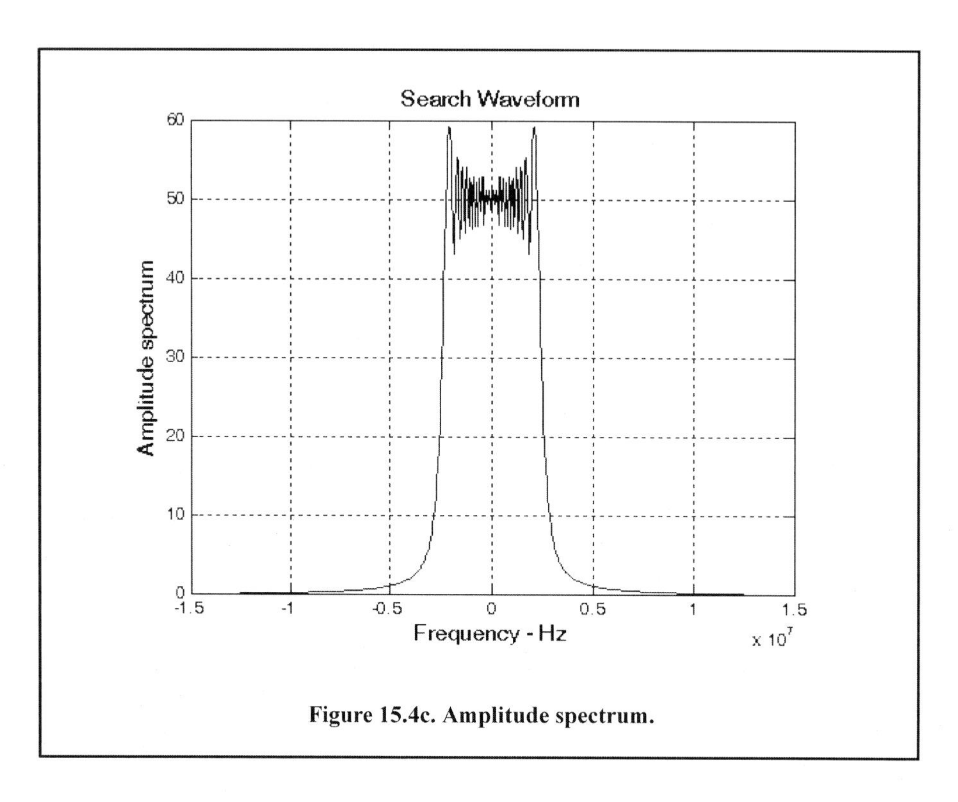

Figure 15.4c. Amplitude spectrum.

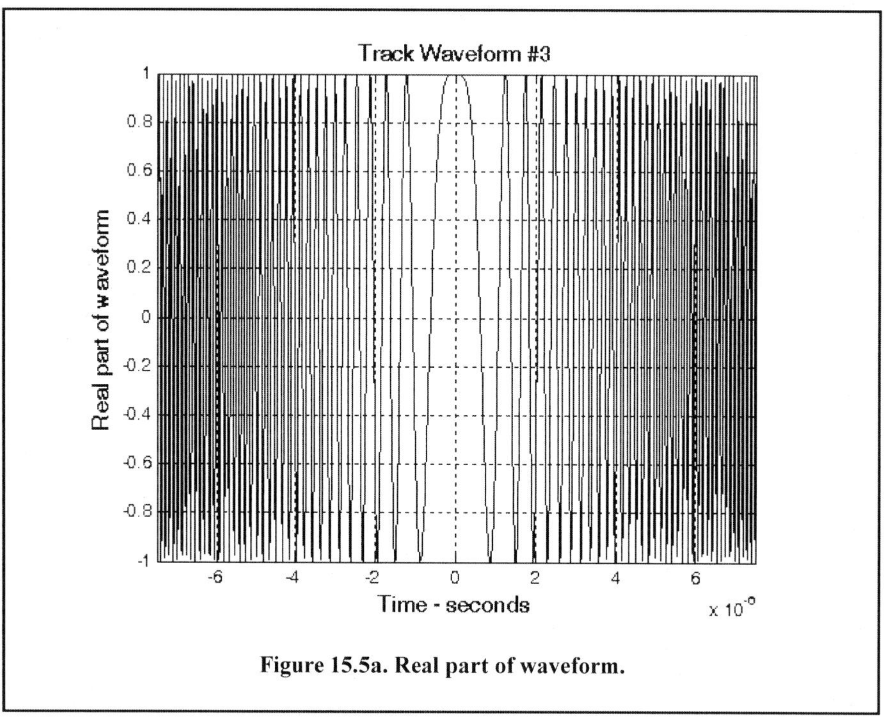

Figure 15.5a. Real part of waveform.

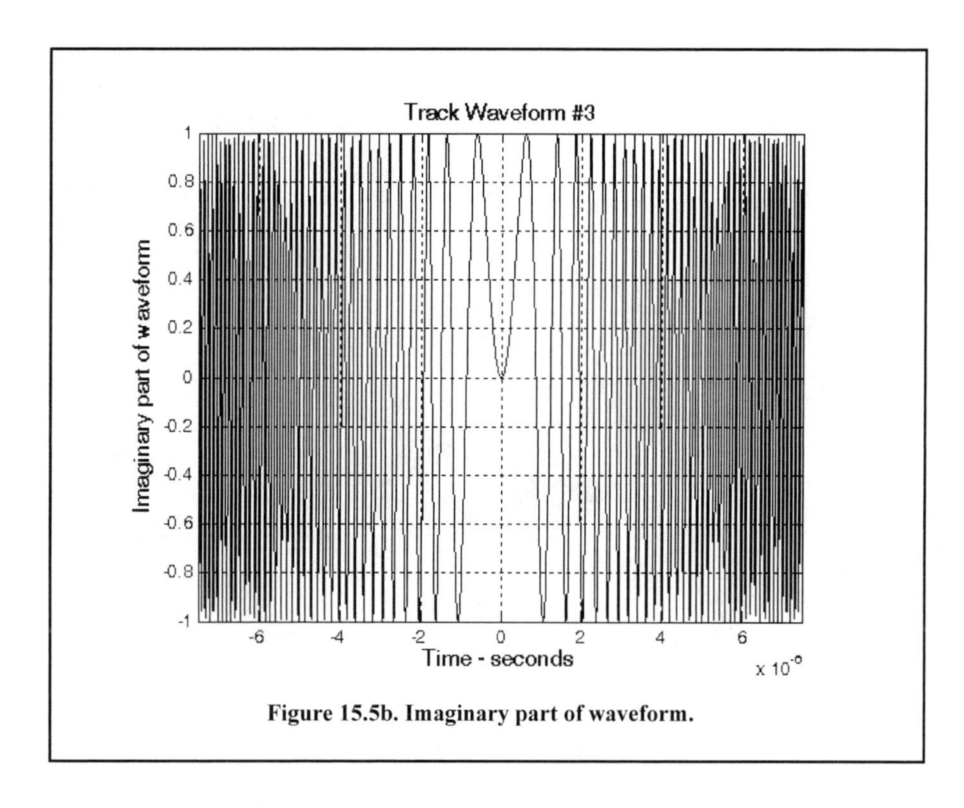

Figure 15.5b. Imaginary part of waveform.

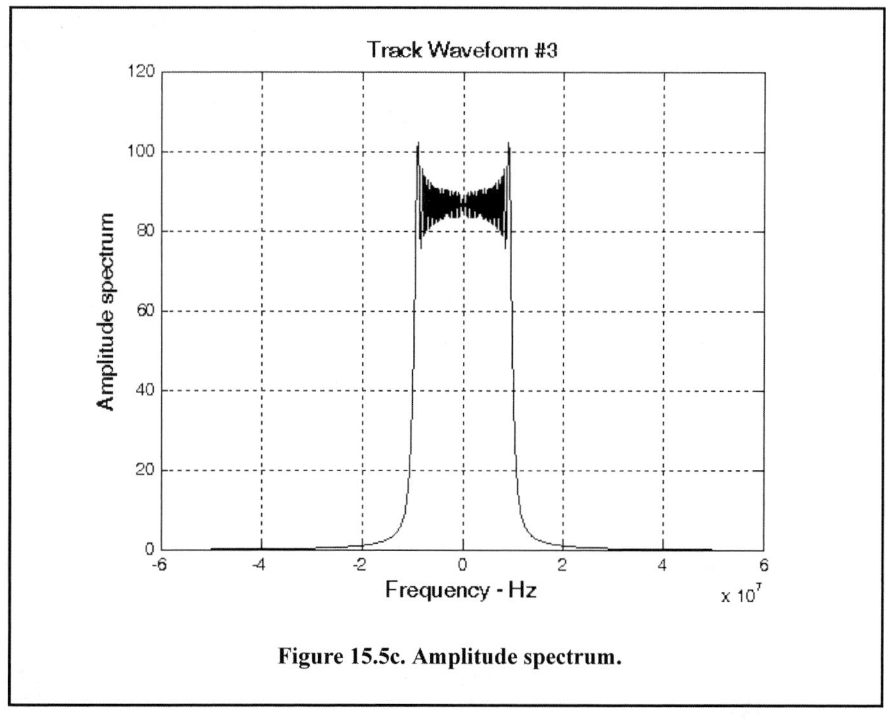

Figure 15.5c. Amplitude spectrum.

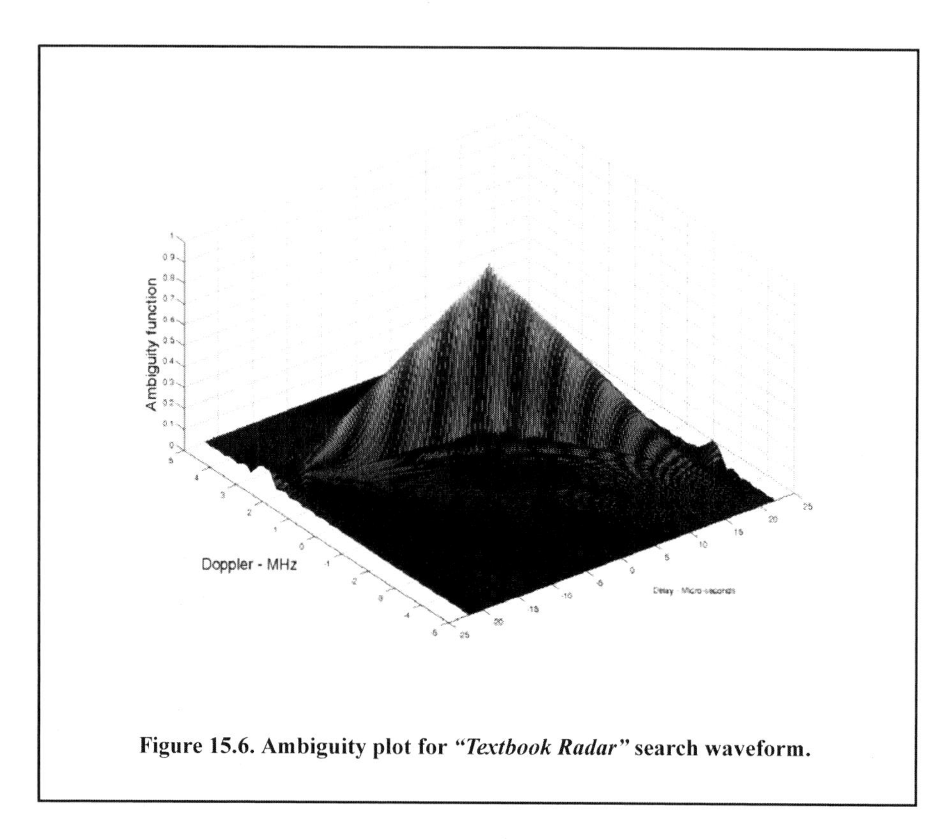

Figure 15.6. Ambiguity plot for *"Textbook Radar"* search waveform.

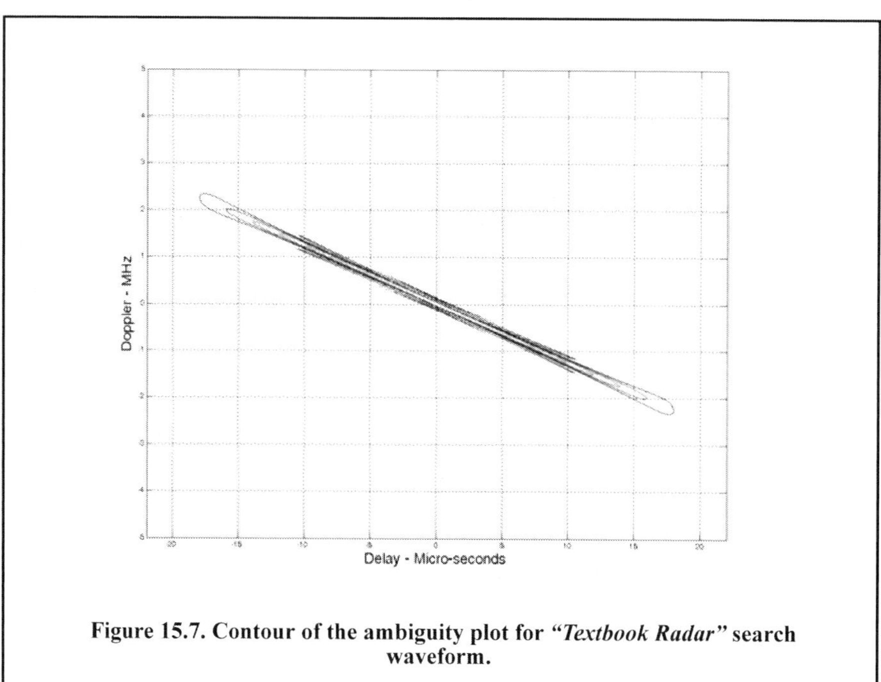

Figure 15.7. Contour of the ambiguity plot for *"Textbook Radar"* search waveform.

15.3. Textbook Radar Design Case Study - Visit #3

Assume that the threat may consist of multiple aircraft and missiles. Show how the matched filter receiver can resolve multiple targets with a minimum range separation of 50 meters. Also verify that the waveforms selected in the previous section are adequate to maintain proper detection and tracking (i.e. provide sufficient SNR).

A Design - Visit #3

It was determined earlier that the pulsed compressed range resolutions during search and track are respectively given by

$$\Delta R_{search} = 30m; \ B_{search} = 5MHz \qquad \text{Eq. (15.35)}$$

$$\Delta R_{track} = 7.5m; \ B_{track} = 20MHz. \qquad \text{Eq. (15.36)}$$

Assume that track is initiated once detection is declared. Aircraft target types are detected at $R^{a}_{max} = 90Km$ while the missile is detected at $R^{m}_{max} = 55Km$. The minimum SNR at these ranges for both target types is $SNR \geq 4dB$ when 4-pulse non-coherent integration is utilized along with cumulative detection. Also, a single pulse option was not desirable since it required prohibitive values for the peak power. At this point one should however take advantage of the increased SNR due to pulse compression. The pulse compression gain, for the selected waveforms, is equal to 10 dB. One should investigate this SNR enhancement in the context of eliminating the need for pulse integration.

The pulsed compressed SNR can be computed using the radar equation,

$$SNR = \frac{P_t \tau' G^2 \lambda^2 \sigma}{(4\pi)^3 R^4 k T_s F L}. \qquad \text{Eq. (15.37)}$$

From the earlier sections, we selected $G = 34.5dB$, $\lambda = 0.1m$, $F = 6dB$, $L = 8dB$, $\sigma_m = 0.5m^2$, $\sigma_a = 4m^2$, and $P_t = 20KW$. The search pulsewidth is $\tau' = 20\mu s$ and the track waveforms are $12.5\mu s \leq \tau'_i \leq 20\mu s$. First consider the missile case. The single pulse SNR at the maximum detection range $R^{m}_{max} = 55Km$ is given by

$$SNR_m = \frac{20 \times 10^3 \times 20 \times 10^{-6} \times (10^{3.45})^2 \times (0.1)^2 \times 0.5}{(4\pi)^3 \times (55 \times 10^3)^4 \times 1.38 \times 10^{-23} \times 290 \times 10^{0.8} \times 10^{0.6}} = . \qquad \text{Eq. (15.38)}$$

$$8.7028 \Rightarrow SNR_m = 9.39dB$$

Alternatively, the single pulse SNR, with pulse compression, for the aircraft is

$$SNR_a = \frac{20 \times 10^3 \times 20 \times 10^{-6} \times (10^{3.45})^2 \times (0.1)^2 \times 4}{(4\pi)^3 \times (90 \times 10^3)^4 \times 1.38 \times 10^{-23} \times 290 \times 10^{0.8} \times 10^{0.6}} = . \qquad \text{Eq. (15.39)}$$

$$9.7104 \Rightarrow SNR_m = 9.87dB$$

Next, consider the matched filter and its replicas and pulsed compressed outputs (due to different waveforms). Assume a receive window of 200 meters during search and 50 meters during track. Figure 15.8 shows the replica and the associated uncompressed and compressed signals (the targets consist of two aircraft separated by 50 meters. Figure 15.9 is similar to Figure 15.8, except it is for track waveform number 4 and the target separation is 20 m.

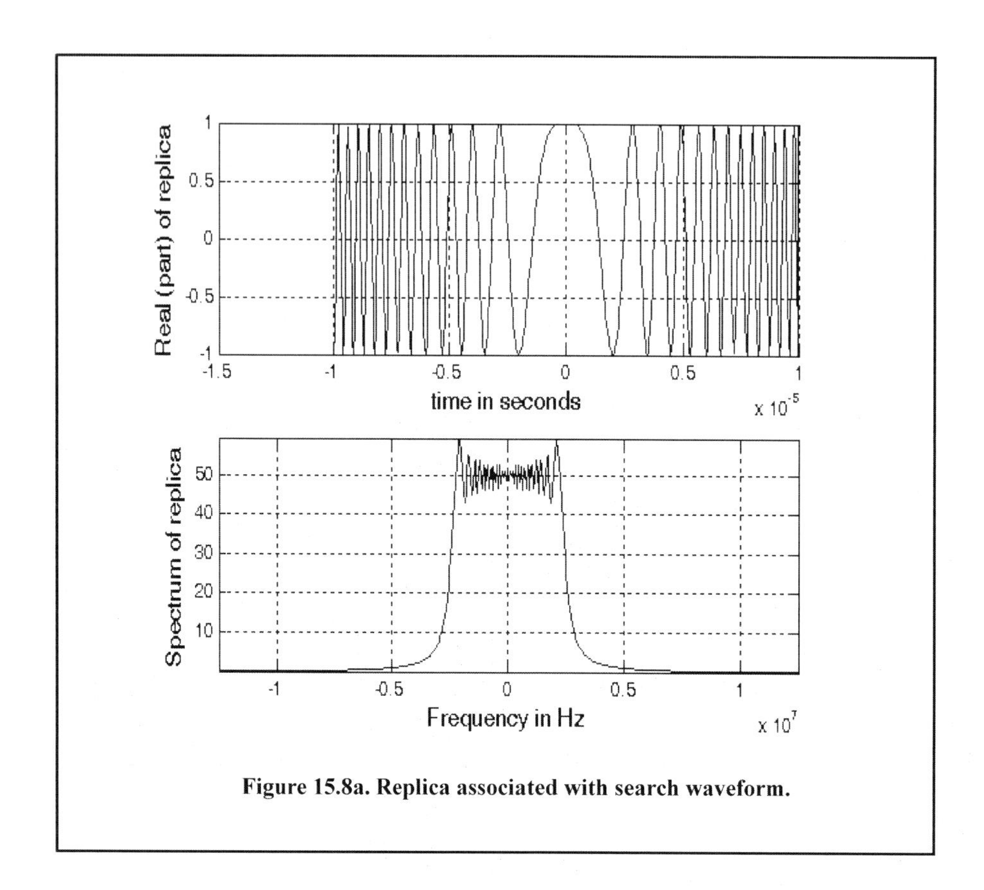

Figure 15.8a. Replica associated with search waveform.

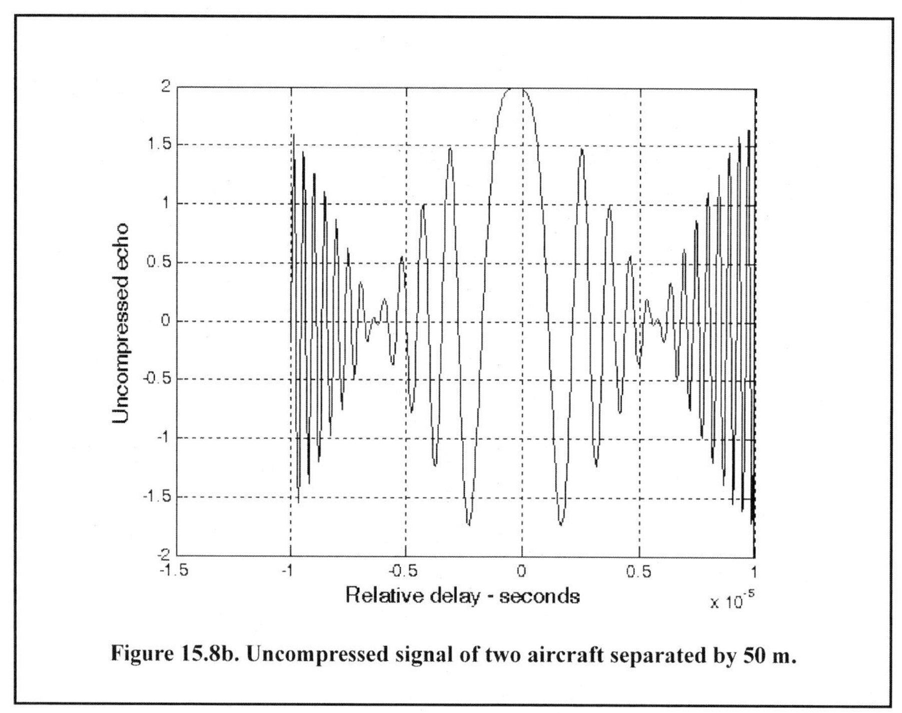

Figure 15.8b. Uncompressed signal of two aircraft separated by 50 m.

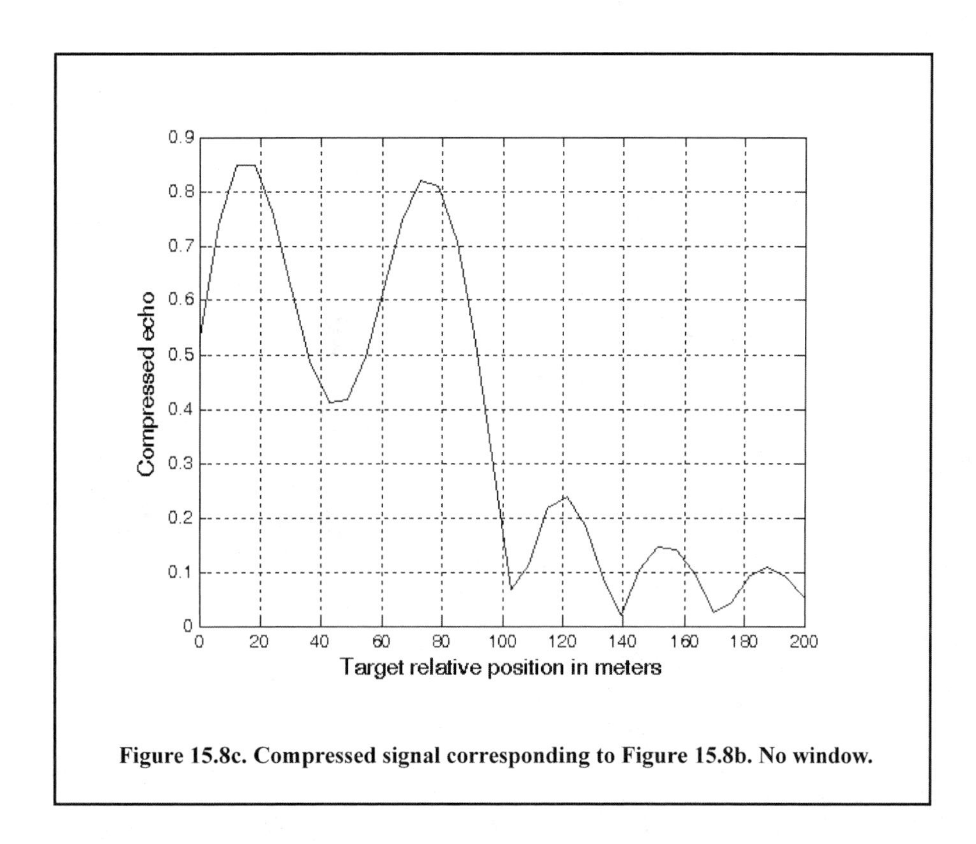

Figure 15.8c. Compressed signal corresponding to Figure 15.8b. No window.

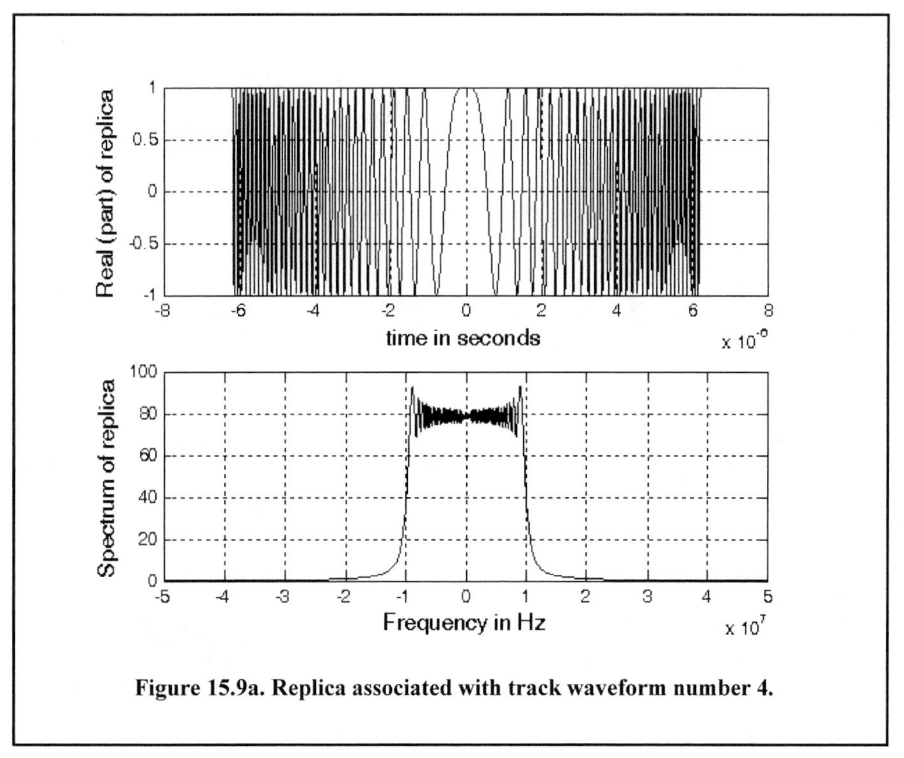

Figure 15.9a. Replica associated with track waveform number 4.

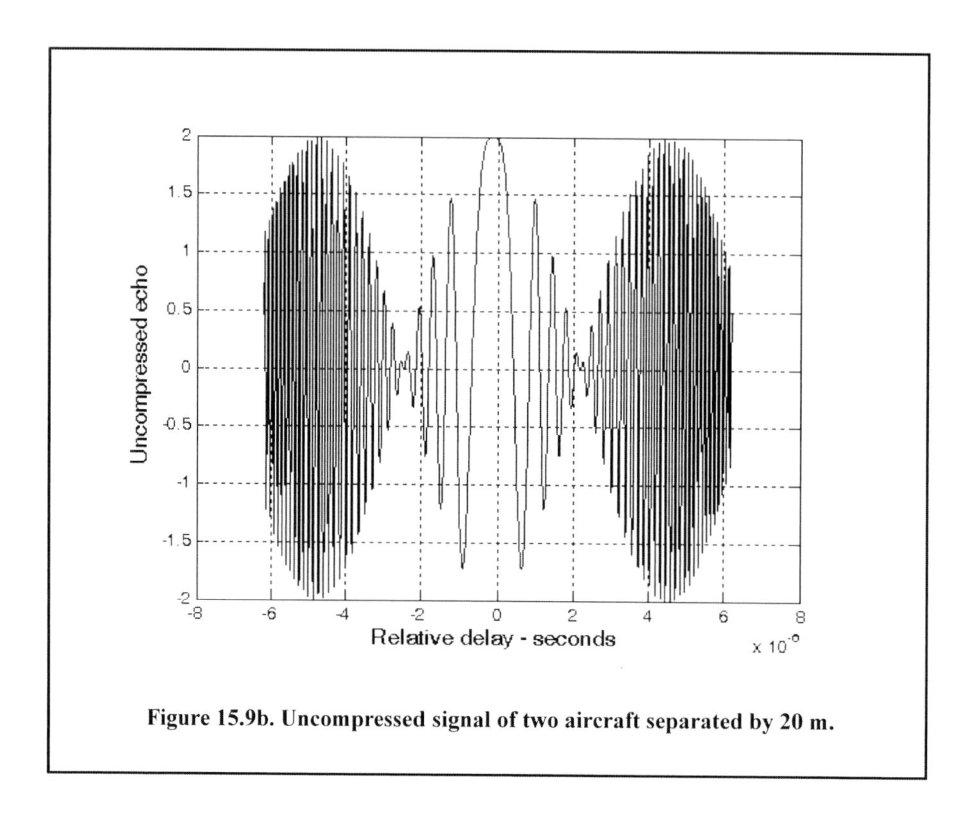

Figure 15.9b. Uncompressed signal of two aircraft separated by 20 m.

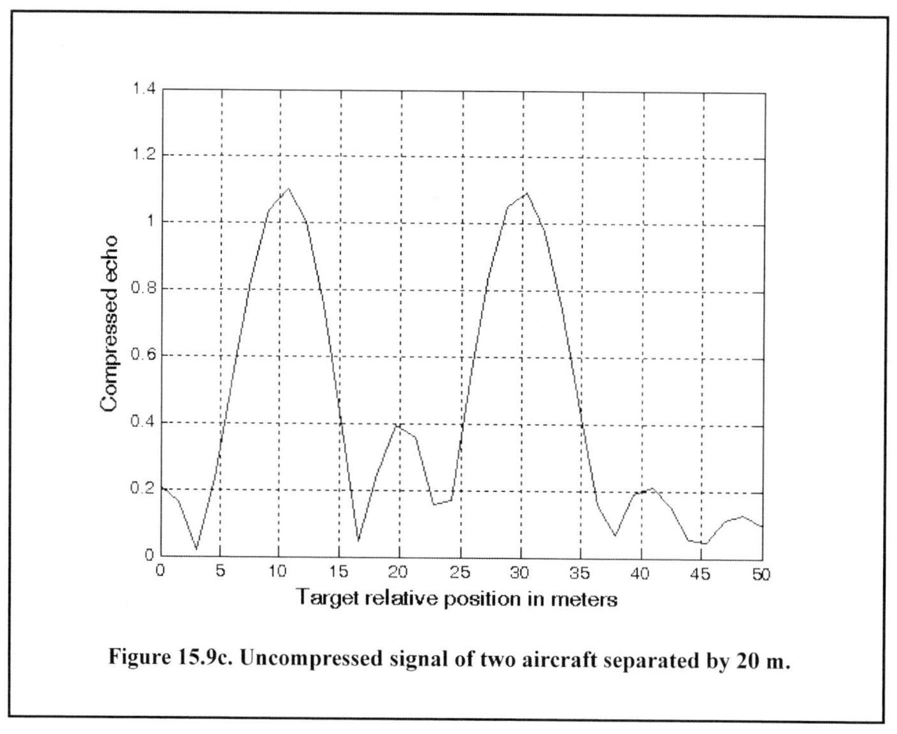

Figure 15.9c. Uncompressed signal of two aircraft separated by 20 m.

15.4. Textbook Radar Design Case Study - Visit #4

Analyze the impact of ground clutter on a *"Textbook Radar"* design case study. Assume

- A Gaussian antenna pattern
- The radar height is 5 meters
- Antenna sidelobe level $SL = -20dB$
- Ground clutter coefficient $\sigma^0 = -15dBsm$.

What conclusions can you draw about the radar's ability to maintain proper detection and track of both targets? Assume a radar height $h_r \geq 5m$.

A Design - Visit #4

From the design process, it is determined that the minimum single pulse SNR required to accomplish the design objectives was $SNR \geq 4dB$ when non-coherent integration (4 pulses) and cumulative detection were used. Factoring in the surface clutter will degrade the SIR. However, one must maintain $SIR \geq 4dB$ in order to achieve the desired probability of detection.

Figure 15.10 shows a plot of the clutter RCS versus range corresponding to *"Textbook Radar"* design requirements.

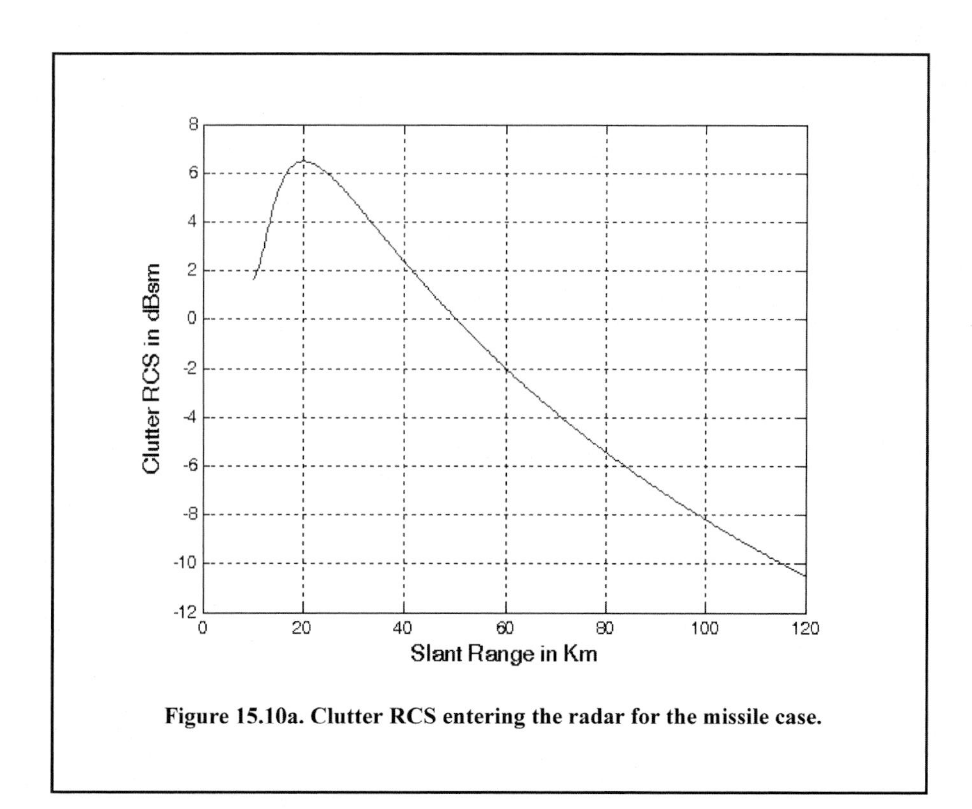

Figure 15.10a. Clutter RCS entering the radar for the missile case.

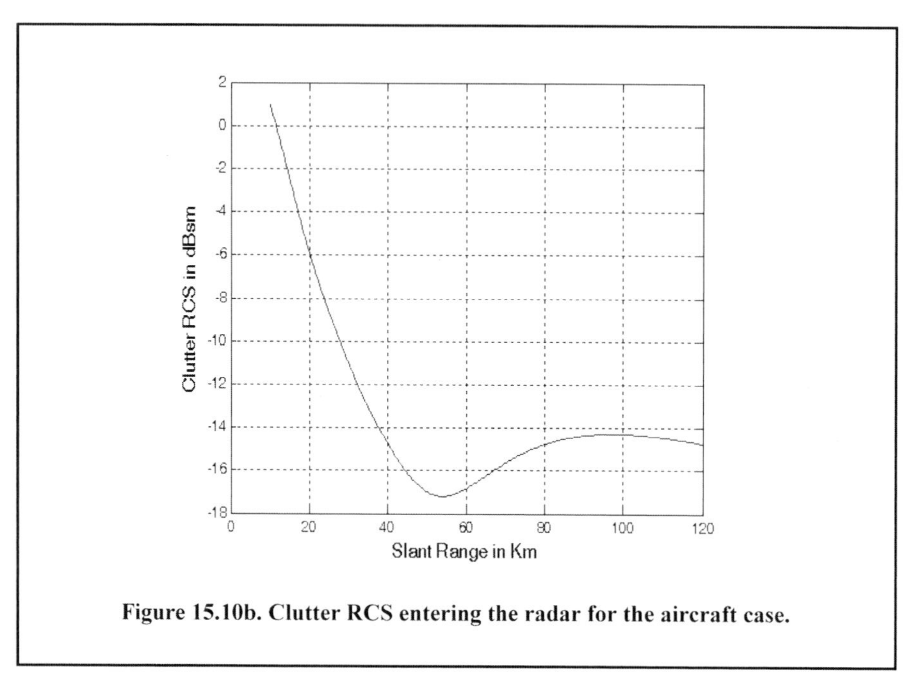

Figure 15.10b. Clutter RCS entering the radar for the aircraft case.

Figure 15.11 shows a plot of the CNR and the SIR associated with the missile. Figure 15.12 is similar to Figure 15.11 except it is for the aircraft case. It is clear from these figures that the required SIR has been degraded significantly for the missile case and not as much for the aircraft case. This should not be surprising, since the missile's altitude is much smaller than that of the aircraft. Without clutter mitigation, the missile would not be detected at all. Alternatively, the aircraft detection is compromised at $R \leq 80Km$.

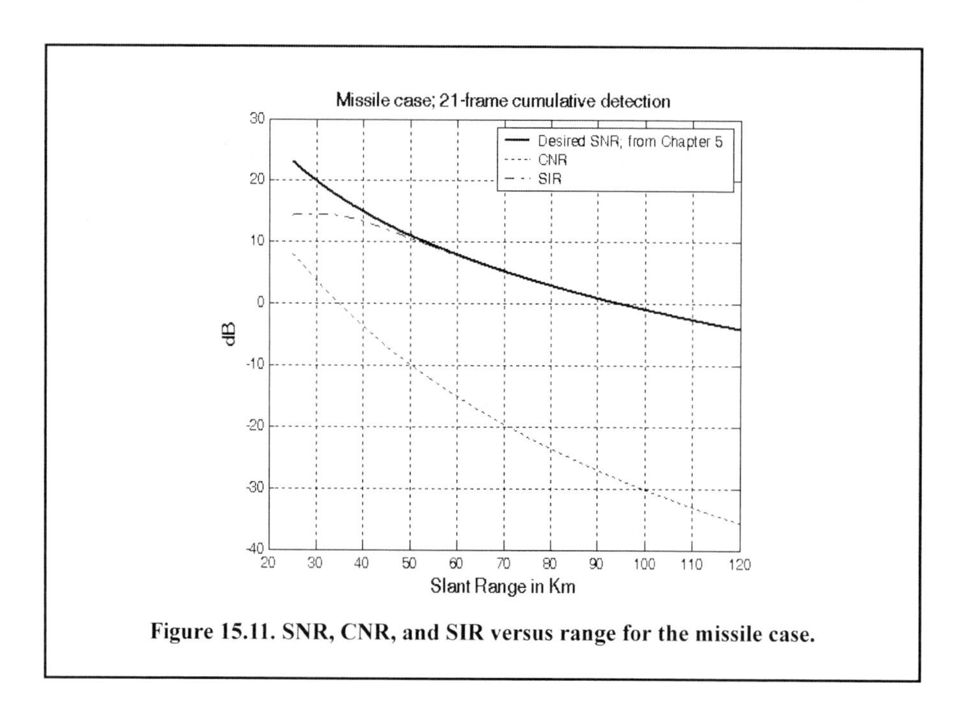

Figure 15.11. SNR, CNR, and SIR versus range for the missile case.

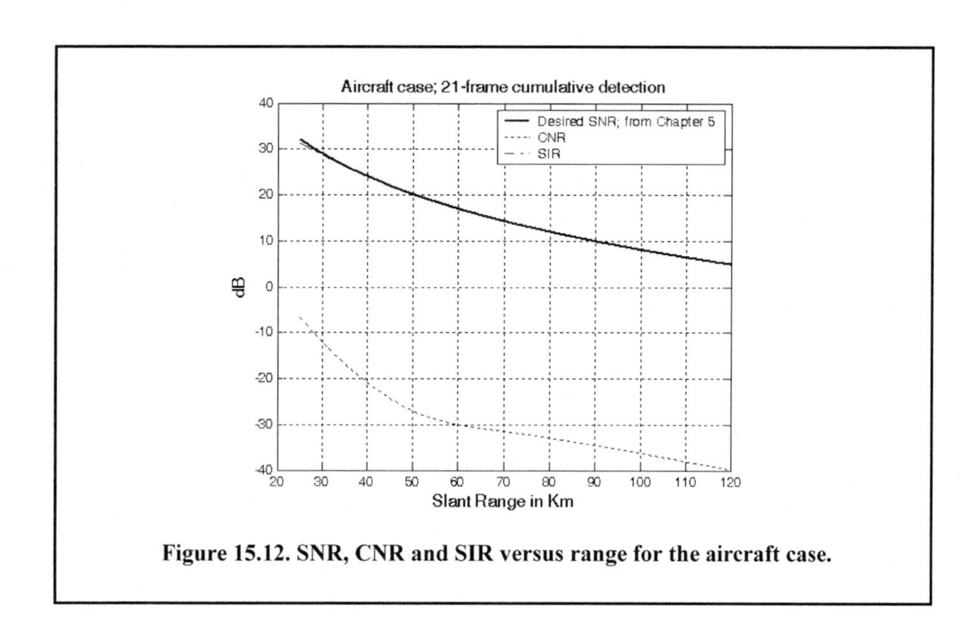

Figure 15.12. SNR, CNR and SIR versus range for the aircraft case.

15.5. Textbook Radar Design Case Study - Visit #5

Assume that the wind rms velocity $\sigma_w = 0.45m/s$. Propose a clutter mitigation process utilizing a 2-pulse and a 3-pulse MTI receiver.

A Design - Visit #5

Earlier, we determined that the wavelength is $\lambda = 0.1m$, the PRF is $f_r = 1KHz$, the scan rate is $T_{scan} = 2s$, and the antenna azimuth 3-dB beamwidth is $\Theta_a = 1.3°$. Thus,

$$\sigma_v = \frac{2\sigma_w}{\lambda} = \frac{2 \times 0.45}{0.1} = 9Hz \qquad \text{Eq. (15.40)}$$

$$\sigma_s = 0.265\left(\frac{2\pi}{\Theta_a T_{scan}}\right) = 0.265 \times \frac{2 \times \pi}{1.32 \times \frac{\pi}{180} \times 2} = 36.136Hz. \qquad \text{Eq. (15.41)}$$

Thus, the total clutter rms spectrum spread is

$$\sigma_t = \sqrt{\sigma_v^2 + \sigma_s^2} = \sqrt{81 + 1305.810} = \sqrt{1386.810} = 37.24Hz. \qquad \text{Eq. (15.42)}$$

The clutter attenuation using a 2-pulse and a 3-pulse MTI are respectively given by

$$I_{2pulse} = 2\left(\frac{f_r}{2\pi\sigma_t}\right)^2 = 2 \times \left(\frac{1000}{2 \times \pi \times 37.24}\right)^2 = 36.531\frac{W}{W} \Rightarrow 15.63dB \qquad \text{Eq. (15.43)}$$

$$I_{3pulse} = 2\left(\frac{f_r}{2\pi\sigma_t}\right)^4 = 2 \times \left(\frac{1000}{2 \times \pi \times 37.24}\right)^4 = 667.247\frac{W}{W} \Rightarrow 28.24dB\,. \qquad \text{Eq. (15.44)}$$

Figures 15.13 and 15.14 show the desired SNR and the calculated SIR using 2-pulse and 3-pulse MTI filters respectively, for the missile case. Figures 15.15 and 15.16 show similar output for the aircraft case. One may argue, depending on the tracking scheme adopted by the radar, that for a tracking radar

$$\sigma_t = \sigma_v = 9Hz \tag{7.45}$$

since $\sigma_s = 0$ for a radar that employs a monopulse tracking option. In this design, we will assume a Kalman filter tracker.

As clearly indicated by these figures, a 3-pulse MTI filter would provide adequate clutter rejection for both target types. However, if we assume that targets are detected at maximum range (90 Km for aircraft and 55 Km for missile) and then are tracked for the rest of the flight, then 2-pulse MTI may be adequate. This is true since the SNR would be expected to be larger during track than it is during detection, especially when pulse compression is used. Nonetheless, in this design a 3-pulse MTI filter is adopted.

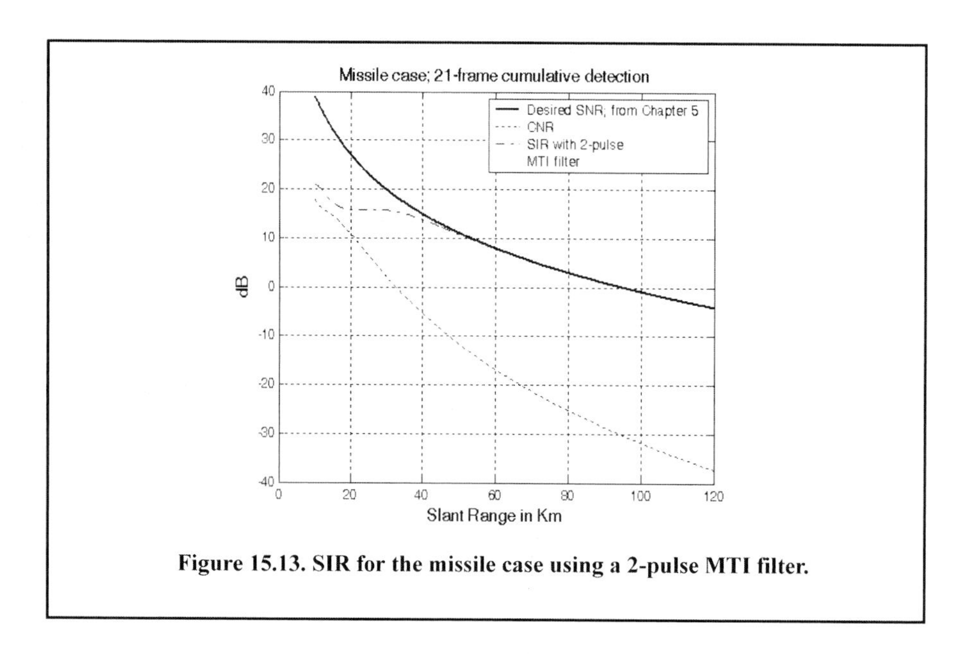

Figure 15.13. SIR for the missile case using a 2-pulse MTI filter.

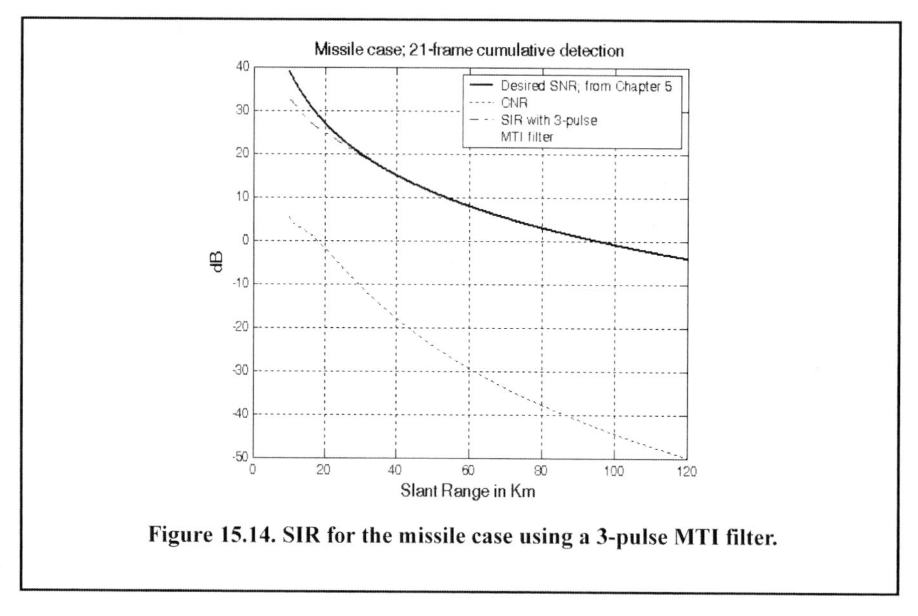

Figure 15.14. SIR for the missile case using a 3-pulse MTI filter.

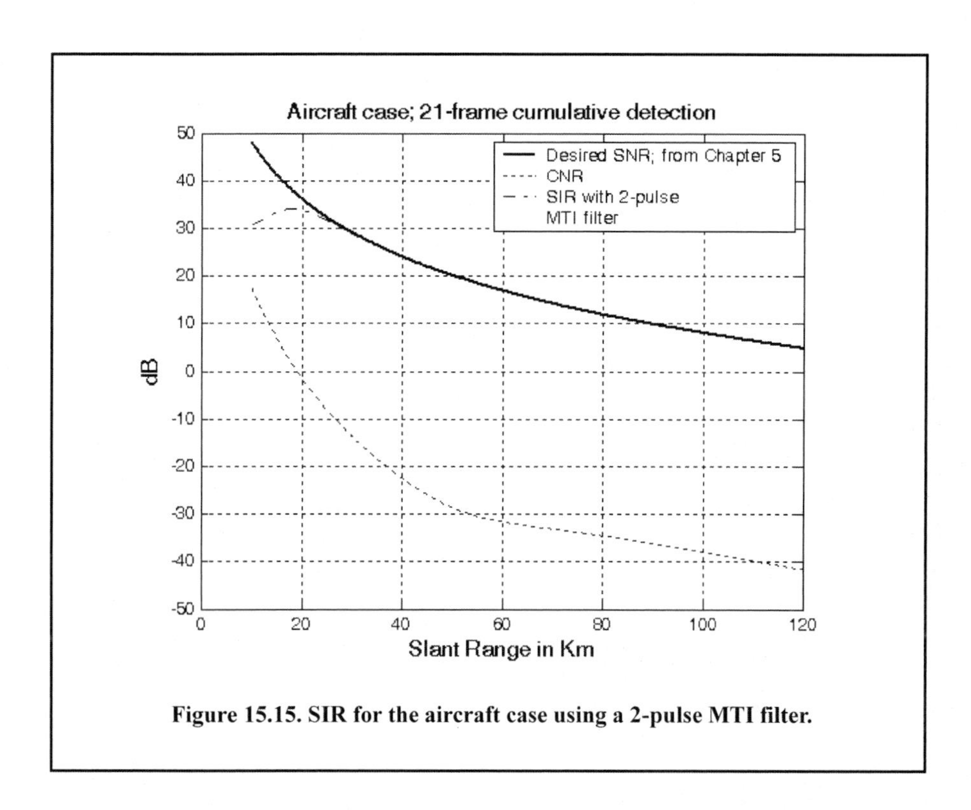

Figure 15.15. SIR for the aircraft case using a 2-pulse MTI filter.

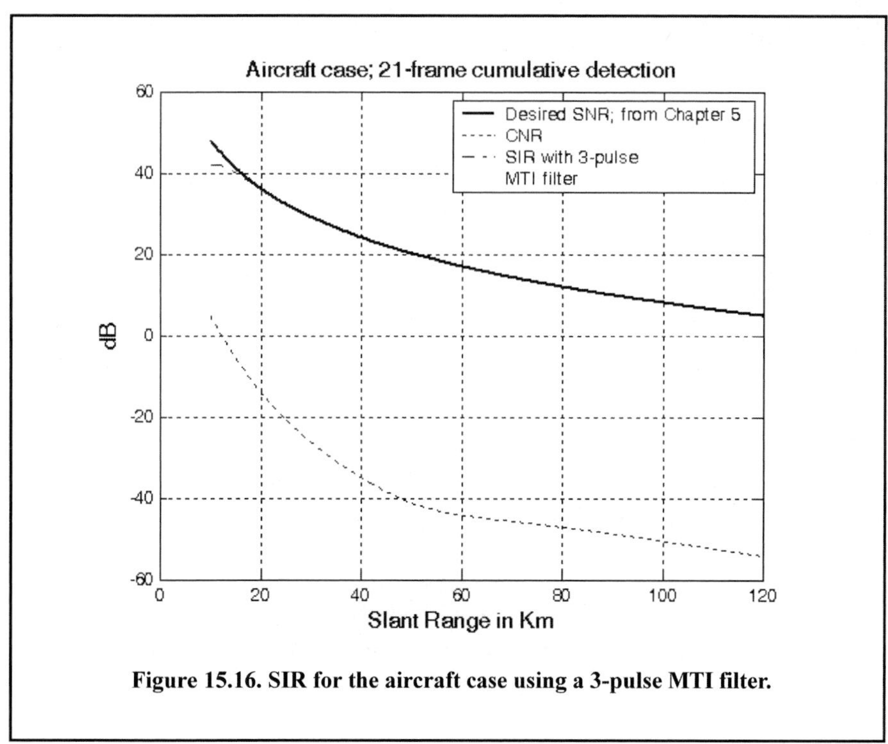

Figure 15.16. SIR for the aircraft case using a 3-pulse MTI filter.

15.6. Textbook Radar Design Case Study - Visit #6

Modify the *"Textbook Radar"* design case study so that a phased array antenna is utilized. For this purpose,

- Modify the design requirements such that the search volume is now defined by $\Theta_e = 10°$ and $\Theta_a \le 45°$
- Assume X-band, if possible
- Design an electronically steered radar (ESR)
- May use non-coherent integration of a few pulses, if necessary
- Size the radar so that it can fulfill this mission
- Calculate the antenna gain, aperture size, missile and aircraft detection range, number of elements in the array, etc.
- All other design requirements are as defined in the previous sections.

A Design - Visit #6

The search volume is

$$\Omega = \frac{10° \times 45°}{(57.296)^2} = 0.1371 \ steradian. \qquad \text{Eq. (15.46)}$$

For an X-band radar, choose $f_o = 9GHz$; then

$$\lambda = \frac{3 \times 10^8}{9 \times 10^9} = 0.0333m. \qquad \text{Eq. (15.47)}$$

Assume an aperture size $A_e = 2.25m^2$; thus

$$G = \frac{4\pi A_e}{\lambda^2} = \frac{4 \times \pi \times 2.25}{(0.0333)^2} = 25451.991 \Rightarrow G = 44dB. \qquad \text{Eq. (15.48)}$$

Assume a square aperture. It follows that the aperture 3-dB beamwidth is calculated from

$$\left(G = \frac{4\pi}{\theta_{3db}^2}\right) \Rightarrow \theta_{3dB} = \sqrt{\frac{4 \times \pi \times 180^2}{25451.991 \times \pi^2}} = 1.3°. \qquad \text{Eq. (15.49)}$$

The number of beams required to fill the search volume is

$$n_b = k_p \frac{\Omega}{(1.3/57.296)^2}\bigg|_{k_p = 1.5} \Rightarrow n_b = 399.5 \Rightarrow choose \ n_b = 400. \qquad \text{Eq. (15.50)}$$

Note that the packing factor k_p is used to allow for beam overlap in order to avoid gaps in the beam coverage. The search scan rate is 2 seconds. Thus, the minimum PRF should correspond to 200 beams per second (i.e. $f_r = 200Hz$). This PRF will allow the radar to visit each beam position only once during a complete scan. Assume that 4-pulse non-coherent integration along with a cumulative detection scheme are required to achieve the desired probability of detection. The corresponding single pulse energy for the missile and aircraft cases are respectively given by

$$E_m = 0.1147 Joules \qquad \text{Eq. (15.51)}$$

$$E_a = 0.1029 Joules. \qquad \text{Eq. (15.52)}$$

However, these values were derived using $\lambda = 0.1m$ and $G = 2827.4$. The new wavelength is $\lambda = 0.0333m$ and the new gain is $G = 25451.99$. Thus,

$$E_m = 0.1147 \times \frac{0.1^2 \times 2827.4^2}{0.0333^2 \times 25452^2} = 0.012765 Joules \qquad \text{Eq. (15.53)}$$

$$E_a = 0.1029 \times \frac{0.1^2 \times 2827.4^2}{0.0333^2 \times 25452^2} = 0.01145 Joules. \qquad \text{Eq. (15.54)}$$

The single pulse peak power that will satisfy detection for both target types is

$$P_t = \frac{0.012765}{20 \times 10^{-6}} = 638.25 W \qquad \text{Eq. (15.55)}$$

where $\tau = 20 \mu s$ is used.

Note that since a 4-pulse non-coherent integration is adopted, the minimum PRF is increased to

$$f_r = 200 \times 4 = 800 Hz \qquad \text{Eq. (15.56)}$$

and the total number of beams is $n_b = 1600$. Consequently the unambiguous range is

$$R_u \leq \frac{3 \times 10^8}{2 \times 800} = 187.5 Km. \qquad \text{Eq. (15.57)}$$

Since the effective aperture is $A_e = 2.25m^2$, then by assuming an array efficiency $\rho = 0.8$ the actual array size is

$$A = \frac{2.25}{0.8} = 2.8125m^2. \qquad \text{Eq. (15.58)}$$

It follows that the physical array sides are $1.68m \times 1.68m$. Thus, by selecting the array element spacing $d = 0.6\lambda$ an array of size 84×84 elements satisfies the design requirements.

Since the field of view is less than $\pm 22.5°$, one can use element spacing as large as $d = 1.5\lambda$ without introducing any grating lobes into the array FOV. Using this option yields an array of size $34 \times 34 = 1156$ elements. Hence, the required power per element is less than $0.6W$.

15.7. Textbook Radar Design Case Study - Visit #7

Modify the *"Textbook Radar"* design case study so that the effects of target RCS fluctuations are taken into account. For this purpose modify the design so that

- The aircraft and missile target types follow Swerling I and Swerling III fluctuations, respectively
- Assume that a $P_D \geq 0.995$ is required at maximum range with $P_{fa} = 10^{-7}$
- You may use either non-coherent integration or cumulative probability of detection
- Finally, you may modify any other design parameters as needed.

A Design - Visit #7

The missile and the aircraft detection ranges were calculated earlier. They are $R_a = 90Km$ for the aircraft and $R_m = 55Km$ for the missile. First determine the probability of detection for each target type with and without the 7-pulse non-coherent integration. These values are

$$SNR\_single\_pulse\_missile = 5.5998\ dB$$
$$SNR\_7\_pulse\_NCI\_missile = 11.7216\ dB$$
$$SNR\_single\_pulse\_aircraft = 6.0755\ dB$$
$$SNR\_7\_pulse\_NCI\_aircrfat = 12.1973\ dB$$

Using the formulas developed earlier in this book (see Chapter 11), yields

$$Pd\_single\_pulse\_missile = 0.013$$
$$Pd\_7\_pulse\_NCI\_missile = 0.9276$$
$$Pd\_single\_pulse\_aircraft = 0.038$$
$$Pd\_7\_pulse\_NCI\_aircraft = 0.8273$$

Clearly in all four cases, there is not enough SNR to meet the design requirement of $P_D \geq 0.995$. Instead, resort to accomplishing the desired probability of detection by using cumulative probabilities. The single frame increments for the missile and aircraft cases are

$$R_{Missile} = scan\ \ rate \times v_m = 2 \times 150 = 300m \qquad \text{Eq. (15.59)}$$

$$R_{Aircraft} = scan\ \ rate \times v_a = 2 \times 400 = 800m\,. \qquad \text{Eq. (15.60)}$$

Single Pulse (Per Frame) Design Option

As a first design option, consider the case where during each frame only a single pulse is used for detection (i.e., no integration). Consequently, if the single pulse detection does not achieve the desired probability of detection at 90 Km for the aircraft or at 55 Km for the missile, then non-coherent integration of a few pulses per frame can then be utilized. Keep in mind that only non-coherent integration can be used in the cases of Swerling type I and III fluctuations.

Assume that the first frame corresponding to detecting the aircraft is 106 Km. This assumption is arbitrary, and it provides the designer with 21 frames. It follows that the first frame, when detecting the missile, is at 61 Km. Furthermore, assume that the SNR at $R = 90Km$ is $(SNR)_{aircraft} = 8.5dB$, for the aircraft case. And, for the missile case, assume that at $R = 55Km$ the corresponding SNR is $(SNR)_{missile} = 9dB$. Note that these values are simply educated guesses, and the designer may be required to perform several iterations in order to accomplish the desired cumulative probability of detection, $P_D \geq 0.995$.

The logic used by this program for calculating the proper probability of detection at each frame and for computing the cumulative probability of detection is described as follows:

1. Assume Swerling V fluctuation and calculate the ith frame-SNR, $(SNR)_i$. For the *"Textbook Radar"* design case study, use $n_P = 1$, $R_0 = 90Km$, and $(SNR_0)_{aircraft} = 8.5dB$. Alternatively use $R_0 = 55Km$ and $(SNR)_{missile} = 9dB$ for

the missile case. Note that the selected SNR values are best estimates or educated guesses, and it may require going through a few iterations before finally selecting an acceptable set.

2. Then calculate the number of frames and their associated ranges.
3. Depending on the fluctuation type calculate the probability of detection for each frame, P_{D_i}.
4. Finally, compute the cumulative probability of detection, P_{D_i}.

Following these steps yields the following cumulative probabilities of detection for the aircraft and missile cases,

$$\begin{aligned} P_{DC_{Missile}} &= 0.99872 \\ P_{DC_{aircraft}} &= 0.99687 \end{aligned} . \qquad \text{Eq. (15.61)}$$

These results clearly satisfy the design requirement of $P_D \geq 0.995$. However, one must re-validate the peak power requirement for the design. To do that, replace the SNR values used earlier in the design by the values adopted in this section (i.e., $(SNR_0)_{aircraft} = 8.5dB$ and $(SNR)_{missile} = 9dB$). It follows that the corresponding single pulse energy for the missile and the aircraft cases are respectively given by

$$E_m = 0.1658 \times \frac{10^{0.9}}{10^{0.56}} = 0.36273 Joules \qquad \text{Eq. (15.62)}$$

$$E_a = 0.1487 \times \frac{10^{0.85}}{10^{0.56}} = 0.28994 Joules . \qquad \text{Eq. (15.63)}$$

This indicates that the stressing single pulse peak power requirement (i.e., missile detection) exceeds $362KW$. This value for the single pulse peak power is high for a mobile ground based air defense radar and practical constraints would require using less peak power. In order to bring the single pulse peak power requirement down, one can use non-coherent integration of a few pulses per frame prior to calculating the frame probability of detection.

Non-Coherent Integration Design Option

The single frame probability of detection can be improved significantly when pulse integration is utilized. One may use coherent or non-coherent integration to improve the frame cumulative probability of detection. In this case, caution should be exercised since coherent integration would not be practical when the target fluctuation type is either Swerling I or Swerling III. Alternatively, using non-coherent integration will always reduce the minimum required SNR.

For this purpose, use $n_P = 4$ and use $SNR = 4dB$ (single pulse) for both the missile and aircraft single pulse SNR[1] at their respective reference ranges, $R_{0_{missile}} = 55Km$ and $R_{0_{aircraft}} = 90Km$. The resulting cumulative probabilities of detection are

1. These values are educated guesses. The designer may be required to go through a few iterations before arriving at an acceptable set of design parameters.

$$P_{DC_{Missile}} = 0.99945$$
$$P_{DC_{aircraft}} = 0.99812 \qquad \text{Eq. (15.64)}$$

which are both within the desired design requirements. It follows that the corresponding minimum required single pulse energy for the missile and the aircraft cases are now given by

$$E_m = 0.1658 \times \frac{10^{0.4}}{10^{0.56}} = 0.1147 Joules \qquad \text{Eq. (15.65)}$$

$$E_a = 0.1487 \times \frac{10^{0.4}}{10^{0.56}} = 0.1029 Joules. \qquad \text{Eq. (15.66)}$$

Thus, the minimum single pulse peak power (assuming the same pulsewidth) is

$$P_t = \frac{0.1147}{1 \times 10^{-6}} = 114.7 KW. \qquad \text{Eq. (15.67)}$$

Note that the peak power requirement can be significantly reduced while maintaining a very fine range resolution when pulse compression techniques are used. Figure 15.17 shows a plot of the SNR versus range for both target types. This plot assumes 4-pulse non-coherent integration.

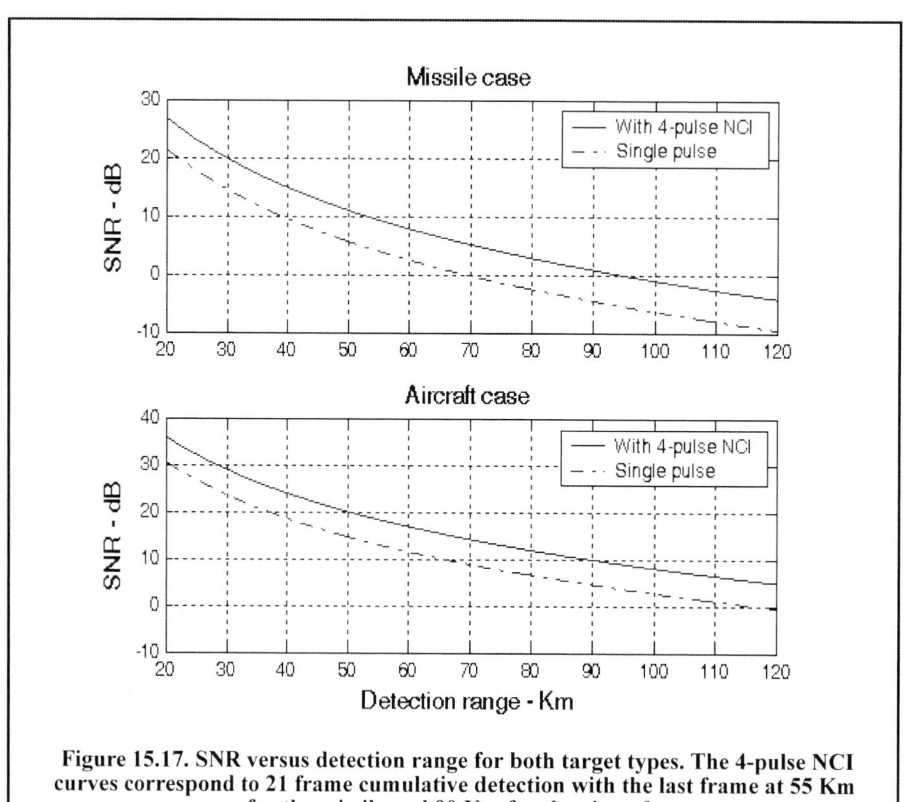

Figure 15.17. SNR versus detection range for both target types. The 4-pulse NCI curves correspond to 21 frame cumulative detection with the last frame at 55 Km for the missile and 90 Km for the aircraft.

Bibliography

[1] Abramowitz, M., and Stegun, I. A., Editors, *Handbook of Mathematical Functions, with Formulas, Graphs, and Mathematical Tables*, Dover Publications, New York, 1970.

[2] Balanis, C. A., *Antenna Theory, Analysis and Design*, Harper & Row, New York, 1982.

[3] Barkat, M., *Signal Detection and Estimation*, Artech House, Norwood, MA, 1991.

[4] Barton, D. K., *Modern Radar System Analysis*, Artech House, Norwood, MA, 1988.

[5] Bean, B. R., and Abbott, R., Oxygen and Water-Vapor Absorption of Radio Waves in the Atmosphere, *Review Geofisica Pura E Applicata (Milano)* 37:127, 1957.

[6] Benedict, T., and Bordner, G., Synthesis of an Optimal Set of Radar Track-While-Scan Smoothing Equations, *IRE Transaction on Automatic Control, Ac-7*, July 1962, pp. 27-32.

[7] Berkowitz, R. S., *Modern Radar: Analysis, Evaluation, and System Design*, John Wiley & Sons, Inc, New York, 1965.

[8] Beyer, W. H., *CRC Standard Mathematical Tables*, 26th Edition, CRC Press, Boca Raton, FL, 1981.

[9] Billetter, D. R., *Multifunction Array Radar*, Artech House, Norwood, MA, 1989.

[10] Blackman, S. S., *Multiple-Target Tracking with Radar Application*, Artech House, Norwood, MA, 1986.

[11] Blake, L. V., *A Guide to Basic Pulse-Radar Maximum Range Calculation. Part-I: Equations, Definitions, and Aids to Calculation*, Naval Res. Lab. Report 5868, 1969.

[12] Blake, L. V., *Curves of Atmospheric-Absorption Loss for Use in Radar Range Calculation*, NRL Report 5601, March 23, 1961.

[13] Blake, L. V., *Radar / Radio Tropospheric Absorption and Noise Temperature*, NRL Report Ad-753 197, October 1972. Distributed by the National Technical Information Service (NTIS).

[14] Blake, L. V., *Radar-Range Performance Analysis*, Lexington Books, Lexington, MA, 1980.

[15] Boothe, R. R., *A Digital Computer Program for Determining the Performance of an Acquisition Radar through Application of Radar Detection Probability Theory*, U.S.

Army Missile Command, Report No. RD-TR-64-2, Redstone Arsenal, Alabama, 1964.

[16] Boothe, R. R., *The Weibull Distribution Applied to the Ground Clutter Backscatter Coefficient*, U.S. Army Missile Command, Report No. RE-TR-69-15, Redstone Arsenal, Alabama, 1969.

[17] Bowman, J. J., Piergiorgio, L. U., and Senior, T. B., *Electromagnetic and Acoustic Scattering by Simple Shapes*, North-Holland Pub. Co, Amsterdam, 1969.

[18] Brookner, E., Editor, *Aspects of Modern Radar*, Artech House, Norwood, MA, 1988.

[19] Brookner, E., Editor, *Practical Phased Array Antenna System*, Artech House, Norwood, MA, 1991.

[20] Brookner, E., *Radar Technology*, Lexington Books, Lexington, MA, 1996.

[21] Burdic, W. S., *Radar Signal Analysis*, Prentice Hall, Englewood Cliffs, NJ, 1968.

[22] Brookner, E., *Tracking and Kalman Filtering Made Easy*, John Wiley & Sons, New York, 1998.

[23] Cadzow, J. A., *Discrete-Time Systems, An Introduction with Interdisciplinary Applications*, Prentice Hall, Englewood Cliffs, NJ, 1973.

[24] Carlson, A. B., *Communication Systems, An Introduction to Signals and Noise in Electrical Communication*, 3rd Edition, McGraw-Hill, New York, 1986.

[25] Carpentier, M. H., *Principles of Modern Radar Systems*, Artech House, Norwood, MA, 1988.

[26] Cook, E. C., and Bernfeld, M., *Radar Signals: An Introduction to Theory and Application*, Artech House, Norwood, MA, 1993.

[27] Costas, J. P., A Study of a Class of Detection Waveforms Having Nearly Ideal Range-Doppler Ambiguity Properties, *Proc. IEEE 72*, 1984, pp. 996-1009.

[28] Crispin, J. W., Jr., and Siegel, K. M, Editors, *Methods of Radar Cross-Section Analysis*, Academic Press, New York, 1968.

[29] Curry, G. R., *Radar System Performance Modeling*, Artech House, Norwood, MA, 2001.

[30] DiFranco, J. V. and Rubin, W. L., *Radar Detection*. Artech House, Norwood, MA, 1980.

[31] Dillard, R. A. and Dillard, G. M., *Detectability of Spread-Spectrum Signals*, Artech House, Norwood, MA, 1989.

[32] Edde, B., *Radar Principles, Technology, Applications*, Prentice Hall, Englewood Cliffs, NJ, 1993.

[33] Elsherbeni, A., Inman, M. J., and Riley, C., Antenna Design and Radiation Pattern Visualization, *The 19th Annual Review of Progress in Applied Computational Electromagnetics*, ACES'03, Monterey, CA, March 2003.

[34] Fehlner, L. F., *Marcum's and Swerling's Data on Target Detection by a Pulsed Radar*, Johns Hopkins University, Applied Physics Lab. Rpt. # TG451, July 2, 1962, and Rpt. # TG451A, September 1964.

[35] Fielding, J. E., and Reynolds, G. D., *VCCALC: Vertical Coverage Calculation Software and Users Manual*, Artech House, Norwood, MA, 1988.

[36] Gelb, A., Editor, *Applied Optimal Estimation*, MIT Press, Cambridge, MA, 1974.

[37] Goldman, S. J., *Phase Noise Analysis in Radar Systems, Using Personal Computers*, John Wiley & Sons, New York, NY, 1989.

[38] Grewal, M. S., and Andrews, A. P., *Kalman Filtering: Theory and Practice Using MATLAB*, 2nd Edition, John Wiley & Sons, New York, 2001.

[39] Hamming, R. W., *Digital Filters*, 2nd Edition, Prentice Hall, Englewood Cliffs, NJ, 1983.

[40] Hirsch, H. L., and Grove, D. C., *Practical Simulation of Radar Antennas and Radomes*, Artech House, Norwood, MA, 1987.

[41] Hovanessian, S. A., *Radar System Design and Analysis*, Artech House, Norwood, MA, 1984.

[42] James, D. A., *Radar Homing Guidance for Tactical Missiles*, John Wiley & Sons, New York, 1986.

[43] Kanter, I., Exact Detection Probability for Partially Correlated Rayleigh Targets, *IEEE Trans, AES-22*, March 1986, pp. 184-196.

[44] Keller, J. B., Geometrical Theory of Diffraction, *Journal Opt. Soc. Amer.*, Vol. 52, February 1962, pp. 116-130.

[45] Klauder, J. R., Price, A. C., Darlington, S., and Albershiem, W. J., The Theory and Design of Chirp Radars, *The Bell System Technical Journal*, Vol. 39, No. 4, 1960.

[46] Knott, E. F., Shaeffer, J. F., and Tuley, M. T., *Radar Cross Section*, 2nd Edition, Artech House, Norwood, MA, 1993.

[47] Lativa, J., Low-Angle Tracking Using Multifrequency Sampled Aperture Radar, *IEEE-AES Trans.*, Vol. 27, No. 5, September 1991, pp. 797-805.

[48] Lee, S. W., and Mittra, R., Fourier Transform of a Polygonal Shape Function and Its Application in Electromagnetics, *IEEE Trans. Antennas and Propagation*, Vol. 31, January 1983, pp. 99-103.

[49] LeFande, R. A., *Attenuation of Microwave Radiation for Paths through the Atmosphere*, NRL Report 6766, Nov. 1968.

[50] Levanon, N., *Radar Principles*, John Wiley & Sons, New York, 1988.

[51] Levanon, N., and Mozeson, E., Nullifying ACF Grating Lobes in Stepped-Frequency Train of LFM Pulses, *IEEE-AES Trans.*, Vol. 39, No. 2, April 2003, pp. 694-703.

[52] Levanon, N., and Mozeson, E., *Radar Signals*, John Wiley-Interscience, Hoboken, NJ, 2004.

[53] Lewis, B. L., Kretschmer, Jr., F. F., and Shelton, W. W., *Aspects of Radar Signal Processing*, Artech House, Norwood, MA, 1986.

[54] Li, J., and Stoica, P., Editors, *MIMO Radar Signal Processing*, John Wiley & Sons, New York, 2009.

[55] Long, M. W., *Radar Reflectivity of Land and Sea*, Artech House, Norwood, MA, 1983.

[56] Lothes, R. N., Szymanski, M. B., and Wiley, R. G., *Radar Vulnerability to Jamming*, Artech House, Norwood, MA, 1990.

[57] Maffett, A. L., *Topics for a Statistical Description of Radar Cross Section*, John Wiley & Sons, New York, 1989.

[58] Mahafza, B. R., *Introduction to Radar Analysis*, CRC Press, Boca Raton, FL, 1998.

[59] Mahafza, B. R., *Radar Systems Analysis and Design Using MATLAB*, 3rd Edition, Taylor & Francis, Boca Raton, FL, 2013.

[60] Mahafza, B. R., *Radar Signal Analysis and Signal Processing Using MATLAB*, Chapman and Hall/CRC, Boca Raton, FL, 2008.

[61] Mahafza, B. R., and Polge, R. J., Multiple Target Detection Through DFT Processing in a Sequential Mode Operation of Real Two-Dimensional Arrays, *Proc. of the IEEE Southeast Conf. '90*, New Orleans, LA, April 1990, pp. 168-170.

[62] Mahafza, B. R., Heifner, L.A., and Gracchi, V. C., Multitarget Detection Using Synthetic Sampled Aperture Radars (SSAMAR), *IEEE-AES Trans.*, Vol. 31, No. 3, July 1995, pp. 1127-1132.

[63] Mahafza, B. R., and Sajjadi, M., Three-Dimensional SAR Imaging Using a Linear Array in Transverse Motion, *IEEE-AES Trans.*, Vol. 32, No. 1, January 1996, pp. 499-510.

[64] Marcum, J. I., A Statistical Theory of Target Detection by Pulsed Radar, Mathematical Appendix, *IRE Trans.*, Vol. IT-6, April 1960, pp. 259-267.

[65] Medgyesi-Mitschang, L. N., and Putnam, J. M., Electromagnetic Scattering from Axially Inhomogeneous Bodies of Revolution, *IEEE Trans. Antennas and Propagation.*, Vol. 32, August 1984, pp. 797-806.

[66] Meeks, M. L., *Radar Propagation at Low Altitudes*, Artech House, Norwood, MA, 1982.

[67] Mensa, D. L., *High Resolution Radar Imaging*, Artech House, Norwood, MA, 1984.

[68] Meyer, D. P., and Mayer, H. A., *Radar Target Detection: Handbook of Theory and Practice*, Academic Press, New York, 1973.

[69] Morchin, W., *Radar Engineer's Sourcebook*, Artech House, Norwood, MA, 1993.

[70] Morris, G. V., *Airborne Pulsed Doppler Radar*, Artech House, Norwood, MA, 1988.

[71] Nathanson, F. E., *Radar Design Principles*, 2nd Edition, McGraw-Hill, New York, 1991.

[72] Navarro, Jr., A. M., *General Properties of Alpha Beta and Alpha Beta Gamma Tracking Filters*, Physics Laboratory of the National Defense Research Organization TNO, Report PHL 1977-92, January 1977.

[73] North, D. O., An Analysis of the Factors Which Determine Signal/Noise Discrimination in Pulsed Carrier Systems, *Proc. IEEE 51*, No. 7, July 1963, pp. 1015-1027.

[74] Oppenheim, A. V., and Schafer, R. W., *Discrete-Time Signal Processing*, Prentice Hall, Englewood Cliffs, NJ, 1989.

[75] Oppenheim, A. V., Willsky, A. S., and Young, I. T., *Signals and Systems*, Prentice Hall, Englewood Cliffs, NJ, 1983.

[76] Orfanidis, S. J., *Optimum Signal Processing, an Introduction*, 2nd Edition, McGraw-Hill, New York, 1988.

[77] Papoulis, A., *Probability, Random Variables, and Stochastic Processes*, 2nd Edition, McGraw-Hill, New York, 1984.

[78] Parl, S. A., New Method of Calculating the Generalized Q Function, *IEEE Trans. Information Theory*, Vol. IT-26, No. 1, January 1980, pp. 121-124.

[79] Peebles, P. Z., Jr., *Probability, Random Variables, and Random Signal Principles*, McGraw-Hill, New York, 1987.

[80] Peebles, P. Z., Jr., *Radar Principles*, John Wiley & Sons, New York, 1998.

[81] Pettit, R. H., *ECM and ECCM Techniques for Digital Communication Systems*, Lifetime Learning Publications, New York, 1982.

[82] Polge, R. J., Mahafza, B. R., and Kim, J. G., *Extension and Updating of the Computer Simulation of Range Relative Doppler Processing for MM Wave Seekers*, Interim Technical Report, Vol. I, prepared for the U.S. Army Missile Command, Redstone Arsenal, Alabama, January 1989.

[83] Polge, R. J., Mahafza, B. R., and Kim, J. G., Multiple Target Detection through DFT Processing in a Sequential Mode Operation of Real or Synthetic Arrays, *IEEE 21st Southeastern Symposium on System Theory*, Tallahassee, FL, 1989, pp. 264-267.

[84] Poularikas, A., and Ramadan, Z. M., *Adaptive Filtering Primer with MATLAB*, Taylor & Francis, Boca Raton, FL, 2006.

[85] Poularikas, A., and Seely, S., *Signals and Systems*, PWS Publishers, Boston, MA, 1984.

[86] Putnam, J. N., and Gerdera, M. B., CARLOS TM: A General-Purpose Three-Dimensional Method of Moments Scattering Code, *IEEE Trans. Antennas and Propagation*, Vol. 35, April 1993, pp. 69-71

[87] Reed, H. R. and Russell, C. M., *Ultra High Frequency Propagation*, Boston Technical Publishers, Inc., Lexington, MA, 1964.

[88] Resnick, J. B., *High Resolution Waveforms Suitable for a Multiple Target Environment*, MS thesis, MIT, Cambridge, MA, June 1962.

[89] Richards, M. A., *Fundamentals of Radar Signal Processing*, McGraw-Hill, New York, 2005.

[90] Rihaczek, A. W., *Principles of High Resolution Radars*, McGraw-Hill, New York, 1969.

[91] Robertson, G. H., Operating Characteristics for a Linear Detector of CW Signals in Narrow-Band Gaussian Noise, *Bell Sys. Tech. Journal*, Vol. 46 April 1967, pp. 755-774.

[92] Rosenbaum, B., *A Programmed Mathematical Model to Simulate the Bending of Radio Waves in the Atmospheric Propagation*, Goddard Space Flight Center Report X-551-68-367, Greenbelt, Maryland, 1968.

[93] Ross, R. A., Radar Cross Section of Rectangular Flat Plate as a Function of Aspect Angle, *IEEE Trans*. AP-14, 1966, p. 320.

[94] Ruck, G. T., Barrick, D. E., Stuart, W. D., and Krichbaum, C. K., *Radar Cross Section Handbook*, Volume 1, Plenum Press, New York, 1970.

[95] Ruck, G. T., Barrick, D. E., Stuart, W. D., and Krichbaum, C. K., *Radar Cross Section Handbook*, Volume 2, Plenum Press, New York, 1970.

[96] Rulf, B., and Robertshaw, G. A., *Understanding Antennas for Radar, Communications, and Avionics*, Van Nostrand Reinhold, 1987.

[97] Scanlan, M. J., Editor, *Modern Radar Techniques*, Macmillan, New York, 1987.

[98] Scheer, J. A., and Kurtz, J. L., Editor, *Coherent Radar Performance Estimation*, Artech House, Norwood, MA, 1993.

[99] Schelher, D. C., *MTI and Pulsed Doppler Radar with MATLAB*, 2nd Edition, Artech House, Norwood MA. 2010.

[100] Shanmugan, K. S., and Breipohl, A. M., *Random Signals: Detection, Estimation and Data Analysis*, John Wiley & Sons, New York, 1988.

[101] Sherman, S. M., *Monopulse Principles and Techniques*, Artech House, Norwood, MA.

[102] Singer, R. A., Estimating Optimal Tracking Filter Performance for Manned Maneuvering Targets, *IEEE Transaction on Aerospace and Electronics, AES-5*, July 1970, pp. 473-483.

[103] Skillman, W. A., *DETPROB: Probability of Detection Calculation Software and User's Manual*, Artech House, Norwood, MA, 1991.

[104] Skolnik, M. I., *Introduction to Radar Systems*, McGraw-Hill, New York, 1982.

[105] Skolnik, M. I., Editor, *Radar Handbook*, 2nd Edition, McGraw-Hill, New York, 1990.

[106] Song, J. M., Lu, C. C., Chew, W. C., and Lee, S. W., Fast Illinois Solver Code (FISC), *IEEE Trans. Antennas and Propagation,* Vol. 40, June 1998, pp. 27-34.

[107] Stearns, S. D., and David, R. A., *Signal Processing Algorithms*, Prentice Hall, Englewood Cliffs, NJ, 1988.

[108] Stimson, G. W., *Introduction to Airborne Radar*, Hughes Aircraft Company, El Segundo, CA, 1983.

[109] Stratton, J. A., *Electromagnetic Theory*, McGraw-Hill, New York, 1941.

[110] Stremler, F. G., *Introduction to Communication Systems*, 3rd Edition, Addison-Wesley, New York, 1990.

[111] Stutzman, G. E., Estimating Directivity and Gain of Antennas, *IEEE Antennas and Propagation Magazine 40*, August 1998, pp. 7-11.

[112] Swerling, P., Probability of Detection for Fluctuating Targets, *IRE Transaction on Information Theory*, Vol. IT-6, April 1960, pp. 269-308.

[113] Taflove, A., *Computational Electromagnetics: The Finite-Difference Time-Domain Method*, Artech House, Norwood, MA, 1995.

[114] Trunck, G. V., *Automatic Detection, Tracking, and Sensor Integration*, NRL Report 9110, June, 1988.

[115] Van Trees, H. L., *Detection, Estimation, and Modeling Theory*, Part I, John Wiley & Sons, New York, 2001.

[116] Van Trees, H. L., *Detection, Estimation, and Modeling Theory*, Part III, John Wiley & Sons, New York, 2001.

[117] Van Trees, H. L., *Optimum Array Processing*, Part IV of *Detection, Estimation, and Modeling Theory*, Wiley & Sons, New York, 2002.

[118] Tzannes, N. S., *Communication and Radar Systems*, Prentice Hall, Englewood Cliffs, NJ, 1985.

[119] Urkowitz, H., *Decision and Detection Theory*, Unpublished Lecture Notes, Lockheed Martin Co., Moorestown, NJ.

[120] Urkowtiz, H., *Signal Theory and Random Processes*, Artech House, Norwood, MA, 1983.

[121] Wehner, D. R., *High Resolution Radar*, Artech House, Norwood, MA, 1987.

[122] Weiner, M. M., Editor, *Adaptive Antennas and Receivers*, Taylor & Francis, Boca Raton, FL, 2006.

[123] White, J. E., Mueller, D. D., and Bate, R. R., *Fundamentals of Astrodynamics*, Dover Publications, New York, 1971.

[124] Ziemer, R. E., and Tranter, W. H., *Principles of Communications, Systems, Modulation, and Noise,* 2nd Edition, Houghton Mifflin, Boston, MA, 1985.

[125] Zierler, N., *Several Binary-Sequence Generators*, MIT Technical Report No. 95, Sept. 1955.

Index

A

B

N

O

P

Q

R

U

V

W

Z